Conservation Science: Sustaining Biodiversity and Species Extinction

Conservation Science: Sustaining Biodiversity and Species Extinction

Edited by Neil Griffin

SYRAWOOD
PUBLISHING HOUSE
New York

Published by Syrawood Publishing House,
750 Third Avenue, 9th Floor,
New York, NY 10017, USA
www.syrawoodpublishinghouse.com

Conservation Science: Sustaining Biodiversity and Species Extinction
Edited by Neil Griffin

International Standard Book Number: 978-1-68286-421-0 (Hardback)

Cataloging-in-publication Data

Conservation science : sustaining biodiversity and species extinction / edited by Neil Griffin.
p. cm.
Includes bibliographical references and index.
ISBN 978-1-68286-421-0
1. Biodiversity conservation. 2. Sustainable biodiversity. 3. Conservation biology. 4. Conservation of natural resources.
5. Nature conservation. I. Griffin, Neil.
QH75 .C66 2017
639.9--dc23

Printed in China.

TABLE OF CONTENTS

PREFACE

The study of protection of biodiversity is called conservation sciences. It is interdisciplinary in nature. This book elucidates new techniques and their applications in a multidisciplinary approach for a better understanding of conservation science. It aims to provide its readers a broad spectrum of topics dealing with care, protection and conservation of species and ecosystem. The text highlights in detail the various processes by which the nature and its resources could be preserved and maintained. This book is a vital tool for all researching or studying this science as it gives incredible insights into emerging trends and concepts. A number of latest researches have been included to keep the readers up-to-date with the global concepts in this area of study.

I am honored to present to you this unique book which encompasses the most up-to-date data in the field. I was extremely pleased to get this opportunity of editing the work of experts from across the globe. I have also written papers in this field and researched the various aspects revolving around the progress of the discipline. I have tried to unify my knowledge along with that of stalwarts from every corner of the world, to produce a text which not only benefits the readers but also facilitates the growth of the field.

Finally, I would like to thank all the contributing authors for their valuable time and contributions. This book would not have been possible without their efforts. I would also like to thank my friends and family for their constant support.

Editor

1

Factors Affecting the Distribution Pattern of Wild Plants with Extremely Small Populations in Hainan Island, China

Yukai Chen, Xiaobo Yang*, Qi Yang, Donghai Li, Wenxing Long, Wenqi Luo

Key Laboratory of Protection and Development Utilization of Tropical Crop Germplasm Resources, Ministry of Education, College of Horticulture and Landscapes, Hainan University, Haikou 570228, Hainan, China

Abstract

Understanding which factors affect the distribution pattern of extremely small populations is essential to the protection and propagation of rare and endangered plant species. In this study, we established 108 plots covering the entire Hainan Island, and measured the appearance frequency and species richness of plant species with extremely small populations, as well as the ecological environments and human disturbances during 2012–2013. We explored how the ecological environments and human activities affected the distribution pattern of these extremely small populations. Results showed that the extremely small populations underwent human disturbances and threats, and they were often found in fragmental habitats. The leading factors changing the appearance frequency of extremely small populations differed among plant species, and the direct factors making them susceptible to extinction were human disturbances. The peak richness of extremely small populations always occurred at the medium level across environmental gradients, and their species richness always decreased with increasing human disturbances. However, the appearance frequencies of three orchid species increased with the increasing human disturbances. Our study thus indicate that knowledge on how the external factors, such as the ecological environment, land use type, roads, human activity, etc., affect the distribution of the extremely small populations should be taken for the better protecting them in the future.

Editor: Ting Wang, Wuhan Botanical Garden, Chinese Academy of Sciences, Wuhan, China, China

Funding: This work was financially supported by grants from the National Natural Science Foundation of China (31060073), the Traditional Chinese medicine resources protection and utilization in some representative area of Hainan province (201207002-03), and the Second national key protected plant resources survey in Hainan province. The funders had no role in study design, data collection and analysis, decision to publish, or preparation of the manuscript.

Competing Interests: The authors have declared that no competing interests exist.

* E-mail: yanfengxb@163.com

Introduction

Plant population is the most basic level of plant ecology, connecting plant individuals with communities and ecosystems. The ecological patterns of a plant population are characterized by its population dynamics, as well as the relationships between the population and the external environments [1]. Many studies of common species and their companion species have been previously conducted, and focused on the population structure, life history, and the relationship between the population and environment, etc [2–4]. But researches concerning the mechanism of endangerment or the threat on the rare and endangered plants are seldom studied [5–8]. These researches, however, become important to biodiversity protection across the world.

Both internal and external factors cause plants to become vulnerable to extinction. Internal factors refer to the plant biological characteristics, including failures in heritability, reproduction, viability, and adaptability [9,10]. External factors include both natural and human factors. Natural factors refer to the ecological environment, including climate, topography, soil, and other biological factors [11,12]. For example, many rare and endangered plants are only distributed in narrow area with a special microclimate. Some of them occur only in temporary habitats, fragmented habitats or "ecological islands" [13,14]. Human factors refer to human disturbances, roads, and land use types. Most of them have become the harmful factors impacting on rare and endangered plants [15–19].

Hainan is the largest island of the Indo Burma Biodiversity Hotspot, and has the best preserved and the most extensive tropical forests in China. It contains rich rare and endangered plant species, and therefore, becomes a priority for species conservation in China [20,21]. But plants on Hainan Island often have a narrow distribution with few individuals, and thus easily become endangered and extinct because of geographical isolation of this Island from the mainland. Additionally, the introduction of alien plants, habitat destruction, and the overexploitation have accelerate the shrink and extinction of these rare and endangered plant populations [22,23].

The State Forestry Administration of China has been carrying out the "Conservation Program for Wild Plants with Extremely Small Populations in China (from 2011 to 2015)". In this plan, 24 wild plant species with extremely small population in Hainan are recorded. The term "extremely small population" refers to a population with a narrow geographical distribution that has been disturbed and constrained by external factors over a long time, and contains less individuals than the minimal number requirements inhibiting from extinction. Some of these plants are qualified as the biological resources important to the country's ecology, science, culture, and economics [24]. For the majority of extremely small populations in China, the main factors causing them susceptible to endangerment or extinction are habitat

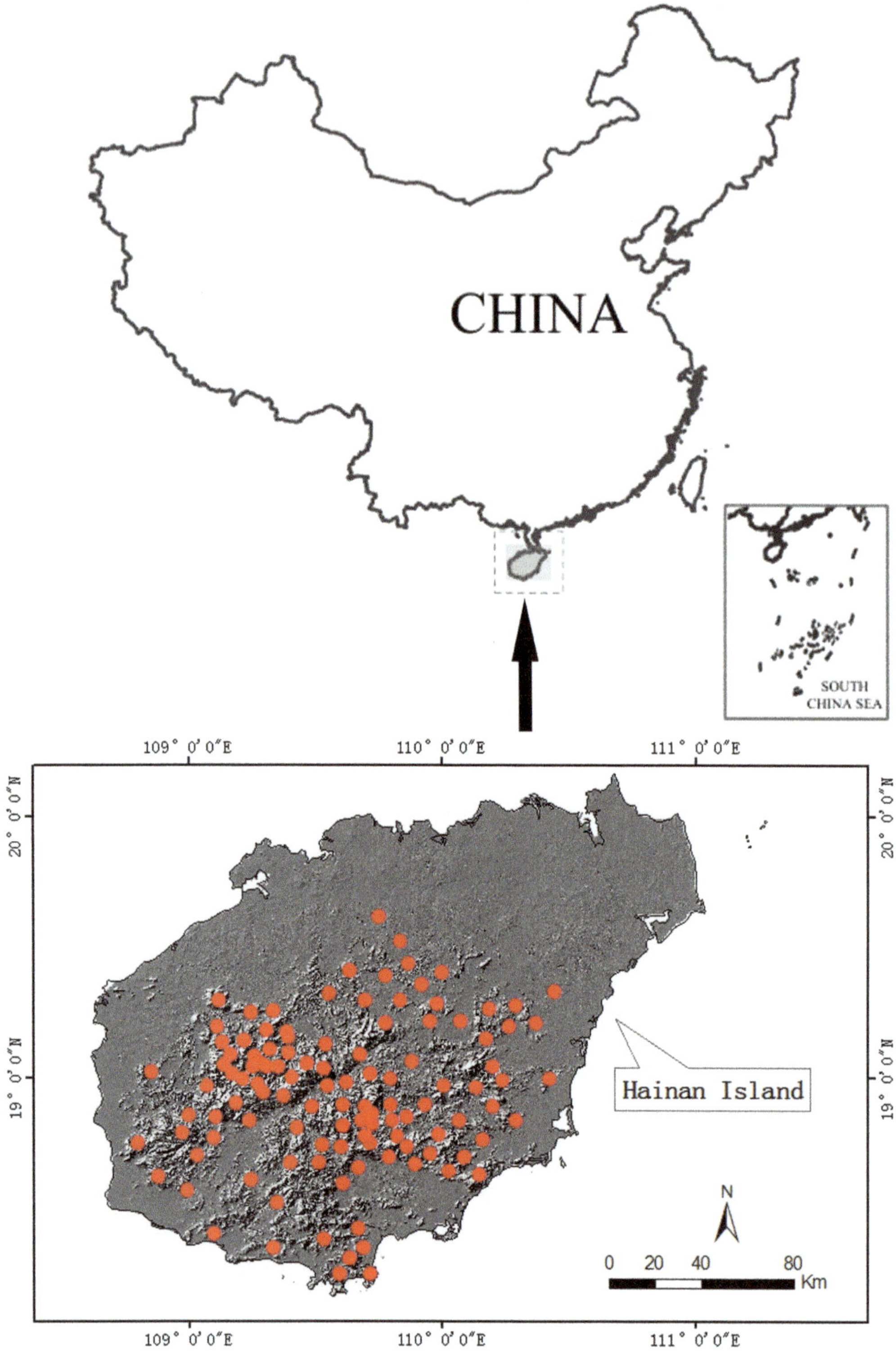

Figure 1. Hainan map and distribution of the 108 sample plots of the field survey. The red circles indicate plots.

destruction, overexploitation, and environmental pollution [25]. Previous each study on the mechanism of endangerment and protection countermeasures for rare and endangered plants has only focused on a single species [6,7,12], leaving open the question of whether there are common mechanisms on the external factors affecting different endangered plants. In this study, we assessed the population sizes and conservation situations of several extremely small populations, studied the characteristics of their habitats, and explored the effect of ecological environment, roads, and human activity on the distribution patterns of these plants. We use our findings to highlight the knowledge gaps in the conservation strategies for these endangered species.

Table 1. Principal component values, eigenvalues, and contribution rate of variables in principal components analysis.

Variable	Principal component		
	1	2	3
Species Richness	0.616	0.077	−0.070
Leaf area index	0.612	0.347	−0.231
Canopy Density	0.812	0.433	−0.083
Annual average temperature (°C)	−0.313	−0.168	−0.465
Average annual precipitation (mm)	0.432	−0.014	0.399
Road Quality	−0.660	0.245	0.224
Distance between Population and Road (m)	0.233	−0.821	0.121
Exposure of rock (%)	−0.591	−0.247	0.031
Altitude	0.721	−0.185	0.105
Slop Aspect	−0.126	0.646	0.386
Slop Gradient	−0.071	−0.041	0.716
Surrounding Population Density	−0.707	0.261	−0.028
Land use type	0.319	0.622	−0.403
Eigenvalue	3.294	1.717	1.395
of variance (%)	27.446	14.307	11.627
Cumulative (%)	27.446	41.753	58.380

Materials and Methods

Study Area

This study was conducted in Hainan Island which is located at the southernmost tip of China (18°10′–20°10′ N, 108°37′–111°03′E) and covers a total area of 33,900 km^2. The island is high at its center and relatively low around its perimeter. The highest peak is Wuzhishan Mountain, with an altitude of 1,867 m (Figure 1). Currently, the island's central and southern mountainous areas contain a rich biodiversity, with many rare and endangered wild plants. However, in the northern and coastal areas, these plant have been almost destroyed by rubber plantations and other forms of agriculture, as well as industrial and urban developments [26,27]. Hainan has a wet tropical monsoon climate, with a distinct wet season from May to October and a dry season from November to April. The annual average temperature is 23°C–25°C, and the annual average rainfall is 1,500–2,500 mm [28].

Field Survey

A total of 108 plots were established covering areas of tropical rainforest, tropical monsoon forest, mangrove forest, tropical coniferous forest, tropical evergreen and semi-evergreen shrubs, semi-natural vegetation, and artificial vegetation. Plots were mainly distributed in the central and southern mountainous areas because of their rich rare and endangered wild plants (Figure 1). Each plot included a 20×20 m-quadrat distributed at the central point, companied with four 20×20 m-quadrates distributed in the four diagonal directions from the central points.

A comprehensive survey was performed to identify plant species names, and measure the all individuals across the entirety of Hainan Island during 2012–2013. The survey of wild plants with extremely small populations in Hainan is permitted and approved by the Ministry of Agriculture of the People's Republic of China, and the Wildlife Protection Bureau of Hainan Province. Our work was conducted with a focus on the protection of wildlife and causing no threats or damage to wildlife, especially to endangered or protected species.

The list of the surveyed plants was in accordance with the species names of "Conservation Program for Wild Plants with Extremely Small Populations in China" [24], including *Cycas changjiangensis*, *Michelia odora*, *Chieniodendron hainanense*, *Ilex kaushue*, *Horsfieldia kingii*, *Paranephelium hainanense*, *Sonneratia hainanensis*, *Lumnitzera littorea*, *Hopea hainanensis*, *Bulbophyllum hainanense*, *Ceratostylis hainanensis*, *Cymbidium eburneum*, *Cymbidium insigne*, *Oxystophyllum changjiangense*, *Dendrobium hainanense*, *Dendrobium sinense*, *Dendrobium strongylanthum*, *Doritis pulcherrima*, *Pinalia quinquelamellosa*, *Dendrolirium tomentosum*, *Gastrochilus acinacifolius*, *Phaius hainanensis*, *Sunipia hainanensis*, and *Thrixspermum odoratum*. We found that four species (*Ilex kaushue*, *Bulbophyllum hainanense*, *Phaius hainanensis*, *Sunipia hainanensis*) that had previously been recorded in Hainan Island were not found in our field survey. Thus, further investigation is required to determine whether these species have disappeared from the wild. Among the remaining 20 species with extremely small populations, we recorded the distributed area, individual number and habitat characteristics (see Appendix S1).

Distribution Patterns

Distribution pattern refers to the appearance frequency and species richness of the extremely small populations. The appearance frequency of each extremely small population was calculated as the ratio of the number of quadrates in which a plant was found to the five quadrates.

Measurement of Affecting Factors

We measured 13 variables which were divided into five types. (1) Topographical: altitude, slope aspect, slope gradient, and exposure of rock. Altitude, slope aspect, and slope gradient were measured using the global positioning system, a compass, and a slope measuring instrument, respectively. And the exposure of rock was estimated by the ratio of the exposed surface stone area to the total area of the quadrates. (2) Climate: annual average

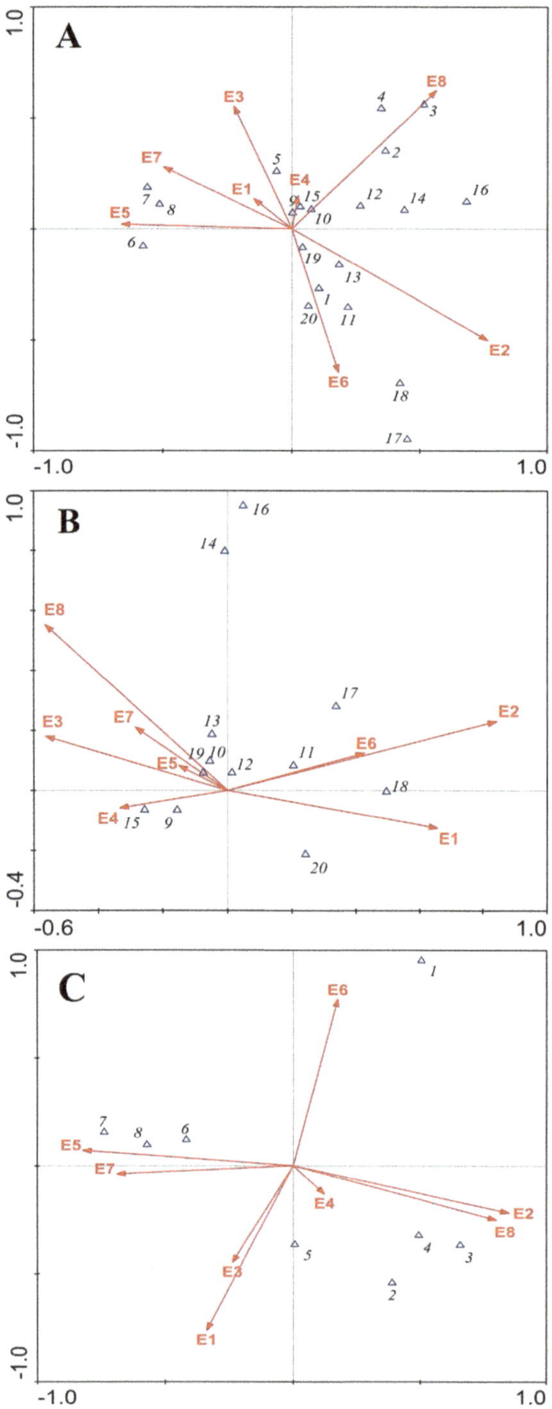

Figure. 2 Canonical correspondence analysis (CCA) between (A), total species; (B), orchid species; (C), non-orchid species with extremely small populations and eight factors. The numbers from 1 to 20 represent the 20 species with extremely small populations, as shown in Appendix S1. E1, canopy density; E2, altitude; E3, slope aspect; E4, slope gradient; E5, road quality; E6, distance between population and road; E7, surrounding population density; E8, land use type.

temperature and average annual precipitation. These variables were extracted for each site from the WorldClim database (http://www.worldclim.org) with a spatial resolution of 0.0083° (approximately 1 km^2 at the equator) [29]. (3) Vegetation: species richness and canopy structure. Species richness was calculated as the mean number of species found in the sample. Canopy structure included both leaf area index and canopy density, which were measured using hemispherical photographs at a height of 1.5 m. Leaf area index, referred to as *LAI* 4 Ring (*LAI*), is the effective leaf area index integrated over the zenith angles 0–60°. Canopy density is the percentage of shelter sky seen from beneath a forest canopy. A hemispherical photograph was taken in each 20×20 m-quadrat using a Nikon 4500 Coolpix Digital camera (Nikon Corporation, Tokyo, Japan) fitted with a Nikon FC-E8 fisheye converter (183° field-of-view, orthographic projection) [30]. The hemispherical images were analyzed with Gap Light Analyzer (GLA) software, Version 2.0 [31]. (4) Road: road quality and distance between population and road. According to the width and the lay material of road, road quality was assigned values of 1, 2, 3, and 4, which represent rural sandstone road (width 2–3 m), rural cement road (width 3–4 m), township road (width 4–5 m), and county road (width 5–6 m), respectively. Distances between population and road were measured using a tape measure and the global positioning system combined with Google Earth. (5) Human activity: surrounding population density and land use type. Surrounding population density was assigned values of 1, 2, 3, and 4, which indicated a depopulated zone, village, town, and suburb, respectively. Land use type was assigned values of 1, 2, 3, and 4, which represent dense forest, sparse woodland or shrubland, wasteland, and plantation or agricultural land, respectively.

Data Analysis

In this study, principal components analysis (PCA) was first used to analyze the 13 variables to reduce the dimension of factors. The top eight variables with absolute eigenvalues more than 0.620 were selected. Then the Canonical correspondence analysis (CCA) and the bivariate correlation analysis were performed to examine the relationship of these selected variables with the appearance frequency of each extremely small population.

We compared the species richness of the extremely small populations that were found at the different levels of the each selected variables using one-way ANOVA.

We divided the altitude values into six levels from 0 to 1800 m: 0–300 m (8 samples), 300–600 m (22 samples), 600–900 m (16 samples), 900–1200 m (44 samples), 1200–1500 m (10 samples), and 1500–1800 m (8 samples), and then analyzed the species similarity between neighboring altitudes using the following Jaccard similarity coefficient index: $C_j = C/(A+B-C)$ [32], where A and B represent the number of extremely small populations found in each neighboring altitude, and C indicates the number of common species found in each neighboring altitude.

We divided the slope aspect into eight ranges: north slope (0°–22.5° and 337.5°–360°, 16 samples), northeast slope (22.5°–67.5°, 16 samples), east slope (67.5°–112.5°, 14 samples), southeast slope (112.5°–157.5°, 11 samples), south slope (157.5°–202.5°, 12 samples), southwest slope (202.5°–247.5°, 8 samples), west slope (247.5°–292.5°, 12 samples), and northwest slope (292.5°–337.5°, 18 samples), and then calculated the frequency for each plant species with extremely small populations occurred at the eight slope aspects. Before the CCA analysis, the slope aspect was cosine transformed, and resulted in values ranged from 0 (south) to 2 (north) [33].

We divided the slope gradient into six levels ranged from 0° to 90°: 0°–15° (15 samples), 15°–30° (34 samples), 30°–45° (40 samples), 45°–60° (9 samples), 60°–75° (4 samples), and 75°–90° (6 samples). Canopy density was divided into six levels from 0.4 to

Table 2. Bivariate correlation analysis between the appearance of the extremely small populations and the measured variables.

Group	Number of distribution sites	Number of individuals	Species	Significant correlation with effecting factors
Orchid species	≤5	≤100	12, 13, 14, 15, 16, 17	E2(0.619*), E7 (−0.790**), E8 (−0.760**)
	≤5	>100	11, 18, 20	E2 (0.607*), E6 (0.515*)
	>5	>100	9, 10, 19	E2 (−0.516*), E1 (−0.575*), E6 (−0.578*), E8 (0.510*)
Non-Orchid species	≤5	≤100	2, 3, 6, 7, 8	E8 (−0.950**), E5 (−0.754**), E6 (0.574*)
	≤5	>100	1	E5 (−0.612*), E7 (−0.456*)
	>5	≤100	5	E6 (0.848**), E7 (−0.602*), E8 (−0.738**)
	>5	>100	4	E2 (0.615*), E6 (0.698*), E8 (−0.753**)

* Significant at the 5% level of probability, ** Significant at the 1% level of probability.
The numbers from 1 to 20 represent the 20 species with extremely small populations, as shown in Appendix S1. E1, canopy density; E2, altitude; E3, slope aspect; E5, road quality; E6, distance between population and road; E7, surrounding population density; E8, land use type.

1.0: 0.4–0.5 (8 samples), 0.5–0.6 (7 samples), 0.6–0.7 (12 samples), 0.7–0.8 (28 samples), 0.8–0.9 (24 samples), and 0.9–1.0 (29 samples). Distance correlation analysis was then performed to calculate the Pearson similarity coefficient, using the presence/absence data of each extremely small population for the six levels of slope gradient or canopy density. Distance between population and road was divided into five levels: 0–200 m (25 samples), 200–400 m (26 samples), 400–600 m (21 samples), 600–800 m (16 samples), and 800–1000 m (20 samples). Road quality, surrounding population density, and land use type were each divided into four levels according to their 1–4 scales.

PCA, bivariate correlation analysis and distance correlation analysis were performed using SPSS 19.0. CCA was performed using CANOCO version 4.5 for Windows, and a Monte Carlo permutation test was conducted with 1000 randomizations for the significance of both the variables and the ordination axes [34]. Figures were created with SigmaPlot 12.0.

Results

PCA of Affecting Factors

PCA showed that the accumulative contribution rate of the three principal components were 58.380% (Table 1). Thus the top eight principal component with absolute eigenvalues more than 0.620 were found. They were road (−0.821), canopy density (0.812), altitude (0.721), slope gradient (0.716), surrounding population density (−0.707), slope aspect (0.646), road quality (−0.660), and land use type (0.622). These eight variables, therefore, are predicted to affect the distribution patterns of wild plants with extremely small populations.

Relationships Between Appearance Frequency and Affecting Factors

CCA showed that there was a significant correlation ($P = 0.002$) between the eight variables and the ordination axes, using the Monte Carlo permutation test. The appearance frequencies of all species were affected by the eight factors, with a decreasing order of importance: altitude, land use type, distance between population and road, road quality, slope aspect, surrounding population density, canopy density, and slope gradient (Figure 2A). The most important factor affecting the appearance frequency of orchid species was altitude (Figure 2B), while the most important factor affecting the appearance frequency of non-orchid species was distance between population and road (Figure 2C).

Among the orchids, six species with less than 5 distribution sites and 100 individuals (see Appendix S1) were significantly and positively correlated with altitude ($P<0.05$), but were significantly negatively correlated with surrounding population density and agricultural activities ($P<0.01$). Three species with less than 5 distribution sites and more than 100 individuals (see Appendix S1) had significant positive correlations with altitude and distance between population and road ($P<0.05$). Three species with more than 5 distribution sites and 100 individuals (see Appendix S1) had significant negative correlations with altitude, canopy density, and distance between population and road ($P<0.05$), but were significantly and positively correlated with agricultural activities ($P<0.05$) (Table 2).

For non-orchid species, five species with less than 5 distribution sites and 100 individuals (see Appendix S1) had significant negative correlations with road quality and agricultural activities ($P<0.01$), but were significantly positively correlated with distance between population and road ($P<0.05$). One species with less than 5 distribution sites and more than 100 individuals (see Appendix S1) had significant negative correlations with road quality and surrounding population density ($P<0.05$). Two species with more than 5 distribution sites had significant positive correlations with altitude, and distance between population and road ($P<0.05$), but had extremely significant negative correlations with agricultural activities ($P<0.01$) (Table 2).

Relationships Between Species Richness and Affecting Factors

The species richness of plant species with extremely small populations decreased with the increasing levels of road quality, agricultural activities, and surrounding population density, but increased with increasing distance of population and road (Figure 3A, 3B). The species richness differed significantly among the five levels of distance between populations and road, as well as the four levels of road quality (Distance: $F_{(4,\ 103)} = 43.43$, $P<0.001$; Road Quality: $F_{(3,\ 104)} = 5.86$, $P = 0.001$; Figure 3A), with largest distance of 800–1000 m, and a rural sandstone road having the peak richness of species. The species richness among the four levels of population density or agricultural activities were significantly different (Population Density: $F_{(3,\ 104)} = 32.55$, $P<0.001$; Agricultural Activities: $F_{(3,\ 104)} = 12.22$, $P<0.001$; Figure 3B), with a depopulated zone or dense forest having the peak richness of species.

The species richness of extremely small populations showed a unimodal curve with the increasing altitude. The species richness among six levels of altitudes were significantly different ($F_{(5,\ 102)} = 3.98$, $P = 0.002$; Figure 3C), with medium altitudes of 900–1200 m having the peak richness of species. The largest Jaccard

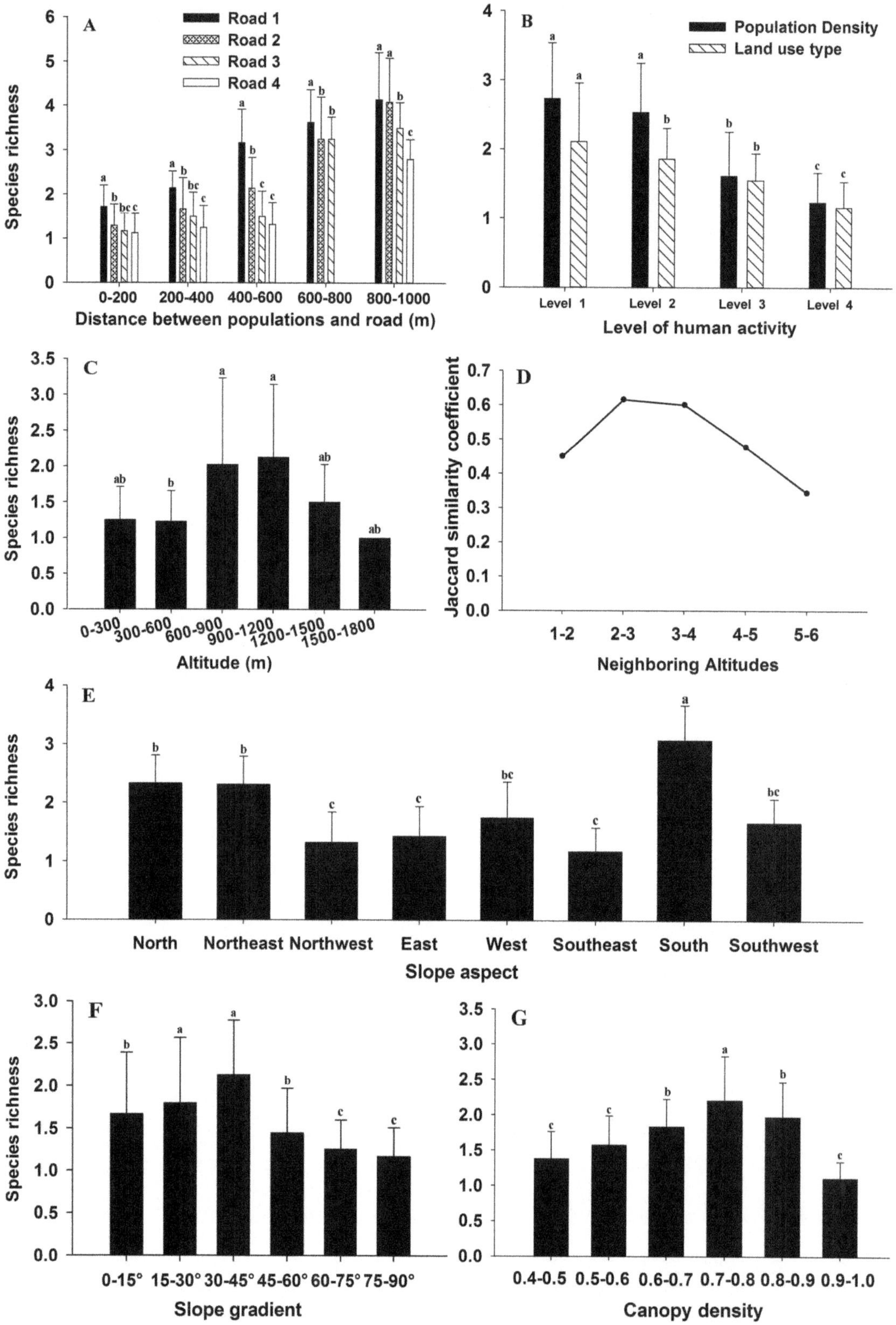

Figure. 3. The relationship between eight factors and the species richness of extremely small populations (mean and standard errors). (A), distance between populations and road (Road 1, Road 2, Road 3, and Road 4 represent rural sandstone road, rural cement road, township road, and county road, respectively); (B), different levels of surrounding population density and land use; (C), different levels of altitudes; (D), coefficient similarity values of neighboring altitudes (the numbers from 1 to 6 represent altitude ranges of 0–300 m, 300–600 m, 600–900 m, 900–1200 m, 1200–1500 m, and 1500–1800 m); (E), different levels of slope aspects; (F), different levels of slope gradients; (G), different levels of canopy density. Different Lowercase letters (a, b, c) indicate significant difference at $p<0.05$.

Table 3. Frequency of extremely small populations appearing on different slope aspects.

Species	North	Northeast	Northwest	East	West	Southeast	South	Southwest
1	0	0	0	0	0	0	0.75	0.25
2	0	0.5	0	0	0	0	0.125	0.375
3	0.444	0.333	0.111	0	0	0.111	0	0
4	0.154	0.076	0	0.154	0.076	0.385	0.076	0.076
5	0.333	0.083	0.167	0.083	0.083	0.083	0.083	0.083
6	0.250	0.375	0	0.125	0.250	0	0	0
7	0	0	0	0	0	0	1.000	0
8	0	0	0	0	0	0	1.000	0
9	0.079	0.105	0.079	0.105	0.105	0.184	0.211	0.132
10	0.222	0.222	0.111	0	0.278	0	0.056	0.111
11	0.182	0.045	0.273	0	0.318	0	0.091	0.091
12	0	0	0	0	0	0.667	0	0.333
13	0	0	0	0.167	0	0	0.167	0.667
14	0.400	0.200	0.400	0	0	0	0	0
15	0	0	0	0	0	0	1.000	0
16	0	0	0	0	0	1.000	0	0
17	0	0	0	0.500	0.500	0	0	0
18	0	0	0	0	0	0.333	0.333	0.333
19	0	0	0.111	0	0	0.333	0.333	0.222
20	0	0.125	0	0.125	0.125	0	0.375	0.250

The numbers from 1 to 20 represent the 20 species with extremely small populations, as shown in Appendix S1.

Table 4. Distance correlation analysis between extremely small populations and different slope gradients.

	1	2	3	4	5	6
1	1.000					
2	0.079	1.000				
3	0.663	0.426	1.000			
4	0.025	0.065	0.583	1.000		
5	0.209	0.161	0.412	0.396	1.000	
6	0.478	0.152	0.189	0.122	0.357	1.000

The numbers from 1 to 6 represent slope gradient ranges of 0°–15°, 15°–30°, 30°–45°, 45°–60°, 60°–75°, and 75°–90°.

similarity coefficient values for the extremely small populations were observed between two neighboring altitudes, including 600–900 m and 900–1200 m, and 900–1200 m and 1200–1500 m (Figure 3D).

The species richness among the eight levels of slope aspects were significantly different ($F_{(7,\ 100)} = 2.16$, $P = 0.012$; Figure 3E), with south aspect having the peak richness of species. Two species (*Chieniodendron hainanense, Dendrobium hainanense*) were found on all slope aspects. Three species (i.e. *Horsfieldia kingii, Paranephelium hainanense, Gastrochilus acinacifolius*) were mainly observed on the north, northeast, and northwest aspects. Seven species (i.e. *Cycas changjiangensis, Thrixspermum odoratum, Doritis pulcherrima, Dendrobium strongylanthum, Cymbidium eburneum, Pinalia quinquelamellosa, Dendrolirium tomentosum*) mainly occurred on the south, southeast, and southwest aspects (Table 3).

The species richness among the six levels of slope gradient were significantly different ($F_{(5,\ 102)} = 3.99$, $P = 0.002$; Figure 3F), with the medium level of slope gradient range from 30° to 45° having the peak richness of species. The slope gradient range from 30° to 45° had a larger Pearson similarity coefficient values with the ranges of 0°–15°, 15°–30°, 45°–60°, and 60–75° respectively, than those between other slope gradient (Table 4).

The species richness among the six levels of canopy density were significantly different ($F_{(5,\ 102)} = 13.34$, $P < 0.001$; Figure 3G), with the medium level of canopy density range 0.7–0.8 having the peak richness of species. The canopy density ranged from 0.7 to 0.8 had a larger Pearson similarity coefficient values with the ranges of 0.4–0.5, 0.5–0.6, 0.6–0.7, and 0.8–0.9 respectively, than those between other canopy density (Table 5).

Discussion

Factors Affecting the Appearance Frequency

As expected, we found that plant species with extremely small populations often lived in a fragmental habitat (Appendix S1). For example, four species (i.e. *Ilex kaushue, Bulbophyllum hainanense, Phaius hainanensis, Sunipia hainanensis*) with extremely small populations were not found in our field survey, and the remaining 20 species underwent human disturbances and threats. Thus the population size of these species are smaller than the minimum number requirements inhibiting from extinction, suggesting that these extremely small populations face the threat of extinction if no protective actions are taken. Our findings have been demonstrated by the view that the main factors causing rare and endangered plants susceptible to extinction are habitat destruction, plant overexploitation and the ecological environment deterioration [25,35,36].

Although there is one or several factors causing plant species endangerment, these factors may be completely different for different species [35]. Therefore, we found that the leading factors changing the appearance frequency of extremely small populations differ among plant species. For example, the dominant factors affecting the appearance frequency of orchid and non-orchid species were altitude and the distance between population and road, respectively (Figure 2B, Figure 2C).

For most of the endangered plants, the direct factors leading them susceptible to extinction are human activities [25,36,37]. This point is proved by our results that the appearance frequencies of 17 species with extremely small populations had significant negative correlations with road quality, agricultural activities, or the surrounding human population density (Table 2). These findings may result from that most of these species have a highly

Table 5. Distance correlation analysis between extremely small populations and different canopy densities.

	1	2	3	4	5	6
1	1.000					
2	0.454	1.000				
3	0.243	0.000	1.000			
4	0.375	0.564	0.342	1.000		
5	0.193	0.152	0.124	0.344	1.000	
6	0.059	0.130	0.243	0.193	0.150	1.000

The numbers from 1 to 6 represent canopy density ranges of 0.4–0.5, 0.5–0.6, 0.6–0.7, 0.7–0.8, 0.8–0.9, and 0.9–1.0.

economic utilization value or excellent ornamental characteristics, and thus that they are often overexploited by local people.

In this study, most orchid species were concentrated in high altitudinal habitats (Table 2). Seed germination and growth of some orchids are dependent on high altitudinal habitats, which not only have favorable temperatures and higher humidity, but also avoid human disturbance [38,39]. Additionally, we found that the appearance frequencies of three orchid species (*Dendrobium hainanense, Oxystophyllum changjiangense, Dendrolirium tomentosum*) were significantly and positively correlated with roads and agricultural activities (Table 2). These species were also concentrated along roadsides and in areas of human activity (Appendix S1), suggesting that the distribution of some endangered orchid species benefit from human activities or roads. Roads usually have larger negative effects on the distribution of native plants than positive effects [40,41]. However, the frequent appearance of the orchid species in this study may probably result from that the survival of these species benefits from road disturbance [42], and that an intermediate level of disturbance can help prevent them from becoming endangered [43].

Factors Affecting Species Richness

Our results showed that the species richness of extremely small populations was always reduced by human activities (Figure 3A, 3B). The main factor causing the decreasing species richness for these species probably being the habitat destruction and fragment because of human disturbance, such as land use type, roads, etc [16,44,45,46]. For example, we found that road quality made a decreasing richness of the extremely small populations.

In this study we found that the peak richness of the small populations always occurred at the medium level across an environmental gradient (Figure 3C, 3F, and 3G), and it also had a larger species similarity values (Figure 3D, Table 4, Table 5). Our results were confirmed by many previous studies [47,48,49,50]. For example, Jiang et al. (2002) and Cui et al. (2006) found that many rare and endangered plants are often concentrated at medium altitudes [51,52]. On the other hand, slope gradient can reflect the soil moisture and its nutrient condition [53,54], the distribution of most rare plants are concentrated in a medium level of slope gradients with favorable soil conditions [55,56]. Also, since the increasing forest canopy can improve the forest microhabitat conditions, and regulate the distribution, composition, and growth of forest subordinate layers [57], the seedlings of most of the rare and endangered plants thus grew better with high abundance under medium canopy densities [58].

Our results showed that the peak richness of extremely small populations always appeared at the south aspect, and that their frequencies were always the largest on sunny or semi-sunny slope aspects (Figure 3E, Table 4), due to their drought tolerance and sun preference. But several other studies have shown that the species richness of endangered plants and their individual numbers is higher on shady and semi-shady slope aspects than those on other aspects, because they prefer to high humidity and low irradiance [55,59,60]. These inconsistent results may account for that the different extremely small populations can adapt to different habitats with different humidity and irradiance. We, therefore, should take knowledge of how the external factors, such as the ecological environment, land use type, roads, human activity, etc., affect the distribution of the extremely small populations for the better protecting them in the future.

Author Contributions

Conceived and designed the experiments: YKC DHL XBY. Performed the experiments: YKC QY WQL. Analyzed the data: YKC WXL. Wrote the paper: YKC WXL.

References

1. Odum EP (1971) Fundamentals of ecology. W B Saunders Co.
2. Brooker RW (2006) Plant–plant interactions and environmental change. *New Phytologist* 171(2): 271–284.
3. Allan E, Weisser W, Weigelt A, Roscher C, Fischer M, et al. (2011) More diverse plant communities have higher functioning over time due to turnover in complementary dominant species. *Eric Allan* 108(41): 17034–17039.
4. Sasaki T, Lauenroth WK (2011) Dominant species, rather than diversity, regulates temporal stability of plant communities. *Oecologia* 166(3): 761–768.
5. Albrecht MA, McCue KA (2010) Changes in demographic processes over long time scales reveal the challenge of restoring an endangered plant. *Restoration Ecology* 18: 235–243.
6. Ren H, Zhang Q, Wang Z, Guo Q, Wang J, et al. (2010) Conservation and possible reintroduction of an endangered plant based on an analysis of community ecology, a case study of *Primulina tabacum* Hance in China. *Plant Species Biology* 25(1): 43–50.
7. García MB, Espadaler X, Olesen JM (2012) Extreme Reproduction and Survival of a True Cliffhanger, The Endangered Plant *Borderea chouardii*.(Dioscoreaceae). *PloS one* 7(9): e44657 doi101371/journalpone0044657
8. Neel MC, Che-Castaldo JP (2013) Predicting Recovery Criteria for Threatened and Endangered Plant Species on the Basis of Past Abundances and Biological Traits. *Conservation Biology* 27(2): 385–397.
9. Lauterbach D, Ristow M, Gemeinholzer B (2011) Genetic population structure, fitness variation and the importance of population history in remnant populations of the endangered plant *Silene chlorantha*. (Willd). Ehrh. (Caryophyllaceae). *Plant Biology* 13(4): 667–777.
10. Amat ME, Silvertown J, Vargas P (2013) Strong Spatial Genetic Structure Reduces Reproductive Success in the Critically Endangered Plant Genus *Pseudomisopates*. *Journal of Heredity* 104(5): 692–703.
11. Levine JM, McEachern AK, Cowan C (2010) Do competitors modulate rare plant response to precipitation change. *Ecology* 91: 130–140.
12. Reckinger C, Colling G, Matthies D (2010) Restoring populations of the endangered plant Scorzonera humilis, influence of site conditions, seed source, and plant stage. *Restoration Ecology* 18(6): 904–913.
13. Mason HL (1946) The edaphic factor in narrow endemism.I.The nature of environmental influences. *Madrono* 8: 209–226.
14. Kruckeberg AR (1985) Biological aspects of endemism in higher plants. *Annual Review of Ecology and Systematics* 16: 447–479.
15. Fiedler PL, Ahouse JJ (1992) Hierarchies of cause, Toward an understanding of rarity in vascular plant species. In Fiedler PL and Jain S. eds. The Theory and Practice of Nature Conservation Preservation and Management. NEW York. Chapman and Hall. pp: 23–47.
16. Reed RA, Johnsonbarnard J, Baker WL (1996) Contribution of roads to forest fragmentation in the Rocky Mountains. *Conservation Biology* 10: 1098–1106.
17. Compilation Group of China's Biodiversity (1998) China's biodiversity, a country study. Beijing, China Environmental Science Press.
18. Forman RTT (2000) Estimate of the area affected ecologically by the road system in the United States. *Conservation Biology* 14: 31–35.
19. Zhang YB, Ma KP (2008) Geographic distribution patterns and status assessment of threatened plants in China. *Biodivers Conserv* 17: 1783–1798.
20. Zang RG, Ding Y (2009) Forest recovery on abandoned logging roads in a tropical montane rain forest of Hainan Island, China. *Acta Oecol* 35: 462–470.
21. IUCN (2013) The IUCN Red List of Threatened Species. Version2013. 1. Available at http://www.iucnredlist.org (accessed 2013 July 2).
22. Frankham R (1997) Do island populations have less genetic variation than mainland populations? *Heredity* 78: 311–327.
23. Francisco-Ortega J, Santos-Guerra A, Kim SC, Crawford DJ (2000) Plant genetic diversity in the Canary Islands, a conservation perspective. *American Journal of Botany* 87: 909–919.
24. State Forestry Administration of China (2011) The saving and conservation program on extremely small populations in China. State Forestry Administration of China.
25. Ren H, Zhang Q, Lu H, Liu H, Guo Q, et al. (2012) Wild Plant Species with Extremely Small Populations Require Conservation and Reintroduction in China. *AMBIO* 41(8): 913–917.

26. Guo Z, Zhang Y, Deegen P, Uibrig H (2006) Economic analyses of rubber and tea plantations and rubber-tea intercropping in Hainan, China. *Agroforestry systems* 66: 117–127.
27. Wang BS, Peng SL, Guo L, Ye YH (2007) Diversity of tropical forest landscape-type in Hainan Island, China. *Acta Ecologica Sinica* 27(5): 1690–1695. (in Chinese with English abstract).
28. Zhang HD (2001) The diversity of the Hainan flora. *Ecologic Science* 20: 1–10. (in Chinese with English abstract).
29. Hijmans RJ, Cameron SE, Parra JL, Jones PG, Jarvis A (2005) Very high resolution interpolated climate surfaces for global land areas International. *Journal of Climatology* 25: 1965–1978.
30. Reid TD, Essery RLH (2013) New methods to quantify canopy structure of leafless boreal birch forest from hemispherical photographs. *Open Journal of Forestry* 3(2): 70–74.
31. Frazer GW, Canham CD, Lertzman KP (1999) Gap Light Analyzer (GLA), Version 2.0, Imaging software to extract canopy structure and gap light transmission indices from true-colour fisheye photographs, users manual and program documentation. Simon Fraser University, Burnaby, British Columbia, and the Institute of Ecosystem Studies. New York: Millbrook.
32. Yuan JF, Hu RY, Shen JH, Zhang L, Zhang XY, et al. (2011) Comparison of Species Composition and Diversity of Four Successional Forest Communities in Zhejiang Province, East China. *Bulletin of Botanical Research* 31(1): 61–66. (in Chinese with English abstract).
33. Beers TW, Dress PE, Wensel LC (1966) Aspect transformation in site productivity research. *J Forestry* 64: 691–692.
34. Lepš J, Šmilauer P (2003) Multivariate Analysis of Ecological Data Using CANOCO. New York: Cambridge University Press.
35. Zhang WH, Zu YG, Liu GB (2002) Population Ecological Characteristics and Analysis on Endangered Cause of Ten Endangered Plant Species. *Acta Ecologica Sinica* 22(9): 1512–1519. (in Chinese with English abstract).
36. Groom MJ, Meffe GK, Carroll CR (2006) Principles of conservation biology. Sunderland: Sinauer.
37. Victor AE, Dold AP (2003) Threatened plants of the Albany Centre of Floristic Endemism, South Africa. *South African Journal of Science* 99 (9–10) : 437–446.
38. Jiang HY, Yang FS, Song XQ (2011) Preliminary Study on artifical sowing of *Dendrobium sinense* in natural habitat in Bawangling Natural Reserve, Hainan Island. *Tropical Forestry* 39(2): 40–43 (in Chinese).
39. Gao JG, Zhang Y, Wu YH, Qiu ZJ, Xu GD, et al. (2012) Conservation strategies for *Ulmus elongata* based on the analysis of biological and ecological factors. *Acta Ecologica Sinica* 32(17): 5287–5298. (in Chinese with English abstract).
40. Forman RTT, Alexander LE (1998) Roads and their major ecological effects. *Annual Review of Ecology and Systematics* 207–231.
41. Gelbard JL, Belnap J (2003) Roads as conduits for exotic plant invasions in a semiarid landscape. *Conservation Biology* 17: 420–432.
42. Spooner PG (2005) Response of acacia species to disturbance by road works in roadside environments in southern new south wales, australia. *Biological Conservation* 122: 231–242.
43. Amat ME, Vargas P, Gómez JM (2013) Effects of human activity on the distribution and abundance of an endangered Mediterranean high-mountain plant (*Erysimum penyalarense*). *Journal for Nature Conservation* 21(5): 262–271.
44. Kareiva P, Wennergren U (1995) Connecting landscape patterns to ecosystem and population processes. *Nature* 373: 299–302.
45. Lenzen M, Lane A, Cooper AW, Williams M (2009) Effects of land use on threatened species. *Conservation Biology* 23(2): 294–306.
46. Watkins RZ, Chen JQ, Pickens J, Brosofske KD (2003) Effects of forest roads on understory plants in a managed hardwood landscape. *Conservation Biology* 17: 411–419.
47. Whittaker RH, Niering WA (1965) Vegetation of the Santa Catalina Mountains, Arizona, a gradient analysis of the south slope. *Ecology* 46: 429–452.
48. Kessler M, Parris BS, Kessler E (2001) A comparison of the tropical montane pteridophyte floras of Mount Kinabalu, Borneo, and Parque Nacional Carrasco, Bolivia. *Journal of Biogeography* 28: 611–622.
49. Kessler M (2002) The elevational gradient of Andean plant endemism, varying influences of taxon-specific traits and topography at different taxonomic levels. *Journal of Biogeography* 29: 1159–1165.
50. Wang GH, Zhou GS, Yang LM, Li ZQ (2003) Distribution, species diversity and life-form spectra of plant communities along an altitudinal gradient in the northern slopes of Qilianshan Mountains, Gansu, China. *Plant Ecology* 165(2): 169–181.
51. Jiang MX, Deng HB, Cai QH (2002) Distribution pattern of rare plants along riparian zone and its implication for conservation in Shennongjia area. *Chinese journal of Applied Ecology* 13(11): 1373–1376. (in Chinese with English abstract).
52. Cui Y, Qi PC, Chen XL (2006) A study on verdical distribution and protection of rare plants in Baishuijiang National Reserve, Gansu. *Guihaia* 26(6): 660–664. (in Chinese with English abstract).
53. Tokuchi N, Takeda H, Yoshida K, Iwatsubo G (1999) Topographical variations in a plant–soil system along a slope on Mt Ryuoh, Japan. *Ecological Research* 14(4): 361–369.
54. Qiu Y, Fu BJ, Wang J, Chen LD (2001) Soil moisture variation in relation to topography and land use in a hillslope catchment of the Loess Plateau, China. *Journal of Hydrology* 240(3–4): 243–263.
55. Song P, Hong W, Wu CZ, Feng L, Lan SR (2005) Topographical Pattern of Dominant Populations of *Pinus taiwanensis* Forest in Wuyi Mountain. *Journal of Mountain Science* 23(5): 631–635.
56. Tong YW, Xiang WH, Wang ZW, Durka W, Fischer M (2013) Effects of topography, neighboring plants and size-dependence of *Machillus thunbergii* on sapling growth and survivorship. *Biodiversity Science* 21(3): 269–277. (in Chinese with English abstract).
57. Martens SN, Breshears DD, Meyer CW (2000) Spatial distributions of understory light along the grassland/forest continuum, effects of cover, height, and spatial pattern of tree canopies. *Ecological Modelling* 126(1): 79–93.
58. Qiao Q, Qin XS, Xing FW, Chen HF, Liu DM (2011) Death causes and conservation strategies of the annual regenerated seedlings of rare plant, *Bretschneidera sinensis*. *Acta Ecologica Sinica* 31(16): 4709–4716. (in Chinese with English abstract).
59. Xiao X, Su ZX, Yan XC (2005) Effects of aspect on distribution pattern of *Taxus chinensis* population in Yele, Sichuan Province, An analysis based on patches information. *Chinese journal of Applied Ecology* 16(6): 985–990. (in Chinese with English abstract).
60. Singh KN, Kumar A, Lal B, Todaria NP (2008) Species Diversity and Population Status of Threatened Plants in Different Landscape Elements of the Rohtang Pass, Western Himalaya. *Journal of Mountain Science* 5(1): 73–83.

2

Ecological Effects of the Invasive Giant Madagascar Day Gecko on Endemic Mauritian Geckos: Applications of Binomial-Mixture and Species Distribution Models

Steeves Buckland[1,2]*, Nik C. Cole[1,3,4], Jesús Aguirre-Gutiérrez[5,6], Laura E. Gallagher[2], Sion M. Henshaw[2], Aurélien Besnard[7], Rachel M. Tucker[4,8], Vishnu Bachraz[2], Kevin Ruhomaun[2], Stephen Harris[1]

1 School of Biological Sciences, University of Bristol, Bristol, United Kingdom, **2** National Parks & Conservation Service, Reduit, Mauritius, **3** Durrell Wildlife Conservation Trust, Jersey, Channel Islands, **4** Mauritian Wildlife Foundation, Vacoas, Mauritius, **5** Naturalis Biodiversity Center, Leiden, The Netherlands, **6** Institute for Biodiversity and Ecosystems Dynamics (IBED), University of Amsterdam, Amsterdam, The Netherlands, **7** Laboratoire de Biogeographie et d'Ecologie des Vertébrés, Centre d'Ecologie Evolutive et Fonctionnelle, Montpellier, France, **8** Department of Animal and Plant Sciences, University of Sheffield, Sheffield, United Kingdom

Abstract

The invasion of the giant Madagascar day gecko *Phelsuma grandis* has increased the threats to the four endemic Mauritian day geckos (*Phelsuma* spp.) that have survived on mainland Mauritius. We had two main aims: (i) to predict the spatial distribution and overlap of *P. grandis* and the endemic geckos at a landscape level; and (ii) to investigate the effects of *P. grandis* on the abundance and risks of extinction of the endemic geckos at a local scale. An ensemble forecasting approach was used to predict the spatial distribution and overlap of *P. grandis* and the endemic geckos. We used hierarchical binomial mixture models and repeated visual estimate surveys to calculate the abundance of the endemic geckos in sites with and without *P. grandis*. The predicted range of each species varied from 85 km^2 to 376 km^2. Sixty percent of the predicted range of *P. grandis* overlapped with the combined predicted ranges of the four endemic geckos; 15% of the combined predicted ranges of the four endemic geckos overlapped with *P. grandis*. Levin's niche breadth varied from 0.140 to 0.652 between *P. grandis* and the four endemic geckos. The abundance of endemic geckos was 89% lower in sites with *P. grandis* compared to sites without *P. grandis*, and the endemic geckos had been extirpated at four of ten sites we surveyed with *P. grandis*. Species Distribution Modelling, together with the breadth metrics, predicted that *P. grandis* can partly share the equivalent niche with endemic species and survive in a range of environmental conditions. We provide strong evidence that smaller endemic geckos are unlikely to survive in sympatry with *P. grandis*. This is a cause of concern in both Mauritius and other countries with endemic species of *Phelsuma*.

Editor: Danilo Russo, Università degli Studi di Napoli Federico II, Italy

Funding: This study was funded by Oregon Zoo, Chicago Herpetological Society and Sigma Xi. SB was supported by the University of Bristol. The funders had no role in study design, data collection and analysis, decision to publish, or preparation of the manuscript.

Competing Interests: The authors have declared that no competing interests exist.

* E-mail: stvsbuckland@yahoo.com

Introduction

Invasive alien species (IAS) are capable of establishing, dispersing and causing harm to indigenous species [1]. They can cause cascading effects in native ecosystems by disrupting trophic interactions and sharing ecological resources with native species [2–4]. Although IAS have led to the extinction of many endemic species [5], they are still being spread around the world. Oceanic islands are at greater risk [6,7] as island endemics have evolved in the absence of predators and do not have anti-predator defence mechanisms [8]. So successful invaders encounter less competition from island endemics and, being genetically more adaptable, tend to expand their niches [9].

Reptile extinctions in Mauritius have mostly been due to the introduction of mammalian invaders, with 69% of its endemic reptile diversity lost since human colonisation at the end of the 16th century [10,11]. Only one terrestrial skink (*Gongylomorphus bojerii fontenayi*, Macchabé skink) and four arboreal day geckos (*Phelsuma cepediana*, blue-tailed day gecko; *Phelsuma guimbeaui*, lowland forest day gecko; *Phelsuma ornata*, ornate day gecko; *Phelsuma rosagularis*, upland forest day gecko) survive on mainland Mauritius. While *P. cepediana* consists of three un-described species [11], they are treated as a single species for this analysis because they cannot be distinguished phenotypically with confidence. A further seven endemic species of reptile survive on seven offshore islets, one of which is *Phelsuma guentheri* (Günther's gecko), a large species of day gecko that co-existed with most, if not all, of the smaller species of *Phelsuma* on mainland Mauritius before the 1800s [12]. However, the recent introduction of *Phelsuma grandis* (giant Madagascar day gecko) is believed to threaten the four surviving smaller species of *Phelsuma* on mainland Mauritius. *P. grandis* was originally introduced in Baie du Tombeau (Figure 1) in the early 1990s through the pet trade and has since been deliberately moved elsewhere. This species can attain a length of 24–30 cm [13], nearly double the length of the four endemic geckos [14].

Although the export of *P. grandis* is controlled through CITES (Appendix II), it has also been introduced to Afghanistan [15],

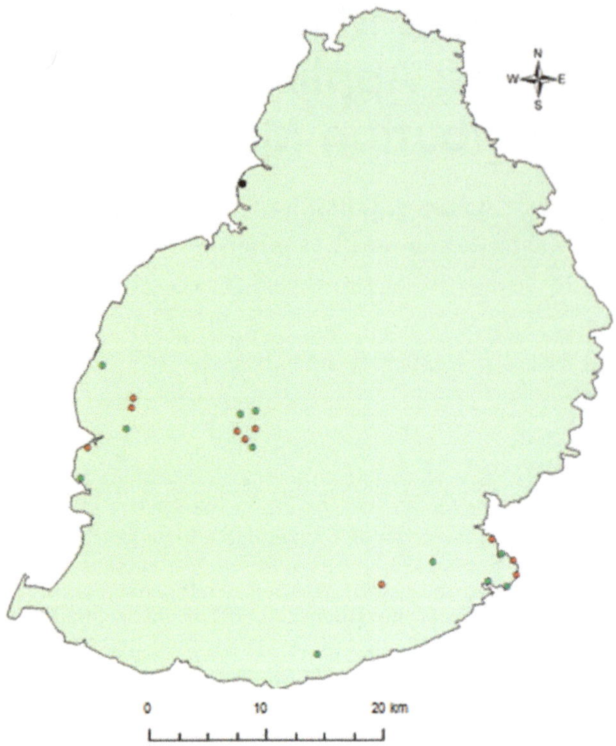

Figure 1. Locations of the ten sites with *Phelsuma grandis* (red) and 11 sites without *Phelsuma grandis* (green) used to estimate the abundance of endemic species of *Phelsuma*. The black dot indicates Baie du Tombeau, the location where *Phelsuma grandis* was first introduced.

Florida [16], Hawaii [17] and Réunion [18]. Whether *P. grandis* has significant impacts in these localities is currently unknown, but it has been observed predating the endemic *Phelsuma borbonica* (gecko vert des hauts) and introduced *Gehyra mutilata* (stump-toed gecko) in Réunion [18]. In Mauritius, *P. grandis* has been observed to predate on geckos such as *P. ornata* and the introduced species *G. mutilata*, *Hemidactylus frenatus* (common house gecko) and *Hemidactylus parvimaculatus* (Sri Lankan house gecko).

Apart from the anecdotal sightings of predation and localised presence of *P. grandis* in Mauritius, our current knowledge of its ecological impact is limited. Therefore, we first investigated the potential distribution of the endemic species of *Phelsuma* in mainland Mauritius and *P. grandis* to determine whether they could share the same spatial distribution. We used Species Distribution Modelling (SDM) to predict the distribution of the endemic geckos and *P. grandis*, adjusted to their physical environment. SDM is widely used in studies of biogeography, global change ecology, conservation biology [19–22] and invasion biology [23–25], and is a valuable management tool [26]. We hypothesised that *P. grandis* is a "generalist" and so is able to occupy a broad climatic and habitat range which would overlap with the four endemic geckos. Second, we surveyed sites with and without *P. grandis* to determine their impact upon the number of endemic geckos. Abundance estimation in many ecological studies is based on counting the number of individuals seen without accounting for probability of detection of the species being studied [27]. Lack of detection can be affected by (i) site-level covariates such as behavioural changes in the presence of a competitor/predator and (ii) observation-level covariates such as climatic conditions, which might influence detection [27–29]. Hence, we used hierarchical binomial mixture modelling (BMM) [30] that implements a metapopulation approach to adjust abundance estimation to the probability of detection [31]. This enabled us to calculate the true abundance (hereafter, abundance) of geckos at each site. BMM has mainly been used in bird studies [32–33], but has also been used to estimate the abundance of amphibians [34–35] and mammals [36]. We hypothesised that *P. grandis* would have a negative impact on the abundance of the four endemic geckos. We then use this information to discuss the management of *P. grandis* in Mauritius and elsewhere.

Methods

Ethical statement

This study was approved by the University of Bristol's Ethical Review Committee (University Investigation Number UB/11/031) and the National Parks and Conservation Service, Ministry of the Agro-Industry, Mauritius.

Predicting the distribution and overlap of endemic geckos and *P. grandis*

Collecting presence data. Articles were published in national newspapers requesting that members of the public report *P. grandis* sightings. We received more than 100 telephone calls giving locations of *P. grandis* across the island. Site visits were conducted to confirm their presence and geographical locations were recorded. Presence data on endemic species of *Phelsuma* were obtained through extensive field surveys in private and public forests between January and March 2010. Surveys were conducted along existing tracks between 07:00 and 19:00; any *Phelsuma* species seen were recorded. Opportunistic observations in urban areas were also collected between January 2010 and August 2012.

Environmental layers. We obtained 19 bioclimatic variables from WorldClim (http://www.worldclim.org/). They represented the annual trends in seasonality in Mauritius over the period 1950 to 2000 [37]. Elevation data were acquired from a Digital Elevation Model obtained from EOSDIS (http://reverb.echo.nasa.gov/r). We derived topographic information such as aspect and slope from the Digital Elevation Model. We also obtained Normalised Difference Vegetation Index (NDVI) data from a 2008 Landsat ETM+ dataset (http://earthexplorer.usgs.gov/).

All the variables were tested for multicollinearity using the pairwise Pearson's correlation test and only the most biologically meaningful ones with correlation values <0.7 were kept for further analyses. The nine retained variables were: BIO3 (isothermality), BIO4 (temperature seasonality), BIO7 (temperature annual range), BIO13 (precipitation of wettest period), BIO19 (precipitation of coldest quarter), aspect, elevation, NDVI and slope. As the bioclimatic variables had a resolution of 1 km^2 and the other four layers had a spatial resolution of 0.0009 km^2, we rescaled the five bioclimatic variables to 0.0009 km^2 to take the advantage of the high resolution of the other layers to produce a map with 2,041,627 grid cells.

Model fitting and evaluation. Prior to analysis, we removed duplicate sightings of each species at the same or different sites within a grid cell. From the original 950 records, 777 presence data remained (*P. cepediana* 265; *P. grandis* 74; *P. guimbeaui* 174; *P. ornata* 220; *P. rosagularis* 44). We selected five algorithms to predict distribution and construct our ensemble models [38]: (i) Maximum Entropy Modelling (MaxEnt) [39]; (ii) Generalised Boosted Model (GBM) [40]; (iii) Random Forest (RF) [41]; (iv) Generalised Linear Models (GLMs) [42]; and (v) Generalised Additive Model (GAM) [43]; for model specifications for these algorithms see [44]. We implemented this particular approach because it is more accurate than using a single algorithm [38].

We randomly divided the original dataset, using 80% to construct the models and 20% to validate their accuracy. We carried out 10 repetition runs to obtain robust estimates of the species distributions, and model accuracy [45] was tested by means of the Area Under the Curve (AUC) of a Receiver Operating Characteristic plot (ROC) [46]. The AUC is a threshold independent measure of accuracy to evaluate the performance of SDMs [38,47,48]. Absence data were needed to evaluate model performance for all the algorithms except MaxEnt. Since only presence data were collected, we randomly generated background pseudo-absences i.e. locations with no sightings of a particular species were selected at random and assigned as absent. The number of pseudo-absences per species was ten times the number of sites with presence data [49].

We considered an AUC<0.7 as a poor model and an AUC>0.9 as a highly accurate model [46]. We generated 50 models per species (i.e. 250 models in total) and models with an AUC>0.7 were selected to construct an ensemble model [44]. The ensemble model corresponded to the median probability of occurrence across the selected models for each grid cell. The median value was chosen because it was less sensitive to outliers than the mean. We converted the continuous predictions (Figure S1) into presence-absence prediction maps (Figures 2 and 3) to carry out further analyses. For this conversion, we applied an optimal probability threshold that maximises the sensitivity and specificity of the created models [50,51]. The presence/absence map was used to project the potential distribution and overlap between each endemic species and *P. grandis*. We also calculated the predicted distribution and overlap between the combined range of the four species of endemic geckos and *P. grandis*. Since we considered *P. guimbeaui* and *P. rosagularis* to be the species most at risk, with only 30 and two known subpopulations respectively, we calculated the distance between the predicted range of *P. grandis* and known subpopulation of *P. guimbeaui* and *P. rosagularis*.

Niche breadth. We examined the environmental breadth indices to determine whether *P. grandis* had a narrow (specialist = 0) or wide (generalist = 1) niche. We used the binary prediction of presence and absence maps (Figures 2 and 3), modelled from the combined influence of all the environmental predictors and occurrence of each species, to estimate niche breadths. We used Levins' niche breadth metrics [52], where a value of 0 was equivalent to only one grid cell being suitable and a value of 1 was equivalent to all grid cells being suitable [53].

Data handling. Data processing, model fitting and projection were conducted in R (R 3.0.1 Development Core Team 2013) BIOMOD2 package [44], while breadth metrics were estimated in ENMTools (http://enmtools.blogspot.co.uk/). SDM predicted range and overlap estimation and GIS manipulations were carried out in ARCGIS 10.1 (ESRI, Redlands, California, USA) and Quantum GIS 1.8 (http://www.qgis.org/en/site/).

Impact of *P. grandis* on the abundance of endemic geckos

Field surveys. We conducted visual estimate surveys (VES) at ten sites where *P. grandis* were present and 11 sites where they were believed absent (Figure 1). We used two types of categorical habitat data: (i) building or non-building and (ii) vegetation type i.e. palm or non-palm. There were four building sites (two with, two without *P. grandis*), four non-palm sites (two with, two without *P. grandis*) and 13 palm sites (six with, seven without *P. grandis*). With the exception of one palm site, each *P. grandis* site was matched with a site without *P. grandis* that had comparable habitat characteristics in terms of area, number of trees and tree diameter at breast height (DBH). Building sites were human dwellings in residential areas along the south-east coast of Mauritius, with an average area of 176 m^2 and 189 m^2 for sites with and without *P. grandis*. Palm sites contained trees from the Arecaceae family. Each consisted of an isolated clump of four to 12 palm trees with a height less than 12 m and occupying an area from 50 m^2 to 100 m^2. Average basal tree coverage (based on DBH) was 0.26 m^2 and 0.21 m^2 in sites with and without *P. grandis*. Non-palm sites were isolated and covered an area of ~400 m^2 with a maximum of four trees up to 20 m high, mainly *Terminalia arguna* (terminalia) and *Ficus benghalensis* (banyan tree). The average basal tree coverage was 10.12 m^2 and 10.00 m^2 in sites with and without *P. grandis*.

To calculate the abundance of endemic geckos, one person slowly walked round each site, scanning every tree and wall surface, counting all the geckos seen and identifying them to species. We used features such as bite marks, size and general colour patterns to avoid repeat counts of the same gecko. This was repeated for 15 minutes each hour during peak activity periods from 08:00 to 11:00 and 14:00 to 18:00, i.e., six times in a day. All 21 sites were surveyed for one day between 16 and 23 May 2011. We assumed that the different time slots and minimum distance between sites (>100 m) would ensure that the sites would be temporally and spatially independent. Temperature was recorded at the start of each count using a Lutron Lm-8000 environmental meter (Lutron Electronic Enterprise Co. Ltd., Taipei, Taiwan). Cloud cover was estimated visually to the nearest 5%. Temperature and cloud cover were included as observation-level covariates to test their respective effects on detection probabilities. Site-level covariates, such as status (presence or absence) of *P. grandis* and habitat types were modelled to investigate their effects on abundance.

Data analyses. The emergence of hierarchical models has decreased the dependence on labour-intensive mark recapture in the estimation of population parameters such as abundance and occupancy [54]. Simulation studies have demonstrated that BMM was robust in the estimation of abundance [55], but dependent upon meeting key assumptions i.e. that the population is closed during the survey and that the sites are spatially and temporally independent [56].

We developed a BMM using two probability distributions:-

$$\mathrm{Ni} \sim \mathrm{dist}(\lambda) \text{ and}$$

$$\mathrm{Yij} \sim \mathrm{Binomial}(\mathrm{Ni}, \mathrm{pij})$$

where Ni = the abundance of geckos in site i; dist = either Poisson, negative binomial or zero-inflated Poisson distribution; λ = mean gecko abundance; Yij = number of geckos recorded at site i during survey j; pij = probability of detecting any gecko at site i during survey j; i = site 1,2,...21; and j = the number of the survey 1,2...6.

An integrated likelihood method as implemented in the unmarked R package [57] was used to calculate the above variables. We found the most parsimonious model by an information-theoretic approach using the corrected Akaike Information Criterion (AICc) in the R package AICcmodavg [58]. The AICc was used because the sample size (n) divided by the number of parameters (K) was less than 40 [59]. Model adequacy of the global model was tested with a parametric bootstrapping chi-squared goodness of fit. Depending on the model adequacy, a Poisson, negative binomial or zero-inflated Poisson probability

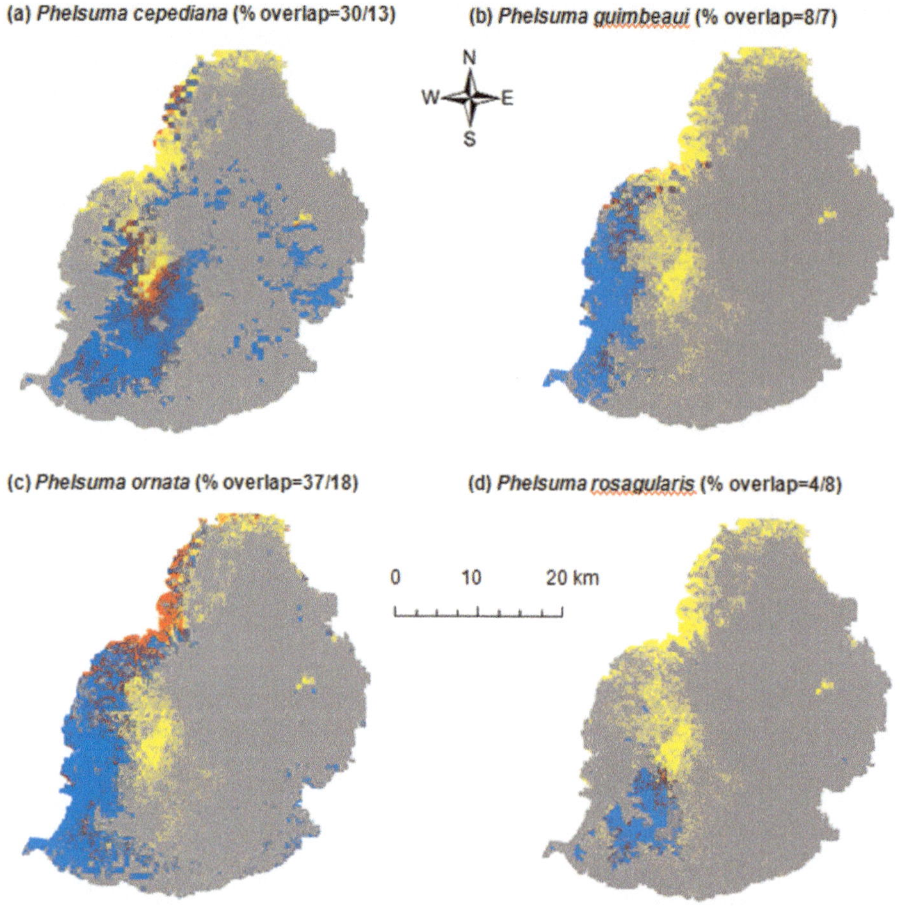

Figure 2. The binary projection and overlap between *Phelsuma grandis* and four endemic species of *Phelsuma*. Each map shows the overlap of *P. grandis* and one of the endemic species. On each map, grey indicates that no species of *Phelsuma* were predicted to be present, yellow shows the predicted range of *Phelsuma grandis*, blue the predicted range of that species of endemic *Phelsuma*, and red areas of predicted overlap between the two species. The first number in each heading is the % overlap of the predicted range of *Phelsuma grandis* with that species of endemic gecko and the second number is the % overlap of the endemic species' predicted range with that of *Phelsuma grandis*. The different maps suggest that *Phelsuma cepediana* (a) and *Phelsuma ornata* (c) will overlap more with *Phelsuma grandis* and thus could be at greater risk.

distribution was used to determine abundance. Only the top model was used to predict the average detection probability and abundance under different covariate conditions.

Results

Species distribution modelling and niche breadths

Only 9 of the 250 SDMs i.e. 7 GLMs, 1 MaxEnt and 1 GAM had an AUC<0.7 and were therefore excluded from the ensemble model. Four ensemble models were considered as excellent with AUC>0.9 and one as a good model with AUC>0.8 (Table 1). *P. cepediana* had the largest predicted suitable range of 376 km^2 (Table 1, Figure 2a). *P. grandis* had a predicted suitable range of 161 km^2 in the centre and along the west to the north-east coast of Mauritius (Table 1, Figure 3). *P. guimbeaui* had a predicted suitable range of 186 km^2 on the west side of the island (Table 1, Figure 2b). *P. ornata* had a predicted suitable range of 342 km^2 along the coast, particularly from the south-west to north-west (Table 1, Figure 2c). *P. rosagularis* had the smallest predicted suitable range, only 85 km^2, mainly in south-west Mauritius (Table 1, Figure 2d).

Between 4% and 37% of the predicted suitable range of *P. grandis* overlapped with each endemic species, whereas 7% to 18% of the predicted suitable range of each endemic species overlapped with *P. grandis* (Figure 2). For *P. grandis*, 60% of its predicted range overlapped with the combined area of suitability for all four endemic species of *Phelsuma*, whereas 15% of the predicted ranges of all the four endemic species overlapped with *P. grandis* (Figure 3). Of the 32 known subpopulations of *P. guimbeaui* and *P. rosagularis*, 13 were within the predicted range of *P. grandis*, four within 50 m, five within 200 m, five within 500 m, three within 1000 m and two within 1500 m.

Niche breadth for the five species of *Phelsuma* varied from 0.140 to 0.652. *P. guimbeaui* and *P. rosagularis* had a narrow specialist niche breadth metric, whereas *P. cepediana*, *P. grandis* and *P. ornata* had a moderate to broad generalist niche breadth metric (Table 2).

Effect of *P. grandis* on the abundance of endemic geckos

We recorded three endemic species, *P. cepediana*, *P. guimbeaui* and *P. ornata*, during visual surveys of the 21 sites. Since endemic geckos were sympatric, and of similar sizes and highly detectable colours, we combined their counts and compared abundance adjusted to imperfect detection probability between sites with and without *P. grandis*. On 126 surveys (21 sites each surveyed 6 times), we recorded 32 sightings of endemic geckos and 346 of *P. grandis* in the invaded sites, and 441 sightings of endemic geckos and no *P.*

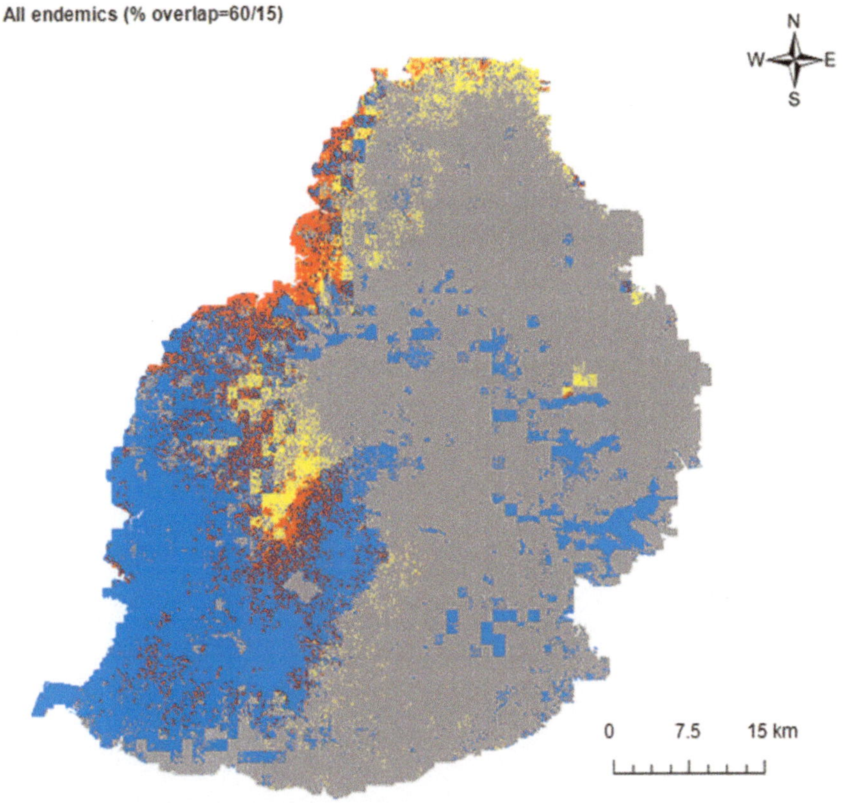

Figure 3. The binary projection and overlap between *Phelsuma grandis* and the combined predicted ranges of the four endemic species of *Phelsuma* on mainland Mauritius. Grey indicates that no species of *Phelsuma* were predicted to be present, yellow shows the predicted range of *Phelsuma grandis*, blue the combined predicted range of all the species of endemic *Phelsuma*, and red areas of predicted overlap. The first number in the heading is the % overlap of the predicted range of *Phelsuma grandis* with the combined predicted ranges of the four species of endemic *Phelsuma*, and the second number is the overlap of the endemic species' combined predicted ranges with the predicted range of *Phelsuma grandis.*

grandis in the control sites. We only detected endemic geckos in six of the ten sites with *P. grandis*.

We used three different statistical distributions to test for over-dispersion. The Poisson distribution showed signs of over-dispersion in the global model and was not used in model fitting. The negative binomial and zero-inflated Poisson distributions passed the goodness of fit test and adequately fitted the abundance model; the zero-inflated Poisson distribution for abundance was selected because it had smaller confidence intervals.

The lowest AICc model (AICc weight of 76%) was chosen as the best fitting model (Table 3). The selected model showed that the abundance of endemic geckos was affected by the presence of *P. grandis*; the abundance of endemic geckos was 13.3 (95% confidence interval 7.0–19.9) and 1.6 (95% confidence interval 0.0–3.2) in sites without and with *P. grandis* respectively (Figure 4). This represented a decline of 89% of endemic geckos in *P. grandis* sites.

Probability of detection was affected by habitat and cloud cover (Table 3). Probability of detection was similar in the three habitats, with the highest detection in building sites (0.530, 95% confidence interval 0.413–0.644), followed by palm (0.465, 95% confidence interval 0.375–0.557) and non-palm sites (0.374, 95% confidence

Table 1. Ensemble model evaluation results, showing the Area Under the Curve (AUC) of the median ensemble model for the five species of *Phelsuma*.

Species	AUC	Threshold	Sensitivity	Specificity	Predicted range (km^2)
P. cepediana	0.888	0.438	84.9	78.3	376
P. grandis	0.973	0.566	98.3	90.1	161
P. guimbeaui	0.960	0.568	89.2	91.1	186
P. ornata	0.951	0.309	97.3	80.7	342
P. rosagularis	0.995	0.515	100.0	95.7	85

Phelsuma cepediana had a good model fit (AUC>0.8), while the other four species had an excellent model fit (AUC>0.9). A probability threshold was used to maximise sensitivity and specificity to produce presence/absence maps to predict the range of each species.

Table 2. Levins' niche breadth metric, where restricted to specific environmental conditions (specialist) = 0 and able to exploit a wide range of environmental conditions (generalist) = 1.

Species	Levins' niche breadth
P. cepediana	0.652
P. grandis	0.497
P. guimbeaui	0.179
P. ornata	0.356
P. rosagularis	0.140

interval 0.284–0.437) (Figure 5a). This is not surprising: building sites had the least vertical diversity, making it relatively easy to spot geckos. The higher detectability between palm versus non-palm sites can be explained by the simple structure of palms with heights less than 12 m, compared to the complex structure of branching non-palm trees with heights up to 20 m. An increase in cloud cover had a negative effect on probability of detection in the three habitat types because the geckos do not bask in cloudy conditions, and so are less likely to be detected as they find refuge in optimal thermo-regulatory spots such as between leaves (Figure 5b).

Discussion

We used SDMs combined with BMM to investigate the ecological impacts of *P. grandis* on the endemic *Phelsuma* community in mainland Mauritius. The large AUC values for the five ensemble models suggest that the predictive maps had high statistical robustness. Though SDMs can give an accurate prediction of the spatial distributions of IAS [24], interpretation of the results need to be treated with caution. Model projections assumed complete random dispersal. However, as gecko movements are restricted by habitat isolation and fragmentation, the predictive maps of all five species could be overestimates [60]. Certainly, for both *P. guimbeaui* and *P. rosagularis*, there are very few known subpopulations within their predicted range. Some studies advocate the use of variables such as species interactions [61], species traits [62] or "natural history" [63] to build SDMs. However, little information was available for these variables, especially for *P. grandis*. BMM model performance using the zero-inflated Poisson distribution showed that our models were robust and had a good fit to the data. We used BMM to account for imperfect detection in the estimation of the abundance of geckos [34,36,57,64], since this can lead to erroneous conclusions [27], especially when invasive and native species co-occur [65]. *Phelsuma* spp. tend to partition their habitats along an axis to reduce competition [66,67], and so one possible response would be for the endemic species to shift along an axis following invasion by *P. grandis*. However, we did not observe any shift in habitat selection by endemic geckos in the presence of *P. grandis*; there were significantly fewer endemic geckos in the presence of *P. grandis*, and endemic geckos were not even detected at four of the ten sites with

Table 3. Model selection results. Abundance was modelled with habitat and status as site-level covariates.

Detection	Abundance	K	AICc	Delta AICc	AICc weight
Cloud+habitat	Status	7	477.4	0.0	0.76
Cloud+status	Status	6	482.0	4.6	0.07
Cloud+habitat+status	Status	8	482.6	5.3	0.05
Habitat+status	Status	7	483.3	5.9	0.04
No covariate	Status	4	484.1	6.8	0.03
Status	Status	5	487.6	10.2	0.00
Habitat+status+temp	Status	8	488.4	11.0	0.00
Cloud+habitat+status+temp	Status	9	488.6	11.2	0.00
Cloud+habitat+status+temp	Habitat+status	11	490.3	12.9	0.00
Cloud	Habitat	6	527.2	49.9	0.00
Habitat	Habitat	7	527.9	50.5	0.00
Cloud+temp	Habitat	7	530.3	52.9	0.00
Cloud+habitat+temp	Habitat	9	531.8	54.4	0.00
Habitat+status	Habitat	8	533.3	55.9	0.00
Cloud+habitat	No covariate	6	537.4	60.0	0.00
Status	Habitat	6	539.0	61.7	0.00
Habitat	No covariate	5	540.9	63.6	0.00
No covariate	No covariate	3	549.9	72.5	0.00
Status	No covariate	4	552.8	75.4	0.00

K = number of parameters used.
Delta AICc = difference between lowest AICc model and model AICc.
AICc weight = model probability among all candidate models.
Detection probability was modelled with observation-level covariates: cloud = cloud percentage cover; habitat = building, non-palm or palm; status = presence or absence of *P. grandis*; and temp = temperature. We used the corrected Akaike Information Criterion (AICc) to determine the best supported model.

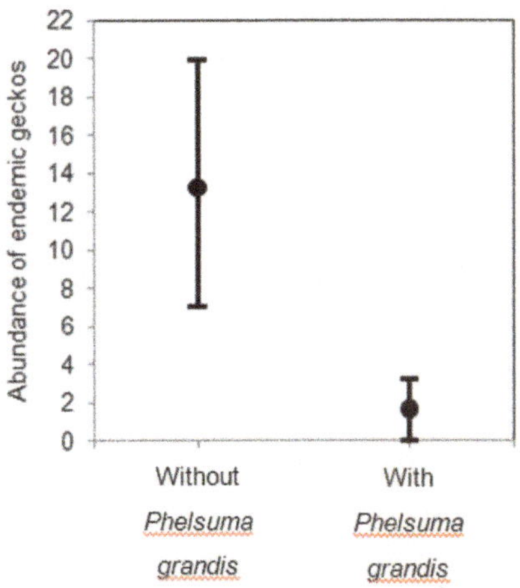

Figure 4. The mean endemic gecko abundance (with 95% confidence intervals) in sites with and without *Phelsuma grandis*. Sites without *Phelsuma grandis* had a high abundance of endemic geckos, those with *Phelsuma grandis* a low abundance of endemic geckos. N = 10 sites with *Phelsuma grandis*, N = 11 sites without *Phelsuma grandis*.

P. grandis. Despite being a snapshot survey, we believe that the same results would have been observed throughout the year.

Characteristics of a generalist invader and the potential threat of invasion

The predictive overlap maps and a relatively large niche breadth suggest that *P. grandis* has the typical attributes of a generalist invader [68], with the ability to persist in a large range of environmental conditions. The predictive maps also suggest that the two commonest species, *P. cepediana* and *P. ornata*, are the most threatened by *P. grandis*. However, the distance between the predicted range of *P. grandis* and known subpopulations of *P. guimbeaui* and *P. rosagularis* were relatively small, suggesting that they could be under more immediate threat from *P. grandis* invasion. Both these endemic species had narrow Levins' niche breadths typical of specialists; they also occupy small restricted ranges such that the potential threat posed by *P. grandis* enhances their vulnerability to extinction [7].

With a relatively large niche breadth, we expected *P. grandis* to have a bigger predicted distribution. For example, the niche breadth of *P. ornata* was 0.356, with a predicted range of 342 km^2, compared to a niche breadth of 0.497 and a predicted range of 161 km^2 for *P. grandis*. One plausible explanation for this apparent disparity is that our presence data do not reflect the full extent of the niche suitability for *P. grandis*, which has not had time to invade all the available niches in the 20 years since its introduction. For instance, in its native range in Madagascar (approximately 1000 km to the west of Mauritius), *P. grandis* has been recorded at elevations up to 900 m [15]. This suggests that the whole altitudinal range of Mauritius, which has a maximum altitude of 828 m, is vulnerable to invasion. However, we had no records of *P. grandis* above 700 m in Mauritius. When an IAS first arrives, it usually utilises habitats similar to its native range [69] and subsequently expands and invades new niches [9,70]. We suspect that eventually *P. grandis* will have a larger range in Mauritius than we have predicted from its early spread and pattern of habitat selection, and so there will be more extensive overlap with the ranges of the endemic species of *Phelsuma*.

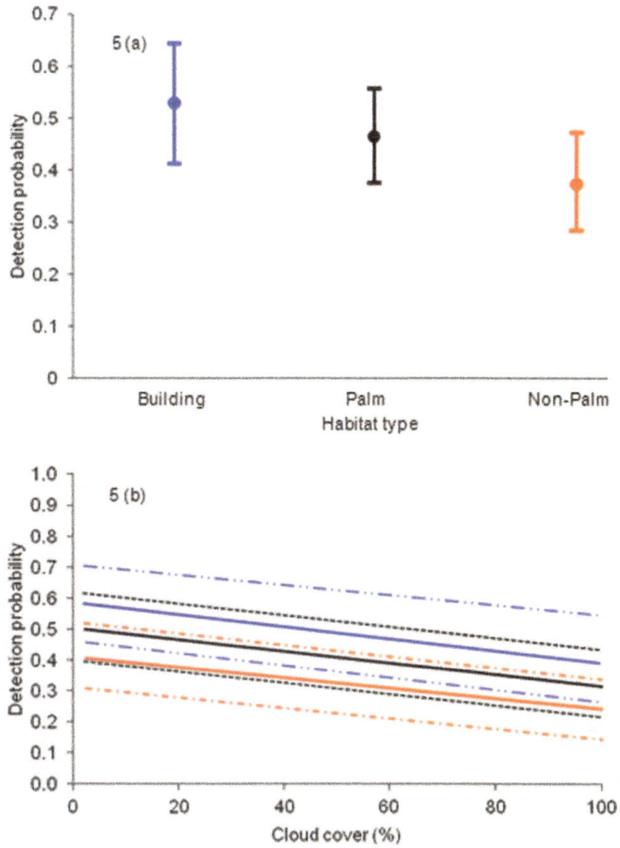

Figure 5. Effects of habitat type and cloud cover in each habitat type on detection probability. Figure 5a shows that the detection probability (with 95% confidence intervals) was similar in the three habitat types, with detection probability slightly higher in building sites than palm followed by non palm sites. Figure 5b shows a general decrease in detection probability (with 95% confidence intervals indicated by broken lines) with an increase in cloud cover in the three habitat types. Blue indicates building sites, black palm sites and red non-palm sites.

The threat to endemic geckos

There was a dramatic decline or total absence of endemic species of *Phelsuma* in the presence of *P. grandis*, suggesting that extirpation follows the arrival of *P. grandis*. Observations by local residents near the study sites suggest that *P. grandis* took less than 12 years to cause the disappearance of the endemic species. Similarly, reports from the public in Baie du Tombeau indicate that, in the two decades since its release, *P. grandis* has colonised the entire 1.8 km^2 suburban region and no *P. cepediana*, *P. guimbeaui* or *P. ornata* have been seen in this area since 2009.

How *P. grandis* is leading to the extirpation of endemic *Phelsuma* populations is unclear, although competitive exclusion and predation are believed to be key drivers of species extinction [71,72]. Field observations suggest that *P. grandis* shares temporal and spatial niches with the endemic geckos and so there is potential for competition and predation. Being larger than the endemic geckos, *P. grandis* is likely to consume larger prey items and thus relax competition for food [73], although day geckos frequently feed on nectar, pollen [74] and tree saps, and so there may be competition with *P. grandis* for these food resources [75]. We often observed aggressive behaviour from *P. grandis* towards endemic geckos at inflorescences and tree-sap foraging sites, suggesting competitive displacement at important food resources. A decrease in egg production [76,77] and/or reduced mating success [77] can occur in the presence of a predator or competitor, leading to a decline in reproductive output and increasing the risk of extinction. Understanding the underlying causes of how *P. grandis* leads to the decline and local loss of endemic *Phelsuma* populations will be key to quantifying the long-term impact and potentially managing systems to mitigate the threats.

P. grandis has also been introduced in nearby Réunion. Two endemic species of day gecko, *P. borbonica* and *P. inexpectata* (Manapany day gecko), are both threatened by the arrival of *P. grandis* [18]. *P. grandis* is widely available in the pet trade and there is the risk of further invasions in countries such as the Comores and Seychelles that have their own endemic species of *Phelsuma.*

Conclusions

With the ever-increasing numbers of invasive species, it is important to decide whether an IAS needs to be eradicated. Ideally, this decision needs to be made before the IAS is well-established. While the eradication of an IAS can help in the restoration of native ecosystems [78], eradication can also cause more damage [79], especially when an IAS has established key ecological functions [9]. However, since the majority of alien species will never be eradicated, alternative management strategies may be more appropriate in many circumstances [80].

P. grandis is a relatively recent introduction to Mauritius. Our data suggest that it is a generalist capable of invading a diversity of Mauritian habitats with dramatic impacts on the endemic *Phelsuma* community. Currently, *P. grandis* is localised in Mauritius, and mostly occurs in private gardens and plantations. However, this poses additional risks since it could be spread by accidental anthropogenic transportation. *P. grandis* is also being actively moved to new locations by locals to control other introduced geckos that are considered to be a messy and noisy nuisance in houses. It is now critical to identify the mechanisms whereby *P. grandis* leads to the loss of local populations of endemic *Phelsuma* to limit further impacts, and to decide if, and how, an eradication programme could be undertaken before *P. grandis* becomes better established in Mauritius. Our data also highlight the importance of banning the importation of *P. grandis* to other countries, particularly those with vulnerable populations of *Phelsuma* and other endemic geckos.

Acknowledgments

We thank Mannickchand Puttoo from the National Parks and Conservation Service, Ministry of the Agro-Industry, Mauritius for support and encouragement; Terry Burke and colleagues of the Molecular Ecology Laboratory of Sheffield University for their support; the Mauritian Wildlife Foundation for constant logistical support; J.C. (Koos) Biesmeijer and Leon Marshall from Naturalis Biodiversity Center, The Netherlands, for help with the SDM analyses; Ralph Budzinski and Mickael Sanchez for all their helpful information; and the National Parks and Conservation Service for permission to undertake this research.

Author Contributions

Conceived and designed the experiments: SB, NCC, SH, VB, LEG, SMH, KR, RMT. Analyzed the data: AB, BMM, SB, JAG, SDM. Wrote the manuscript: SB SH NCC. Data collection: SB NCC LEG SMH RMT.

References

1. Mack RN, Simberloff D, Lonsdale WM, Evans H, Clout M, et al. (2000) Biotic invasions: causes, epidemiology, global consequences, and control. Ecol Appl 10: 689–710. doi:10.2307/2641039.
2. O'Dowd DJ, Green PT, Lake PS (2003) Invasional 'meltdown' on an oceanic island. Ecol Lett 6: 812–817. doi:10.1046/j.1461-0248.2003.00512.x.
3. Charles H, Dukes JS (2007) Impacts of invasive species on ecosystem services. In: Nentwig DW, editor. Biological invasions. Berlin: Springer. pp. 217–237.
4. Kurle CM, Croll DA, Tershy BR (2008) Introduced rats indirectly change marine rocky intertidal communities from algae- to invertebrate-dominated. Proc Natl Acad Sci USA 105: 3800–3804. doi:10.1073/pnas.0800570105.
5. Clavero M, García-Berthou E (2005) Invasive species are a leading cause of animal extinctions. Trends Ecol Evol 20: 110. doi:10.1016/j.tree.2005.01.003.
6. Case TJ, Bolger DT (1991) The role of introduced species in shaping the distribution and abundance of island reptiles. Evol Ecol 5: 272–290.
7. Clavero M, Brotons L, Pons P, Sol D (2009) Prominent role of invasive species in avian biodiversity loss. Biol Conserv 142: 2043–2049. doi:10.1016/j.biocon.2009.03.034.
8. Atkinson IAE (2001) Introduced mammals and models for restoration. Biol Conserv 99: 81–96. doi:10.1016/S0006-3207(00)00189-0.
9. Pearson DE (2009) Biological invasions on oceanic islands: implications for island ecosystems and avifauna. In: Chae HY, Choi CY, Nam HY, editors. Seabirds in danger: invasive species and conservation of island ecosystem, pp. 3–16. Available: http://www.fs.fed.us/rm/pubs_other/rmrs_2009_pearson_d003.pdf. Accessed 16 June 2013.
10. Arnold EN (2000) Using fossils and phylogenies to understand evolution of reptile communities on islands. Bonn Zool Monogr 46: 309–324.
11. Austin JJ, Arnold EN, Jones CG (2004) Reconstructing an island radiation using ancient and recent DNA: the extinct and living day geckos (*Phelsuma*) of the Mascarene islands. Mol Phylogenet Evol 31: 109–122. doi:10.1016/j.ympev.2003.07.011.
12. Cole N. (2009) A field guide to the reptiles and amphibians of Mauritius. Vacoas, Mauritius: Mauritian Wildlife Foundation. 81 p.
13. Glaw F, Vences M (2007) Field guide to the amphibians and reptiles of Madagascar. Köln: Vences & Glaw. 436 p.
14. Vinson J, Vinson J-M (1969) The saurian fauna of the Mascarene Islands. Mauritius Inst Bull 4: 203–320.
15. Ratsoavina F, Glaw F, Rakotondrazafy NA (2011) *Phelsuma grandis*. IUCN Red List of Threatened Species. Version 2012.2. Available: http://www.iucnredlist.org/details/193490/0. Assessed 25 April 2013.
16. Krysko KL, Hooper AN, Sheehy CM (2003) The Madagascar giant day gecko, *Phelsuma madagascariensis grandis* Gray 1870 (Sauria: Gekkodinae): a new established species in Florida. Flo Scient 66: 222–225.
17. Goldberg SR, Bursey CR, Kraus F (2010) Helminth records for the Madagascan giant day gecko, *Phelsuma grandis* (Gekkonidae) from Hawaii. Occas Pap Bernice P Bishop Mus 108: 49–52.
18. Sanchez M, Gandar A (2010) Etat des lieux de la population introduite à Manapany-les-Bains du grand gecko vert malgache, *Phelsuma grandis* Gray 1870. Reunion: Association Nature Océan Indien unpublished report. 26 p.
19. Carvalho SB, Brito JC, Pressey RL, Crespo E, Possingham HP (2010) Simulating the effects of using different types of species distribution data in reserve selection. Biol Conserv 143: 426–438. doi:10.1016/j.biocon.2009.11.010.
20. Doko T, Fukui H, Kooiman A, Toxopeus AG, Ichinose T, et al. (2011) Identifying habitat patches and potential ecological corridors for remnant Asiatic black bear (*Ursus thibetanus japonicus*) populations in Japan. Ecol Model 222: 748–761. doi:10.1016/j.ecolmodel.2010.11.005.
21. Bosso L, Rebelo H, Garonna AP, Russo D (2013) Modelling geographic distribution and detecting conservation gaps in Italy for the threatened beetle *Rosalia alpina*. J Nat Conserv 21: 72–80. doi:10.1016/j.jnc.2012.10.003.
22. Virkkala R, Heikkinen RK, Fronzek S, Leikola N (2013) Climate change, northern birds of conservation concern and matching the hotspots of habitat suitability with the reserve network. PLoS One 8(5): e63376. doi:10.1371/journal.pone.0063376.
23. Beaumont LJ, Gallagher RV, Thuiller W, Downey PO, Leishman MR, et al. (2009) Different climatic envelopes among invasive populations may lead to underestimations of current and future biological invasions. Divers Distrib 15: 409–420. doi:10.1111/j.1472-4642.2008.00547.x.
24. Ficetola GF, Thuiller W, Padoa-Schioppa E (2009) From introduction to the establishment of alien species: bioclimatic differences between presence and reproduction localities in the slider turtle. Divers Distrib 15: 108–116. doi:10.1111/j.1472-4642.2008.00516.x.

25. Ficetola GF, Maiorano L, Falcucci A, Dendoncker N, Boitani L, et al. (2010) Knowing the past to predict the future: land-use change and the distribution of invasive bullfrogs. Global Change Biol 16: 528–537. doi:10.1111/j.1365-2486.2009.01957.x.
26. Gallien L, Münkemüller T, Albert CH, Boulangeat I, Thuiller W (2010) Predicting potential distributions of invasive species: where to go from here? Divers Distrib 16: 331–342. doi:10.1111/j.1472-4642.2010.00652.x.
27. Kéry M (2010) Introduction to WinBUGS for ecologists: a Bayesian approach to regression, ANOVA, mixed models and related analyses. Amsterdam: Elsevier. 302 p.
28. MacKenzie DI, Nichols JD, Lachman GB, Droege S, Royle JA, et al. (2002) Estimating site occupancy rates when detection probabilities are less than one. Ecology 83: 2248–2255. doi.org/10.1890/0012-9658(2002)083[2248:ESORWD]2.0.CO;2.
29. De Wan AA, Sullivan PJ, Lembo AJ, Smith CR, Maerz JC, et al. (2009) Using occupancy models of forest breeding birds to prioritize conservation planning. Biol Conserv 142: 982–991. doi:10.1016/j.biocon.2008.12.032.
30. Royle JA (2004) N-mixture models for estimating population size from spatially replicated counts. Biometrics 60: 108–115. doi:10.1111/j.0006-341X.2004.00142.x.
31. Kéry M, Royle JA, Plattner M, Dorazio RM (2009) Species richness and occupancy estimation in communities subject to temporary emigration. Ecology 90: 1279–1290. doi.org/10.1890/07-1794.1.
32. Royle JA, Dawson DK, Bates S (2004) Modeling abundance effects in distance sampling. Ecology 85: 1591–1597. doi.org/10.1890/03-3127.
33. Kéry M (2008) Estimating abundance from bird counts: binomial mixture models uncover complex covariate relationships. Auk 125: 336–245. doi:10.1525/auk.2008.06185.
34. Dodd CK, Dorazio RM (2004) Using counts to simultaneously estimate abundance and detection probabilities in a salamander community. Herpetologica 60: 468–478. doi:10.1655/03-60.
35. McKenny HC, Keeton WS, Donovan TM (2006) Effects of structural complexity enhancement on eastern red-backed salamander (*Plethodon cinereus*) populations in northern hardwood forests. Forest Ecol Manage 230: 186–196. doi:10.1016/j.foreco.2006.04.034.
36. Graves TA, Kendall KC, Royle JA, Stetz JB, Macleod AC (2011) Linking landscape characteristics to local grizzly bear abundance using multiple detection methods in a hierarchical model. Anim Conserv 14: 652–664. doi:10.1111/j.1469-1795.2011.00471.x.
37. Hijmans RJ, Cameron SE, Parra JL, Jones PG, Jarvis A (2005) Very high resolution interpolated climate surfaces for global land areas. Int J Climatol 25: 1965–1978. doi:10.1002/joc.1276.
38. Aguirre-Gutiérrez J, Carvalheiro LG, Polce C, van Loon EE, Raes N, et al. (2013) Fit-for-purpose: species distribution model performance depends on evaluation criteria - Dutch hoverflies as a case study. PloS One 8(5): e63708. doi:10.1371/journal.pone.0063708.
39. Phillips SJ, Anderson RP, Schapire RE (2006) Maximum entropy modeling of species geographic distributions. Ecol Model 190: 231–259. doi:10.1016/j.ecolmodel.2005.03.026.
40. Ridgeway G (1999) The state of boosting. Comp Sci Stat 31: 172–181.
41. Breiman L (2001) Random forests. Mach Learn 45: 5–32.
42. McCullagh P, Nelder JA (1989) Generalized linear models. London: Chapman and Hall. 511 p.
43. Hastie T, Tibshirani R (1999) Generalized additive models. London: Chapman and Hall. 335 p.
44. Thuiller W, Lafourcade B, Engler R, Araújo MB (2009) BIOMOD - a platform for ensemble forecasting of species distributions. Ecography 32: 369–373. doi:10.1111/j.1600-0587.2008.05742.x.
45. Barbet-Massin M, Jiguet F, Albert CH, Thuiller W (2012) Selecting pseudo-absences for species distribution models: how, where and how many? Methods Ecol Evol 3: 327–338. doi:10.1111/j.2041-210X.2011.00172.x.
46. Swets JA (1988) Measuring the accuracy of diagnostic systems. Science 240: 1285–1293. doi:10.1126/science.3287615.
47. Fielding AH, Bell JF (1997) A review of methods for the assessment of prediction errors in conservation presence/absence models. Environ Conserv 24: 38–49.
48. Thuiller W, Lavorel S, Araújo MB (2005) Niche properties and geographical extent as predictors of species sensitivity to climate change. Global Ecol Biogeogr 14: 347–357. doi:10.1111/j.1466-822X.2005.00162.x.
49. Thuiller W, Georges D, Engler R (2013) Package "biomod2". Species distribution modeling within an ensemble forecasting framework. Available: http://cran.r-project.org/web/packages/biomod2/biomod2.pdf. Accessed 25 August 2013.
50. Jiménez-Valverde A, Lobo JM (2007) Threshold criteria for conversion of probability of species presence to either-or presence-absence. Acta Oecol 31: 361–369. doi:10.1016/j.actao.2007.02.001.
51. Liu C, White M, Newell G (2013) Selecting thresholds for the prediction of species occurrence with presence-only data. J Biogeogr 40: 778–789. doi:10.1111/jbi.12058.
52. Levins R (1968) Evolution in changing environments: some theoretical explorations. Princeton: Princeton University Press. 132 p.
53. Pike DA (2013) Climate influences the global distribution of sea turtle nesting. Global Ecol Biogeogr 22: 555–566. doi:10.1111/geb.12025.
54. Royle JA, Dorazio RM (2006) Hierarchical models of animal abundance and occurrence. J Agric Biol Envir Stat 11: 249–263. doi:10.1198/108571106X129153.
55. Dorazio RM (2007) On the choice of statistical models for estimating occurrence and extinction from animal surveys. Ecology 88: 2773–2782. doi:10.1890/07-0006.1.
56. Mazerolle MJ, Bailey LL, Kendall WL, Royle JA, Converse SJ, et al. (2007) Making great leaps forward: accounting for detectability in herpetological field studies. J Herpetol 41: 672–689.
57. Fiske I, Chandler R, Miller D, Royle A, Kéry M (2010) Models for data from unmarked animals. R package version 0.8–6. Available: http://ftp.ctex.org/mirrors/CRAN/web/packages/unmarked/. Accessed 29 June 2013.
58. Mazerolle MJ (2011) AICcmodavg-package. R Package Version 131. Available: http://www.inside-r.org/packages/cran/AICcmodavg/docs/AICcmodavg. Accessed 29 June 2013.
59. Burnham KP, Anderson DR (2002) Model selection and multimodel inference: a practical information-theoretic approach. New York: Springer. 488 p.
60. Sinclair SJ, White MD, Newell GR (2010) How useful are species distribution models for managing biodiversity under future climates? Ecol Soc 15(1): 8. http://www.ecologyandsociety.org/vol15/iss1/art8/.
61. Araújo MB, Luoto M (2007) The importance of biotic interactions for modelling species distributions under climate change. Global Ecol Biogeogr 16: 743–753. doi:10.1111/j.1466-8238.2007.00359.x.
62. Evangelista PH, Kumar S, Stohlgren TJ, Jarnevich CS, Crall AW, et al. (2008) Modelling invasion for a habitat generalist and a specialist plant species. Divers Distrib 14: 808–817. doi:10.1111/j.1472-4642.2008.00486.x.
63. Rödder D, Schmidtlein S, Veith M, Lötters S (2009) Alien invasive slider turtle in unpredicted habitat: a matter of niche shift or of predictors studied? PLoS One 4(11): e7843. doi:10.1371/journal.pone.0007843.
64. Chandler RB, King DI (2011) Habitat quality and habitat selection of golden-winged warblers in Costa Rica: an application of hierarchical models for open populations. J Appl Ecol 48: 1038–1047. doi:10.1111/j.1365-2664.2011.02001.x.
65. Cayuela H, Besnard A, Joly P (2013) Multi-event models reveal the absence of interaction between an invasive frog and a native endangered amphibian. Biol Invasions 15: 2001–2012. doi:10.1007/s10530-013-0427-x.
66. Harmon LJ, Harmon LL, Jones CG (2007) Competition and community structure in diurnal arboreal geckos (genus *Phelsuma*) in the Indian Ocean. Oikos 116: 1863–1878. doi:10.1111/j.2007.0030-1299.15958.x.
67. Noble T, Bunbury N, Kaiser-Bunbury CN, Bell DJ (2011) Ecology and co-existence of two endemic day gecko (*Phelsuma*) species in Seychelles native palm forest. J Zool 283: 73–80. doi:10.1111/j.1469-7998.2010.00751.x.
68. Williamson MH (1996) Biological invasions. London: Chapman and Hall. 244 p.
69. Bomford M, Kraus F, Barry SC, Lawrence E (2009) Predicting establishment success for alien reptiles and amphibians: a role for climate matching. Biol Invasions 11: 713–724. doi:10.1007/s10530-008-9285-3.
70. Fitzpatrick MC, Weltzin JF, Sanders NJ, Dunn RR (2007) The biogeography of prediction error: why does the introduced range of the fire ant over-predict its native range? Global Ecol Biogeogr 16: 24–33. doi:10.1111/j.1466-8238.2006.00258.x.
71. Losos JB (2000) Ecological character displacement and the study of adaptation. Proc Natl Acad Sci USA 97: 5693–5695. doi:10.1073/pnas.97.11.5693.
72. Schoener TW, Spiller DA, Losos JB (2001) Predators increase the risk of catastrophic extinction of prey populations. Nature 412: 183–186. doi:10.1038/35084071.
73. Pianka ER, Pianka HD (1976) Comparative ecology of twelve species of nocturnal lizards (Gekkonidae) in the Western Australian desert. Copeia 1976: 125–142. doi:10.2307/1443783.
74. Hansen DM, Kiesbüy HC, Jones CG, Müller CB (2007) Positive indirect interactions between neighboring plant species via a lizard pollinator. Am Nat 169: 534–542. doi:10.1086/511960.
75. Krysko KL, Hooper AN (2006) *Phelsuma madagascariensis grandis* (Madagascar giant day gecko). Nectarivory; potential pollination. Herpetol Rev 37: 226.
76. Rummel JD, Roughgarden J (1985) Effects of reduced perch-height separation on competition between two *Anolis* lizards. Ecology 66: 430–444. doi:10.2307/1940392.
77. Schoener TW, Spiller DA, Losos JB (2002) Predation on a common *Anolis* lizard: can the food-web effects of a devastating predator be reversed? Ecol Monogr 72: 383–407. doi.org/10.1890/0012-9615(2002)072[0383:POACAL]2.0.CO;2.
78. McGeoch MA, Butchart SHM, Spear D, Marais E, Kleynhans EJ, et al. (2010) Global indicators of biological invasion: species numbers, biodiversity impact and policy responses. Divers Distrib 16: 95–108. doi:10.1111/j.1472-4642.2009.00633.x.f.
79. Bergstrom DM, Lucieer A, Kiefer K, Wasley J, Belbin L, et al. (2009) Indirect effects of invasive species removal devastate World Heritage Island. J Appl Ecol 46: 73–81. doi:10.1111/j.1365-2664.2008.01601.x.
80. Carroll SP (2011) Conciliation biology: the eco-evolutionary management of permanently invaded biotic systems. Evol Appl 4: 184–199. doi:10.1111/j.1752-4571.2010.00180.x.

3

The Lilliput Effect in Colonial Organisms: Cheilostome Bryozoans at the Cretaceous–Paleogene Mass Extinction

Caroline E. Sogot[1,2]*, Elizabeth M. Harper[1], Paul D. Taylor[2]

1 Department of Earth Sciences, University of Cambridge, Cambridge, United Kingdom, **2** Department of Earth Sciences, Natural History Museum, London, United Kingdom

Abstract

Consistent trends towards decreasing body size in the aftermath of mass extinctions – Lilliput effects – imply a predictable response among unitary animals to these events. The occurrence of Lilliput effects has yet to be widely tested in colonial organisms, which are of particular interest as size change may potentially occur at the two hierarchical levels of the colony and the individual zooids. Bryozoans are particularly useful organisms in which to study colonial size response as they have well-defined zooids. Additionally, a number of analyses of present-day bryozoans have shown that zooid size reflects local environmental conditions, most notably seawater temperature and possibly also food supply. Following the hypothesised decline in primary productivity at the Cretaceous–Paleogene (K–Pg) mass extinction, it is predicted that bryozoan zooid size should decline in the early Paleogene, resulting in a Lilliput effect. To test this prediction, zooid size was compared across the K–Pg boundary at the assemblage level and also within 4 surviving genera. Analysis of 59 bryozoan species from assemblages on either side of the K–Pg boundary showed no significant change in zooid length. Zooid size was also measured in 98 Maastrichtian colonies and 162 Danian colonies belonging to four congeneric species. Only one of these genera showed a significant size decrease across the K–Pg boundary, the other three maintaining constant zooidal lengths, widths and areas. Additionally, the sizes of 210 Maastrichtian colonies and 163 Danian colonies did not show consistent size decrease across the K–Pg boundary in these same species, although maximum colony size did decline in three out of four genera. Furthermore, this lack of consistent size change is uniform between two distinct biogeographical regions, Denmark and the southeastern USA.

Editor: Richard J. Butler, University of Birmingham, United Kingdom

Funding: This work was supported by a Natural Environment Research Council studentship to CES. The funders had no role in study design, data collection and analysis, decision to publish, or preparation of the manuscript.

Competing Interests: The authors have declared that no competing interests exist.

* E-mail: csogot@gmail.com

Introduction

The Lilliput effect [1] describes a decrease in body size following mass extinctions and is often thought to reflect an ecophenotypic response to environmental changes at these times. If this phenomenon occurs in all taxa and across all mass extinctions, it means that organisms have reacted predictably to these events, regardless of their proximate cause [2]. Such predictability has the potential to assist our understanding of future responses by organisms to contemporary ecological disturbances [3]. Recent studies, however, have questioned the ubiquity of the Lilliput effect [4–7], prompting discussion of its importance.

The original definition of the 'Lilliput effect' originated from the observation of diminutive colonies among some graptolite species that survived the Late Silurian biotic crises [1]. The term has become modified over time; it is often used in a more general sense for any examples of small-sized, post-extinction organisms, as opposed to size change within lineages crossing mass extinctions [8–9]. As such, it is unclear how often the Lilliput effect, in terms of its original more restricted definition, really occurs [10]. Within-lineage size decrease has been studied less often as it ideally demands a reliable phylogeny [3]. Although patterns of within-lineage size change have been observed to be more variable than those reported for higher taxa [8], a Lilliput effect has been reported to occur in some instances at lower taxonomic levels (Table 1).

There are numerous Lilliput analyses for an array of unitary (solitary) organisms but very few for colonial organisms [10], despite the original concept being based on graptolites [1]. Colonial organisms are important components of modern benthic ecosystems (e.g. reef corals) and their size response has potential implications for other taxa dependent upon them. Understanding such reactions to changing environments is therefore of particular importance at the present day. Size changes documented for colonial organisms are of further interest as they can occur at two hierarchical levels: the colony and its constituent individual modules (zooids). Colony size reduction could potentially be the result of smaller zooids and/or fewer component zooids [1]. Paralleling solitary organisms, it might also be expected that the individual zooids would experience a reduction in size. However, zooid size change in colonial organisms has yet to be analysed across a mass extinction boundary.

The aim of this study is to gain new perspectives on the Lilliput effect by investigating changes across the Cretaceous–Paleogene (K–Pg) mass extinction in bryozoan size at the two hierarchical levels of the colony and the zooid. Cheilostome bryozoans lend themselves particularly well to studies of size change as their box-like zooids retain a fixed size after they are budded [11], and the

Table 1. Summary of within-lineage size trends across mass extinction intervals in invertebrate species or genera.

Size trends	Taxa		Mass extinction	Reference
Size decrease	Graptolites	*Pristiograptus dubius parvus**	Upper Silurian	[1]
		*Pristiograptus dubius tumescens**		
		*Saetograptus leintwardinensis**		
	Brachiopods	?*Tethyochonetes* sp.		[54][a]
		Paryphella orbicularis		
		*'Lingula'**		[9]
	Gastropod	*Bellerophon**		
	Bivalve	*Pseudomytiloides**	Pliensbachian- Toarcian	[50][b]
	Ammonite	*Dactylioceras**		
	Brachiopods	*Deliella**		
		Strophomena (and *Katastrophomena*)*	End-Ordovician	[8]
		*Skenidioides**		
		*Coolinia**		
No change	Belemnites	*Passaloteuthis milleri*	Pliensbachian- Toarcian	[55][b]
		Passaloteuthis bisulcata		
Initial decrease, then increase	Brachiopods	*Paracraniops*	End-Ordovician	[8]
		Brevilamnulella		
Size increase	Brachiopods	*Triplesia*	End-Ordovician	[8]
		Eospirifer		

Only taxa that cross the extinction boundaries are considered.
* described in the original reference as a 'Lilliput effect'.
[a]Other brachiopod species analysed in this study show varying trends prior to the Permo–Triassic boundary: two species decrease and two increase in size prior to the boundary. However, only the species listed in the above table cross over the Permo-Triassic boundary itself.
[b]Other taxa show variable size trends in this study, but do not cross the boundary itself. Those that do cross the boundary might also show variable size trends leading up to or beyond the boundary, e.g. *Passaloteuthis bisulcata* significant increases in size at the start of the extinction interval.

perimeters of the typically sheet-like encrusting colonies are well defined, allowing precise measurement of colony area.

The K–Pg event has long been associated with a significant decline in primary productivity [12], which is considered a key driver of the Lilliput effect [2]. It is therefore hypothesised that survivors of the K–Pg mass extinction should exhibit smaller body size than their pre-extinction relatives [13–17]. In the case of bryozoans, this post-extinction size decrease can be expected at both the colony- and the zooid-level. Decreases in zooid size and colony size have been observed in cheilostome bryozoans during the latest Maastrichtian in Denmark and are thought to represent unstable and unfavourable environmental conditions, with low planktonic productivity, prior to the K–Pg boundary [18]. The study here extends this analysis across the K–Pg boundary to establish whether size reduction continued into the early Danian.

Materials and Methods

Geological Setting

No specific permissions were required for fieldwork, which was carried out on public land. Field studies did not involve endangered or protected species.

Cheilostome bryozoan size is assessed in two regions – southeastern USA and Denmark – to account for biogeographical variation. Although Maastrichtian bryozoan faunas occur in several regions, Paleocene faunas are much rarer [19], making it difficult to compare bryozoans across the K–Pg boundary beyond the two regions studied here [20].

In the southeastern USA, encrusting cheilostome specimens were collected from seven localities across Georgia, Alabama and Mississippi (Figure 1). Fieldwork was undertaken on public land and no protected or valuable specimens were collected. This area was a shallow marine shelf at a palaeolatitude of approximately 33°N in the Late Cretaceous and Early Paleocene [21]. Sediments represent deltaic and nearshore marine settings [22], and occur in mixed carbonate and siliclastic sequences [23]. Most of the bryozoans encrusted 'oyster' shells (*Exogyra*, *Pycnodonte*, etc.) and were collected from the Maastrichtian Prairie Bluff Chalk and the Danian Clayton Formation.

In Denmark, specimens were collected from three localities at Stevns Klint, 45 km south of Copenhagen (Figure 1). No specific permissions were required and no protected or valuable specimens were collected. Stevns Klint was situated at approximately 45°N in the Late Cretaceous and Early Paleocene [24]. The K–Pg sections here consist of virtually pure chalk, contrasting with the more siliciclastic sediments of the southeastern USA. At Stevns Klint, cheilostome colonies mostly encrust echinoid tests, particularly *Echinocorys*. Maastrichtian specimens are from the Højerup Member whereas Danian specimens are from the Paleogene Korsnæb Member overlying the intervening Danian Cerithium Limestone which contains few bryozoans.

Zooid size and colony size were measured in encrusting sheet-like cheilostome bryozoans. Zooid size changes were analysed at both the assemblage level and within surviving clades to provide two different comparisons of size trends across the K–Pg boundary.

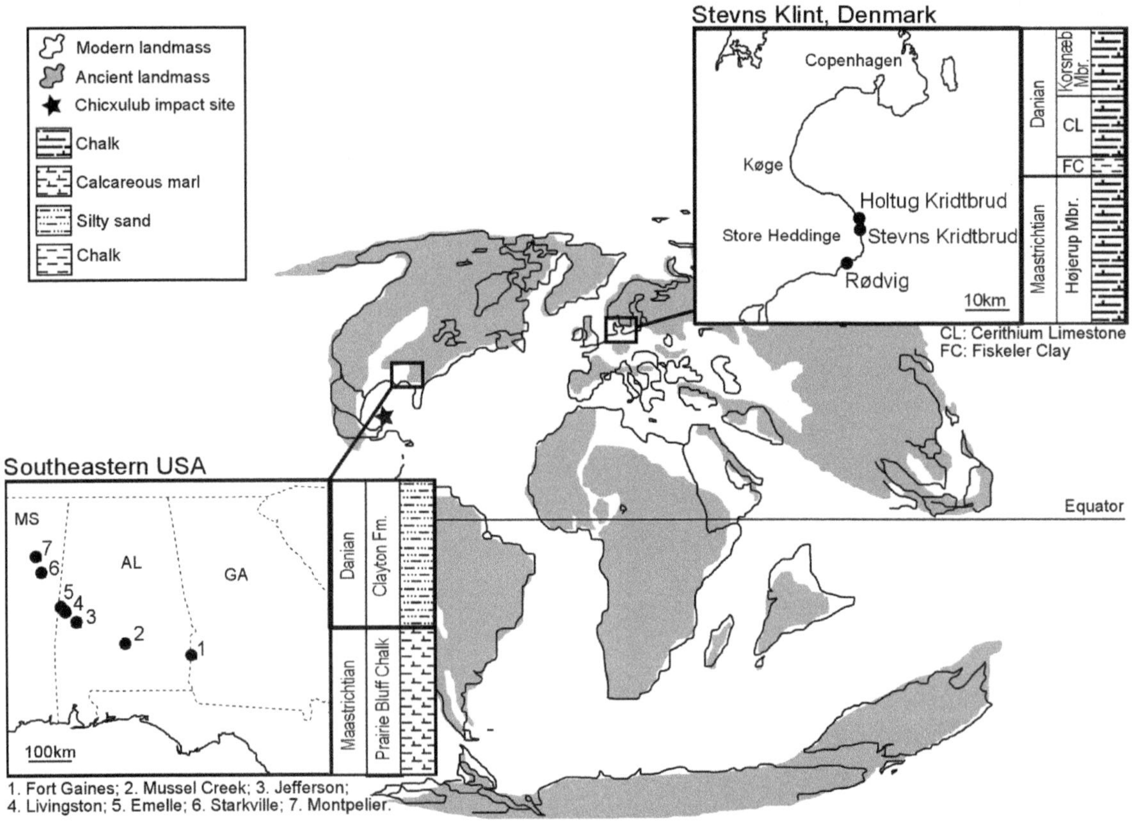

Figure 1. Locality maps and stratigraphical logs for the two study regions, Denmark and the southeastern USA. Stratigraphical formations and members are shown to highlight the units from which specimens were collected for analysis. Global palaeomap adapted from: http://scotese.com/K/t.htm.

Within-genus colony size

To minimise the potential confounding influence of phylogenetic differences, and in keeping with the original definition of the Lilliput effect, the sizes of closely related congeneric species were compared before and after the K–Pg boundary (Figure 2). Although some generic names traditionally used for these species in the Cretaceous and Paleocene species in the USA differ [25–26], the selected pairs of species from either side of the K–Pg boundary are here considered to be congeneric and, for *Pliophloea subcornuta*, conspecific. The lack of a phylogenetic study of these bryozoans, which would have identified sister groups, forced this selection of these taxa, which are based on morphological similarities.

Colony sizes were compared across the K–Pg boundary in two congeneric pairs from the USA and two from Denmark (Figure 2). Every colony of these species present in the collections was measured. This meant that colonies were sampled from multiple localities within each region. In total, the areas of 136 colonies from the USA (72 Maastrichtian, 64 Danian) and 237 colonies from Denmark (138 Maastrichtian, 99 Danian) were measured (Tables 2, 3).

Photographs were taken of each colony and their areas were measured using ImageJ software (http://rsbweb.nih.gov/ij/). The mean, minimum and maximum colony area was determined for each species, before and after the K–Pg boundary. The coefficient of variation (CV) was also calculated by dividing the standard deviation of colony sizes for each species by the mean colony size.

The perimeters of encrusting cheilostome colonies are generally well defined. However, preservational losses make it impossible to measure the original sizes of all colonies as breakage of colony edges means that measured colony areas underestimate the true area. However, as there is no evidence that breakage differed systematically between Maastrichtian and Danian samples, this factor is believed not to bias the results.

The hierarchical nature of the dataset – colonies, localities and stratigraphical intervals – lends itself to nested ANOVA analysis [27–30]. A general linear model was used to conduct the analysis, as this is an ANOVA method that allows for non- matching sized data sets. This method allows the comparison of size at different levels: (1) variation between colonies to test for variation within species from the same environment and time; (2) variation between localities, to test for variation between colonies from a given stratigraphical horizon; and (3) variation between formations to evaluate variation across the K–Pg boundary. Nested ANOVA therefore allows assessment of whether variation is indeed between formations on opposite sides of the K–Pg boundary, removing these other potentially confounding factors. It also allows a comparison of the proportion of variation that each factor contributes overall, by calculating the percentage variance component.

Within-genus zooid size

The same species as those analysed for colony size were also assessed for zooid size. Zooid sizes were therefore compared across the K–Pg boundary in two congeneric pairs from the USA and two pairs from Denmark. All selected species have well defined zooids, are abundant and generally show good preservation. Specimens collected in the field were supplemented by material

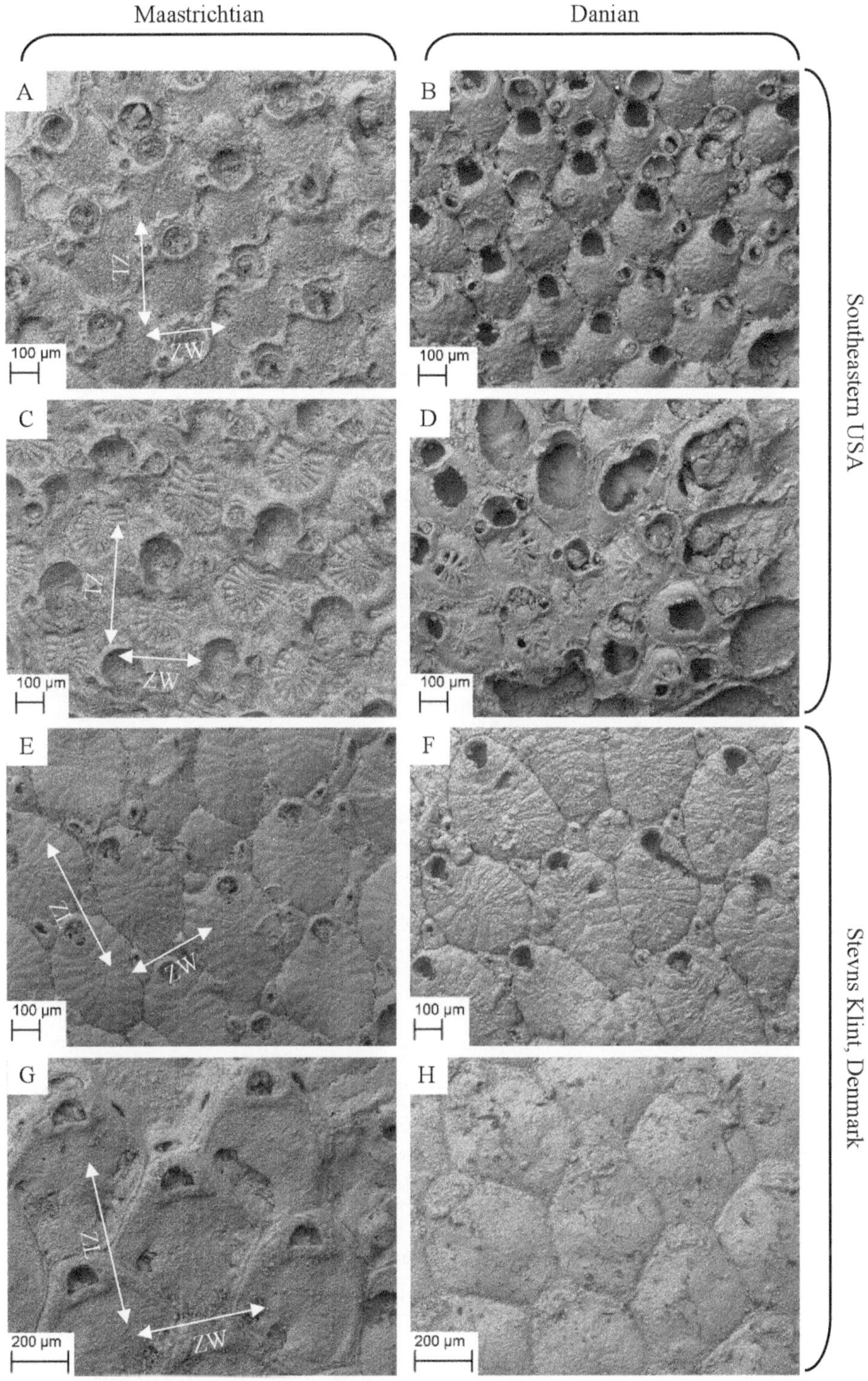

Figure 2. SEM images of species used in this study. Left-hand column shows Maastrichtian specimens, right-hand column Danian specimens. Rows are congeneric (or conspecific) pairs. Examples of maximum zooid length (ZL) and maximum zooid width (ZW) measurements are shown on the Maastrichtian specimens. A. *Balantiostoma nomas* (Shaw, 1967); B. *Balantiostoma midwayanica* (Canu and Bassler, 1920); C. *Tricephalopora larwoodi* (Shaw, 1967); D. *Tricephalopora levigatum* (Canu and Bassler, 1920); E. *Pliophloea subcornuta* (Berthelsen, 1962); F. *Pliophloea subcornuta*; G. *Stichomicropora* sp. 1; H. *Stichomicropora* sp. 2.

from the collection of Canu and Bassler (1920) at the Smithsonian Institution, Washington D.C. (USNM), including type specimens, the SEM images of which also allowed confirmation of taxonomic identifications. Colonies were selected from two localities within each region to include the influence of local environmental differences (Tables 2, 3).

Colonies were selected for analysis using guidelines adapted from previous studies of zooid size in bryozoans [11,31]: (1) the

Table 2. Number of (A) colonies measured, (B) zooids measured and (C) species, from which zooids were measured in the USA.

Formation	**Prairie Bluff**				**Clayton**			**Total**		
Locality	**M.**	**L.**	**E.**	**J.**	**F. G.**	**M.C.**	**S.**	**Prairie Bluff**	**Clayton**	**Total**
A) Taxon	**Number of colonies**									
Balantiostoma	-	4	6	17	-	38	16	27	54	81
Tricephalopora	-	15	2	28	-	2	8	45	10	55
Total								**72**	**64**	**136**
B) Taxon	**Number of zooids**									
Balantiostoma	30	40	-	-	50	40	-	70	90	160
Tricephalopora	60	50	-	-	20	10	-	110	30	140
Total								**180**	**120**	**300**
C) All species	**Number of species**									
Total								**20**	**2**	**22**

Locality abbreviations: M. = Montpelier; L. = Livingston; E = Emelle; J = Jefferson; F. G. = Fort Gaines; M. C. = Mussel Creek; S = Starkville.

colony had to be of reasonable size, with at least ten complete autozooids beyond the primary zone of astogenetic change around the colony origin [32]; (2) the colony had to have encrusted a flat, regular surface, allowing accurate measurement of undistorted zooids; (3) colonies were chosen that occurred on different substrates wherever possible to reduce any substrate-associated biases; and (4) the colony must have been relatively well-preserved and have zooids with well-defined outlines.

For comparisons of zooid size, multiple colonies of the same genus were selected from each locality and stratigraphic horizon to allow for genotypic variation between colonies. Zooids were measured in 30 colonies from the USA (18 Maastrichtian, 12 Danian) and 23 colonies from Denmark (8 Maastrichtian, 15 Danian). These were the only colonies that fulfilled the selection criteria outlined above. SEM images were taken of the selected colonies, either at the Natural History Museum, London (NHMUK), or at the USNM. An area that displayed at least 10 'normal' autozooids, beyond the primary zone of astogenetic change, was selected for scanning.

From each colony, 10 autozooids were randomly selected by numbering each zooid in the SEM image and then randomly selecting 10 of these, excluding: (1) polymorphic zooids such as kenozooids or avicularia; (2) zooids from the primary zone of astogenetic change [32]; and (3) zooids of abnormal size or shape due to physical damage or biotic interactions. In some instances only ovicellate zooids were available for measurement, in which case the ovicells were omitted from the zooidal measurements.

The maximum length and maximum width of each selected zooid was measured using ImageJ software. Zooid area was approximated by multiplying zooid length and width [11]. In total, the length, width, and area of 530 individual zooids were used for this study. Data were normally distributed and a nested ANOVA analysis was therefore conducted, with the addition of intracolonial comparisons to test for size variation between genetically identical individual zooids within each colony and account for measurement error.

Assemblage level zooid size

A total of 59 species were identified in the assemblage collections, 22 from the USA (20 Maastrichtian, 2 Danian) and 37 from Denmark (25 Maastrichtian, 12 Danian) (Tables 2, 3).

Table 3. Number of (A) colonies measured, (B) zooids measured and (C) species, from which zooids were measured in Denmark.

Member	**Højerup**			**Korsnæb**			**Total**		
Locality	**H.**	**S.**	**R.**	**H.**	**S.**	**R.**	**Højerup**	**Korsnæb**	**Total**
A) Taxon	**Number of colonies**								
Pliophloea subcornuta	67	5	-	-	24	37	72	61	133
Stichomicropora	30	7	29	7	23	8	66	38	104
Total							**138**	**99**	**237**
B) Taxon	**Number of zooids**								
Pliophloea subcornuta	50	-	-	-	30	80	50	110	160
Stichomicropora	10	-	20	-	40	-	30	40	70
Total							**80**	**150**	**230**
C) All species	**Number of species**								
Total							**25**	**12**	**37**

Locality abbreviations: H. = Holtug Kridtbrud; S. = Stevns Kridtbrud; R. = Rødvig.

Table 4. Nested ANOVA results for colony area.

Congeneric	Mean area (mm^2) (SD)		Min. area (mm^2)		Max. area (mm^2)		F-ratios		% variance component[a]		
	K	Pg	K	Pg	K	Pg	Fm.	Loc.	Fm.	Loc.	Col.
Balantiostoma nomas	7.74 (7.24)	8.44 (10.20)	0.84	0.23	32.27	62.39	0.35	7.69***	0	35.21	64.79
Tricephalopora larwoodi	17.87 (23.85)	15.39 (15.89)	0.72	1.74	115.89	53.48	1.35	0.56	3.07	0	96.93
Pliophloea subcornuta	43.76 (74.82)	38.61 (50.06)	0.27	1.25	524.07	259.59	0.02	2.16	0	5.71	94.29
Stichomicropora	54.24 (84.59)	17.07 (17.61)	0.57	0.91	54.24	17.07	4.27	0.50	5.55	0	94.45

* = $p<0.05$; ** = $p<0.01$; *** = $p<0.001$.
[a]Negative estimates of variance components associated with non-significant F-ratios were assigned zero values in keeping with standard practice [26].

For each species in these assemblages, a well-preserved colony was selected using the guidelines discussed above. If more than one colony within a species was present, the colony with the best preservation was selected for analysis. A light microscope with a graticule was used to measure the length of 10 randomly selected autozooids from each colony, as for the within-lineage study. Only the length was measured in this instance as environmental conditions have been shown to be most influential on this parameter [27,33,34] and length is easier to determine accurately under a light microscope than is width.

As before, a nested ANOVA was applied to the zooid length measurements to test for variation within colonies, between colonies, and between stratigraphical intervals.

Results

Within-genus colony size

Mean colony size does not change significantly across the K–Pg boundary for the pairs of species analysed in this study (Table 4, Figure 3, Table S1). *Stichomicropora* showed the greatest variation in colony size, with a decrease of 37.2 mm^2 ($p = 0.07$). The mean colony sizes of *Tricephalopora* and *Pliophloea subcornuta* also decreased across the boundary, by 2.5 mm^2 and 5.2 mm^2 respectively, but these changes are not statistically significant ($p = 0.26$ and $p = 0.89$, respectively). There is also no significant change in colony size between localities for these three genera, suggesting that locality is not a confounding factor (Table 4).

The mean colony area of *Balantiostoma* increased across the K–Pg boundary by 0.7 mm^2, but again this change is not significant ($p = 0.60$). There is, however, significant variation between localities for all *Balantiostoma nomas* colonies ($p<0.05$), which contributes to 35% of the total variance for this species.

Maastrichtian colonies exhibit greater maximum sizes for three of the pairs studied (Figure 3), with maximum colony size decreasing by 50–69% across the K–Pg boundary. However, the opposite trend occurs in *Balantiostoma*, where the largest colony is observed in the Paleocene Clayton Formation, with a 93% increase in maximum colony size when compared to the Prairie Bluff Chalk. Minimum colony size decreases across the K–Pg boundary for *Balantiostoma*, but increases across the boundary for the other taxa studied here (Figure 3).

Colony sizes for these taxa are highly variable, as is shown by their coefficients of variation (CV) (94–171%) and high contributions to the total variation between colonies (65–97%) (Table 4).

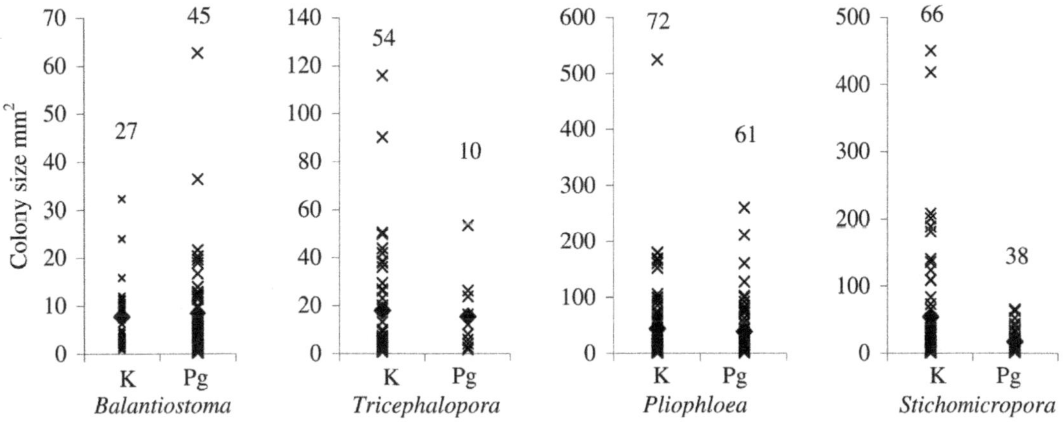

Figure 3. Mean, minimum and maximum colony size for taxa. Colony sizes of taxa analysed from the Maastrichtian Prairie Bluff Chalk/Højerup Member (K) and the Paleogene Clayton Formation/Korsnæb Member (Pg). The total number of colonies measured is indicated above each data point. $p>0.05$ across the K–Pg boundary for each pair.

Table 5. Nested ANOVA results for zooid length.

Congeneric	Mean length (μm) (SD)		F-ratios			% variance component[a]			
	K	Pg	Fm.	Loc.	Col.	Fm.	Loc.	Col.	Zooids
Balantiostoma nomas	348 (36)	278 (43)	13.79	1.49	12.63***	57.13	2.90	21.49	18.48
Tricephalopora larwoodi	446 (64)	355 (52)	3.98	3.59	29.54***	46.99	19.55	24.78	8.68
Pliophloea subcornuta	452 (37)	423 (44)	0.65	1.94	14.47***	0	11.77	50.64	37.59
Stichomicropora	560 (69)	491 (32)	n/a[b]	0.00	25.62***	45.12	0	39.03	15.85

* = $p<0.05$; ** = $p<0.01$; *** = $p<0.001$.
[a]Negative estimates of variance components associated with non-significant F-ratios were assigned zero values in keeping with standard practice [26].
[b]Denominator of the F-test is zero.

There is greater variation in the sizes of *Tricephalopora*, *Pliophloea subcornuta* and *Stichomicropora* colonies prior to the boundary, but the variation in *Balantiostoma* colonies is greater in the Danian (CV 121%) than the Maastrichtian (CV 94%).

Within-genus zooid size

After intra- and inter-colony variation and locality variations were taken into consideration by the nested ANOVA analysis, none of the taxa studied displayed a significant change in zooid length across the K–Pg boundary (Table 5, Figure 4, Table S2). Although mean zooid length declined from the Maastrichtian to the Danian by 29–92 μm (i.e. 6–21%), these decreases were not significant between formations. This variation in size between formations, however, accounted for a significant proportion of the overall variation (45–57%) in all taxa except *Pliophloea subcornuta* (0%) (Table 5). Standard deviations for all taxa in each formation were 32–69 μm around the mean zooid size. There was no consistent directional change in standard deviation, i.e. zooid size was neither more nor less variable on either side of the K–Pg boundary. For all studied taxa, maximum zooid lengths were found in samples from the Maastrichtian, whereas minimum zooid lengths were all recorded from the Danian samples.

There was no significant difference in zooid length between localities and this factor contributed the least amount of total variation (<20%). There was only minor intracolonial variation in zooid length (9–18%) for taxa other than *Pliophloea subcornuta*. However, between colonies there was significant variation in zooid length for all taxa studied ($p<0.001$) and this accounted for 21–51% of the total variation.

Balantiostoma displayed a decrease in zooid width across the K–Pg boundary ($p=0.03$), with a mean decline of 44 μm (18%) (Figure 4). All other taxa measured showed no significant variation

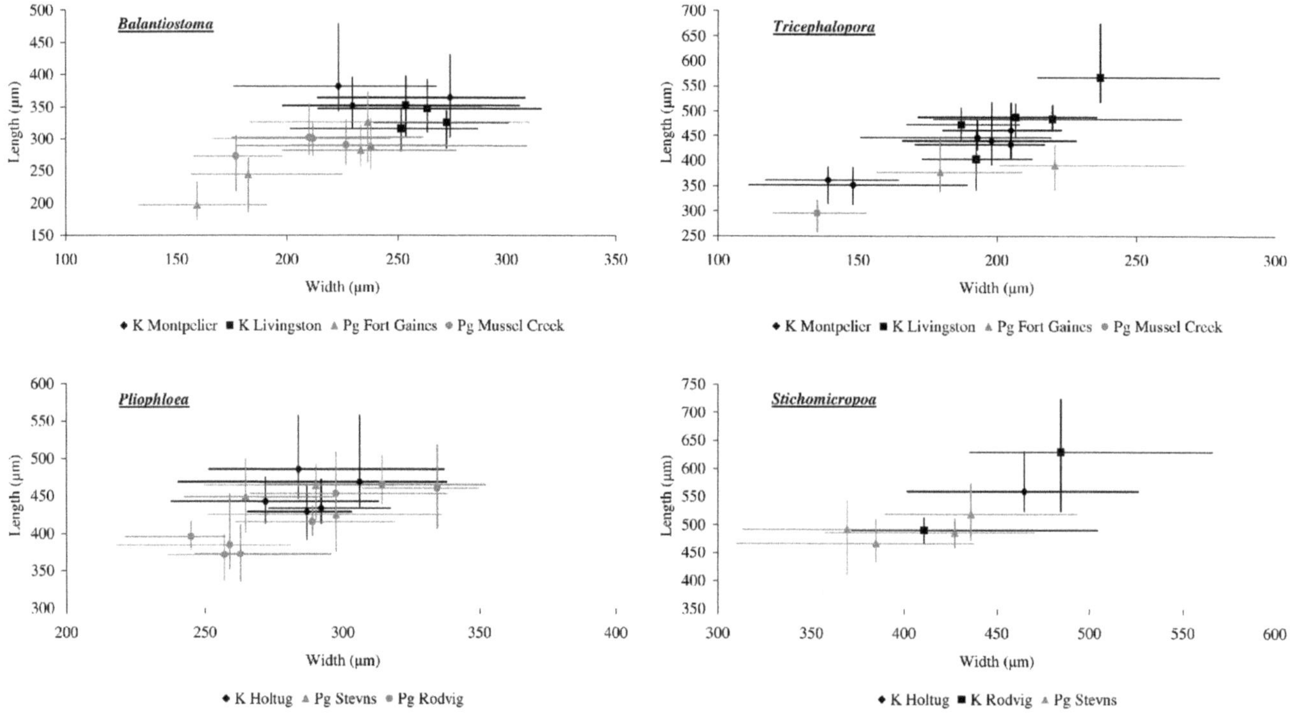

Figure 4. Mean zooid width and length for all colonies. *Balantiostoma*: K = Cretaceous Prairie Bluff Chalk (*Balantiostoma nomas*); Pg = Paleogene Clayton Formation (*B. midwayanica*). ***Tricephalopora***:K = Cretaceous Prairie Bluff Chalk (*Tricephalopora larwoodi*); Pg = Paleogene Clayton Formation (*T. levigatum*). *Pliophloea*:K = Cretaceous Højerup Member; Pg = Paleogene Korsnæb Member. *Stichomicropora*: K = Cretaceous Højerup Member; Pg = Paleogene Korsnæb Member. Error bars show minimum and maximum width and length measured for each colony.

in zooid width between formations, with mean size differences of 15–49 μm (2–11%) (Table 6, Figure 4). Standard deviations remained reasonably constant at 26–50 μm.

Maximum zooid width was observed in the Maastrichtian for all taxa, excluding *Pliophloea subcornuta*, for which a maximum width of 393 μm was observed in the Korsnæb Member compared to a value of 364 μm in the Højerup Member. Similarly, minimum zooid width was always observed in the Danian, except for *Tricephalopora* which had a minimum width of 111 μm in the Prairie Bluff Chalk compared to a value of 120 μm in the Clayton Formation.

As was the case for zooid length, there was no significant variation between localities in zooid width, and *Tricephalopora* is the only genus to contribute to the total variance component for localities (33%). Between 26 and 46% of total variance was intracolonial. Variation between colonies was significant for all species ($p<0.001$) and contributed a higher component of the total variance than that seen for zooid length (27–52%).

Zooid area results are similar to those obtained for zooid length and zooid width, which is unsurprising given that zooid area is a function of these two parameters. *Balantiostoma* was the only taxon to show a decrease in zooid area ($p<0.001$) at the K–Pg, with the average area of zooids declining by 0.03 mm^2 (33%) from 0.09 mm^2 to 0.06 mm^2 (Table 7). The other three taxa showed a decrease in mean area of 0.01–0.06 mm^2 (7–26%) across the boundary, which was not significant. There is no significant variation between localities for zooid area, but variation between colonies accounts for between 21% and 68% of the total variance.

Assemblage level zooid size

There is no significant change in the length of bryozoan zooids across the K–Pg boundary for species in assemblages from the USA or Denmark (Figure 5, Table S3). In the USA, the overall mean zooid size for bryozoans actually increased by 46 μm across the K–Pg boundary, but this is not statistically significant ($p=0.5$). In Denmark, overall mean zooid size decreased by 109 μm ($p=0.05$, which is the level of significance). There is also significant variation in zooid sizes between different species (colonies) within each formation (81%) (Table 8).

In both regions, the longest zooids occur in Maastrichtian formations, with *Conopeum spissamentum* in the Prairie Bluff Chalk (550 μm) and *Ellisina simplex* in the Højerup Member (1299 μm) having the greatest mean zooid lengths. *Ellisina simplex* also exhibits the maximum zooid length observed overall in the Højerup Member (1408 μm), but this taxon is not observed in the Korsnæb Member assemblage studied here, where the maximum zooid length is 693 μm and occurs in the measured "*Membranipora*" sp. colony. In the USA, the maximum zooid length in the Prairie Bluff Chalk is observed in a *Tricephalopora larwoodi* colony measured for this analysis (627 μm). The longest zooid present in the Clayton Formation is also in a *Tricephalopora* colony (576 μm). Maximum zooid length therefore appears to decrease across the K–Pg boundary in both regions. The shortest zooid lengths are also observed in the Maastrichtian assemblage for the USA (*Dionella rinsbergi*: 224 μm) and Denmark (*Cryptostomella* sp.: 310 μm). The minimum zooid lengths in the Danian of the USA (*Balantiostoma midwayanica*: 320 μm) and Denmark (*Pliophloea subcornuta*: 347 μm) are longer than the minimum zooid lengths measured in the corresponding Maastrichtian assemblages (Figure 5), although this may be an artefact of the smaller number of species measured in the Danian.

The only species that crosses the boundary and exhibits a size increase is *Cryptostomella* sp. in Denmark. The remaining 6 species that are present in the assemblages on either side of the boundary appear to decrease in zooid length or maintain a relatively constant zooid length across the K–Pg boundary.

Discussion

There is no significant difference in either zooid size or colony size for any of the analysed cheilostome bryozoan species pairs across the K–Pg boundary, with the exception of *Balantiostoma* which exhibits a significant decease in zooid width and zooid area (but not zooid length) across the boundary, and *Stichomicropora* which decreases slightly in colony size across the boundary. Zooid size in species of *Tricephalopora*, *Pliophloea* and *Stichomicropora* do not vary significantly between formations, nor does colony size in *Balantiostoma*, *Tricephalopora* and *Pliophloea*. In a more general sense, zooid size does not change significantly across the K–Pg boundary when all bryozoan species are considered together in the USA, although there is a slight decrease in zooid size in the Danish assemblage studied here. Variations between localities do not appear to affect colony or zooid size, but there are highly significant between colony variations in zooid size for all taxa in all zooid parameters measured.

Table 6. Nested ANOVA results for zooid width.

Congeneric	Mean width (μm) (SD)		F-ratios			% variance component[a]			
	K	Pg	Fm.	Loc.	Col.	Fm.	Loc.	Col.	Zooids
Balantiostoma nomas	253	208	24.26*	0.41	8.63***	38.13	0	26.78	35.09
	(33)	(39)							
Tricephalopora larwoodi	194	179	0.89	3.51	17.11***	0	32.64	41.56	25.80
	(34)	(40)							
Pliophloea subcornuta	288	283	n/a[b]	0.01	12.17***	1.83	0	51.81	46.36
	(27)	(33)							
Stichomicropora	454	404	n/a[b]	0.13	10.24***	36.67	0	30.41	32.92
	(50)	(45)							

* = $p<0.05$; ** = $p<0.01$; *** = $p<0.001$.

[a]Negative estimates of variance components associated with non-significant F-ratios were assigned zero values in keeping with standard practice [26].

[b]Denominator of the F-test is zero.

Table 7. Nested ANOVA results for zoid area.

Congeneric	Mean area (mm^2)(SD)		F-ratios			% variance component[a]			
	K	Pg	Fm.	Loc	Col.	Fm.	Loc.	Col.	Zooids
Balantiostoma nomas	0.09	0.06	285.96***	0.06	12.12***	60.16	0.00	20.98	18.86
	(0.01)	(0.02)							
Tricephalopora larwoodi	0.09	0.06	2.03	3.34	35.31***	23.52	27.08	38.25	11.15
	(0.02)	(0.02)							
Pliophloea subcornuta	0.01	0.01	1.23	0.46	22.86***	1.06	0.00	67.88	31.05
	(0.01)	(0.02)							
Stichomicropora	0.03	0.02	n/a[b]	0.02	61.62***	48.89	0.00	38.53	12.58
	(0.05)	(0.03)							

* $=p<0.05$; ** $=p<0.01$; *** $=p<0.001$.
[a]Negative estimates of variance components associated with non-significant F-ratios were assigned zero values in keeping with standard practice [26].
[b]Denominator of the F-test is zero.

Colony size across the K–Pg boundary

Mean colony size in the majority of cheilostome bryozoans analysed here does not change significantly across the K–Pg boundary, although the mean size of *Stichomicropora* colonies does decrease slightly. This lack of change contrasts with previously reported decreases in colony size at other biotic crises among graptolites [1] and corals [35] and implies that not all colonial organisms exhibit a Lilliput effect.

Colony size in living bryozoans has been shown to be dependent on both quantity and quality of food supply [36–38]. As suspension feeders, it was expected that bryozoans would exhibit a decrease in size following the hypothesised primary productivity crash of the K–Pg event. In free-living bryozoans, colony size has been used as a measure of the favourability of environmental conditions based on the fact that larger colonies have a greater number of feeding and reproductive zooids [18]. A pre-extinction decrease in bryozoan colony size in *Lunulites pseudocretacea* in the late Maastrichtian of Nye Kløv, Denmark was attributed to low primary productivity levels [18]. In contrast, no evidence for a significant decrease in colony size across the K–Pg boundary itself at Stevns Klint is apparent in the *Pliophloea subcornuta* colonies studied here, although *Stichomicropora* does exhibit a slight but insignificant decrease in colony size.

Colony size is a function of zooid size and the number of constituent zooids. In the majority of bryozoan species studied here the size of the zooids remained more or less constant across the K–Pg boundary, indicating that there must also be limited variation in the number of zooids per colony for *Tricephalopora* and *Pliophloea subcornuta*. The decrease observed in zooid size (area) in *Balantiostoma*, however, implies that the average number of zooids per colony must increase in order to maintain the constant average colony size observed for this genus. This suggests that, for *Balantiostoma*, the number and size of zooids are more plastic than is total colony area.

The size of the substrates encrusted by bryozoans may place an upper bound on colony size. It is therefore necessary to determine whether colony size responds indirectly to environmental variations or instead adapts to substrate size, which itself may vary according to environmental factors. Although anecdotal observations show a decrease in substrate size across the K–Pg boundary for the material analysed here, and echinoid size has previously been shown to decrease across the K–Pg boundary [13], the limited size response observed in the colonies studied here suggests that colony size does not simply scale to substrate size.

There is, however, a clear decrease in the maximum size of colonies across the K–Pg boundary, for all studied taxa apart from *Balantiostoma*. This apparent decline in maximum colony size indicates that, whilst the mass extinction appears to have had negligible influence on the mean size of these colonies, the extremely large colony sizes reached in the Maastrictian were not attained in the Danian.

A possible reason for the lack of mean colony size decrease found in this study is that sampling across the K–Pg boundary may have been over a timescale that extended beyond the duration of any Lilliput effect, which is in the order of a few hundred thousand years [1]. The duration of the size response has important implications as it can help to determine whether Lilliput effects are short-term ecological responses or represent long-term resettings of evolutionary history [3]. The samples used in this study were not collected from sediments deposited immediately post-extinction (Figure 1) and may have failed to detect a short-term Lilliput effect: size may have returned back to 'normal' pre-extinction level by the time of the Paleocene bryozoans used for this study. In particular, the Danish specimens come from the Korsnæb Member as too few bryozoans are present in the Cerithium Limestone immediately above the K–Pg boundary. Likewise, a depositional hiatus underlying the sampling levels for the Clayton Formation bryozoans from the USA that were used in this study is likely [39].

Zooid size across the K–Pg boundary

Decrease in zooid width and area of Balantiostoma across the K–Pg boundary. Of the genera studied here, only *Balantiostoma* showed significant changes in size across the K–Pg boundary, with a decrease in both zooid width and area. To interpret the cause of this size decrease, it is necessary to consider the standard phenotypic responses of bryozoan zooids to unfavourable conditions [1]. A reduction in primary productivity has been hypothesised as the cause for small organism size following the K–Pg event [13,17], although an array of other environmental stresses, such as greenhouse warming, shallowing seas and the restriction of basins, as well as volcanic activity during the late Maastrichtian, have been cited as possible causes for a trend towards small-sized foraminifera [40]. Alternative environmental factors suggested to induce Lilliput effects at mass

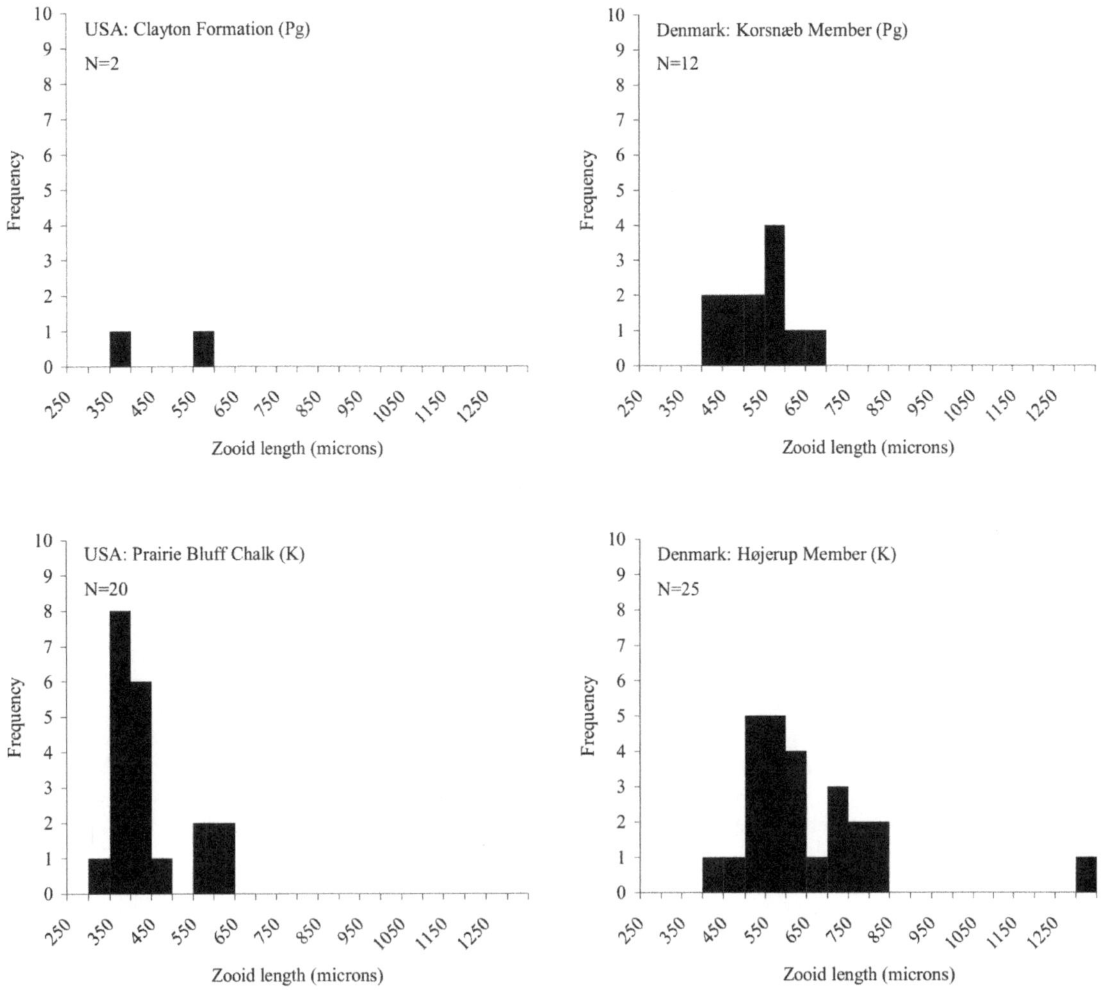

Figure 5. Mean zooid length of all species present in the studied samples. Histograms showing the lengths of zooids from the Maastrichtian Prairie Bluff Chalk/Højerup Member (K) and the Paleogene Clayton Formation/Korsnæb Member (Pg). The total number of species is indicated. $p>0.05$ across the K–Pg boundary in the USA and Denmark in a comparison of overall bryozoan zooid size.

extinctions elsewhere in the geological record include anoxia, and changes in temperature and salinity levels [2].

Environmental influences on zooid size in living cheilostome bryozoans have been investigated using laboratory and field experiments [31,41]. Temperature is often identified as the most important single environmental influence on zooid size, with an inverse relationship between size and temperature both within colonies and within species [27,33,37,41–47]. This relationship is believed to be a taxon-independent function of metabolism at different temperatures [44]. Therefore, the zooid size-temperature trend should not have altered through geological time [11]. Although there is some evidence that a zooid size-temperature relationship may occur between congeneric species [48], applying this relationship to changes in zooid size between related species across the K–Pg is more difficult to justify given our current state of knowledge. For example, it seems unlikely that only *Balantiostoma* among the genera studied would have experienced temperature-related zooid size reduction, especially as the studied colonies are from the same localities as *Tricephalopora* colonies that showed no change in zooid size.

Alternative environmental parameters that have been shown to influence zooid size include salinity and oxygen [37]. Based on studies of other groups, the likeliest driver of any zooid decrease at the K–Pg boundary is the decline in primary productivity hypothesised to have occurred at this time [12]. However, contradictory results have been published on the effect of food quality and quantity on zooid size in modern bryozoans [37,42,45]. Since food supply has sometimes been shown to have no influence on zooid size, it has been suggested that its impact is slight and overprinted by temperature [46]. However, a positive correlation between zooid size and both food quality and quantity in laboratory experiments on *Electra pilosa* has been observed [38]. Possible explanations for mixed results in food studies on zooid size include different methodologies [38], the fact that laboratory experiments do not reflect the true complexities of the natural world [45], and a non-uniform response of different bryozoan species to changes in food supply [49]. The implication of zooid size decrease only occurring in *Balantiostoma* may be that not all bryozoan genera were similarly affected by the hypothesised primary productivity decline across the K–Pg boundary.

The lack of change in zooid length in *Balantiostoma* at the K–Pg underlines the importance of obtaining multiple measurements of body size and understanding the implications that such parameters may have on the physiology and ecology of the studied organisms. In *Balantiostoma*, the variation in zooid width but not length contrasts with previous bryozoan zooid size studies that have

Table 8. Nested ANOVA results for zoid length of all species.

Region	Mean length (μm)(SD)		F-ratios		% variance component[a]		
	K	Pg	Fm.	Col.	Fm.	Col.	Zooids
USA	386 (87)	432 (95)	0.48	129.15***	11.78	81.86	6.36
Denmark	599 (181)	490 (83)	4.08	219.18***	15.44	80.86	3.81

* = $p<0.05$; ** = $p<0.01$; *** = $p<0.001$.

found zooid length to be the most sensitive size parameter to environmental changes [27,33,34]. It has been suggested that zooid width is determined mainly by the position of the zooid in the colony, particularly with respect to row bifurcations, whereas zooid length is more dependent on environmental influences [34,37]. Further knowledge of the relationship between these size parameters would help to understand why most change in *Balantiostoma* is in width rather than length. At the assemblage-level, additional analysis of trends in zooid width might therefore also aid further insight into bryozoan size response.

Lack of consistent zooid size change. The results of this study contradict the Lilliput hypothesis, which predicts significant size decrease across the K–Pg boundary in all species. None of the four analysed species show significant changes in zooid length at this boundary, and there is also no significant change in zooid width or area for three of these species. Furthermore, a more general analysis of zooid size in assemblages does not show a significant trend to smaller size after the K–Pg mass extinction in the USA, in contrast to previous whole assemblage size analyses, which constitute the majority of Lilliput studies [2,50], although a slight decrease in zooid size was detected in the Danish assemblages.

However, size change at mass extinction horizons is likely to be complex and it cannot be expected that all clades will respond in the same manner, or that all mass extinctions will have the same effects [10]. The definition of the Lilliput effect allows for more variety in the response of organisms than is often supposed [8]; in Urbanek's original study of the Lilliput effect [1], size decrease was not consistent across all species, and studies of within-lineage size across mass extinction events have since obtained mixed results (Table 1). Furthermore, the difference in size trends between *Balantiostoma* and the three other bryozoan genera studied here demonstrates that even quite closely related taxa do not always respond identically [8].

As with colony size, it is possible that the timescale of any Lilliput effect on bryozoan zooid size was too short to be detected in this study. It is also possible that zooid sizes prior to the K–Pg event were smaller than 'normal'. Pre-extinction dwarfing has previously been observed in other taxa [51,52], and an observed shift to smaller bryozoan zooid size prior to the K–Pg boundary in Denmark was attributed to a change in temperature [18]. These results imply that 'unfavourable' environmental conditions affected zooid size leading up to the mass extinction, potentially 're-setting' size trends prior to the K–Pg boundary rather than across the K–Pg boundary itself. This trend corresponds with a suggestion that the biotic responses to the K–Pg event were not instantaneous and that environmental perturbations predated the K–Pg boundary [53]. However, the size decrease observed in *Balantiostoma* indicates that at least one species experienced a change on the timescale used in this study and the decline in assemblage-level zooid size in Denmark also suggests an appropriate timescale was analysed. It is unclear, therefore, whether these results indicate lack of a size response, or different timings of size responses between species. Analysis of additional colonies extending a further time distance from the boundary are therefore necessary in order to establish the true extent and influence of the K–Pg event on bryozoan zooid size.

Genotypic variation. In all parameters measured for all species studied, there are highly significant variations in zooid size between colonies, underlying similarities found at the formation level. These variations may be inferred as genotypic in origin, which is known to be responsible for a large amount of zooid size variation observed within species of living bryozoans [37,38,42]. Therefore, caution must be taken when inferring environmental factors influencing bryozoan zooid size [48]; mean zooid size can often vary more between genotypes of the same species than over time or between environments [46]. Genotypic effects can be expected to be even greater among closely related, congeneric species, like those analysed in the current study. However, the nature of the nested ANOVA analysis demonstrates that the differences and similarities at the formation level (i.e. across the K–Pg boundary) occur despite size variation within and between colonies from each formation.

Biogeographical and regional trends

The two regions from which the bryozoans were collected contrast markedly in their depositional environments, latitude, distance from the Chicxulub impact size, and types of hard substrates available. The fact that both show a similar lack of change in zooid size across the boundary indicates that such stasis is more likely to be a global than a regional pattern [20]. If localised variations had been influential on bryozoan size, greater variation in size trends between the two regions would be expected. For example, it has previously been shown that colonies of a modern cheilostome species living on different substrate types differ in size [34], but a similar pattern is not evident in the current study.

For within-lineage analyses, excluding *Balantiostoma* zooid size and *Stichomicropora* colony size, there is no significant difference between zooid or colony sizes of any other studied species at the locality level. This implies that size patterns override local differences in environments, with no regional variation detected in this study. Additionally, analysis of two species from the USA and two species from Denmark shows a similar lack of zooid size response in these biogeographical regions. The only genus to display a significant decrease in zooid size is *Balantiostoma*, but as *Tricephalopora* colonies from the same localities failed to show a parallel trend, this indicates that the change was taxon-specific. There is also a difference in zooid size trends between regions at the assemblage level, with Danish cheilostome zooid sizes decreasing slightly across the boundary while the USA sizes remained static.

Conclusions

1) Bryozoan colony size and zooid size do not change significantly across the K–Pg boundary in two widely separated biogeographical regions, Denmark and the south-eastern USA. The influence of locality on colony size is variable and is significant only for *Balantiostoma* from the SE USA among the four genera studied.
2) Although *Stichomicropora* did show a slight decrease in colony size, the size of colonies for the three other genera studied

here remained static across the K–Pg boundary, contradicting the predicted decline reflecting a Lilliput effect and some previous studies of colony size change in other colonial groups across mass extinctions [1,34].

3) Zooid size generally remained stable in assemblages across the K–Pg boundary in the USA, which is in contrast to size decreases observed at the assemblage-level in solitary organisms in other Lilliput studies. In Denmark, however, a slight decline in zooid size was observed.
4) Zooid size change across the K–Pg boundary in pairs of congeneric species varied according to species, which is consistent with the original Lilliput study of graptolites [1] and more recent analyses at lower taxonomic levels [8]. Three out of the four studied genera showed no significant change in zooid length, width, or area across the K–Pg boundary, whereas zooid width and area were found to decrease in *Balantiostoma*.
5) The slight and non-universal size reduction observed in the bryozoans studied here at both the colony- and zooid-levels across the K–Pg boundary shows that these colonial animals do not exhibit a classic and predictable Lilliput effect, despite the fact that various studies of present-day bryozoans have shown that size is responsive to localised environmental changes.
6) The possibility that size change in bryozoans across the K–Pg boundary occurred over a shorter interval than sampled in this study cannot be ruled out.
7) This caveat apart, the lack of size change in the majority of bryozoans studied here at the K–Pg mass extinction indicates that a universal Lilliput effect occurring across all taxa at all mass extinction events cannot be supported. Instead, it seems probable that different organisms have responded differently in terms of body size to different mass extinctions.

Supporting Information

Table S1 Colony sizes for within-genus analysis.

Table S2 Zooid sizes for within-genus analysis.

Table S3 Measurements of zooid length for assemblage level analysis.

Acknowledgments

We thank Sten Jakobsen, Mark Wilson, Megan Innis, George Phillips and Jon Bryan for their guidance and assistance in the field and JoAnn Sanner for scanning electron microscopy of some bryozoan collections held at the Smithsonian Institution, Washington DC. Comments provided by the two journal reviewers, Richard Twitchett and Alexander Nützel, greatly improved the final manuscript.

Author Contributions

Conceived and designed the experiments: CES EMH PDT. Performed the experiments: CES. Analyzed the data: CES. Contributed reagents/materials/analysis tools: CES EMH PDT. Wrote the paper: CES EMH PDT.

References

1. Urbanek A (1993) Biotic crises in the history of Upper Silurian graptoloids: a palaeobiological model. Hist Biol 7: 29–50.
2. Twitchett RJ (2007) The Lilliput effect in the aftermath of the end-Permian extinction event. Palaeogeogr Palaeoclim Palaeoecol 252: 132–144.
3. Wade BS, Twitchett RJ (2009) Extinction, dwarfing and the Lilliput effect. Palaeogeogr Palaeoclim Palaeoecol 284: 1–3.
4. Nützel A (2005) A new Early Triassic gastropod genus and the recovery of gastropods from the Permian/Triassic extinction. Acta Palaeontol Pol 50: 19–24.
5. Nützel A (2005) Recovery of gastropods in the Early Triassic. Comptes Rendus Palevol 4: 501–515.
6. McGowan AJ, Smith AB, Taylor PD (2009) Faunal diversity, heterogeneity and body size in the Early Triassic: testing post-extinction paradigms in the Virgin Limestone of Utah, USA. Aust J Earth Sci 56: 859–872.
7. Brayard A, Nützel A, Stephen DA, Bylund KG, Jenks J, et al. (2010) Gastropod evidence against the Early Triassic Lilliput effect. Geology 38: 147–150.
8. Huang B, Harper DAT, Zhan R, Rong J (2010) Can the Lilliput Effect be detected in the brachiopod faunas of South China following the terminal Ordovician mass extinction? Palaeogeogr Palaeoclim Palaeoecol 285: 277–286.
9. Metcalfe B, Twitchett RJ, Price-Lloyd N (2011) Changes in size and growth rate of "Lilliput" animals in the earliest Triassic. Palaeogeogr Palaeoclim Palaeoecol 308: 171–180.
10. Harries PJ, Knorr PO (2009) What does the "Lilliput Effect" mean? Palaeogeogr Palaeoclim Palaeoecol 284: 4–10.
11. O'Dea A, Okamura B (2000) Intracolony variation in zooid size in cheilostome bryozoans as a new technique for investigating palaeoseasonality. Palaeogeogr Palaeoclim Palaeoecol 162: 319–332.
12. Arthur MA, Zachos JC, Jones DS (1987) Primary productivity and the Cretaceous/Tertiary boundary event in the oceans. Cretac Res 8: 43–54.
13. Smith AB, Jeffery CH (1998) Selectivity of extinction among sea urchins at the end of the Cretaceous period. Nature 392: 69–71.
14. Jablonski D, Raup D (1995) Selectivity of end-Cretaceous marine bivalve extinctions. Science 268: 389–391.
15. Jeffery CH (2001) Heart urchins at the Cretaceous/Tertiary boundary: a tale of two clades. Paleobiology 27: 140–158.
16. Lockwood R (2005) Body size, extinction events, and the early Cenozoic record of veneroid bivalves: a new role for recoveries? Paleobiology 31: 578–590.
17. Aberhan M, Weidemeyer S, Kiessling W, Scasso R, Medina F (2007) Faunal evidence for reduced productivity and uncoordinated recovery in Southern Hemisphere Cretaceous-Paleogene boundary sections. Geology 35: 227–230.
18. O'Dea A, Håkansson E, Taylor PD, Okamura B (2011) Environmental change prior to the K–T boundary inferred from temporal variation in the morphology of cheilostome bryozoans. Palaeogeogr Palaeoclim Palaeoecol 308: 502–512.
19. Macleod N, Rawson PF, Forey PL, Banner FT, Boudagher-Fadel MK, et al. (1997) The Cretaceous–Tertiary biotic transition. J Geol Soc Lond 154: 265–292.
20. Sogot CE, Harper EM, Taylor PD (2013) Biogeographical and ecological patterns in bryozoans across the Cretaceous–Paleogene boundary: Implications for the phytoplankton collapse hypothesis. Geology 41: 631–634.
21. Bryan JR, Jones DS (1989) Fabric of the Cretaceous-Tertiary marine macrofaunal transition at Braggs, Alabama. Palaeogeogr Palaeoclim Palaeoecol 69: 279–301.
22. Schulte P, Speijer RP (2009) Late Maastrichtian-Early Paleocene sea level and climate changes in the Antioch Church Core (Alabama, Gulf of Mexico margin, USA): A multi-proxy approach. Geol Acta 7: 11–34.
23. Mancini EA, Puckett TM, Tew BH, Smith CC (1995) Upper Cretaceous Sequence Stratigraphy of the Mississippi-Alabama Area. Gulf Coast Assoc Geol Soc Trans 45: 377–384.
24. Anderskouv K, Damholt T, Surlyk F (2007) Late Maastrichtian chalk mounds, Stevns Klint, Denmark – Combined physical and biogenic structures. Sediment Geol 200: 57–72.
25. Canu F, Bassler RS (1920) North American Early Tertiary Bryozoa. Smithson Inst United States Natl Mus Bulletin 106: 897pp.
26. Taylor PD, McKinney FK (2006) Cretaceous Bryozoa from the Campanian and Maastrichtian of the Atlantic and Gulf Coastal Plains, United States. Scr Geol 132: 1–346.
27. Okamura B, Bishop JDD (1988) Zooid size in cheilostome bryozoans as an indicator of relative palaeotemperature. Palaeogeogr Palaeoclim Palaeoecol 66: 145–152.
28. Sokal RR, Rohlf FJ (1995) Biometry. 3rd ed. New York: W. H. Freeman and Company. 850 p.
29. Grafen A, Hails R (2002) Modern statistics for the life sciences. Oxford UK: Oxford University Press. 351 p.
30. Dytham C (2003) Choosing and using statistics – a biologist's guide. 2nd ed. Oxford UK: Blackwell. 248 p.
31. Okamura B, O'Dea A, Knowles T (2011) Bryozoan growth and environmental reconstruction by zooid size variation. Mar Ecol Prog Ser 430: 133–146.
32. Boardman RS, Cheetham AH, Cook PL (1969) Intracolony variation and the genus concept in Bryozoa. Proceedings of the North American Paleontological Convention. Chicago. pp. 294–320.

33. Lombardi C, Cocito S, Occhipinti-Ambrogi A, Hiscock K (2006) The influence of seawater temperature on zooid size and growth rate in *Pentapora fascialis* (Bryozoa: Cheilostomata). Mar Biol 149: 1103–1109.
34. Yagunova EB, Ostrovsky N (2008) Encrusting bryozoan colonies on stones and algae: variability of zooidal size and its possible causes. J Mar Biol Assoc United Kingd 88: 901–908.
35. Kaljo D (1996) Diachronous recovery patterns in Early Silurian corals, graptolites and acritarchs. Biotic Recovery from Mass Extinction Events. Geological Society Special Publication. Vol. 102. pp. 127–133.
36. Winston JE (1976) Experimental culture of the estuarine ectoproct *Conopeum tenuissimum* from Chesapeake Bay. Biol Bull 150: 318–335.
37. O'Dea A, Okamura B (1999) Influence of seasonal variation in temperature, salinity and food availability on module size and colony growth of the estuarine bryozoan *Conopeum seurati*. Mar Biol 135: 581–588.
38. Hageman SJ, Needham LL, Todd CD (2009) Threshold effects of food concentration on the skeletal morphology of the bryozoan *Electra pilosa* (Linnaeus, 1767). Lethaia 42: 438–451.
39. Savrda CE (1993) Ichnosedimentologic evidence for a noncatastrophic origin of Cretaceous–Tertiary boundary sands in Alabama. Geology 21: 1075–1078.
40. Keller G, Abramovich S (2009) Lilliput effect in late Maastrichtian planktic foraminifera: Response to environmental stress. Palaeogeogr Palaeoclim Palaeoecol 284: 47–62.
41. Knowles T, Taylor PD, Williams M, Haywood AM, Okamura B (2009) Pliocene seasonality across the North Atlantic inferred from cheilostome bryozoans. Palaeogeogr Palaeoclim Palaeoecol 277: 226–235.
42. Hunter E, Hughes R (1994) The influence of temperature, food ration and genotype on zooid size in *Celleporella hyalina* (L.). Biology and palaeobiology of bryozoans. Fredensborg: Olsen & Olsen. pp. 83–86.
43. O'Dea A, Okamura B (2000) Life history and environmental inference through retrospective morphometric analysis of bryozoans: a preliminary study. J Mar Biol Assoc United Kingd 80: 1127–1128.
44. O'Dea A (2003) Seasonality and zooid size variation in Panamanian encrusting bryozoans. J Mar Biol Assoc United Kingd 83: 1107–1108.
45. Amui-Vedel AM, Hayward PJ, Porter JS (2007) Zooid size and growth rate of the bryozoan *Cryptosula pallasiana* Moll in relation to temperature, in culture and in its natural environment. J Exp Mar Biol Ecol 353: 1–12.
46. O'Dea A, Rodriguez F, Romero T (2007) Response of zooid size in *Cupuladria exfragminis* (Bryozoa) to simulated upwelling temperatures. Mar Ecol 28: 315–323.
47. Knowles T, Leng MJ, Williams M, Taylor PD, Sloane HJ, et al. (2010) Interpreting seawater temperature range using oxygen isotopes and zooid size variation in *Pentapora foliacea* (Bryozoa). Mar Biol 157: 1171–1180.
48. Kuklinski P, Taylor PD (2008) Are bryozoans adapted for living in the Arctic? Va Mus Nat Hist Spec Publ 15: 101–110.
49. Berning B (2007) The Mediterranean bryozoan *Myriapora truncata* (Pallas, 1766): a potential indicator of (Palaeo-) environmental conditions. Lethaia 40: 221–232.
50. Fraiser ML, Bottjer DJ (2004) The Non-Actualistic Early Triassic Gastropod Fauna: A Case Study of the Lower Triassic Sinbad Limestone Member. Palaios 19: 259–275.
51. He W, Shi GR, Feng Q, Campi MJ, Gu S, et al. (2007) Brachiopod miniaturization and its possible causes during the Permian–Triassic crisis in deep water environments, South China. Palaeogeogr Palaeoclim Palaeoecol 252: 145–163.
52. Wade BS, Olsson RK (2009) Investigation of pre-extinction dwarfing in Cenozoic planktonic foraminifera. Palaeogeogr Palaeoclim Palaeoecol 284: 39–46.
53. Tobin TS, Ward PD, Steig EJ, Olivero EB, Hilburn IA, et al. (2012) Extinction patterns, δ18 O trends, and magnetostratigraphy from a southern high-latitude Cretaceous–Paleogene section: Links with Deccan volcanism. Palaeogeogr Palaeoclim Palaeoecol 350–352: 180–188.
54. He WH, Twitchett RJ, Zhang Y, Shi GR, Feng QL, et al. (2010) Controls on body size during the Late Permian mass extinction event. Geobiology 8: 391–402.
55. Morten SD, Twitchett RJ (2009) Fluctuations in the body size of marine invertebrates through the Pliensbachian–Toarcian extinction event. Palaeogeogr Palaeoclim Palaeoecol 284: 29–38.

4

Inferring Gene Family Histories in Yeast Identifies Lineage Specific Expansions

Ryan M. Ames[1]*, Daniel Money[2], Simon C. Lovell[1]

1 Computational and Evolutionary Biology, Faculty of Life Sciences, The University of Manchester, Manchester, United Kingdom, 2 Department of Ecology and Evolutionary Biology, University of Kansas, Lawrence, Kansas, United States of America

Abstract

The complement of genes found in the genome is a balance between gene gain and gene loss. Knowledge of the specific genes that are gained and lost over evolutionary time allows an understanding of the evolution of biological functions. Here we use new evolutionary models to infer gene family histories across complete yeast genomes; these models allow us to estimate the relative genome-wide rates of gene birth, death, innovation and extinction (loss of an entire family) for the first time. We show that the rates of gene family evolution vary both between gene families and between species. We are also able to identify those families that have experienced rapid lineage specific expansion/contraction and show that these families are enriched for specific functions. Moreover, we find that families with specific functions are repeatedly expanded in multiple species, suggesting the presence of common adaptations and that these family expansions/contractions are not random. Additionally, we identify potential specialisations, unique to specific species, in the functions of lineage specific expanded families. These results suggest that an important mechanism in the evolution of genome content is the presence of lineage-specific gene family changes.

Editor: Maria Anisimova, Swiss Federal Institute of Technology (ETH Zurich), Switzerland

Funding: This work is funded by BBSRC grant BB/I020489/1. The funders had no role in study design, data collection and analysis, decision to publish, or preparation of the manuscript.

Competing Interests: The authors have declared that no competing interests exist.

* E-mail: ryan.ames@manchester.ac.uk

Introduction

The creation of new genetic content in the form of new genes is a key component of genome evolution. New genes can arise through a variety of mechanisms including gene duplication, retroposition, horizontal gene transfer and *de novo* origination [1]; however, gene duplication has been recognised, since the 1970s, as the most prevalent source of new genes [2]. Indeed, in a comparison of the relative contributions of these gene creation mechanisms, gene duplication was shown to have produced ~80% of new genes in several *Drosophila* species [3].

Rates of gene gain by duplication have been shown to be high in a variety of species, and it has been suggested that in eukaryotes 50% of genes are expected to duplicate at least once every 35–350 million years [4]. In *Drosophila* the rate of gene gain has been estimated to be in the range of 5 to 11 genes every million years [3] and as high as 17 genes every million years [5]. However, current genome content is a balance between the rate of gene gain and the rate at which genes are lost. Indeed, the most common fate of duplicate genes is expected to be nonfunctionalisation [4], which may be followed by removal from the genome. That the size of the genome appears to be constant over time is probably due to the high rate of gene gain coupled with a high rate of gene loss. It is therefore important to consider the rates of both gene gain and gene loss independently in order to accurately understand genome evolution.

Reconstructing histories of gene families is currently an active area of research, and there have been several methods developed, along with some genome-wide studies of the evolution of gene families [6–11]. Inferences of gain and loss events along the human lineage of the mammalian phylogeny suggests that, between humans and chimpanzees, the complement of genes differs more than the sequences of orthologous nucleotides [12], leading to the argument that a "revolving door" of gain and loss leads to large differences between the genomes of humans and chimpanzees. Likewise, in comparisons of gene families in several species of *Drosophila* [5] large numbers of genes are both gained and lost with over 40% of families varying in size. Genome-wide studies of gene family evolution have also been conducted in species sets containing yeast [13–15]. In all species analysed a large turnover of genes is seen, and specific functions are associated with these changes.

Many different methods have been developed to study gene family evolution. Novel reconstruction algorithms [16,17] have been used to infer the histories of gene families. These algorithms a incorporate gain, loss and horizontal gene transfer and utilise phyletic profiles and a species tree. Parsimony [13] and weighted parsimony [9] have both been used to infer the evolution of gene families. Although these methods are quick and can be applied to genome-wide studies, they cannot account for multiple events (gain and subsequent loss) on a single branch.

As an alternative to parsimony methods that make use of phyletic profiles and a single species tree there are many methods that make use of gene trees produced from protein families. These methods aim to reconcile the gene trees with the single species tree and in the process of reconciliation infer the gain and loss events. Several software tools have been developed for gene/species tree reconciliation [18–22]. As with the parsimony methods these tools are quick, can be applied to whole genomes and have been shown to be accurate, although they may be affected by bias in some cases [23,24]. More recently, a range of reconciliation methods [6,8,15,25,26] have used probabilistic models or Bayesian methods to infer the gene trees and for gene/species tree reconciliation. These methods have shown to be both accurate and applicable to whole genome analyses.

The problem of inferring gene family histories has also been addressed with the development of likelihood-based methods that make use of probabilistic models [7,9,10,12,14]. These methods do not rely on individual gene trees but instead use a probabilistic model to infer the evolutionary histories of gene families based on a species phylogeny and phyletic profiles for each family. These methods have been shown to be accurate, applicable to genome-wide analyses and can infer multiple events on a single branch. As these methods are dependent on an underlying model, the biological realism of these models is important for their accuracy, and so, much research is focused on producing more biologically relevant models.

We have previously produced a method with several models of gene gain and loss, allowing for the inference of gene family histories, on a whole genome level, with variable duplication branch lengths and rates of evolution among families [9]. However, the model used in our previous study did not capture all the biological complexity of gene family evolution. We have considered innovation events such as *de novo* gene gain and horizontal gene transfer that may be much more widespread than originally thought, particularly in yeast [27]. However, ourselves, and many others have not considered the complete removal of a gene family by pseudogenisation. To gain a more complete understanding of these processes more biologically accurate models are needed.

Here, we apply a new evolutionary model, implemented in DupliPHY-ML [9], to infer the evolutionary histories of gene families from 9 yeast species, using data from the Génolevures project [28]. Our model, BDIE, allows the estimation of the relative rates of gene birth (new gene gain in existing families), death (gene loss by pseudogenisation), innovation (*de novo* gene gain or horizontal gene transfer creating a new family) and extinction (loss of a complete gene family) in yeast gene families on a whole genome scale. We are able to identify families that have experienced lineage specific expansions in the yeast species and link these to specific functions. We are also able to identify functions that appear to be repeatedly associated with expanding families in multiple lineages indicating that these functions may provide common adaptations in these yeast. In addition, we can also identify functions linked with expanding families in a single lineage suggesting that lineage specific gene family evolution plays a key role in specialisation.

Results

Rates of gene family evolution

We used gene family data from the Génolevures project [28]. Here, the families represent phylogenetic groups of genes related by homology and identified by the similarity between protein-coding genes from all the species. There are a total of 4,578 families (after removal of some families - see methods) with an average size of 9.72 genes and a largest family size, in any single taxa, of 54 members. We also looked at the average number of members in each taxa per family (Figure 1), which shows that in most families the each species contains <5 members. Lists of family membership can be found in File S1.

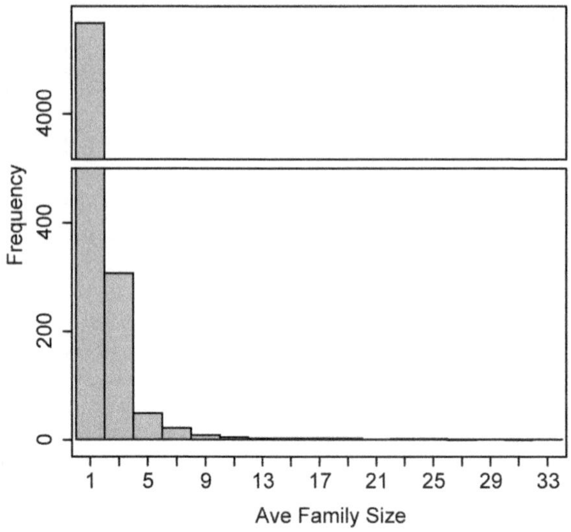

Figure 1. The distribution of average family sizes in the Génolevures data.

A key feature of DupliPHY-ML, is the ability to estimate the rates of gene loss (*d*), gene family extinction (*e*) and the innovation (*i*) of new gene families relative to the rate of gene gain (*b*). In this analysis the rate of gene gain (birth) refers to gene duplication in an existing gene family and gene loss (death) refers to the loss of a gene in a family with 2 or more members. Thus, extinction refers to the loss of a final member of a gene family and innovation refers to the gain of a new gene family by processes such as horizontal gene transfer or *de novo* gain. In our data set we find that $b=1.0$, $d=4.05$, $e=0.59$ and $i=0.05$. The estimates indicate that the rate of gene loss through pseudogenisation is much higher than the rate of gene duplication. This is a surprising result; although it is possible that genomes in at least some of these species are shrinking, it is perhaps more likely to indicate some artefacts from the data generation. We also find that the rate of extinction of a gene family is much lower than the rate of gene loss. The rate of new family formation is extremely low suggesting that in yeast, new gene gain by horizontal gene transfer or *de novo* gain is rare.

Using DupliPHY-ML to infer the ancestral history of gene families, estimates of ancestral family sizes are shown in File S2. DupliPHY-ML also allows us to infer how fast or slow individual families are evolving in terms of gains and losses over the phylogeny. For example, families with a fast evolutionary rate will have had many duplication and loss events in their inferred history. We see that the majority of families are evolving very slowly with only a minority of families displaying rapid evolution (Figure 2).

Data quality may affect the estimation of these rates. It is apparent that both the rates of gene loss and extinction are relatively high compared to the rate of gene gain. If some of the genome sequences for these species are not complete then this missing data may be construed as loss by DupliPHY-ML, which would act to erroneously increase the relative rates of loss and extinction. Indeed, the majority of gene families are very small (Figure 1) and regions with zero sequence coverage in a subset of species may erroneously appear to be extinction events. The effect of such missing sequences would lead to an increase in the extinction rate. However, we note that the rate of gene loss is much higher than that of extinction suggesting that even with

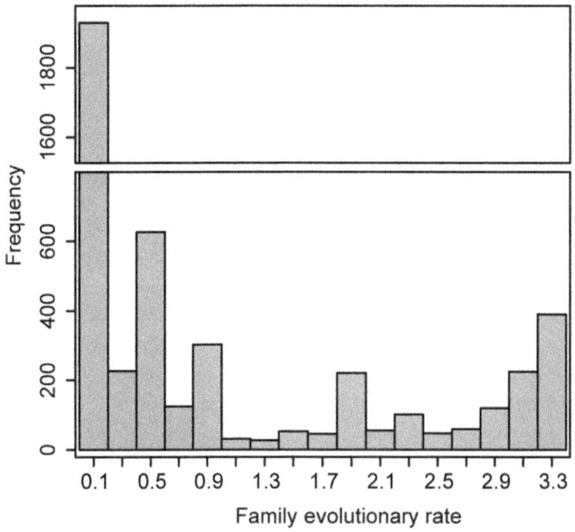

Figure 2. The rate of family evolution in the Génolevures data estimated by DupliPHY-ML as the mean posterior rate with gamma-distributed rates across families. Here a high estimated rate indicates a family that is evolving quickly and has undergone many duplication and loss events in its history. The average rate is 1.

these potential effect of missing data, extinction still seems rare compared to gene family contraction.

Validating the estimated rates of gene family evolution

As the estimated rates of gain and loss may be affected by both the model and the data used we attempted to validate our results. We see that on the Génolevures data our BDIE with gamma-distributed rates across families (+G) we estimate a very high rate of gene loss. In order to check that this is not a consequence of our model choice we repeated the analysis using our birth-death-innovation model [9]. The estimated rates for all the models can be found in Table S1. When we compare BDIE+G and BDI+G we see that the estimates of extinction are similar at 0.05 and 0.08 for the BDIE+G and BDI+G models respectively. The rate of death for the BDIE+G model is higher than that for the BDI+G model. This is likely due to the addition of the extinction parameter. As we think that extinction is a relatively rare event (only changes in size from 1 to 0 cause innovation and all other family size decreases are explained by death), and these rates are averaged over all families, by including extinction we remove rare events from this average, and as a consequence, the death rate increases.

We see similar patterns when we compare the the BDIE and BDI models. As with the gamma models the estimated rate of death is higher in the BDIE model for the reasons discussed above. The estimated rates of innovation are similar, 0.48 and 0.70 for BDIE and BDI respectively. Interestingly, in the non-gamma models there is a large increase in the estimates of the rate of innovation. We have seen that under the gamma model small families and families where one or more species has zero members tend to evolve faster. Our data also consists of many small families. Since the innovation parameter in the BDIE model describes going from 0 to 1 gene in a family, we expect that the innovation parameter will have the greatest effect on smaller families. This suggests that small families may need to be handled differently, which is not possible with the BDIE and BDI models. Further study of this result is warranted as it is suggestive of current models being misspecified.

In order to to test the effect of clustering parameters on the estimates of the rates of birth and death we generated 4 new gene families sets using Tribe-MCL [29] and a range of values to affect cluster granularity. The estimated rates for each clustering can be found in Table S2. As we increased the cluster granularity we identified more gene families and the average size of the gene families decreased. We find that with decreasing family size we see an increase in the estimated rate of death. We speculate that this may be due to the fact that as the death rate increases relative to the birth rate the stationary distribution of family size becomes more skewed towards smaller families and so better fits the distribution we see with larger granularity.

In order to quantify the effect of data quality on our results we generated a new set of gene families excluding *S. cerevisiae*, which has the highest quality sequence with the most annotated genes. When we repeat the DupliPHY-ML analysis we find that $b=1.0$, $d=2.89$, $e=0.25$ and $i=0.08$. The relative rates estimated without *S. cerevisiae* change from those estimated using all the species. Most notably, the rates of gene loss and extinction are reduced indicating that there is a higher rate of gene gain inferred in this analysis compared to our original analysis. Although the values of these estimates change, the overall pattern of these estimates remains the same.

We have used several methods of validating the estimated rates of gene family evolution. In each case by altering the gene family evolution model or the input data we see variation in the estimated rates. In many cases this variation can be explained. Importantly, in all cases the trends of the results remain the same; The rate of gene loss is higher than the rate of gene birth, the rate of extinction is lower than that of birth and innovation appears to be a rare event.

Gene family evolution over the phylogeny

DupliPHY-ML estimates the average number of gain and loss events per gene family on each branch. These events, represented as branch lengths, are useful for identifying lineages, including ancestral lineages, that have a high turnover of genes in gene families. It should be noted that model misspecification may lead to very short internal branches followed by longer branches at the tip of the tree. In the species phylogeny (Figure 3) we see a long internal branch after divergence from the outgroup *Y. lipolytica*. Notably, *Y. lipolytica* has a branch length of zero suggesting there are no lineage specific events on this branch. This is unlikely to be an accurate estimate, since *Y. lipolytica* would be expect to have at least a few branch-specific events. The long internal branch length from the outgroup indicates a large turnover of genes on this branch and recapitulates the larger genetic distance between the out groups and more closely related species in the tree. Other long internal branches may represent expansions or contractions of gene families that are common to multiple species within the phylogeny. There are short branch lengths leading to *S. kluyveri* and *K. thermotolerans* (Figure 3), suggesting that these species have a very similar evolutionary history and have not had much lineage-specific evolution of gene families.

Terminal branches with long lengths represent lineage specific turnover of genes in gene families. Species with long branches may represent many gains of genes in a family and/or many losses. *D. hansenii* and *E. gossypii* both have longer terminal branches than the other species, indicating a large turnover of genes in these species. Since gains, and potentially losses, of genes in families may indicate some adaption to a new environment or evolution after speciation. In order to identify signals of environmental adaptation

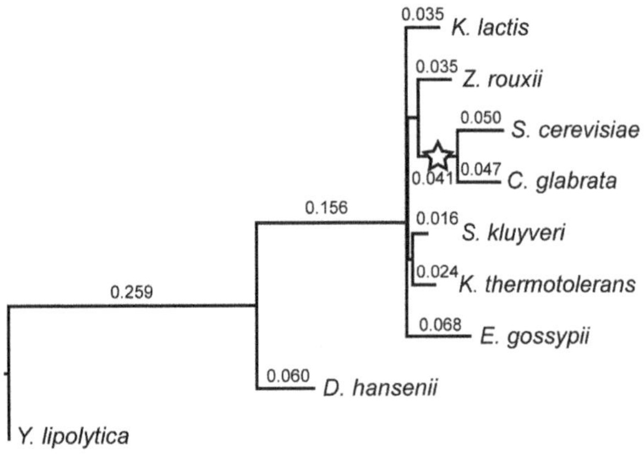

Figure 3. The phylogeny for the Génolevures data. Branch lengths represent the average number of gain and loss events per gene family as inferred by DupliPHY-ML. The branch on which the whole-genome duplication occurred is indicated with a star.

we first need to identify significant lineage specific expansions and contractions of gene families.

Lineage specific expansions of gene families

We used the net change in family size on each terminal branch compared with the net change on all other branches to identify those families with significant lineage specific expansion. We specifically chose the net change in order to identify those families that have undergone expansion and those genes have been retained, rather than identifying families with just a high turnover of genes. We have identified a large number of lineage specific expanding gene families (Table 1). In *S. cerevisiae* and *D. hansenii* we can identify more than 50 such families. The ranges of expanding family sizes also varies greatly within and between species. In all species with identified expanding families we observe families with only 2 members identified as expanding. In these cases it may be that all other species lack the family or have single member families (singletons) and the expansion to 2 members is lineage specific and significant by our definition.

The observed patterns of evolution of gene families, including both lineage-specific expansions and slowly-evolving families, may be due to either selection or neutral evolution. If the process of expansion is neutral we expect the newly introduced genes to nonfunctionalise and later be removed. If selection is acting to increase family size in specific families we would expect that these families would be associated with specific functions that may aid adaptation, and, in particular, may be relatable to known environmental adaptation routes in yeast. Additionally, in the presence of selection we expect that some of these functions may be common to multiple species. In order to determine whether selection is operating, we determined the functional classes of genes where families are changing rapidly.

In order to identify the functions relating to the lineage specific expanding gene families we used the Gene Ontology (GO) [30]. First, we identified enriched GO slim terms for each of the expanding families (File S3). We then visualised the GO slim terms on a Voronoi treemap and highlighted the enriched terms. These visualisations allow us to detect similarities and differences in types of gene families that are expanding in these species.

The species analysed show a wide variance in enrichment of GO terms (Figure 4). Most species show few enriched GO terms in expanding families. *S. cerevisiae* shows the most enrichment with some terms enriched in multiple expanding gene families. Enriched *S. cerevisiae* terms include those related to transport of amino acids and carbohydrates as well as enzyme activity such as isomerase, helicase and hydrolase activity. Transport of carbohydrates and amino acids are commonly enriched GO terms and are also enriched in the expanding families of *D. hansenii* and *K. thermotolerans*. Finally, we can identify expanding families that might be indicative of adaptation to a specific environment given the family's functional annotation; *Z. rouxii* is the only species to show enrichment for terms relating to the response to chemical stimuli and response to oxidative stress.

The observation that many of the expanding families have similar functional annotation suggests that many of the expansions we observe are driven by selection rather than random turnover of genes, or systematic gaps in sequence data. The common identification of nutrient uptake terms enriched in expanding gene families suggests that the expansion of gene in nutrient transporter families may be a common response to new environments or selection pressures. In this scenario following speciation which may have been precluded by a change in environmental conditions, selection for more efficient or varied uptake of nutrients from the environment may result in the observed expansions. In contrast, the identification of families showing unique functional annotation, such as the response to oxidative stress in *Z. rouxii*, suggests that we are also able to detect species specific adaptations.

Interestingly, we can identify rapidly expanding families of unknown function. As not all genes from these species can be

Table 1. Lineage specific gene family expansions in the Génolevures data.

Taxa	Num Genes	Num families	Ave size	Min size	Max size
S. kluyveri	5321	19	5.21	2	24
K. lactis	5076	19	4.15	2	12
K. thermotolerans	5092	15	5.93	2	22
Y. lipolytica	6448	0	0.00	0	0
C. glabrata	5203	13	5.46	2	13
Z. rouxii	4991	20	6.15	2	21
S. cerevisiae	6663	53	6.86	2	33
E. gossypii	4715	12	3.50	2	6
D. hansenii	6272	51	7.45	3	54

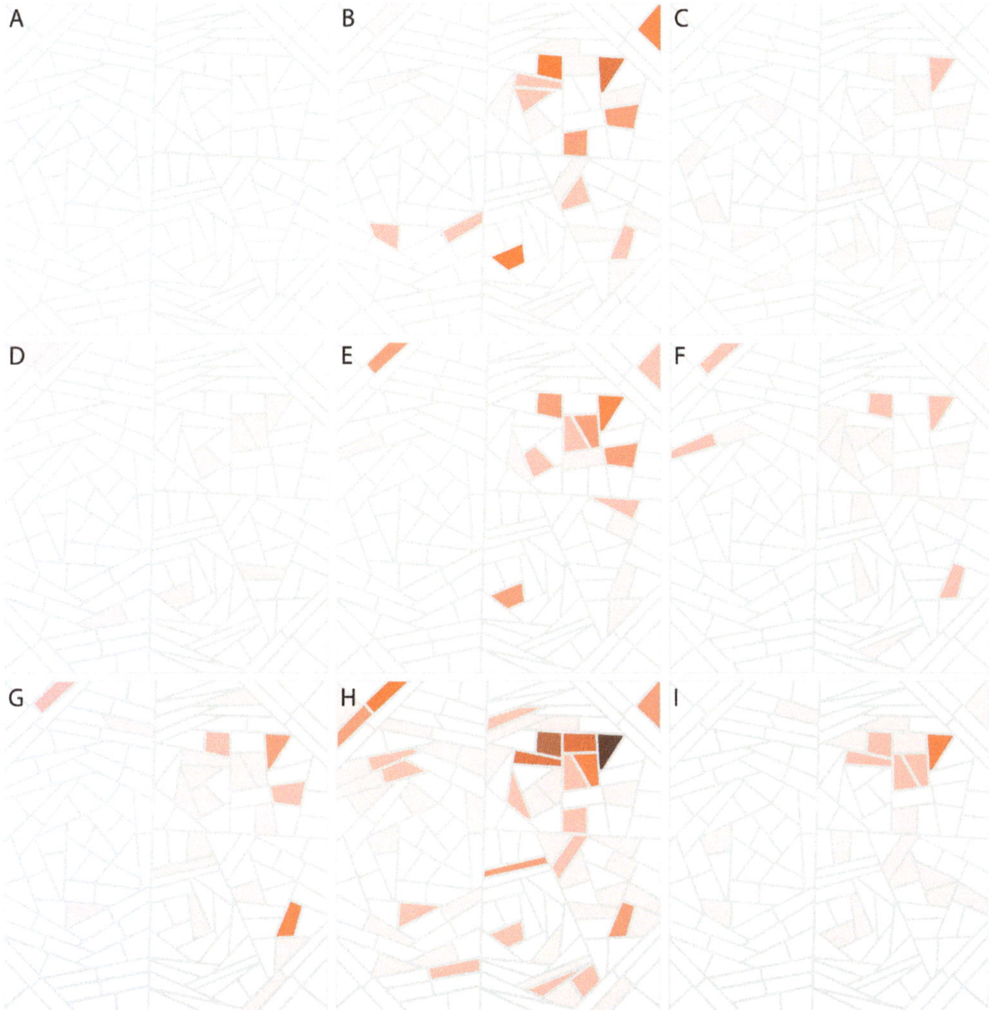

Figure 4. Functional enrichment of lineage specific expanding gene families for GO slim terms. Each cell represents a GO slim term from the biological process, molecular function or cellular component ontology. The positioning of the cells is determined by a term overlap metric so that more similar terms are positioned close together. Cells are shaded if one or more expanding gene family shows enrichment for that term. The intensity of the shading represents the number of gene families showing enrichment for that term. The most intense colour shows > 10 families are enriched. Each panel represents a species from the Génolevures data set: *Y. lipolytica* (A), *D. hansenii* (B), *E. gossypii* (C), *K. lactis* (D), *S. kluyveri* (E), *K. thermotolerans* (F), *Z. rouxii* (G), *S. cerevisiae* (H) and *C. glabrata* (I). Lists of enriched terms in expanding families are available in File S3

annotated with *S. cerevisiae* identifiers, our data contain many genes with no known functions. By annotating these genes with an "unknown" function, we can test to see whether genes with this unknown function are enriched in any rapidly expanding families. Indeed, we find enrichment for unknown functions in all species except *S. cerevisiae* and *Y. lipolytica*, which has no rapidly expanding families. The presence of these expanding families suggests that there are species specific adaptations using novel functions. We also identify two families enriched for the unknown function that are rapidly expanding in multiple species. Family GL3C0410 is enriched for the unknown function in *C. glabrata*, *D. hansenii* and *S. kluyveri*. Family GL3R0575 is also enriched for the unknown function in *D. hansenii* and *K. lactis*. The same family expanding in multiple species may indicate that the function represented by these families is important to multiple species.

Lineage specific contractions of gene families

In addition to looking at families showing rapid expansion we also looked at families showing significant contraction along a terminal branch. Contraction of gene families may signify the removal of non-adaptive functions or a streamlining of the genome and may result from specific adaptations. We find fewer families experiencing significant contraction than families experiencing expansion. *E. gossypii* has the largest number of contracting families with 28. Interestingly, *D. hansenii*, which has more than 50 expanding families only contains a single significantly contracting family suggesting the *D hansenii* genome has recently undergone an increase in size.

As before we used the GO to identify the functions of the contracting gene families. We find several terms common to multiple species again suggesting that species specific losses may reflect common adaptations. *S. cerevisiae*, *E. gossypii*, *S. kluyveri* and *K. thermotolerans* all show contractions of gene families related to the golgi apparatus. Furthermore, *C. glabrata*, *E. gossypii*, *S. kluyveri*, *K. lactis* and *Z. rouxii* all show enrichment for transmembrane transport terms in contracting families. Full lists of the GO terms associated with contracting families can be found in File S4.

Whole-genome duplication

Two of the species in our data set (*S. cerevisiae* and *C. glabrata*) have a whole-genome duplication (WGD) in their evolutionary past [31,32]; the branch on which this event occurred is indicated by a star on Figure 3. Subsequent to this duplication, a large number of duplicate genes were lost [32,33]. In common with all previously-published methods, our BDIE model does not incorporate any parameters that represent whole genome duplication. The presence of whole-genome duplication therefore has the potential to give rise to biases in the results, especially since the subset of genes retained after the WGD is enriched for specific functions [33,34].

There are three possible patterns of gene loss that may have occurred subsequent to WGD. First, a gene duplicated by WGD may be lost in both daughter species; in this case neither the duplication nor loss would be present in our input data. Second, a gene duplicated by WGD may be retained in both daughters. DupliPHY should assign this duplication to the WGD branch, leading to a large number of gains, and so long branch length in Figure 3. Third, a duplicated gene may be differentially lost in one daughter species and retained in the other. In this case the only signal for the duplication is the presence of a gene in one species. Here the duplication event may be assigned to the terminal branch rather than the ancestral WGD branch and then inferring a loss in the other species.

The first case is invisible to all methods of ancestral reconstruction that are based on analysis of extant genomes, including ours. However, we see that on the WGD branch of the phylogeny 4,161 gene families do not appear to change size. Since, by definition, all gene families must change size during a WGD, this figure gives an indication of how many events are invisible to our method and similar methods. We see some evidence of case two (retention in both species). There are 285 gene gains (in 263 families) on the WGD branch, and only 172 losses (in 152 families). This pattern is the reverse of that seen throughout the tree as a whole, where gene loss is four times more likely than gene gain.

In order to investigate the possible causes of bias that arise from case 3 (differential retention being miss-assigned), we determined how many of the gains on the *S. cerevisiae* and *C. glabrata* lineages have arisen via WGD. We used the assignment of duplications to either the WGD set (data from Byrne and Wolfe [35]) or small-scale duplication set (SSD, data from Hakes *et al* [36]). We find that, for *S. cerevisiae* 64 of the lineage specific gene gains arise from the WGD event, 203 from SSD events, with 97 gains unknown. For *C. glabrata* only 2 genes arise from WGD, 23 from SSD and 46 genes arise from unknown origins. The rapidly expanding families that contain WGDs for both *S. cerevisiae* and *C. glabrata* can be found in File S5. As the method of Hakes *et al* is conservative in assigning SSDs and much research has focussed on identifying WGDs, we believe that the majority of the "unknown" gains are likely to be SSDs. Although it is still possible that some of these genes may be unannotated WGDs or may have arisen from innovation events such as horizontal gene transfer between species. Since there are many more WGDs than SSDs (551 and 272 gene pairs, respectively), the observed pattern is unlikely to be an artefact arising from differing number of duplications in the input set. We conclude, therefore, that although the whole-genome duplication must have affected the pattern of observed duplications, particularly for the *S. cerevisiae* and *C. glabrata* lineage-specific duplications, the effect is unlikely to change our conclusions substantially.

In addition we also repeated the inference of duplicate gene family histories excluding any gene families in the Génolevures data that contained a known WGD from Byrne and Wolfe [35]. This data set contained 4,064 families with an average size of 8.88 genes. Here we find that $b = 1.0$, $d = 5.09$, $e = 1.01$ and $i = 0.05$. Overall the relative rates show the same pattern as the original analysis. The rate of loss is high compared to duplication and innovation is rare. Notably, the relative rates of both gene loss and gene family extinction have increased compared to the original analysis. This suggests that most genes retained from the WGD were inferred as gains in our analysis and the removal of these families has increased the estimation of the rate of loss and extinction relative to the rate of gene gain.

Discussion

Here we have used the software DupliPHY-ML to infer the evolutionary history of gene families in a range of yeast species and present evidence that suggests lineage specific gene family expansions lead to species-specific functional adaptations. We have implemented a new model with parameters for birth, death, innovation and extinction.

We note that in our model the birth parameter represents the retention of a gene after duplication rather than simply the occurrence of a duplication event. The observed birth rate is therefore a combination of the underlying rate of duplication, and the various factors that affect fixation rate, including selection. Although we may define the biological event represented by birth differently from previous work the model is similar to that of many other birth-death models used to study the evolution of gene families [6,8,14]. We make this distinction because it may be expected that the underlying duplication rate would correlate with gene family size, since large families contain more genes that are candidates for duplication. However, there is clear evidence for selection to remove or silence recently duplicated genes [4]. The combination of these factors means that the size of families are not well described by a simple model where birth rate increases linearly with family size [37,38]. There is, however, a weak correlation between the rate of change and family size; this conclusion holds for a range of species, including yeast. It is therefore likely that our model could be improved by the addition of a suitably weak correlation parameter.

DupliPHY-ML estimates the rate of gene loss, gene family extinction and gene family innovation relative to the rate of gene gain. The rates we estimate from the data show that the rate of gene loss is much higher than the rate of gene gain. Pseudogenisation followed by gene loss is thought to be the most common fate of duplicate genes [39]. The rate of innovation is much lower than that of the rate of gene gain by duplication within an existing family. Indeed, *de novo* gain of genes seems to be very rare, with only a handful of reported cases in yeast [40,41], *Drosophila* [3] and human [42]. We identified innovation events in our data by looking for examples of gene family gain within our phylogeny *i.e.* going from 0 to 1 or more members along a branch. The gene *BSC4* has been reported as a *de novo* gene in *S. cerevisiae* and has been associated with DNA repair during the stationary phase in *S. cerevisiae* when shifted to a nutrient-poor environment [41]. Here we find it is a member of a singleton family, and therefore, as arising from an innovation event. Hall *et al.* [40] identified two genes, *BDS1* and *URA1*, in *S. cerevisiae* that originated by horizontal gene transfer. In our data we also identify *BDS1*, an aryl- and alkyl-sulfatase gene required for the use of specific sulfates as sulphur sources, as arising from an innovation event. Interestingly, this gene appears to be present in both *S. cerevisiae* and *K. thermoltolerans*, suggesting that this gene may have been introduced to these genomes by separate horizontal transfer events. *URA1*, a dihydroorotate dehydrogenase involved in the in the *de novo* biosynthesis of pyrimidines, is not identified in our data as originating from an innovation event as all species studied here contain a gene with similarity to *URA1*, meaning the gene family was present at the root of our phylogeny.

Recently, Carvunis *et al* [27] described a process of gene formation from non-genic sequences through the formation of proto-genes that later evolve into genes. The finding of a significant number of proto-genes (~ 1900) in the yeast genome suggests that the rate of innovation identified in this study may be well below the true level of *de novo* gene gain. Interestingly, we find evidence of 1, 292 innovation events across all 8 species used in this study. We point out that although are estimates of the rate of innovation are much smaller than that of gene duplication we still find a large number of cases of innovation in our data. Additionally, it is possible that the exclusion of singleton families from this analysis has led to the low estimate of the rate of innovation of genes.

As with innovation, we find that gene family extinction is rare compared to the loss of a single gene in a family with more than one member. In mammals the rate of extinctions is lower than both expansions and contractions [12]. It has been suggested that gene family extinction may occur when there are shifts in nutrient availability, meaning specialised families are no longer necessary for growth [43].

We note that the estimated rates of death and extinction appear to be artificially high. Although the pairwise comparisons of the rates of birth/innovation and death/extinction make biological sense there are alternative explanations for the high estimates of death and extinction. Firstly, there are differences in the set of genes identified in these species that may have arisen because of low sequencing coverage or errors in gene identification. We have demonstrated that the estimated rates are not overly affected by the presence of one well sequenced species by removing *S. cerevisiae*. When *S. cerevisiae* is removed from the analysis the relative rates estimated by DupliPHY-ML are affected. In particular the estimated rate of loss is reduced from 4.05 to 2.89. These estimates of loss may be artificially high because repeated lineage specific losses of ancestral genes may inflate the estimated rate of loss. Alternatively, the genomes may be shrinking from some large ancestral genome at, or before, the last common ancestor of the set of species used in our analysis. Finally, large differences in the number of identified genes (Table 1) across all species may explain the high estimates of loss. CAFE 3 [11] attempts to correct for missing data but these types of corrections could not be used for this study as we have no accurate estimate of the amount of data missing from these genomes.

We have repeated much of our analysis to validate our estimates of the rates of birth, death, innovation and extinction by removing families, species, running different models and using data generated with different clustering methods. In some cases we find that the estimates of these parameters vary depending on the data or the model used. This indicates that the model used in this study still does not capture the complete biological processes of gene family evolution and suggests that further research of these models is warranted. Importantly, for all of the validation the trends in the estimates of these parameters remains the same and so, we can draw biological conclusions from our results.

Extended internal or terminal branches may represent areas of the phylogeny that have undergone accelerated gene family evolution in response to some selection pressure. In the species phylogeny *D. hansenii* and *E. gossypii* show the longest terminal branch lengths. These long branches may well be representative of adaptation to their respective environments. Indeed, these yeasts appear to occupy very specialised environments with *D. hansenii* being a cryotolerant, halotolerant marine yeast often found in cheese [44] and *E. gossypii* being a cotton pathogen [45]. Alternatively, long branch lengths may represent some stochastic turnover of genes in areas of the genome that experience high rates of gene gain or loss, such as subtelomeric regions [46,47].

The DupliPHY-ML method allows the determination of rates of gene gain and loss in a branch-specific manner. Families with low rates of changes may be tightly constrained by selection, with changes in the numbers of members having a deleterious effect. By contrast, the fast evolving families may represent those that are expanding or contracting in response to selection from the environment. Indeed, Demuth *et al.* [12] identified a set of rapidly evolving gene families in a mammalian phylogeny and showed that these families were associated with the same biological functions as quickly evolving genes and regulatory regions. Similarly, gene families that are evolving quickly in yeast may have important biological functions if there is selection pressure from the environment to increase the membership in these families. Alternatively, these families may be experiencing stochastic turnover of genes. In order to distinguish between evolution by adaptive or neutral processes we first needed to identify gene families that show significant expansion in a lineage. Families showing significant expansion on the terminal branches of the phylogeny represent lineage specific examples and may provide insight into the genetic mechanisms that lead to species specific adaptations.

We find that the ASP family is rapidly expanding; the ASP genes are asparaginases involved in the catabolism of alternative nitrogen sources [48,49]. This family is shown to be expanding on the lineage leading to *S. cerevisiae* which is consistent with previous analyses [50]. Other *S. cerevisiae* lineage specific expansions show an expansion of two of the major subtelomeric gene families. There are expansions in the seripauperin multigene (*PAU*) family and the *DUP240* integral membrane protein family. The *DUP240* family is a collection of nonessential genes that have been linked with membrane trafficking processes [51]. Subtelomeric regions have been shown to contain many duplicated genes that may be a consequence of higher rates of recombination in these regions [46,47]. Interestingly, in *S. cerevisiae* some genes gained in these families have relocated to internal sites on chromosomes [50] suggesting that duplication events that increase gene family membership may be coupled with relocation events.

The use of GO [30] gives an overview of the types of genes in expanding gene families. There is a wide range of variation in functional annotation, through enrichment of GO terms, in the expanding families. Several species show very little enrichment whereas *S. cerevisiae* shows wide ranging enrichment. The species in these data are sufficiently diverged that it is possible to detect differences in the types of families that are expanding in these species (Figure 4). We can identify unique functions associated with families expanding in a single species that might indicate lineage specific adaptation. A single expanding family in the species *Z* shows enrichment for response to chemical stimuli and response to oxidative stress. This family is comprised of 5 genes, 3 of which have high similarity to *S. cerevisiae CTA1* and 2 with similarity to *CTT1*, suggesting lineage specific duplication of these genes. *CTA1* is a catalase A involved in hydrogen peroxide detoxification and is important during the oxidative stress response [52]. *CTT1* is a catalase T and is also involved in hydrogen peroxide detoxification and is known to be induced under oxidative stress conditions [52,53]. In *C. albicans* it has been shown that trehalose mobilisation is important for tolerance of hydrogen peroxide [54]. *Z. rouxii* is well known to be tolerant to salt and osmotic stress [55,56] and it has been suggested that the high expression of trehalose synthesis genes under non-stress conditions [57] may be important in a range of stress responses [58]. It may be the case that *Z. rouxii's* naturally high rate of trehalose synthesis makes the species tolerant of a wide range of stresses and the expansion of a family associated with hydrogen peroxide resistance is a complementary adaptation to a specific environmental stress. Additionally, the presence of enrichment for an "unknown" GO term in expanding gene families suggests that there are also species specific adaptations using novel functions.

It is also possible to identify common functions associated with families expanding in multiple species. As these expansions are lineage specific and thus, independent, any common functional enrichment of these families may represent repeated selection for a specific adaptation. Many species contain lineage specific expanding families that are associated with carbohydrate transport and metabolism as well as amino acid transport. *S. cerevisiae* is well adapted to the uptake of sugars and has a range of transporters for different sugars [59,60]. It seems that there has been expansion in the sugar transporter family in *S. cerevisiae* through a combination of whole genome and tandem duplication possibly to facilitate the evolution of aerobic fermentation [61,62]. Indeed, we also see a large expansion of a sugar transporter family in *S. cerevisiae. D. hansenii* also shows expansion of families containing carbohydrate transporters and has been shown to be able to transport and utilise a range of sugars including hexoses and pentoses [63]. The prevalence of sugar transporter family expansion across several species of yeast may reflect a general adaptation strategy to environments containing an array of sugars with different genes amplified to fine-tune a yeast's metabolism to a specific sugar. Indeed, *K. lactis* which is found mostly in dairy products has acquired the ability to utilise lactose. We find that few *HXT* genes have been retained [62] and no expanding families show enrichment for sugar transporters.

Interestingly, we can also identify gene families that are contracting and the functions represented by these families. As with expanding families it is possible to detect different families with the same functional annotation contracting in multiple species. These functions tend to be associated with general transmembrane transporter activity and exocytosis (File S4). It has been argued that loss of genes may be adaptive [64] by the removal of non-adaptive functions and streamlining of the genome. It is tempting in this case to speculate that the losses shown here, coupled with gains of specific transporters, represent a restructuring of the uptake capabilities of the yeast in response to a new environment. Thus contractions as well as expansions may play a role in species specific adaptations.

The evolutionary histories of gene families can provide important insights into the past and present adaptations of species. We have inferred the evolutionary histories of all gene families across a variety of species, using the most sophisticated published models. We find that families evolve at a variety of rates and that these rates vary on different branches of the phylogeny. This variability suggests that rates are likely to be dictated by a range of selection pressures that act on a particular species. Additionally, we have identified families that show significant expansions on the terminal branches of the phylogeny. These families may be experiencing high rates of expansion because of neutral mechanisms or alternatively, may be under selection for increased membership. Indeed, we have identified several families showing rapid lineage specific expansion that are located in subtelomeric regions of the genome that are known to undergo regular recombination events. We also see that functional annotations associated with these expanding families suggest that there is selection for expansion of stress response and sugar transporter families. These expansions contribute to species differences and their individual functional specialisations.

Methods and Materials

Genomic and gene familiy data

Annotated genome sequences for 9 species of hemiascomycete yeast were downloaded from the Génolevures project [28]. These species were *Eremothecium gossypii, Candidia glabrata, Debaryomyces hansenii var. hansenii, Kluyveromyces lactis var. lactis, kluyveromyces thermotolerans, Saccharomyces cerevisiae, Saccharomyces kluyveri, Yarrowia lipolytica* and *Zygosaccharomyces rouxii.* These specific species were selected because they have previously been used to generate a set of gene families using a consistent and comprehensive method [65]. For all species we used the predicted ORFs and protein sequences identified by the sequencing project. Genes were annotated using BLAST [66] and the *S. cerevisiae* annotated genome as a reference. An E-value cutoff of 1×10^{-4} was used to call annotations in the BLAST results.

We used gene families from the Génolevures data as previously identified [65]. Briefly, the authors aligned the proteomes of the yeast species using BLAST [66] and the Smith-Waterman algorithm. Following this Tribe-MCL [29] was used to cluster the dat set using a range of inflation coefficients to produce clusters at different granularities. Finally, the authors use consensus clustering and an election algorithm to compare clusters and categorise them as robust, consensus, multiple choice or unique. In order to use the most reliable predictions of gene families we used those classified as "robust" or "consensus" families. Additionally, we removed families where any single taxa contains more than 75 members. We did this to make the analysis computationally tractable but it might also be the case that extremely large families are under specific selection pressures that may not be well defined by our current models. We note that this removes some notoriously complicated families such as polyproteins and repeat domains as well as single gene families which represent unique genes in the tree.

The specific clustering parameters used to generate the Génolevures data and our selection of "robust" and "consensus" families, may have introduced some bias into the identified families. In order to assess any bias in the data performed our own clustering on the raw Génolevures data. Here we used the protein sequences available for each species from the Génolevures website to perform an all against all BLAST search. The resulting network of BLAST hits was then clustered using Tribe-MCL [29] with a range of inflation values that control cluster granularity. Tribe-MCL was run with 4 inflation values of 1.4, 2, 4 and 6. For each of the resulting sets of gene families we ran DupliPHY-ML with the BDIE+G model (described below). We note that for computational tractability we still removed families where any single taxa contains more than 75 members and removed any singleton families. This analysis allows us to determine the effect of clustering on our results.

Phylogenetic tree

A phyogenetic species tree is necessary to infer the evolutionary histories of gene families. DupliPHY-ML does not use gene trees to infer events but instead will infer gain and loss events on the species phylogeny. Here, we used a subset of the cladogram presented in [67] to get the phylogenetic tree structure. Branch lengths were estimated for this structure using Baseml of the PAML package [68]. We used MUSCLE [69] to align the coding sequences of 2324 common genes between all species (based on

BLAST annotation described above), excluding any genes that appeared to have more than one copy in any species, and these alignments were concatenated. The list of common genes used in this step can be found in File S6. Baseml was run using the general time reversible model with no molecular clock to infer branch lengths. These branch lengths were used to test for significantly expanding or contracting gene families after the inference of gene family histories.

Inferring the evolutionary history of gene families

Given the identified gene families and phylogenetic trees for these data sets we then used DupliPHY-ML [9] to infer the evolutionary history of each gene family. These histories were inferred under the Birth-Death, Innovation and Extinction (BDIE) model with gamma-distributed rates across families. This model is similar to the Birth-Death-Innovation (BDI) model described previously [9], with the addition of an extinction parameter. DupliPHY-ML estimates each of the parameters from the available gene family data. Here the rate of gene birth corresponds to the rate of gene gain by duplication in existing families. The rate of gene death simulates the loss of a duplicate gene in a family with more than 1 member. Innovation represents the gain of a novel gene family through *de novo* gene gain or horizontal gene transfer and extinction represents the loss of a gene family from the tree. Once a gene family has been lost, it may only arise again through an innovation event. The BDIE model has an instantaneous rate matrix, **Q**, defined by equation 1.

$$Q_{i,j} = \begin{cases} b \text{ if } j-i=1 \text{ and } i \neq 0 \text{ (birth)} \\ d \text{ if } i-j=1 \text{ and } j \neq 0 \text{ (death)} \\ h \text{ if } i=0 \text{ and } j=1 \text{ (innovation)} \\ k \text{ if } j = 0 \text{ and } i=1 \text{ (extinction)} \\ 0 \text{ if} |i-j| > 1 \text{ (maximum one event)} \end{cases} \quad (1)$$

We also supplied DupliPHY-ML with a set of unobservable families. Unobservable families represent potential phyletic patterns that can occur in these species but have been purposely removed from the analysis. In this case these are families that contain only a single gene. DupliPHY-ML corrects for these cases using the method of Felsenstein [70].

In order to validate the BDIE model we repeated the analysis of the Génolevures data using the previously described BDI model [9]. We also ran these models without gamma-distributed rates across families. All models were run on the same dataset and with all other run options kept constant.

To identify lineage specific gene family expansions we compared the change in number of members of an inferred gene family on an internal node to the observed gene family size the tips of the tree. The change in the number of members was normalised by the terminal branch length to control for greater divergence time allowing for higher turnover. If this change was positive and greater than the mean expansion or contraction on all other branches in the tree we used a Wilcoxon Rank Sum test to check for a significant expansion. We controlled for false discovery rate using the method of Benjamini and Hochberg [71]. All tests with $P < 0.05$ after false discovery rate correction were taken as the set of rapidly expanding families. We note that by simply calculating the change in family membership from an ancestral node to a child node is likely to underestimate the true number of changes along a branch as some genes may be gained and lost on that branch. This simplification means that the calculations of the amount of change along a branch are likely to be an underestimate and so our results may be conservative. It also means that we focus on those gene gains that have been retained in the extant species.

We also identified lineage specific gene family contractions. Much like for expanding families we looked for the change in number of family members on each branch of the tree normalised by the branch length. If the change was negative (i.e. loss) and the number of losses was greater than the mean number of events on all other branches we used a Wilcoxon Rank Sum test to check for a significant expansion. We controlled for false discovery rate using the method of Benjamini and Hochberg [71]. All tests with $P < 0.05$ after false discovery rate correction were taken as the set of rapidly contracting families.

Controlling for genome annotation and whole genome duplicates

The presence of genomes with better quality sequence and annotation may affect our analysis by artificially altering the estimated rates of gain and loss. In order to determine the effects of variable genome quality on our analysis we removed *S. cerevisiae* and repeated the analysis. To create a new set of gene families that exclude *S. cerevisiae* we first performed an all against all BLAST [66] search of all remaining species. We then used Tribe-MCL [29] with an inflation parameter of 1.2 to cluster the BLAST similarity network and generate a new set of gene families. Finally, we repeated the inference of gene family history using DupliPHY-ML as described above.

We also note that the new model of genome evolution (and all other existing models) doesn't have a parameter for large scale duplication events. As the phylogeny used in this study is known to contain a whole genome duplication event [32] we controlled for the effect of whole duplication on our estimates of gain and loss by removing all families that contained a known whole genome duplicate. After the removal of these families we repeated the DupliPHY-ML analysis as described above.

Functional annotation, enrichment and visualisation of gene families

In order to assign function to the expanding gene families we used the slim version of the Gene Ontology (GO) [30]. As GO terms are associated with genes via *S. cerevisiae* identifiers we annotated all genomes with an *S. cerevisiae* identifier (see above). We note that some genes that have no GO term annotation and as a consequence some expanding gene families or some members of these families have no functional annotation. There are also cases of genes in some species having no associated gene identifier from *S. cerevisiae*, in these cases we have assigned these genes a pseudo "unknown" GO term. This allows us to identify families that are enriched for an unknown function and may perform a novel function not performed by *S. cerevisiae*. We then used the GO slim term associations available at the *Saccharomyces* Genome Database (SGD) to assign GO terms to genes.

To test for enrichment of a GO term in a gene family we used Fisher's exact test. Here, the sample was all annotated genes in a gene family and the background population was all the annotated genes in the yeast genome. The sample successes therefore were all genes in the gene family annotated with the specific GO term and the population successes were all genes in the genome annotated with the specific GO term. All P-values were false discovery rate corrected using the method described by Benjamini and Hochberg [71] with a significance cutoff of $P < 0.05$. This analysis was applied to both the rapidly expanding and contracting gene families.

In order to visually compare functional enrichment, the Term Overlap (TO) metric of the GLASS software (available at http://www.bioinformatics.ic.ac.uk/glass/) was used to determine the pairwise distances between GO slim terms. A tree-structure was generated from these pairwise distances using the neighbour-joining algorithm implemented in Quicktree [72]. The tree structure was then represented in two dimensions using Voronoi Treemaps [73,74], implemented with GLASS [75]. In this visualisation each cell represents a GO slim term, whose location within the panel is determined by the TO distance to all other terms. A cell is coloured if one of more expanding gene families are enriched for that GO term, with the intensity of the colour indicating the number of gene families enriched for that term.

Supporting Information

File S1 This file lists all of the Génolevures families including the identified present in each family for each species.

File S2 The ancestral reconstructions for each family analysed with DupliPHY-ML. The first line of the file is a newick string representation of the yeast phylogeny used in the study. The rest of the file is a tab delimited list of gene family ID and the number of members present at each node of the tree.

File S3 The functional enrichment of GO slim terms for rapidly expanding families in each species. The file lists the enriched terms, the ontology within GO slim and the raw and corrected P-values.

File S4 The functional enrichment of GO slim terms for rapidly contracting families in each species. The file lists the enriched terms, the ontology within GO slim and the raw and corrected P-values.

File S5 Expanding families in *S. cerevisiae* and *C. glabrata* that contain whole genome duplicates.

File S6 The list of 2324 common genes used to infer the branch lengths on the yeast phylogeny.

Table S1 Comparison of parameter estimates of 4 DupliPHY-ML models run on the Génolevures data.

Table S2 The effects of gene family data on the estimation of rates under the BDIE+G model.

Author Contributions

Conceived and designed the experiments: RMA. Performed the experiments: RMA. Analyzed the data: RMA. Contributed reagents/materials/analysis tools: RMA DM. Wrote the paper: RMA SCL.

References

1. Long M, Betrán E, Thornton K, Wang W (2003) The origin of new genes: glimpses from the young and old. Nature Reviews Genetics 4: 865–875.
2. Ohno S (1970) Evolution by Gene Duplication. Springer, New York.
3. Zhou Q, Zhang G, Zhang Y, Xu S, Zhao R, et al. (2008) On the origin of new genes in Drosophila. Genome Research 18: 1446–1455.
4. Lynch M, Conery J (2000) The evolutionary fate and consequences of duplicate genes. Science 290: 1151–1155.
5. Hahn M, Han M, Han S (2007) Gene family evolution across 12 Drosophila genomes. PLoS Genetics 3: e197–e209.
6. Akerborg O, Sennblad B, Arvestad L, Lagergren J (2009) Simultaneous Bayesian gene tree reconstruction and reconciliation analysis. Proceedings of the National Academy of Sciences of the United States of America 106: 5714–5719.
7. Csuros M (2010) Count: evolutionary analysis of phylogenetic profiles with parsimony and likelihood. Bioinformatics 26: 1910–1912.
8. Rasmussen MD, Kellis M (2011) A bayesian approach for fast and accurate gene tree reconstruction. Molecular Biology and Evolution 28: 273–290.
9. Ames R, Money D, Ghatge V, Whelan S, Lovell S (2012) Determining the evolutionary history of gene families. Bioinformatics 28: 48–55.
10. Librado P, Vieira F, Rozas J (2012) Badirate: estimating family turnover rates by likelihood-based methods. Bioinformatics 28: 279–281.
11. Han MV, Thomas GW, Lugo-Martinez J, Hahn MW (2013) Estimating gene gain and loss rates in the presence of error in genome assembly and annotation using cafe 3. Molecular Biology and Evolution: In Press.
12. Demuth J, Bie T, Stajich J, Cristianini N, Hahn M (2006) The evolution of mammalian gene families. PLoS ONE 1: e85.
13. Krylov DM, Wolf YI, Rogozin IB, Koonin EV (2003) Gene loss, protein sequence divergence, gene dispensability, expression level, and interactivity are correlated in eukaryotic evolution. Genome Research 13: 2229–2235.
14. Hahn M, De Bie T, Stajich J, Nguyen C, Cristianini N (2005) Estimating the tempo and mode of gene family evolution from comparative genomic data. Genome Research 15: 1153–1160.
15. Wapinski I, Pfeffer A, Friedman N, Regev A (2007) Natural history and evolutionary principles of gene duplication in fungi. Nature 449: 54–61.
16. Kunin V, Ouzounis C (2003). GeneTRACE-reconstruction of gene content of ancestral species.
17. Mirkin B, Fenner T, Galperin M, Koonin E (2003) Algorithms for computing parsimonious evolutionary scenarios for genome evolution, the last universal common ancestor and dominance of horizontal gene transfer in the evolution of prokaryotes. BMC Evolutionary Biology 3: 2.
18. Page R (1998) GeneTree: comparing gene and species phylogenies using reconciled trees. Bioinformatics 14: 819–820.
19. Chen K, Durand D, Farach-Colton M (2000) NOTUNG: A Program for Dating Gene Duplications and Optimizing Gene Family Trees. Journal of Computational Biology 7: 429–447.
20. Durand D, Halldórsson B, min Vernot B (2006) A Hybrid Micro-Macroevolutionary Approach to Gene Tree Reconstruction. Research in Computational Molecular Biology: 9th Annual International Conference, RECOMB 2005, Cambridge, MA, USA, May 14–18, 2005: Proceedings.
21. Vernot B, Stolzer M, Goldman A, Durand D (2008) Reconciliation with non-binary species trees. Journal of Computational Biology 15: 981–1006.
22. Wehe A, Bansal MS, Burleigh JG, Eulenstein O (2008) Duptree: a program for large-scale phylogenetic analyses using gene tree parsimony. Bioinformatics 24: 1540–1541.
23. Górecki P (2004) Reconciliation problems for duplication, loss and horizontal gene transfer. Proceedings of the eighth annual international conference on Resaerch in computational molecular biology: 316–325.
24. HahnM(2007) Bias in phylogenetic tree reconciliation methods: implications for vertebrate genome evolution. Genome Biology 8: R141–R149.
25. Arvestad L, Berglund AC, Lagergren J, Sennblad B (2003) Bayesian gene/species tree reconciliation and orthology analysis using mcmc. Bioinformatics 19: i7–i15.
26. Larget BR, Kotha SK, Dewey CN, Ané C (2010) Bucky: gene tree/species tree reconciliation with bayesian concordance analysis. Bioinformatics 26: 2910–2911.
27. Carvunis AR, Rolland T, Wapinski I, Calderwood MA, Yildirim MA, et al. (2012) Proto-genes and de novo gene birth. Nature: In Press.
28. Sherman D, Martin T, Nikolski M, Cayla C, Souciet J, et al. (2009) Genolevures: protein families and synteny among complete hemiascomycetous yeast proteomes and genomes. Nucleic Acids Research 37: D550–D554.
29. Enright A, Van Dongen S, Ouzounis C (2002) An efficient algorithm for large-scale detection of protein families. Nucleic acids research 30: 1575–1584.
30. Ashburner M, Ball C, Blake J, Botstein D, Butler H, et al. (2000) Gene ontology: tool for the unification of biology. The Gene Ontology Consortium. Nature Genetics 25: 25–29.
31. Wolfe K, Shields D (1997) Molecular evidence for an ancient duplication of the entire yeast genome. Nature 387: 708–713.
32. Kellis M, Birren B, Lander E (2004) Proof and evolutionary analysis of ancient genome duplication in the yeast *Saccharomyces cerevisiae*. Nature 428: 617–624.
33. Scannell D, Byrne K, Gordon J, Wong S, Wolfe K (2006) Multiple rounds of speciation associated with reciprocal gene loss in polyploid yeasts. Nature 440: 341–345.
34. Conant G, Wolfe K (2008) Turning a hobby into a job: how duplicated genes find new functions. Nature Reviews Genetics 9: 938–950.
35. Byrne K, Wolfe K (2005) The Yeast Gene Order Browser: combining curated homology and syntenic context reveals gene fate in polyploid species. Genome research 15: 1456–1461.

36. Hakes L, Pinney J, Lovell S, Oliver S, Robertson D (2007) All duplicates are not equal: the difference between small-scale and genome duplication. Genome Biology 8: R209–R222.
37. Karev G, Wolf Y, Rzhetsky A, Berezovskaya F, Koonin E (2002) Birth and death of protein domains: A simple model of evolution explains power law behavior. BMC Evolutionary Biology 2: 18–34.
38. Novozhilov AS, Karev GP, Koonin EV (2006) Biological applications of the theory of birth-anddeath processes. Briefings in bioinformatics 7: 70–85.
39. Lynch M, Force A (2000) The probability of duplicate gene preservation by subfunctionalization. Genetics 154: 459–473.
40. Hall C, Brachat S, Dietrich FS (2005) Contribution of horizontal gene transfer to the evolution of saccharomyces cerevisiae. Eukaryotic cell 4: 1102–1115.
41. Cai J, Zhao R, Jiang H, Wang W (2008) De novo origination of a new protein-coding gene in Saccharomyces cerevisiae. Genetics 179: 487–496.
42. Knowles D, McLysaght A (2009) Recent de novo origin of human protein-coding genes. Genome Research 19: 1752–1759.
43. Demuth J, Hahn M (2009) The life and death of gene families. BioEssays 31: 29–39.
44. Fleet G (1990) Yeasts in dairy products. Journal of Applied Microbiology 68: 199–211.
45. Ashby S, Nowell W (1926) The fungi of stigmatomycosis. Annals of Botany: 69–84.
46. Ames RM, Rash BM, Hentges KE, Robertson DL, Delneri D, et al. (2010) Gene duplication and environmental adaptation within yeast populations. Genome Biology and Evolution 2: 591–601.
47. Brown C, Murray A, Verstrepen K (2010) Rapid expansion and functional divergence of subtelomeric gene families in yeasts. Current Biology 20: 895–903.
48. Dunlop P, Meyer G, Ban D, Roon R (1978) Characterization of two forms of asparaginase in *Saccharomyces cerevisiae*. Journal of Biological Chemistry 253: 1297–1304.
49. Sinclair K, Warner J, Bonthron D (1994) The asp1 gene of Saccharomyces cerevisiae, encoding the intracellular isozyme of l-asparaginase. Gene 144: 37–43.
50. Gordon J, Byrne K, Wolfe K (2009) Additions, losses, and rearrangements on the evolutionary route from a reconstructed ancestor to the modern *Saccharomyces cerevisiae* genome. PLoS Genetics 5: e1000485–e1000499.
51. Despons L, Wirth B, Louis V, Potier S, Souciet J (2006) An evolutionary scenario for one of the largest yeast gene families. Trends in Genetics 22: 10–15.
52. Cohen G, Fessl F, Traczyk A, Rytka J, Ruis H (1985) Isolation of the catalase a gene of *Saccha-romyces cerevisiae* by complementation of the cta1 mutation. Molecular and General Genetics 200: 74–79.
53. Lee J, Godon C, Lagniel G, Spector D, Garin J, et al. (1999) Yap1 and skn7 control two specialized oxidative stress response regulons in yeast. Journal of Biological Chemistry 274: 16040–16046.
54. Alvarez-Peral F, Zaragoza O, Pedreno Y, Argüelles J (2002) Protective role of trehalose during severe oxidative stress caused by hydrogen peroxide and the adaptive oxidative stress response in *Candida albicans*. Microbiology 148: 2599–2606.
55. Onishi H (1960) Studies on osmophilic yeasts. Bulletin of the agricultural chemical society of japan 24: 126–130.
56. D'Amore T, Crumplen R, Stewart G (1991) The involvement of trehalose in yeast stress tolerance. Journal of Industrial Microbiology & Biotechnology 7: 191–195.
57. Kwon H, Yeo E, Hahn S, Bae S, Kim D, et al. (2003) Cloning and characterization of genes encoding trehalose-6-phosphate synthase (tps1) and trehalose-6-phosphate phosphatase (tps2) from *Zygosaccharomyces rouxii*. FEMS Yeast Research 3: 433–440.
58. Smits G, Brul S (2005) Stress tolerance in fungi—to kill a spoilage yeast. Current Opinion in Biotechnology 16: 225–230.
59. Lagunas R (1993) Sugar transport in Saccharomyces cerevisiae. FEMS Microbiology Letters 104: 229–242.
60. Ozcan S, Johnston M (1999) Function and Regulation of Yeast Hexose Transporters. Microbiology and Molecular Biology Reviews 63: 554.
61. Brown C (1998). Multiple duplications of yeast hexose transport genes in response to selection in a glucose-limited environment.
62. Lin Z, Li W (2011) Expansion of hexose transporter genes was associated with the evolution of aerobic fermentation in yeasts. Molecular Biology and Evolution 28: 131–142.
63. Nobre A, Lucas C, Leão C (1999) Transport and utilization of hexoses and pentoses in the halotolerant yeast *Debaryomyces hansenii*. Applied and environmental microbiology 65: 3594–3598.
64. Olson M (1999) When less is more: gene loss as an engine of evolutionary change. American journal of human genetics 64: 18–23.
65. Nikolski M, Sherman D (2007) Family relationships: should consensus reign?—consensus clustering for protein families. Bioinformatics 23: e71–e76.
66. Altschul S, Gish W, Miller W, Myers E, Lipman D (1990) Basic local alignment search tool. Journal of Molecular Biology 215: 403–410.
67. Dujon B (2010) Yeast evolutionary genomics. Nature Reviews Genetics 11: 512–524.
68. Yang Z (2007) PAML 4: phylogenetic analysis by maximum likelihood. Molecular Biology and Evolution 24: 1586–1591.
69. Edgar R (2004) MUSCLE: multiple sequence alignment with high accuracy and high throughput. Nucleic Acids Research 32: 1792–1797.
70. Felsenstein J (1992) Phylogenies from restriction sites: A maximum-likelihood approach. Evolution 46: 159–173.
71. Benjamini Y, Hochberg Y (1995) Controlling the false discovery rate: a practical and powerful approach to multiple testing. Journal of the Royal Statistical Society Series B (Methodological) 57: 289–300.
72. Howe K, Bateman A, Durbin R (2002) Quicktree: building huge neighbour-joining trees of protein sequences. Bioinformatics 18: 1546–1547.
73. Balzer M, Deussen O (2005) Exploring relations within software systems using treemap enhanced hierarchical graphs. In: Visualizing Software for Understanding and Analysis, 2005. VISSOFT 2005. 3rd IEEE International Workshop on. IEEE, pp. 1–6.
74. Balzer M, Deussen O, Lewerentz C (2005) Voronoi treemaps for the visualization of software metrics. In: Proceedings of the 2005 ACM symposium on Software visualization. ACM, pp. 165–172.
75. Ames RM, MacPherson JI, Pinney JW, Lovell SC, Robertson DL (2013) Modular biological function is most effectively captured by combining molecular interaction data types. PLoS ONE 8: e62670.

5

Habitat Fragmentation and Species Extirpation in Freshwater Ecosystems; Causes of Range Decline of the Indus River Dolphin (*Platanista gangetica minor*)

Gill T. Braulik[1,2,3]*, Masood Arshad[2], Uzma Noureen[2], Simon P. Northridge[1]

1 Sea Mammal Research Unit, Scottish Oceans Institute, University of St. Andrews, St. Andrews, Fife, United Kingdom, **2** World Wildlife Fund-Pakistan, Lahore, Pakistan, **3** Wildlife Conservation Society, Zanzibar, United Republic of Tanzania

Abstract

Habitat fragmentation of freshwater ecosystems is increasing rapidly, however the understanding of extinction debt and species decline in riverine habitat fragments lags behind that in other ecosystems. The mighty rivers that drain the Himalaya - the Ganges, Brahmaputra, Indus, Mekong and Yangtze - are amongst the world's most biodiverse freshwater ecosystems. Many hundreds of dams have been constructed, are under construction, or are planned on these rivers and large hydrological changes and losses of biodiversity have occurred and are expected to continue. This study examines the causes of range decline of the Indus dolphin, which inhabits one of the world's most modified rivers, to demonstrate how we may expect other vertebrate populations to respond as planned dams and water developments come into operation. The historical range of the Indus dolphin has been fragmented into 17 river sections by diversion dams; dolphin sighting and interview surveys show that river dolphins have been extirpated from ten river sections, they persist in 6, and are of unknown status in one section. Seven potential factors influencing the temporal and spatial pattern of decline were considered in three regression model sets. Low dry-season river discharge, due to water abstraction at irrigation barrages, was the principal factor that explained the dolphin's range decline, influencing 1) the spatial pattern of persistence, 2) the temporal pattern of subpopulation extirpation, and 3) the speed of extirpation after habitat fragmentation. Dolphins were more likely to persist in the core of the former range because water diversions are concentrated near the range periphery. Habitat fragmentation and degradation of the habitat were inextricably intertwined and in combination caused the catastrophic decline of the Indus dolphin.

Editor: Tom Gilbert, Natural History Museum of Denmark, Denmark

Funding: World Wildlife Fund-Pakistan -(http://www.wwfpak.org/) funded the interview surveys with support through the Pakistan Wetlands Project. The US Marine Mammal Commission (www.mmc.gov) provided funding to GB for reporting and analysis in grant number E4047595. The funders had no role in study design, data collection and analysis, decision to publish, or preparation of the manuscript.

Competing Interests: The authors have declared that no competing interests exist.

* Email: gillbraulik@downstream.vg

Introduction

Fresh waters are experiencing declines in biodiversity far greater than those in the most affected terrestrial ecosystems [1,2]. Dam construction has dramatically increased habitat fragmentation and degradation in freshwaters, which is likely to have incurred a large unredeemed extinction debt [3]. However, this debt is not yet well quantified or understood, as metapopulation ecology in freshwaters has lagged behind similar work in other habitats, such as tropical forests [4,5,6]. A fundamental, yet unanswered question for conservation biology is how rapidly freshwater species disappear from river fragments and which factors influence the extinction of freshwater species in habitat patches. We investigate this issue using the example of the highly fragmented Indus River system and its endemic, endangered freshwater cetacean.

The great rivers that drain the Himalaya are amongst the world's most biodiverse freshwater ecosystems, but they are increasingly under threat as the emerging nations of China, India, and Pakistan and the countries of Southeast Asia scramble to harness hydropower and provide water for expanding agrarian economies, in the midst of increasing water scarcity and climatic uncertainty [7,8,9]. Many hundreds of new dams and other water development projects are planned or under construction in the region, including mainstem and tributary mega-dams, run-of-the-river hydroelectric schemes, irrigation barrages, and inter-basin water transfers [10,11]. The Himalayan region will soon have the highest concentration of dams in the world [12]. The combined effects of these activities are predicted to cause rapid and escalating hydrological change and habitat fragmentation that will negatively impact riverine biodiversity and ecosystem services [9,13].

The freshwater cetaceans that inhabit the largest Himalayan rivers, the Indus, Ganges, Brahmaputra, Yangtze, Mekong, and Ayeyarwady, collectively form one of the world's most endangered groups of mammals, each listed as endangered or critically endangered on the IUCN RedList [14], and, in the case of the Yangtze River dolphin (*Lipotes vexillifer*), the species has probably been extinct since the mid-2000's [15]. River dolphins are iconic species that can serve as charismatic flagships for conservation of freshwater ecosystems; but they are poorly understood and increasingly threatened. This is exemplified by the demise of the Yangtze River dolphin which disappeared so quickly that a comprehensive evaluation of the causes of its decline was not conducted until after it was presumed extinct [16].

We examine the pattern, and causes of range decline of the Indus dolphin (*Platanista gangetica minor*), an obligate freshwater cetacean endemic to the Indus River system, which, after more than 150 years of barrage (gated-dam) construction and removal of water to feed irrigated agriculture, is one of the most fragmented and modified rivers in the world. The distribution of the Indus dolphin was carefully documented in the 1870's, (just prior to the first major barrage being constructed) and at that time the dolphin inhabited the entire lower Indus River system from the delta with the Indian Ocean, to the foothills of the Himalayas in what is now India and Pakistan [17]. By the early 1990's, Indus dolphins had undergone an 80% reduction in range, having been extirpated from the upper and lower reaches of the Indus and four of the largest tributaries [18]. They are now confined to five contiguous 'river sections' on the Indus mainstem in Pakistan, separated by barrages, and in the Beas River, in India (Fig. 1) [19,20]. Details of when dolphins were extirpated from different parts of their former range are vague, and the causes not clearly understood. However, the construction of twenty irrigation barrages between 1886 and 1971 (gated-dams used for water diversion) that fragmented the dolphins historical range into 17 river sections (numbered 1–17 on Fig. 1), and large-scale water abstraction for irrigation rendering many sections of river almost dry for many months, certainly played a role (see Fig. 2) [19,21].

In this paper we document the spatial and temporal dynamics of the Indus River dolphin range decline, and then use a series of regression models to determine the causes of the spatial pattern of decline, the timing of subpopulation extirpation, and the speed of subpopulation disappearance after habitat fragmentation. Greater understanding of how the Indus dolphin has responded to the presence of dams and water diversions within its habitat demonstrates how we may expect other vertebrate populations to respond as planned dams and water developments come into operation elsewhere. The results provide important and relevant insights into factors influencing species extinction in fragmented riverscapes.

Methods

Last Dolphin Sighting Date

The status of each of the six extant dolphin subpopulations is fairly well understood [19,22], but there is little information on when or why dolphins disappeared from the 11 river sections where dolphins are presumed extirpated (Fig. 1). In 2007, we compiled historical dolphin sighting dates and locations from areas in which dolphins are believed to have been extirpated, by conducting community interviews, and a review of historical literature. GB received clearance from the Pakistan Home Department for the survey, and as there was no appropriate ethics or review board in Pakistan to provide approval no other permits were required. Depending on location, dolphins are believed to have disappeared between 20 and 80 years ago, so the short structured interviews targeted elderly riverside inhabitants old enough to have seen dolphins in their lifetimes. The objective of the interviews was explained to potential interviewees and they provided their verbal, rather than written, consent to participate as the majority were illiterate. All interviews were anonymous. A calendar of significant local historical events was compiled to assist informants in recalling dates correctly. A last dolphin sighting date (LDSD) was allocated to each river section based on the most recent dolphin sighting that we identified. We did not attempt to identify the exact extinction date of each subpopulation [23], but used the LDSD as a general indicator of when dolphins disappeared [24]. Inexact dates were assigned to 5 year intervals, for example, if the date was early 1970s, a date of 1972 was assigned, if it was mid 1970s, 1975 was designated, and if it was late 1970s the date used was 1978.

Identifying the Causes of Range Decline

The following seven explanatory variables that may have contributed to the Indus dolphin range decline were determined for each of the 17 sections of the dolphin's former range:

1. *Fragmentation date* - The year that each river section was created; assigned as the date that the second of the two bounding barrages became operational. For the river section upstream of Harike barrage in India, the isolation date was assigned as the completion date of Hussainiwala barrage which is located only 30 km downstream of Harike.
2. *River length* - The number of river kilometres between two barrages.
3. *Proximity to range edge* - The distance along the river's course from the former dolphin distributional limit recorded by Anderson [17], to the barrage located closest to the range core.
4. *Size of river* - The mean annual discharge in Million Acre Feet reported for each river prior to implementation of the Indus Water Treaty in 1960 [25]. This illustrates original river size prior to large-scale water abstraction.
5. *Confluences* - The number of river confluences within each river section was included as an indicator of favourable habitat, as Indus dolphins occur with higher frequency at confluences [26].
6. *River slope* - The slope within each river section was calculated as the drop in elevation between the up and downstream barrages, or upstream range limit in the case of peripheral segments, divided by the length of river. Slope exerts a direct effect on flow velocity and sediment transport and therefore may influence dolphin habitat.
7. *Dry season river discharge* - River discharge data were obtained for all twelve barrages and two dams north of Guddu on the Indus River system in Pakistan for the period July 2008 to April 2013 (4 years 9 months). The daily discharge below Guddu and Kotri barrages was obtained from October 2010 to April 2013 (~2 ½ years), and discharge below Sukkur barrage was obtained from April 1994 to January 2000, and October 2010 to April 2013 (8 years 3 months). Occasional missing data were interpolated. It is very low flows that are likely to adversely impact dolphins, therefore median daily discharge during the dry season (1st October to 31st March) was determined using the years for which there was an entire dry season's data. Mean monthly discharge was available above and below Harike and Hussainiwala barrages in India from January 2009 to December 2011, and for these two river sections the median of the mean monthly dry season discharge was used. The number of years of data available differed according to barrage but the temporal discharge pattern was predictable and similar across years in each location.

Generalised Linear Models (GLMs) and a survival analysis were used with the seven explanatory variables described above as predictors of the continued presence of river dolphins in each of the 17 river sections. Generalised Additive Models were used in

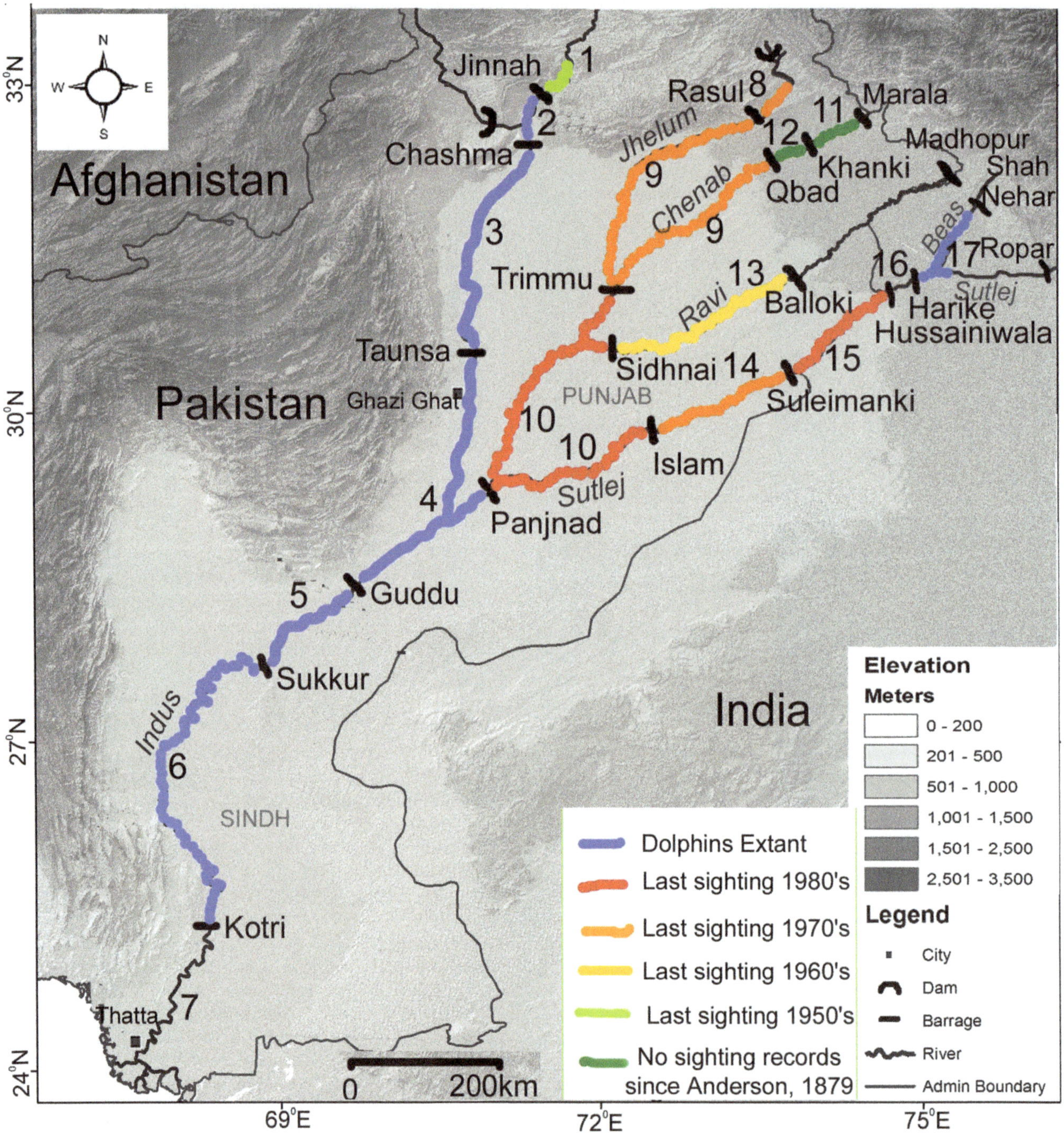

Figure 1. Map of the lower Indus River system. Rivers and barrages are named, and each river section is numbered and coloured according to whether river dolphins are extant, or the approximate date that they were extirpated (see Table A1 for details).

the initial data exploration to visually investigate whether the relationship between the predictor and explanatory variables was linear, and, which type of transformation could be used to best account for non-linearity. Three sets of models were developed, with objectives summarised below:

Spatial pattern of dolphin persistence. The objective of the first set of models was to identify which factors best explained the observed geographic pattern of range decline. The presence or absence of dolphins in each river section was modelled using a GLM and a binomial error distribution, with presently extant populations coded as 1 and extirpated populations coded as 0. The best fitting models were then used to predict the probability that dolphins are still present in the Harike-Hussainiwala river section on the India-Pakistan border (Fig. 1: no. 16), where dolphin presence is unknown.

Temporal pattern of decline. The second model set included only sections where dolphins have been extirpated and examined which factors influenced when dolphins disappeared. The number of years since dolphins were sighted (as of 2013) was

Figure 2. Aerial photograph of Sukkur Barrage. Image shows the seven canals diverting water out of the river, and demonstrates the dramatically reduced flows downstream (river flow direction right to left).

the response variable modelled using a GLM with a quasi-Poisson error distribution.

Time to extirpation. The third model set used a survival function to investigate which factors influenced the speed with which dolphin populations were extirpated following their isolation between barrages. We used the Kaplan–Meier estimate of survival which allows for the inclusions of censored data, in this instance allowing for the inclusion of river sections where dolphins have not yet been extirpated, as well as those where dolphins have disappeared. Each river section was qualified with a status assignment, where 1 = dolphins extirpated, and 0 = dolphins extant [27]. The time to extirpation was calculated as the number of years between the fragmentation date and either the LDSD if dolphins have been extirpated, or the current year, 2013, where they are still present. Time to extirpation and status together were the predictor variable (the Kaplan-Meier survivorship object) modelled using the 'survreg' function in the survival library of the program R [28]. Both an exponential and a Weibull error distribution were tested, and the Weibull distribution was selected as it provided a significantly better fit to the data (ΔAIC 15.17) [27].

All models were implemented using the program R 2.15.1 [28]. Logit, probit and cloglog link functions were included in global models and the logit function, which resulted in the best fit, applied. Variance Inflation Factors (VIFs) that demonstrate the degree of collinearity between variables were generated from the maximal models and collinear variables removed until VIF scores were less than five [27]. Three two-way interactions that described potentially meaningful relationships between variables (fragmentation date and dry season discharge, fragmentation date and river length, and river length and proximity to range edge) were included, as well as second and third order polynomials of significant variables. The binomial and survival models were simplified using backwards stepwise selection based on Akaike's Information Criteria (AIC). Quasi-Poisson models were selected on the basis of quasi-AIC (QAIC) scores, and non-significant terms were sequentially dropped based on their levels of significance. Models separated by at least two AIC/QAIC points were assumed to be significantly different [29]. Goodness of fit for the GLMs was measured by determining the proportion of the total deviance explained by the final model. Model plots were examined for non-normality of errors, statistical independence of observations, heteroscedasticity and influential points [30].

Results

Pattern of Range Decline

Historical dolphin sightings were obtained for all river sections formerly occupied by dolphins except for the area downstream of Kotri Barrage to the delta (Fig. 1: no. 7) and the stretch from Harike to Hussainiwala barrage (Fig. 1: no. 16) which is close to the India-Pakistan border. Our focus on retired fishermen and on areas from which dolphins have already been extirpated meant that the communities were forth-coming with information because it was not regarded by them as sensitive. However, there were very few elderly members of each fishing community and our pool of available informants was consequently small (n = 57). 79% of informants were, or had been, full-time commercial fishermen or contractors and the remainder were part-time subsistence fishermen. There was no significant difference in the age of informants interviewed at each barrage (GLM $p = 0.7$) but older individuals were significantly more likely to have seen dolphins than younger informants (GLM $p < 0.01$). We found no evidence that dolphins persist in any of the Indus tributaries in Pakistan. Of 17 sections of river, dolphins are extant in six, have been extirpated from ten, and in one border area that could not be surveyed (no. 16), dolphin presence or absence is unknown (Fig. 1).

Causes of Range Decline

For each river section, dolphin presence or absence, estimated LDSD, and the time to extirpation were compiled along with the physical characteristics and these data included in each of the models (Table S1). On the upper Chenab River (Fig1:,no 11 & 12) the LDSD is that reported by Anderson in 1879. Time to extirpation was not calculated for these two sections reflecting the lack of recent sighting evidence and uncertainty of the extirpation date.

Where dolphins are still extant the median monthly dry season river discharge averaged 30,830 cusecs (ranged 7,224–47,040 cusecs), as compared to an average of 8,022 cusecs (range 0–38,000 cusecs) in locations from which dolphins have been extirpated. In general, sections of river where dolphins are still present were fragmented by barrages later, are further from the range periphery, are of longer length, have a shallower slope and greater dry season discharge than river sections where dolphins are no longer found.

Sixteen river sections, 6 where dolphins are extant and ten where they have been extirpated, were included in the spatial GLM models. The VIFs generated from the full model indicated that river discharge and slope were collinear. Slope was considered to be less important than discharge in explaining dolphin range decline because it has not changed substantially in hundreds of years, and it was therefore removed from further candidate models. The final model that best explained the observed spatial pattern in Indus dolphin range decline retained the explanatory variables dry season river discharge and distance from range edge (Table 1). The probability that an Indus dolphin subpopulation is still extant increases with increasing distance from the range edge, and with increasing dry season river discharge (Fig. 3). The final spatial model predicted only a 2.6% probability that dolphins are still extant in the Harike-Hussainiwala river section in India that was not surveyed for dolphins. This is not unexpected given that this section has very low dry season discharge and is near the periphery of the dolphin's range, both factors that increase the likelihood of subpopulation extirpation. Although the linear arrangement of river segments might suggest a lack of statistical independence, there was no clumping of residuals according to the geographic position of the river segments, and the independence of observations was further shown by the Durbin-Watson test of correlated errors ($p = 0.24$).

The variables that described the temporal pattern of Indus dolphin subpopulation extirpation were the same as those that influenced the spatial pattern of decline: dry season river discharge and distance from former range limit (Table 2). Within areas from which dolphins have disappeared, they were extirpated earlier from river sections where discharge was lower, and from those sections located near the periphery of the subspecies former range (Fig. 3).

In river sections where dolphins have been extirpated, the mean time from fragmentation to the LDSD was 50 years (SD = 23, range = 9–76). For river sections where dolphins are still extant, the mean time from subpopulation isolation to present (2013) was 57 years (SD = 15, range = 42–86). Thirteen river sections were included in the survival model (number 2, 10, 12 and 16 were excluded because of missing data) and the slope parameter was included. The final survival model retained four variables: median dry season river discharge, isolation date, length of river section and slope. Dolphin subpopulations were extirpated more quickly in sections with low dry season river discharge. Subpopulations persisted longer where the river slope is more gentle (e.g. in the lower reaches) and those that were isolated between barrages a long time ago persisted for longer than those in more recently

Table 1. Summary of spatial range decline model output.

Model	*AIC*	*Δ AIC*	*% explained Deviance*	*n*	*Deviance*						
					Q	**Range**	**Is. Date**	**L**	**Conf**	**Size**	**USD: Is. Date**
1	18.26	1.82	32.6	1	6.91	-					
2	**16.44**	-	**50.7**	**2**	**7.88**	**2.84**	-	-	-	-	-
3	17.72	1.28	54.1	3	7.88	2.84	0.72	-	-	-	-
4	19.67	3.23	54.3	4	7.88	2.84	0.68	0.08	-	-	-
5	21.64	5.20	54.4	5	7.88	2.84	0.42	0.31	0.07	-	-
6	23.64	7.20	54.5	6	7.88	2.84	0.42	0.31	0.07	-	0.001
7	25.16	8.72	56.7	7	7.88	2.84	0.79	0.31	0.07	0.0001	0.11

n = number of covariates, Is. Date = Isolation Date, L = Length of river section, Range = Distance from range edge, Size = River size, Conf = confluences, Q = River discharge, USD: Is. Date = Interaction between Isolation Date and River Discharge. Model in bold was the final selected model.

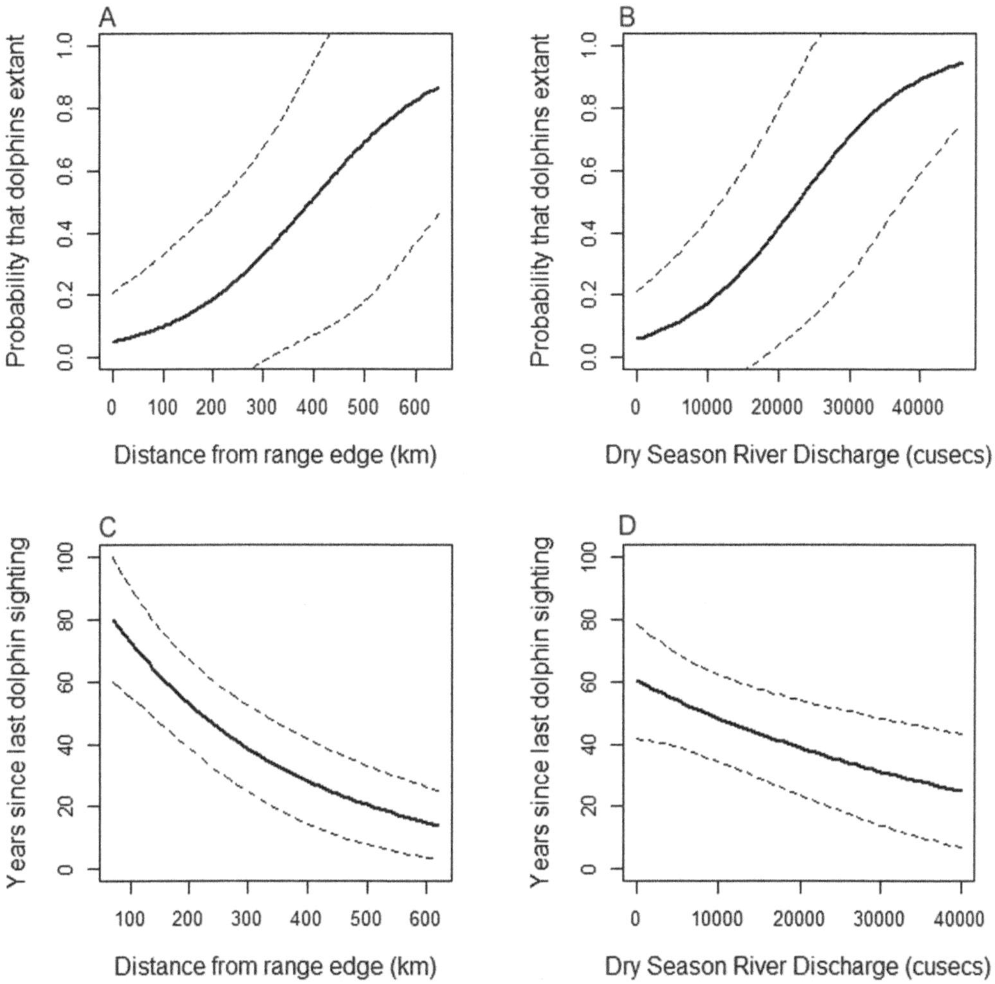

Figure 3. Significant relationships retained in GLM models of the causes of the spatial and temporal pattern of Indus dolphin decline. The figure demonstrates the probability that an Indus dolphin subpopulation is extant according to A) proximity to the edge of the former range and B) median dry season river discharge, and the relationship between the number of years since a dolphin was sighted and C) distance from the historical distributional limit, and D) median dry season river discharge.

subdivided river sections. Fifty years after Indus dolphins were isolated between barrages there is a less than a 50% chance that they will still be extant, and after 100 years this probability drops to 37%.

Discussion

Model evaluation

The river discharge data used in these models were from the last ten years but they explained well the pattern of dolphin decline that occurred decades ago. Although river discharge varies from year to year, and has generally declined, the relative discharge among barrages (e.g. the spatial relationship) has remained constant with the same locations consistently reporting high (e.g. the upper Indus) and low discharge (e.g. Indus tributaries) over time. Therefore, the assumption implicit in this analysis that the present spatial pattern of discharge reflects that present during the period leading up to dolphin extirpation is not unreasonable.

As for terrestrial habitats, such as forest fragments, we would have expected to see a relationship between species extinction and habitat patch size [5]. In fact, length of river section was one of the first variables to be excluded in candidate models. This may be because only the current configuration of 17 comparatively small habitat fragments were included in the models, and we did not consider the progression of escalating habitat fragmentation and concomitant diminishing fragment size over time. To investigate this, we constructed an additional model considering all 33 river sections that have existed since the onset of barrage construction (Table A2) in a binomial GLM, with dolphins recorded as extirpated (0) or still extent (1) in each river section at the point it was further subdivided. Explanatory variables were 1) Length of river section, 2) Isolation Date, and 3) End Date, taken as the year a new barrage was completed resulting in the sections further subdivision. When considering the entire history of habitat fragmentation, the models showed that dolphins were significantly more likely to be extirpated in smaller fragments ($p<0.05$), and that this relationship was independent of fragment creation date or duration. The link between species extirpation and habitat fragment size has been clearly established in terrestrial habitats for several species groups [5] but this is one of the first studies to show a similar relationship in riverscapes. It underlines the great importance of maintaining large sections of intact river habitat to sustain tropical aquatic biodiversity.

Table 2. Summary of temporal range decline model outputs.

Model	*AIC*	*Δ AIC*	*% explained Deviance*	*n*	*Deviance*						
					Q	**Range**	**Is. Date**	**L**	**Conf**	**Size**	**USD*Is. Date**
1	17.57	1.36	56.75	1	-	115.36	-	-	-	-	-
2	**16.21**	-	**76.76**	**2**	**40.67**	**115.36**	-	-	-	-	-
3	17.80	1.59	79.17	3	40.67	115.36	-	-	-	-	4.91
4	17.54	1.33	92.62	4	52.06	115.36	-	-	-	6.66	14.20
5	19.50	3.29	92.85	5	52.06	115.36	-	-	6.60	6.66	8.11
6	21.46	5.25	93.12	6	52.06	115.36	0.75	-	6.60	6.66	7.9
7	23.39	7.18	93.49	7	52.06	115.36	0.60	0.33	9.16	6.66	5.89

n = number of covariates, Is. Date = Isolation Date, L = Length of river section, Range = Distance from range edge, Size = River size, Conf. = confluences, Q = River discharge. Model in bold was the final selected model.

River discharge and distance from range periphery provided a good fit to the range decline data, explaining more than 76% of the deviance in the temporal model and 50% in the spatial model. However, three other aspects that may have also have played a role in the dolphins decline were not included as explanatory variables because of a lack of suitable data. These are a) water quality, b) incidental capture in fishing gear and c) hunting. The possible contributions of these to the Indus dolphin decline are discussed below.

The magnitude of surface water pollution in Pakistan has increased at a dramatic rate over the last ten years and more than 90% of industrial and municipal effluents enter water courses untreated [31]. The Indus tributaries flow through the industrial and agricultural heartland of Pakistan and are more polluted than the Indus River itself which has a greater dilution capacity and passes through remoter areas. There has been no systematic monitoring of river water quality that could provide data for this analysis. However, dolphins had already been extirpated from most areas prior to significant declines in water quality which occurred in the 1980s and 90s, and this asynchronous timing indicates that pollution was not primarily responsible for the dolphins' decline.

Mortality from accidental capture in fishing gear is considered to be the greatest threat to most cetacean populations [32]. In the past, the Indus River main channel was not intensively fished because the water was too swift for easy manoeuvrability of oar-powered boats, and instead fishing focused on side channels and adjacent pools that are rarely used by dolphins [33]. Since 2010 changing fishing practices in Sindh Province have led to an increase in dolphin mortality, however, prior to this there are very few records of incidental capture of dolphins in fishing gear and this is not likely to be a large factor in the decline of the Indus dolphin.

Indus dolphins were killed for food, oil and medicine until the late 1970s when the animal became legally protected [17,34]. Information on dolphin hunting is sparse and un-quantified and records refer only to hunting on the Indus River, where dolphins are still extant. Although it is possible that dolphins were hunted throughout the river system, there is no evidence that this was so, and the fact that they persist in the places that hunting is reported to have been intense, and have disappeared from places where hunting was not reported, suggests that this was unlikely to have been the cause of the subspecies' decline. However, the timing of reported hunting coincides with the period of decline and without more information, it is not possible to completely discount the role of hunting.

For the majority of the year the gates on all barrages are lowered to divert water into canals, and the physical opening is sufficiently small that it would be difficult or impossible for dolphins to pass through the gates and between different sections of river. It has been hypothesized that there may be consistent or frequent movements of dolphins through some barrages and between subpopulations [21,35]. It has also been theorised that due to the high water velocity and turbulence often found within the barrage gates it would be more likely for animals to move down-, rather than up-stream, and that this would lead to the downstream migratory attrition of upstream subpopulations [18,36]. The only evidence of this was obtained in 2009, when, during the annual canal maintenance period, which is one of the brief periods in the year when no water is diverted and barrage gates are fully open for several weeks, one radio-tagged Indus dolphin was recorded to move through the gates on Sukkur barrage in both an up- and down-stream direction (WWF-Pakistan unpublished). Each barrage is quite different in terms of design, location and operation and one dolphin moving through Sukkur barrage does not prove that this occurs at other barrages, or that it is a regular occurrence. Therefore, for the purposes of this study, and in the absence of evidence to the contrary, we assume no significant migration between dolphin subpopulations. It is possible that future research may demonstrate that dolphins do move across some, or all, barrages with regularity and if the movement of individuals is primarily in one direction this would be another important factor to consider in the extinction dynamics. The pattern of subpopulation persistence near the range core, and earlier extirpation of subpopulations near the range periphery could perhaps be partially explained by the consistent downstream migration of animals, but again this pattern is disrupted by the presence of dolphins in the Beas River.

Causes of Indus dolphin range decline

The clear result of this study was the relationship between low dry season river discharge and the decline of the Indus dolphin. Reduced flows directly impact dolphins by reducing the physical space available to them, reducing average water velocity and depth and increasing water temperatures. Flow regulation is also likely to indirectly impact river dolphins due to declines in fish diversity, the dominance of generalist fish species, and increased success of invasive species [37,38]. The dampened flood peaks typically associated with dams and diversions reduce the frequency, extent and duration of floodplain inundation that determines how long fish can gain access to nursery habitat and food. Water abstraction also exacerbates and concentrates existing anthropogenic threats, for example, increasing the concentration of nutrients and pollutants, and concentrating dolphins into deep pools that are also important areas for fishing, thereby increasing the chances of negative human interactions [39]. The altered hydrological regime on the Indus River has likely reduced the complexity of hydrologic and geomorphologic habitat and ultimately also diminished its carrying capacity and ability to support large numbers of aquatic megafauna. To preserve aquatic biodiversity, river management is needed that focuses on restoring both the timing and duration of flood pulses, as well as on maintaining critical minimum flows in the dry season.

The persistence of dolphins in the Beas River, India is likely to be due to the presence of constant water supplies little depleted by diversions. Dolphins in the Beas River occur in an isolated habitat fragment as the river downstream is virtually dry, and only connected with the rest of the river system for a few weeks each year during the monsoon floods. This demonstrates that in the presence of sufficient water, and an absence of other threats, river dolphins can persist for decades even in relatively small fragments of habitat near the periphery of their range. This subpopulation is of conservation importance, as all other Indus dolphins occur in a single river, the Indus, and are therefore at risk from environmentally correlated catastrophic events [40]. However, based on the historical pattern of decline, Indus dolphins are most likely to disappear in the future from locations with low river discharge located closer to the range periphery, meaning that dolphins in the Beas (close to range periphery with moderate discharge) and between Sukkur and Kotri Barrages (with low discharge located a moderate distance from the former range edge) are most at risk.

The date of habitat fragmentation was not selected by any of the models as a strong predictor of whether dolphins are still present. However depleted river discharge and habitat fragmentation by barrages are inextricably intertwined as barrages are responsible for diverting water, and they are a physical barrier that greatly impedes or prevents the dispersal of dolphins out of impacted river reaches.

Population extirpation, core habitat and conservation

Contraction of geographic range is one of the principal characteristics exhibited by declining or threatened species [41,42]. In general, at the periphery of a species geographic range, populations occupy less favourable habitat and occur at lower and more variable densities. Therefore, as a species becomes endangered it is expected that its geographic range will contract inwards, and that populations will persist in the range core until the final stages of decline [43]. However, for many endangered mammals the pattern of range decline is instead dictated by the spread of factors driving the decline, with those populations last impacted, regardless of their location, persisting longer than those that were historically large [41,43]. The range of the Indus dolphin has also contracted inwards, and dolphins persist primarily in what is assumed to be the former range core or higher density area, however this is likely to be because the greatest threat, water extraction, is concentrated in the periphery of the subspecies range. This conclusion is supported by the continued persistence of the Beas River population at the range limit. However, that animals naturally occur at lower density in upstream areas and smaller rivers and are therefore more vulnerable near the range limit is also certainly a factor. The spatial pattern of Indus River dolphin decline is very different from the gradual decline in abundance described for the Yangtze River dolphin [16].

One of the greatest challenges in conservation science involves disentangling the relative contributions of multiple factors in the decline of species, especially when causes interact or vary spatially and temporally with importance [44]. Nevertheless, the primary factor identified in these models (i.e. low dry season discharge due to water diversion at barrages) is well supported, and is also the most salient for informing current ecosystem management. In the Mekong River, numerous stressors such as fisheries bycatch, hunting and habitat destruction have reduced the resident Irrawaddy dolphin (*Orcaella brevirostris)* population to less than 100 individuals [45]. The results of our study suggests that habitat fragmentation and/or flow disruption associated with the many proposed new dams on the Mekong are likely to further drive Irrawaddy dolphin decline potentially leading to local extinction. The Indus dolphin range decline is probably most prescient for the related Ganges dolphin (*Platanista gangetica gangetica*) which occurs in the neighbouring Brahmaputra and Ganges River systems and is subject to the same threats. The Ganges River is fragmented by barrages and flow is severely reduced in many areas, and the Brahmaputra River system is the focus of massive hydropower development [46,47]. The range of the Ganges dolphin has begun to decline especially at the upper limit of the distributional range [48], and if water development continues as planned the range of the dolphin is expected to continue to shrink towards larger habitat fragments with higher discharge that are primarily in downstream locations. The results of this study suggest that other vertebrate populations in other large rivers, such as the Amazon, Orinoco and Ayeyarwady, will also respond with dramatic declines in range when dams and other water developments that reduce discharge, fragment habitat and change the hydrological regime are constructed.

The amount of habitat fragmentation and level of water withdrawals from rivers in Pakistan is extreme, negatively affecting human communities, eroding the delta, destroying fisheries and concentrating pollutants. This study indicates that if water development plans in South Asia and the wider Himalayan region proceed as currently proposed [10,13]and follow the pattern demonstrated by the Indus, the resulting habitat fragmentation and flow disruption will likely cause large declines in resident freshwater cetaceans and other freshwater dependent species. Healthy rivers are of great importance to communities, and it is critical that where developments are planned, environmental flow and impact assessments be conducted that balance human requirements for irrigation water and power with the habitat requirements of the aquatic ecosystem that are vital to humans.

Acknowledgments

We are grateful to the Office of Executive Engineers at Harike and Hussainiwala Canal Divisions and Ropar Headworks, India for allowing us river discharge data. A. Khan and R. Garstang facilitated the fieldwork. Valuable reviews were provided by P. Hammond, M. Lonergan, S. Turvey and G. Ryan.

Author Contributions

Conceived and designed the experiments: GTB. Performed the experiments: GTB MA UN. Analyzed the data: GTB SPN. Contributed to the writing of the manuscript: GTB MA UN SPN.

References

1. Sala OE, Chapin FS, Armesto JJ, Berlow R, Bloomfield J, et al. (2000) Global biodiversity scenarios for the year 2100. Science 287.
2. Dudgeon D, Arthington AH, Gessner MO, Kawabata Z-I, Knowler DJ, et al. (2006) Freshwater biodiversity: importance, threats, status and conservation challenges. Biological Reviews 81: 163–182.
3. Tilman D, May RM, Lehman CL, Nowak MA (1994) Habitat destruction and the extinction debt. Nature 371: 65–66.
4. Strayer DL, Dudgeon D (2010) Freshwater biodiversity conservation: recent progress and future challenges. Journal of the North American Benthological Society 29: 344–358.
5. Gibson L, Lynam AJ, Bradshaw CJA, He F, Bickford DP, et al. (2013) Near-complete extinction of native small mammal fauna 25 years after forest fragmentation. Science 341: 1508–1510.
6. Prugh LR, Hodges KE, Sinclair ARE, Brashares JS (2008) Effect of habitat area and isolation on fragmented animal populations. PNAS 105: 20770–20775.
7. FAO (2012) Coping with water scarcity. An action framework for agriculture and food security. Rome, Italy: FAO Water Reports No.38.
8. World Water Assessment Programme (2012) The United Nations World Water Development Report 4. Managing water under uncertainty and risk. Paris: UNESCO.
9. Dudgeon D (2000) Large-scale hydrological changes in tropical Asia: prospects for riverine biodiversity. Bioscience 50: 793–806.
10. Dutta AP (2010) Reservoir of Dams. Down to Earth May 1–15: 32–39.
11. Verma S, Kampman DA, van der Zaag P, Hoekstra AY (2009) Going against the flow: A critical analysis of inter-state virtual water trade in the context of India's National River Linking Program. Physics and Chemistry of the Earth 34: 261–269.
12. Dharmadhikary S (2008) Mountains of Concrete: Dam building in the Himalayas. Berkeley, CA: International Rivers. 48 p.
13. Ziv G, Baran E, Nam S, Rodríguez-Iturbe I, Levin SA (2012) Trading-off fish biodiversity, food security, and hydropower in the Mekong River basin. PNAS www.pnas.org/cgi/doi/10.1073/pnas.1201423109.
14. IUCN (2013) IUCN Red List of Threatened Species. version 2013.2 <http://www.iucnredlist.org> Downloaded on 6 January 2014.
15. Turvey ST, Pitman RL, Taylor BL, Barlow J, Akamatsu T, et al. (2007) First human-caused extinction of a cetacean species? Biology Letters 3: 537–540.
16. Turvey ST, Barrett LA, Hart T, Collen B, Yujiang H, et al. (2010) Spatial and temporal extinction dynamics in a freshwater cetacean. Proceedings of the Royal Society B 277: 3139–3147.
17. Anderson J (1879) Anatomical and zoological researches: comprising an account of the zoological results of the two expeditions to western Yunnan in 1868 and 1875 and a monograph of the two cetacean genera *Platanista* and *Orcella*. Bernard Quaritch, Piccadilly, London.
18. Reeves RR, Chaudhry AA, Khalid U (1991) Competing for water on the Indus Plain: Is there a future for Pakistan's river dolphins? Environmental Conservation 18: 341–349.
19. Braulik GT (2006) Status assessment of the Indus River dolphin, *Platanista gangetica minor*, March-April 2001. Biological Conservation 129: 579–590.
20. Behera SK, Nawab A, Rajkumar B (2008) Preliminary investigations confirming the occurrence of Indus River dolphin *Platanista gangetica minor* in River Beas, Punjab, India. Journal of the Bombay Natural History Society 105: 90–126.
21. Reeves RR, Leatherwood S (1994) Dams and river dolphins: can they co-exist? Ambio 23: 172–175.
22. Braulik GT, Bhatti ZI, Ehsan T, Hussain B, Khan AR, et al. (2012) Robust abundance estimate for endangered river dolphin subspecies in South Asia. Endangered Species Research 17: 201–215.
23. Collen B, Purvis A, Mace GM (2010) When is a species really extinct? Testing extinction inference from a sighting record to inform conservation assessment. Diversity and Distributions 16: 755–764.
24. Butchart SHM, Stattersfield AJ, Brooks TM (2006) Going or gone: defining 'Possibly Extinct' species to give a truer picture of recent extinctions. Bulletin-British Ornithologists Club 126: 7–24.
25. IUCN (2011) Pakistan Water Gateway: Gateway to water information about Pakistan. www.waterinfo.net.pk.
26. Braulik GT, Reichert AP, Ehsan T, Khan S, Northridge SP, et al. (2012) Habitat use by a freshwater dolphin in the low-water season. Aquatic Conservation: Marine and Freshwater Ecosystems 22: 533–546.
27. Crawley MJ (2007) The R Book. Chichester, England: John Wiley & Sons Ltd. 951 p.
28. R Development Core Team (2012) R: A language and environment for statistical computing. Vienna, Austria: R Foundation for Statistical Computing.
29. Burnham KP, Anderson DR (2002) Model selection and multi- model inference: a practical information-theoretic approach. 2nd edition. New York: Springer-Verlag. 496 p.
30. Fox J (2008) Applied regression analysis and Generalised Linear Models. 2nd edition. Los Angeles: Sage Publications, Inc. 665 p.
31. Directorate of Land Reclamation Punjab (2007) Surface water quality monitoring plan. Lahore: Irrigation and Power Department, Government of Punjab. 27 p.
32. Read AJ (2008) The looming crisis: interactions between marine mammals and fisheries. Journal of Mammalogy 89: 541–548.
33. Khan H (1947) A fishery survey of River Indus. Journal of the Bombay Natural History Society 46: 529–535.
34. Pilleri G (1972) Field observations carried out on the Indus dolphin *Platanista indi* in the Winter of 1972. Investigations on Cetacea 4: 23–29.
35. Braulik GT (2012) Conservation ecology and phylogenetics of the Indus River dolphin (*Platanista gangetica minor*). St. Andrews, UK. 267: University of St. Andrews.
36. Reeves RR (1991) Conservation of the bhulan (blind river dolphin) in the Punjab. Natura. 3–22.
37. Nilsson C, Reidy CA, Dynesius M, Revenga C (2005) Fragmentation and flow regulation of the world's large river systems. Science 308: 405–408.
38. Xenopoulos MA, Lodge DM (2006) Going with the flow: using species-discharge relationships to forecast losses in fish biodiversity. Ecology 87: 1907–1914.
39. Kelkar N, Krishnaswamy J, Choudhary S, Sutaria D (2010) Coexistence of fisheries with river dolphin conservation. Conservation Biology 24: 1130–1140.
40. Gilpin ME (1990) Extinction in finite metapopulations living in correlated environments. In: Shorrocks B, Swingland IR, editors. Living in a patchy environment. New York: Oxford University Press. 177–187.
41. Channell R, Lomolino MV (2000) Trajectories to extinction: spatial dynamics of the contraction of geographical ranges. Journal of Biogeography 27: 169–179.
42. Simberloff D (1986) The proximate causes of extinction. In: Raup DM, Jablonski D, editors. Patterns and processes in the history of life. Berlin: Springer-Verlag. 259–276.
43. Lomolino MV, Channell R (1995) Splendid isolation: patterns of geographic range collapse in endangered mammals. Journal of Mammalogy 76: 335–347.
44. Johnson PTJ, McKenzie VJ, Peterson AC, Kerby JL, Brown J, et al. (2010) Regional decline of an iconic amphibian associated with elevation, land-use change and invasive species. Conservation Biology 25: 556–566.
45. Ryan GE, Dove V, Trujillo F, Doherty Jr PF (2011) Irrawaddy dolphin demography in the Mekong River: an application of mark-resight models. Ecosphere 2: 1–15.
46. Bashir T (2010) Ganges river dolphin (Platanista gangetica) seeks help. Current Science 98: 287–288.
47. Baruah D, Hazarika LP, Bakalial B, Borah S, Dutta R, et al. (2012) A grave danger for the Ganges dolphin (Platanista gangetica, Roxburgh) in the Subansiri River due to a large hydroelectric project. Environmentalist 32: 85–90.
48. Sinha RK, Smith BD, Sharma G, Prasad K, Choudhury BC, et al. (2000) Status and distribution of the Ganges Susu (*Platanista gangetica*) in the Ganges River system of India and Nepal. In: Reeves RR, Smith BD, Kasuya T, editors. Biology and Conservation of Freshwater Cetaceans in Asia. Gland, Switzerland & Cambridge, U.K.: IUCN. 54–61.

6

Clade Age and Diversification Rate Variation Explain Disparity in Species Richness among Water Scavenger Beetle (Hydrophilidae) Lineages

Devin D. Bloom[1,2], Martin Fikáček[3,4], Andrew E. Z. Short[1,2*]

1 Department of Ecology & Evolutionary Biology, University of Kansas, Lawrence, Kansas, United States of America, **2** Division of Entomology, Biodiversity Institute, University of Kansas, Lawrence, Kansas, United States of America, **3** Department of Entomology, National Museum, Prague, Czech Republic, **4** Department of Zoology, Faculty of Sciences, Charles University, Prague, Czech Republic

Abstract

Explaining the disparity of species richness across the tree of life is one of the great challenges in evolutionary biology. Some lineages are exceptionally species rich, while others are relatively species poor. One explanation for heterogeneity among clade richness is that older clades are more species rich because they have had more time to accrue diversity than younger clades. Alternatively, disparity in species richness may be due to among-lineage diversification rate variation. Here we investigate diversification in water scavenger beetles (Hydrophilidae), which vary in species richness among major lineages by as much as 20 fold. Using a time-calibrated phylogeny and comparative methods, we test for a relationship between clade age and species richness and for shifts in diversification rate in hydrophilids. We detected a single diversification rate increase in Megasternini, a relatively young and species rich clade whose diversity might be explained by the stunning diversity of ecological niches occupied by this clade. We find that Amphiopini, an old clade, is significantly more species poor than expected, possibly due to its restricted geographic range. The remaining lineages show a correlation between species richness and clade age, suggesting that both clade age and variation in diversification rates explain the disparity in species richness in hydrophilids. We find little evidence that transitions between aquatic, semiaquatic, and terrestrial habitats are linked to shifts in diversification rates.

Editor: Diego Fontaneto, Consiglio Nazionale delle Ricerche (CNR), Italy

Funding: MF received funding from the Czech Science Foundation grant P506/12/P096 and from the Ministry of Culture of the Czech Republic (DKRVO 2014/13, National Museum, 00023272). The funders had no role in study design, data collection and analysis, decision to publish, or preparation of the manuscript.

Competing Interests: The authors have declared that no competing interests exist.

* E-mail: aezshort@ku.edu

Introduction

One of the most remarkable and pervasive patterns on Earth is the uneven distribution of species richness among clades. Indeed, some clades such as beetles are astoundingly species rich, while others such as monotremes are species poor. While there has long been interest in the disparity of species richness across the tree of life [1–3], recent advances in comparative methods have made investigating the underlying causes tractable [4]. However, the causes of disparity in species richness among lineages remain controversial [5,6].

One intuitive explanation for disparity in species richness is that species rich clades are older, and thus have had more time to accumulate diversity than younger clades [7]. This scenario assumes that constant rates of lineage diversification over time result in a predictive positive relationship between clade age and species richness. Disparity in species richness among clades may also result from differences in net diversification rates (speciation minus extinction) among lineages [8]. Diversification rate differences can result from both intrinsic factors (e.g., key innovations) and extrinsic factors (e.g., habitat shifts), or ecological limits (density-dependence) on clade diversity [9,10]. Widely varying net diversification rates are expected to weaken a clade age-species richness relationship [10], or even decouple age-diversity correlations completely [11].

Studies investigating the disparity of species richness among clades have been met with mixed results [5,7]. Some studies have failed to detect a relationship between clade age and species richness in groups including plants [12,13], birds [14], and squamates [8,15]. Other studies have detected a positive relationship between clade age and species richness in groups such as turtles [16], geckos [17], and diving beetles [18]. In a landmark meta-analysis study, McPeek and Brown [7] analyzed 163 species-level phylogenies and found a positive relationship between clade age and species richness and concluded that species richness in most clades is explained by clade longevity. More recently, using a time-tree of Eukaryotes, Rabosky et al. [5] found no relationship between clade age and species richness across all multi-cellular organisms. Intriguingly, when Rabosky et al. [5] analyzed 12 major subgroups independently (e.g., gymnosperms, mammals), beetles were the only group that showed a positive clade age-species richness relationship. However, when a more densely sampled coleopteran data set was analyzed this positive clade age species relationship did not hold [5]. The generality of this pattern within Coleoptera remains unknown because there have been few studies explicitly investigating the roles of clade age and diversification rates in determining species richness patterns within diverse beetle groups.

Here we investigate clade age and diversification rate in water scavenger beetles (Hydrophilidae). Hydrophilids are an excellent group for investigating the processes that determine species richness patterns because they are a diverse group with over 3000 described species and a nearly global distribution [19], and they show a huge disparity in species richness across major lineages. As their name implies, many hydrophilid species occupy aquatic habitats, such as small ponds, stream margins, and wetlands. However, hydrophilids have also diversified across a remarkable array of semiaquatic (or "intermediate") habitats including waterfalls and seeps, and terrestrial habitats such as dung, flowers, and forest litter (Figs. 1 and 2). Hydrophilids are thought to have repeatedly transitioned between these varied habitats [19–21], however the relative number and frequency of habitat transitions is largely unknown.

Habitat has previously been show to influence diversification rates in aquatic beetles, although most studies have focused on microhabitat differences such as lotic and lentic environments [22,23]. Macrohabitat transitions in beetles have received far less attention, but studies on other taxonomic groups have demonstrated that macrohabitat and ecological dynamics in general can influence diversification rates [24–29]. Thus transitions between aquatic and terrestrial habitats may play an important role in determining diversity patterns in hydrophilids. Indeed, the largest terrestrial clade (Sphaeridiinae+Rygmodinae) contains approximately 35% of all hydrophilid species, suggesting that a shift between aquatic and terrestrial habitats may have been a key event [sensu 30,31] that promoted an increase in diversification rates. However, investigations of aquatic-terrestrial habitat shifts across major lineages of insects and the role of these habitat transitions in determining diversity patterns remain understudied.

The absence of a comprehensive time-calibrated phylogeny for hydrophilids has precluded efforts to disentangle the evolutionary processes that explain species richness patterns in this group. Here we use an extensive set of recently revised fossil taxa and relaxed molecular clock analyses to estimate diversification times for water scavenger beetles. We integrate our phylogeny with data on species diversity based on detailed taxonomic expertise of the group and data on habitat preferences of particular taxa largely based on our direct observations in the field. We use comparative methods to investigate the roles of clade age and among-lineage diversification rate variation in determining patterns of species richness in water scavenger beetles, and explore the influence of transitions between aquatic and terrestrial habitats in driving diversification rates.

Methods

To determine divergence times for hydrophilids we used a six-gene molecular data set of 151 species that included all major lineages of Hydrophilidae [19], and ran a Bayesian relaxed clock analysis with eight fossil calibrations in the program BEAST v1.7.2 [32]. We used the following fossils to calibrate the tree (see online supplementary material for more details on fossil ages and the calibration schemes): *Protochares brevipalpis* (Late Jurassic, Australia) and *Baissalarva hydrobioides* (Early Cretaceous, Russia) [33] *Hydrobius titan* (actually belonging to the genus *Sperchopsis*, Late Eocene,

	Amphiopini	Berosini	Paracymus group	Laccobius group	Hydrobiusini	Hydrophilini	Chaetarthriini	Enochrini	Acidocerini	Rygmodinae	Omicrini	Coelostomatini	Protosternini	Sphaeridiini	Megasternini	
Lentic	●	●	●	●	●	●	●	●	●			●			●	Aquatic
Lotic		●	●	●	●	●	●	●	●	●		●				
Hygropetric		●	●	●	●		●	●	●	●		●				Semiaquatic
Wet Sand							●									
Phytotelmata							●				●	●			●	
Forest Litter	●						●	●	●	●	●	●	●		●	Terrestrial
Moss			●		●		●					●			●	
Myrmecophilus											●		●		●	
Beach Wrack															●	
Subcortical											●	●	●		●	
Dung										●				●	●	
Floral										●						
Fungi															●	

Figure 1. Distribution of habitat types across the major clades of water scavenger beetles.

Figure 2. A selection of habitat diversity within the Hydrophilidae. A) Lentic habitat in a vegetated marsh, B) Lotic habitat along a mountain stream, C) Coastal dune pond, D) Detrital forest pool, E) *Anacaena* sp. (Chaetarthriinae) inside a drained bromeliad tank, F) *Oocyclus* sp. (Laccobiini: *Laccobius* group) on a wall seepage, G) *Pelosoma* sp. (Megasternini) emerging from a *Heliconia* inflorescence, H) wet sand habitat along a creek, I) *Dactylosternum* sp. (Coelostomatini) under the bark of a freshly cut tree, J) *Chimaearocyon shimadai* (Megasternini) in the brood chamber of *Pheidole* ants, photo credit: Taku Shimada, K) *Rygmodus* sp. (Rygmodinae) on flowers, photo credit Richard Leschen, L) *Nitidulodes* sp. (Megasternini) on an aroid inflorescence. All photos by A.E.Z. Short unless otherwise indicated.

USA; Fikáček et al., unpubl. data); *Limnoxenus olenus* (Latest Oligocene, France) [34]; *Anacaena paleodominica* from Dominican amber (Early Miocene, Dominican Republic) [35], *Helochares* (*Hydrobaticus*) sp. and *Cercyon* sp. from Baltic amber (Eocene, Europe) (Fikáček, unpubl. data and [36]), and *Helophorus paleosibiricus* (Early Cretaceous, Russia) [37]. We used an uncorrelated lognormal tree prior and a birth-death prior for rates of cladogenesis. The dataset was partitioned by gene with partitions unlinked and a GTR model with gamma-distributed rate heterogeneity used for each partition. We ran two analyses for 100 million generations, sampling every 1,000th generation. We verified convergence of parameter estimates and that effective sample sizes were >200 for all parameters using Tracer 1.5 [38]. We combined runs using LogCombiner v1.6.1 [38] and the maximum credibility tree was generated in TreeAnnotator v1.6.1 [38].

To identify shifts in diversification rate we used MEDUSA [39], a comparative method that combines taxonomic and phylogenetic information to fit diversification models using stepwise addition and Akaike Information Criterion (AIC). We accounted for missing species by incorporating our species richness estimates for each major hydrophilid lineage, and pruned the tree to the most terminal clade for which species richness could be confidently estimated (see electronic supplementary material). MEDUSA uses maximum likelihood to fit birth-death, Yule, or a mixed (both birth-death and Yule) diversification models beginning with a single rate model, and using stepwise addition to add models with increasing complexity (i.e. additional rate shifts). Rate shift models are compared using AICc, with more complex models being added until the AICc threshold is no longer met and the single most likely model is selected. Due to the difficulty of estimating extinction rates from molecular data [40], we implemented all three options (birth-death, Yule, and mixed models).

We determined the expected species richness of a clade given a net diversification rate (using background rate from MEDUSA), a relative extinction rate, and clade age [13] using the R package Geiger [41]. We determined the 95% confidence intervals (CI) for models incorporating high ($e = 0.90$) and low ($e = 0.0$) extinction rates. The estimated number of species for clades was plotted with the expected diversity estimates to identify clades that have significantly high or low richness given their respective ages.

We used phylogenetic generalized least-squares regression and standard linear regression to test for a relationship between clade age and log-transformed species richness values using the stem clade ages (some clades were represented by a single representative preventing the use of crown ages) from our hydrophilid time tree and current figures for species richness for each major lineage compiled from literature. It is well known that species-rich lineages of insects harbor much greater diversity than is presently described [42]. To account for this undescribed diversity, we combined our taxonomic expertise to estimate expected species richness values for each major lineage and repeated the analyses (see Table S1 in File S1).

To explore the relative number of transitions between aquatic, semiaquatic and terrestrial habitats we conducted ancestral character reconstruction. Our taxon sampling does not allow us to determine the absolute number of transitions between these macrohabitats, but we can assess where across the entire hydrophilid tree transitions have occurred, and couple this with MEDUSA determined diversification rate shifts to explore a possible relationship between habitat type and diversification rates. We coded aquatic, semiaquatic, and terrestrial habitats as discrete, unordered character states. All character reconstructions were conducted on the maximum clade credibility tree from our BEAST analyses. We used maximum likelihood (ML) in Mesquite v2.6 [43] to reconstruct ancestral character states under the Mk model [44].

Results

We recovered a topology for Hydrophilidae in our relaxed clock analysis that is consistent with Short & Fikáček [19] (Fig. 3, Figures S1, S2, and S3, and Table S2 in File S1). Our divergence time analyses indicate a Late Triassic origin of modern Hydrophilidae (214.1 Ma). Divergence of most major clades (subfamilies and tribes) took place in the Jurassic, with only the two youngest tribes (Megasternini and Sphaeridiini) diverging in the Early Cretaceous (127.5 Ma). Strikingly, one of the youngest clades, Megasternini, is the most speciose major clade of Hydrophilidae.

Our MEDUSA analyses selected a Yule (pure birth) model as the best-fit diversification model. We found a single net diversification rate increase that occurred within the exceptionally diverse Megasternini ($r = 0.053$) relative to the background rate for hydrophilids ($r = 0.032$) (Fig. 4). We did not detect any rate decreases across the hydrophilid tree. However, we found that Amphiopini had fewer species than expected (95% CI) under both high and low extinction rate models (Fig. 4). Meanwhile, Megasternini had higher than expected diversity under a model with low extinction rates, but was within the 95% confidence interval of expected species richness under a high extinction rate.

Using phylogenetic least squares regression and standard linear regression we found no relationship between clade age and (estimated) species richness across the full hydrophilid tree (PGLS $p = 0.603$, linear regression $p = 0.520$, Fig. 4). However, the removal of Megasternini and Amphiopini (see above) from the dataset results in a significant positive relationship between clade age and species richness (PGLS $p = 0.017$, linear regression $p = 0.009$, Fig. 4). We repeated the same analyses using the number of currently described hydrophilids for each major lineage; our major findings were consistent regardless of which species richness values we used (Tables S3, S4, S5 in File S1), therefore we report and discuss the results from the estimated values.

Our ancestral character reconstructions indicate that hydrophilids were ancestrally aquatic (Fig. 5). We infer at least three independent transitions from aquatic to terrestrial habitats, and eight independent transitions from aquatic to intermediate (or semiaquatic) habitats. We infer two secondary returns of terrestrial lineages to aquatic environment. Terrestrial lineages evolved relatively early in the hydrophilid tree, the first instance occurring approximately 171 Ma (Rygmodinae + Sphaeridiinae) and the most recent 80 Ma (*Grodum*-lineage of *Anacaena*). We did not recover any instances where intermediate habitats were a transitional step between aquatic and terrestrial environments.

Discussion

Triassic Origins of the Hydrophilidae

Our results show that the origin of Hydrophilidae (214 Ma) considerably predates the estimate found by Hunt et al. [45] for the entire superfamily Hydrophiloidea (175±23.4 Ma), and is more congruent with age estimates for the Hydrophiloidea found by McKenna & Farrell [46] (229–225 Ma). Our molecular age estimates for Early-Middle Jurassic origins of most major hydrophilid clades are consistent with the fossil record because several modern hydrophiloid families were already well established by the Late Jurassic [21,47] and Hydrophilidae was worldwide in distribution by the Late Jurassic [48]. Thus, our study provides the most robust temporal framework to date for interpreting the diversification of hydrophiloid water beetles.

Is Disparity in Species Richness Explained by Clade Age or Diversification Rate?

Our study suggests that both clade age and among lineage diversification rate differences explain the disparity of species richness among hydrophilid clades. Our MEDUSA analyses revealed a single increase in diversification rate that occurred in Megasternini. We also found that Amphiopini has fewer species that expected (95% CI) under both high and low extinction rate models given a constant diversification rate (Fig. 4), suggesting this group has unusually low diversity for its relatively old age. However, detecting only a single rate increase suggests that diversification rate variation alone does not explain the disparity in species richness. Our regression analyses show that when hydrophilids are analyzed as a whole, there is a positive, but non-significant relationship between clade age and species richness. When Amphiopini and Megasternini are excluded from the regression analyses (PGLS and standard linear regression), the remaining clades show a significant positive relationship between clade age and species richness (Fig. 4). Thus only two clades, Amphiopini, a relatively old species poor clade, and Megasternini, a relatively young species rich clade, account for the lack of a positive relationship between clade age and species richness. The remaining diversity of hydrophilids shows that species richness can be predicted by clade age; older clades have more species due to greater amounts of time to accumulate lineage diversity.

It is not clear whether net diversification rate or clade age is the predominate factor in determining the pervasive disparity of species richness across the tree of life [5–7]. Our study suggests that both factors play a role. However, the majority of diversity within hydrophilids is explained by clade age, supporting the study by McPeek and Brown [7], which showed a strong correlation between clade age and species richness across a wide range of animal taxa. While such studies on beetles are limited to date, there is some evidence that species richness in diving beetles is correlated with clade age [18], suggesting a more general pattern among Coleoptera. When analyzing a time tree representing all Eukaryotes, beetles were one of the very few groups that Rabosky et al. [5] found to exhibit a correlation between clade age and species richness. However, when Rabosky et al. [5] further analyzed beetles using the data set from Hunt et al. [45] they no longer detected a significant relationship between age and

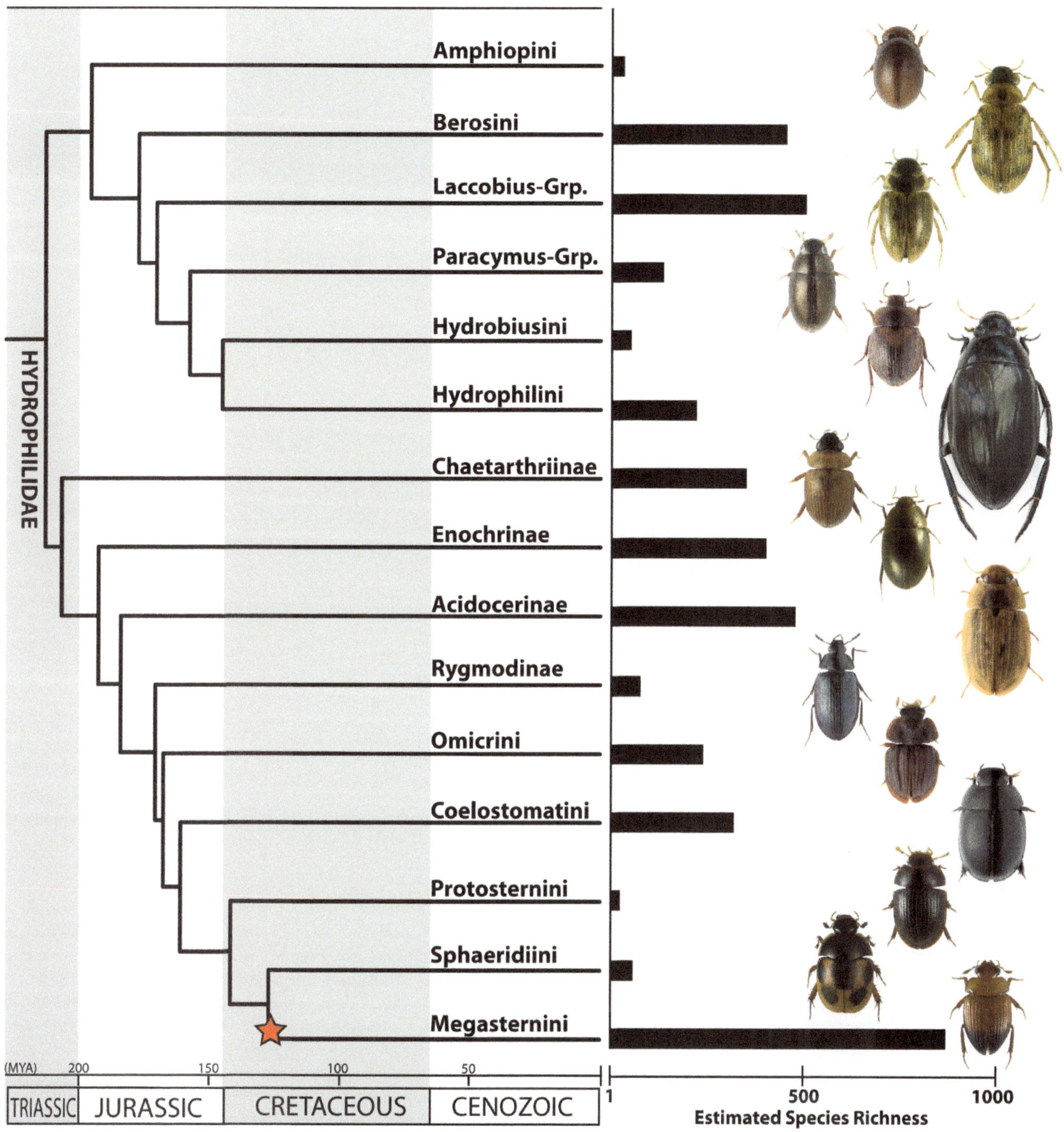

Figure 3. Time-calibrated phylogeny for major lineages of Hydrophilidae, along with estimated species richness values. The star indicates the location of the diversification rate increase determined by our MEDUSA analysis.

richness. It is possible that, much like our results, the latter outcome is driven by a few young exceptionally species rich clades, a few old exceptionally species poor clades, or a combination of the two. If this is the case then beetle diversity may be largely explained by low extinction rates [49] and the ability persisted for a remarkably long time [7,45].

Macroecological Shifts and Diversification Rate

Our results support the hypothesis that water scavenger beetles were ancestrally aquatic and have repeatedly shifted between aquatic and terrestrial habitats [20]. However, our results do not support so-called intermediate (semiaquatic) habitats as a transitional step between terrestrial and aquatic states (Fig. 5). Interestingly, transitions from fully aquatic to semiaquatic habitats seem to occur more frequently (eight transitions) than transitions between fully aquatic and terrestrial habitats (five transitions), or between terrestrial habitats and intermediate habitats (zero transitions). This suggests that some transitions such as aquatic to semiaquatic habitats are relatively easy for hydrophilids, but that there are much stronger constraints [50–53] on other types of

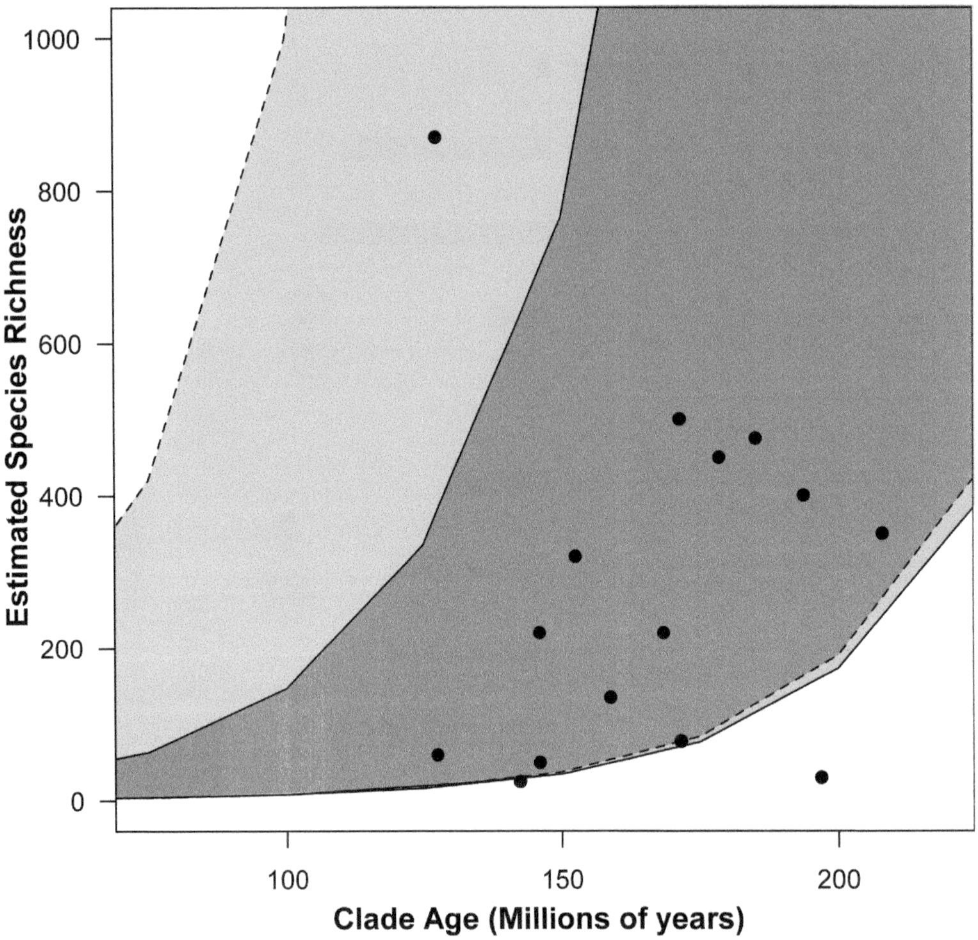

Figure 4. Relationship between clade age and estimated species richness in Hydrophilidae. The shaded regions show 95% confidence interval of the expected diversity under a low (e = 0, solid lines) and high (e = 0.9, dashed lines) extinction.

transitions such as re-invading fully aquatic habitats from either semiaquatic or terrestrial habitats. Ribera [54] argued that lentic habitats (e.g., ponds & lakes) select for generalists that are pre-adapted to invade other niches, whereas lotic habitats (e.g., rivers and streams) select for specialists that are unlikely to undergo habitat transitions. Our results suggest that semiaquatic habitats (e.g. seeps, waterfalls, and phytotelmata) also represent highly specialized adaptive peaks, rather than an intermediate stage between aquatic and terrestrial environment. Once a lineage invades this specialized niche it is difficult leave for other regions of the adaptive landscape [55,56].

We did not detect a diversification rate shift directly associated with any transition between major habitat types (Figs. 3 & 5). Instead, the single rate shift we detected by our MEDUSA analyses followed a major shift from an aquatic-dominant to a terrestrial-dominant lineage by nearly 45 million years (Fig. 5). Furthermore, terrestriality evolved at least two other times (and likely many more in lineages not sampled) in hydrophilid clades that did not experience a diversification rate shift (Fig. 5). This suggests that transitioning into terrestrial habitats did not immediately trigger rapid diversification. However, it is possible that the transition to a terrestrial environment may not immediately spur an increase in diversification, but it sets the stage for a diversification rate shift to occur [57–59] by providing the opportunity for rapid diversification in some clades but not others. This might explain the lag time between the origin of terrestriality and the diversification rate shift and why not all terrestrial lineages experienced a rate shift. Whether a trait or evolutionary transition spurs diversification is contingent on interactions with other organisms, traits and the environment [57]. Thus, invading terrestrial habitats may promote diversification, but only in conjunction with other factors. It is also possible that macrohabitat differences do not directly influence net diversification rates to the degree that the use of microhabitats do. If this is the case, it will be necessary to take a more fine scale approach to delineate which terrestrial niches confer elevated rates of diversification.

Why is Megasternini so Diverse?

With over 540 described species and an estimated 870 species, the Megasternini is remarkably diverse compared to other major hydrophilid clades. This diversity is ultimately the result of an exceptional increase in net diversification rates compared to other hydrophilids (Figs. 3 & 4, supplementary materials). It is difficult to determine if the diversification rate shift is due to an increase in speciation, a decrease in extinction, or both. However, the observed diversity of Megasternini was within the expected diversity for a clade of that age with a high rate of extinction (Fig. 4). One interpretation of this result is that extinction simply has not had enough time to reduce the diversity of this clade (i.e.,

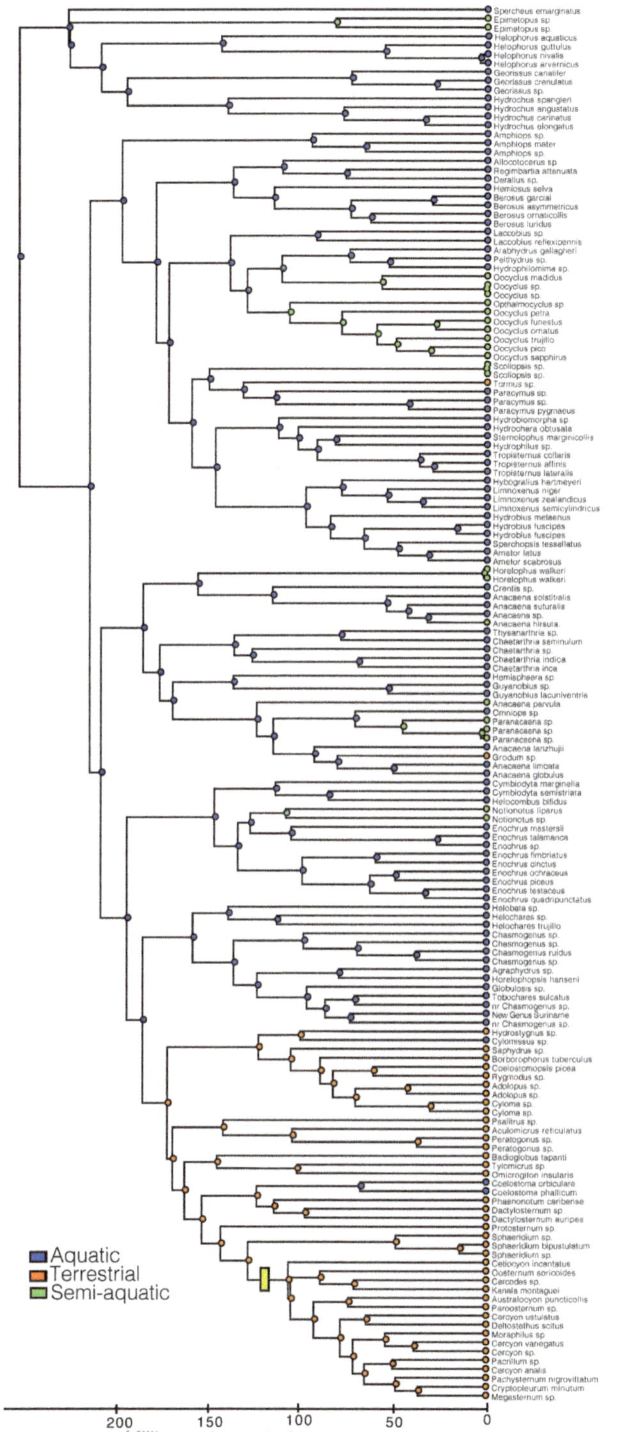

Figure 5. Ancestral character reconstructions of habitat transitions on the time-calibrated phylogeny of Hydrophilidae. The pie charts at nodes indicate maximum likelihood support for ancestral states. A yellow bar indicates the location of the diversification rate shift determined by MEDUSA.

the pull of the present). However, the selection of a (Yule) pure-birth model suggests extinction rates may not be the driving factor, and that the increase in net diversification is due to increased speciation rates.

Though our results do not implicate a transition from aquatic to terrestrial environments as a cause for explosive diversification within the family, the exceptional species diversity of Megasternini may have an ecological explanation. Megasternini occupies a remarkable diversity of niches within the terrestrial environment and are found in a broader array of habitats than most other hydrophilid lineages (Figs. 2 & 5), such as *Heliconia* inflorescences, leaf litter, ant nests, and mammal dung. For example, the tribe is the only lineage to have significant radiations of myrmecophilous or beach wrack taxa [60,61]. As habitat has been shown to play an important role in lineage diversification [22,25,50], it is possible that the increase in net diversification rate in Megasternini (either a decrease in extinction rate, increase in speciation rate, or both) is linked to the broad spectrum of habitats found in this lineage. Habitats can differ in diversification rates due to various parameters that are associated with that habitat such as the relative presence of barriers, corresponding geographic range size, or some other property linked to habitat [25,62–64]. Alternatively, repeated habitat shifts may have circumvented diversity-dependent regulations of clade growth. The primary mechanism behind diversity-dependence is interspecific competition [11]; the basic concept is that closely related species will compete for limited resources, which in turn reduces speciation and increases extinction and results in a characteristic slowing of lineage accumulation over time [11]. It is possible that repeated transitions between microhabitats allowed Megasternini to escape diversity-dependence by repeatedly presenting ecological opportunity for diversification [9,65–70]. Disentangling which of these mechanisms best explains the remarkable species richness in Megasternini will require dense taxonomic sampling of this clade and to utilize diversification models that explicitly estimate speciation and extinction rates for particular habitats (character states) [71], and to fit diversity dependent diversification models [11,72].

Why is Amphiopini Species Poor?

Despite its early Jurassic origin, the Amphiopini is significantly more species-poor than expected under both high and low extinction scenarios given a constant rate (Fig. 4). Our MEDUSA analysis does not detect an exceptional slowdown in diversification rate, and thus begs the question of why the tribe has so few species. One explanation may be that just as the Megasternini exhibits remarkable niche breadth, the ecological diversity within the Amphiopini is atypically narrow for the family. Amphiopini is commonly found in lentic habitats in Africa, Asia, and northern Australia [73–75], and represented by a single species found in leaf litter from Madagascar [76]. Lentic habitats are frequently more isolated and more temporary over evolutionary time scales than lotic habitats [54]; these traits have been shown to convey higher rates of dispersal and lower rates of diversification compared to lotic habitats in other aquatic beetle lineages [22]. Thus Amphiopini's principal specialization in lentic niches may partly explain its relatively low diversity.

Another factor contributing to low species richness in Amphiopini may be its relatively restricted geographic range. Amphiopini is notable in being one of only two major lineages of Hydrophilidae to be absent from the New World. Interestingly, the only tribe that is less diverse (though younger) than the Amphiopini–the Protosternini–also has a relatively restricted geographic range compared to other hydrophilid lineages. It may be that while the diversification rate of Amphiopini has remained similar to the rest of the family, the smaller geographical scale over which it has diversified has limited its absolute species richness.

Acknowledgments

We thank Matt Davis for providing R scripts and analytical advice. Tiffany Schriever provided statistical advice and help with R. Joseph Brown graciously offered guidance using MEDUSA. Richard Glor, Matt Davis, Leo Smith, Daniel Rabosky, and Ignacio Ribera provided critical feedback that greatly improved this study.

Author Contributions

Conceived and designed the experiments: DDB MF AEZS. Performed the experiments: DDB. Analyzed the data: DDB. Contributed reagents/ materials/analysis tools: DDB MF AEZS. Contributed to the writing of the manuscript: DDB MF AEZS.

References

1. Farrell BD (1998) "Inordinate Fondness" Explained: Why Are There So Many Beetles? Science 281: 555–559.
2. Hutchinson GE (1959) Homage to Santa Rosalia or Why Are There So Many Kinds of Animals? The American Naturalist 93: 145–159.
3. Fisher AG (1960) Latitudinal variations in organic diversity. Evolution 14: 64–81.
4. Pyron RA, Burbrink FT (2013) Phylogenetic estimates of speciation and extinction rates for testing ecological and evolutionary hypotheses. Trends in ecology & evolution (Personal edition) 28: 729–736.
5. Rabosky DL, Slater GJ, Alfaro ME (2012) Clade Age and Species Richness Are Decoupled Across the Eukaryotic Tree of Life. PLOS Biol 10: e1001381.
6. Wiens BJohn J (2011) The Causes Of Species Richness Patterns Across Space, Time, And Clades And The Role Of "Ecological Limits". The Quarterly Review of Biology 86: 75–96.
7. McPeek MA, Brown JM (2007) Clade age and not diversification rate explains species richness among animal taxa. The American Naturalist 169: E97–E106.
8. Rabosky DL, Donnellan SC, Talaba AL, Lovette IJ (2007) Exceptional among-lineage variation in diversification rates during the radiation of Australia's most diverse vertebrate clade. Proceedings of the Royal Society B-Biological Sciences 274: 2915–2923.
9. Rabosky DL (2009) Ecological limits and diversification rate: alternative paradigms to explain the variation in species richness among clades and regions. Ecology Letters 12: 735–743.
10. Rabosky DL (2010) Primary Controls on Species Richness in Higher Taxa. Systematic Biology 59: 634–645.
11. Rabosky DL (2013) Diversity-Dependence, Ecological Speciation, and the Role of Competition in Macroevolution. Annual Review of Ecology, Evolution, and Systematics 44: null.
12. Ricklefs RE, Schwarzbach AE, Renner SS (2006) Rate of lineage origin explains the diversity anomaly in the world's mangrove vegetation. American Naturalist 168: 805–810.
13. Magallon S, Sanderson MJ (2001) Absolute diversificatoin rates in Angiosperm clades. Evolution 55: 1762–1780.
14. Ricklefs RE (2003) Global diversification rates of passerine birds. Proceedings of the Royal Society of London Series B: Biological Sciences 270: 2285–2291.
15. Ricklefs RE, Losos JB, Townsend TM (2007) Evolutionary diversification of clades of squamate reptiles. Journal of Evolutionary Biology 20: 1751–1762.
16. Stephens PR, Wiens JJ (2003) Explaining Species Richness from Continents to Communities: The Time-for-Speciation Effect in Emydid Turtles. The American Naturalist 161: 112–128.
17. Gamble T, Bauer AM, Colli GR, Greenbaum E, Jackman TR, et al. (2011) Coming to America: multiple origins of New World geckos. Journal of Evolutionary Biology 24: 231–244.
18. Ribera I, Vogler AP, Balke M (2008) Phylogeny and diversification of diving beetles (Coleoptera: Dytiscidae). Cladistics 24: 563–590.
19. Short AEZ, Fikáček M (2013) Molecular Phylogeny, Evolution, and Classification of the Hydrophilidae (Coleoptera). Systematic Entomology 38: 723–752.
20. Bernhard D, Schmidt C, Korte A, Fritzsch G, Beutel RG (2006) From terrestrial to aquatic habitats and back again – molecular insights into the evolution and phylogeny of Hydrophiloidea (Coleoptera) using multigene analyses. Zoologica Scripta 35: 597–606.
21. Fikáček M, Prokin AA, Angus RB, Ponomarenko AG, Yue Y, et al. (2012) Phylogeny and the fossil record of the Helophoridae reveal Jurassic origin of modern hydrophiloid lineages (Coleoptera: Polyphaga). Systematic Entomology 37: 420–447.
22. Ribera I, Barraclough TG, Vogler AP (2001) The effect of habitat type on speciation rates and range movements in aquatic beetles: inferences from species-level phylogenies. Molecular Ecology 10: 721–735.
23. Ribera I, Foster GN, Vogler AP (2003) Does habitat use explain large scale species richness patterns of aquatic beetles in Europe? Ecography 26: 145–152.
24. Vrba ES (1987) Ecology in relation to speciation rates: some case histories of Miocene-Recent mammal clades. Evolutionary Ecology 1: 283–300.
25. Bloom DD, Weir JT, Piller KR, Lovejoy NR (2013) Do freshwater fishes diversify faster than marine fishes? A test using state-dependent diversification analyses and molecular phylogenetics of New World Silversides (Atherinopsidae). Evolution 67: 2040–2057.
26. Benton MJ (2009) The Red Queen and the Court Jester: Species diversity and the role of biotic and abiotic factors through time. Science 323: 728–732.
27. Alfaro ME, Santini F, Brock CD (2007) Do reefs drive diversification in marine teleosts? Evidence from the pufferfish and their allies (Order Tetraodontiformes). Evolution 61: 2104–2126.
28. Sequeira AS, Lanteri AA, Albelo R, Bhattacharya S, Sijapati M (2008) Colonization history, ecological shifts and diversification in the evolution of endemic Galapagos weevils. Molecular Ecology 17: 1089–1107.
29. McPeek MA (2007) The macroevolutionary consequences of ecological differences among species. Palaeontology 50: 111–129.
30. Moore BR, Donoghue MJ (2009) A Bayesian approach for evaluating the impact of historical events on rates of diversification. Proceedings of the National Academy of Sciences of the United States of America 106: 4307–4312.
31. Donoghue MJ (2008) A phylogenetic perspective on the distribution of plant diversity. Proceedings of the National Academy of Sciences 105: 11549–11555.
32. Drummond AJ, Suchard MA, Xie D, Rambaut A (2012) Bayesian phylogenetics with BEAUti and the BEAST 1.7. Molecular Biology and Evolution.
33. Fikáček M, Prokin A, Yan E, Yue Y, Wang B, et al. (2014) Modern hydrophilid clades present and widespread in the Late Jurassic and Early Cretaceous (Coleoptera: Hydrophiloidea: Hydrophilidae). Zoological Journal of the Linnean Society 170: 710–734.
34. Fikáček M, Prokop J, Nel A (2010) Fossil water scavenger beetles of the subtribe Hydrobiusina (Coleoptera: Hydrophilidae) from the Upper Oligocene locality of Aix-en-Provence. Acta Entomologica Musei Nationalis Pragae 50: 445–458.
35. Fikáček M, Engel MS (2011) An aquatic water scavenger beetle in Early Miocene amber from the Dominican Republic (Coleoptera: Hydrophilidae). Annales Zoologici 61: 621–628.
36. Kubisz D (2000) Fossil beetles (Coleoptera) from Baltic amber in the collection of the Museum of Natural History of ISEA in Kraków. Polish Journal of Entomology 69: 225–230.
37. Fikáček M, Prokin AA, Angus RB, Ponomarenko AG, Yue Y, et al. (2012) Revision of Mesozoic fossils of the helophorid lineage of the superfamily Hydrophiloidea (Coleoptera: Polyphaga). Acta Entomologica Musei Nationalis Pragae 52: 89–127.
38. Drummond AJ, Rambaut A (2007) BEAST: Bayesian evolutionary analysis by sampling trees. BMC Evolutionary Biology 7: 214.
39. Alfaro ME, Santini F, Brock C, Alamillo H, Dornburg A, et al. (2009) Nine exceptional radiations plus high turnover explain species diversity in jawed vertebrates. Proceedings of the National Academy of Sciences of the United States of America 106: 13410–13414.
40. Rabosky DL (2010) Extinction rates should not be estimated from molecular phylogenies. Evolution 6: 1816–1824.
41. Harmon LJ, Weir JT, Brock CD, Glor RE, Challenger W (2008) GEIGER: investigating evolutionary radiations. Bioinformatics 24: 129–131.
42. Hamilton Andrew J, Basset Y, Benke Kurt K, Grimbacher Peter S, Miller Scott E, et al. (2010) Quantifying Uncertainty in Estimation of Tropical Arthropod Species Richness. The American Naturalist 176: 90–95.
43. Maddison WP, Maddison DR (2011) Mesquite: a modular system for evolutionary analysis. 2.75 ed.
44. Pagel M (1999) The maximum likelihood approach to reconstructing ancestral character states of discrete characters on phylogenies. Systematic Biology 48: 612–622.
45. Hunt T, Bergsten J, Levkanicova Z, Papadopoulou A, John OS, et al. (2007) A comprehensive phylogeny of beetles reveals the evolutionary origins of a superradiation. Science 318: 1913–1916.
46. McKenna D, Farrell B (2009) Beetles (Coleoptera). In: Hedges SB, Kumar S, editors. The Timetree of Life: Oxford University Press. 278–289.

47. Prokin AA (2009) New water scavenger beetles (Coleoptera: Hydrophilidae) from the Mesozoic of Mongolia. Paleontological Journal 43: 660–663.
48. Fikáček M, Prokin A, Yan E, Yue Y, Wang B, et al. (in press) Modern hydrophilid clades present and widespread in the Late Jurassic and Early Cretaceous (Coleoptera: Hydrophiloidea: Hydrophilidae). Zoological Journal of the Linnean Society.
49. Labandeira CC, Sepkoski JJ (1993) Insect diversity in the fossil record. Science 261: 310–315.
50. Betancur-R R, Ortí G, Stein AM, Marceniuk AP, Alexander Pyron R (2012) Apparent signal of competition limiting diversification after ecological transitions from marine to freshwater habitats. Ecology Letters 15: 822–830.
51. Bloom DD, Lovejoy NR (2012) Molecular phylogenetics reveals a pattern of biome conservatism in New World anchovies (Family Engraulidae). Journal of Evolutionary Biology 25: 701–715.
52. Crisp MD, Arroyo MTK, Cook LG, Gandolfo MA, Jordan GJ, et al. (2009) Phylogenetic biome conservatism on a global scale. Nature 458: 754–U790.
53. Vermeij GJ, Dudley R (2000) Why are there so few evolutionary transitions between aquatic and terrestrial ecosystems? Biological Journal of the Linnean Society 70: 541–554.
54. Ribera I (2008) Habitat constraints and the generation of diversity in freshwater macroinvertebrates. In: Lancaster J, Briers RA, editors. Aquatic Insects: Challenges to Populations. UK: CAB International. 289–311.
55. Martin CH, Wainwright PC (2013) Multiple Fitness Peaks on the Adaptive Landscape Drive Adaptive Radiation in the Wild. Science 339: 208–211.
56. Svensson EI, Calsbeek R (2012) The Adaptive Landscape in Evolutionary Biology. Oxford: Oxford University Press.
57. de Queiroz A (2002) Contingent Predictability in Evolution: Key Traits and Diversification. Systematic Biology 51: 917–929.
58. Bininda-Emonds ORP, Cardillo M, Jones KE, MacPhee RDE, Beck RMD, et al. (2007) The delayed rise of present-day mammals. Nature 446: 507–512.
59. Wainwright PC, Smith WL, Price SA, Tang KL, Sparks JS, et al. (2012) The Evolution of Pharyngognathy: A Phylogenetic and Functional Appraisal of the Pharyngeal Jaw Key Innovation in Labroid Fishes and Beyond. Systematic Biology 61: 1001–1027.
60. Fikáček M, Short AEZ (2010) A revision of the Neotropical genus Sacosternum Hansen (Coleoptera: Hydrophilidae: Sphaeridiinae). Zootaxa 2538: 1–37.
61. Smetana A (1978) Revision of the subfamily Sphaeridiinae of America north of Mexico (Coleoptera: Hydrophilidae). Memoirs of the Entomological Society of Canada 105: 1–292.
62. Jablonski D (2008) Species selection: Theory and data. Annual Review of Ecology, Evolution, and Systematics 39: 501–524.
63. Jablonski D, Hunt G (2006) Larval ecology, geographic range, and species survivorship in Cretaceous mollusks: Organismic versus species-level explanations. American Naturalist 168: 556–564.
64. Rabosky DL, McCune AR (2010) Reinventing species selection with molecular phylogenies. Trends in Ecology & Evolution 25: 68–74.
65. Rabosky DL (2009) Ecological limits on clade diversification in higher taxa. American Naturalist 173: 662–674.
66. Schluter D (2000) The Ecology of Adaptive Radations: Oxford University Press.
67. Yoder JB, Clancey E, Roches SD, Eastman JM, Gentry L, et al. (2010) Ecological opportunity and the origin of adaptive radiations. Journal of Evolutionary Biology 23: 1581–1596.
68. Burbrink FT, Pyron RA (2010) How Does Ecological Opportunity Influence Rates of Speciation, Extinction, and Morphological Diversification in New World Ratsnakes (Tribe Lampropeltini)? Evolution 64: 934–943.
69. Pyron RA, Burbrink FT (2012) Extinction, ecological opportunity, and the origins of global snake diversity. Evolution 66: 163–178.
70. Parent CE, Crespi BJ (2009) Ecological Opportunity in Adaptive Radiation of Galapagos Endemic Land Snails. The American Naturalist 174: 898–905.
71. Maddison WP, Midford PE, Otto SP (2007) Estimating a binary character's effect on speciation and extinction. Systematic Biology 56: 701–710.
72. Rabosky DL, Lovette IJ (2008) Density-dependent diversification in North American wood warblers. Proceedings of the Royal Society B: Biological Sciences 275: 2363–2371.
73. Minoshima Y, Hayashi M (2012) The first instar larva of Hydrobius pauper Sharp (Coleoptera, Hydrophilidae). Elytra, NS 2: 279–284.
74. Short AEZ, Fikáček M (2011) World catalogue of the Hydrophiloidea (Coleoptera): additions and corrections II (2006–2010). Acta Entomologica Musei Nationalis Pragae 51: 83–122.
75. Watts CHS (1998) Revision of Australian Amphiops Erichson, Allocotocerus Kraatz and Regimbartia Zaitzev (Coleoptera: Hydrophilidae). Records of the South Australian Museum 30: 93–106.
76. Short AEZ (2009) Description of Micramphiops gen. n. from Madagascar (Coleoptera: Hydrophilidae). Koleopterologische Rundschau 79: 189–195.

7

Molecular Phylogenetics and Temporal Diversification in the Genus *Aeromonas* Based on the Sequences of Five Housekeeping Genes

J. Gaspar Lorén[1], Maribel Farfán[1,2]*, M. Carmen Fusté[1,2]

1 Departament de Microbiologia i Parasitologia Sanitàries, Facultat de Farmàcia, Universitat de Barcelona, Barcelona, Spain, **2** Institut de Recerca de la Biodiversitat (IRBio), Universitat de Barcelona, Barcelona, Spain

Abstract

Several approaches have been developed to estimate both the relative and absolute rates of speciation and extinction within clades based on molecular phylogenetic reconstructions of evolutionary relationships, according to an underlying model of diversification. However, the macroevolutionary models established for eukaryotes have scarcely been used with prokaryotes. We have investigated the rate and pattern of cladogenesis in the genus *Aeromonas* (*γ-Proteobacteria*, *Proteobacteria*, Bacteria) using the sequences of five housekeeping genes and an uncorrelated relaxed-clock approach. To our knowledge, until now this analysis has never been applied to all the species described in a bacterial genus and thus opens up the possibility of establishing models of speciation from sequence data commonly used in phylogenetic studies of prokaryotes. Our results suggest that the genus *Aeromonas* began to diverge between 248 and 266 million years ago, exhibiting a constant divergence rate through the Phanerozoic, which could be described as a pure birth process.

Editor: Mark R. Liles, Auburn University, United States of America

Funding: This work was supported by projects from the Ministerio de Educación y Ciencia, Spain [CGL2008-03281/BOS] and from the University of Barcelona [ARZ00F01]. The funders had no role in study design, data collection and analysis, decision to publish, or preparation of the manuscript.

Competing Interests: The authors have declared that no competing interests exist.

* E-mail: mfarfan@ub.edu

Introduction

Speciation is a central topic in evolutionary science and has been the focus of an enormous amount of research, especially during the last 20 years [1–3]. Traditionally, the speciation models and the speciation and extinction rates in a population were determined by analysing the fossil record data, but this is not available for all species, is restricted to the last 600 million years and, in the case of prokaryotes, is scarce and confined to very few taxa. The recent expansion of molecular phylogenetics has provided a useful approach to overcoming this problem.

As tools such as DNA sequencing, genomics and proteomics become feasible for larger samples, it has been possible to analyse diversification patterns from molecular data. Phylogenetic trees inferred from molecular sequences, particularly those including all the living species in a higher taxonomic group, provide an indirect record of speciation events that have led to present day species [1].

Since Nee et al. [4] proposed a method to estimate both speciation and extinction rates of a lineage from phylogenies reconstructed from contemporary taxa, several other methods mainly based on birth-death models have been developed [5]–[7]. In the simplest of these models, the birth and death rates of lineages remain constant through time. However, rates of species origination and extinction can vary over time during evolutionary radiations and among lineages [8], [9]. Therefore, several authors have developed methods to estimate changes in diversification rates through time and across lineages from phylogenetic data of extant species [4], [10], [11]. All these methods have potential applications in the study of speciation and extinction processes in organisms with few or no existent fossil records, such as prokaryotes, although a major problem is the difficulty in estimating divergence times. Phylogenetic trees derived from DNA sequences only contain information about the relative timing of reconstructed speciation events (i.e. branch lengths of these trees represent the evolution rate multiplied by the elapsed time).

Since the seminal papers of Zuckerkandl and Pauling [12] and Kimura [13] molecular dating has been based on the molecular clock hypothesis of a constant chronological rate of sequence change [14]. This approach has been regularly challenged by results obtained using datasets from a variety of organisms, ranging from bacteria to primates, which show considerable departures from clocklike evolution and constant rate variation among lineages, and it has become clear that the strict molecular clock hypothesis is not biologically realistic [15]. This implies that although it is possible to infer phylogenies from molecular sequences, it is not possible to estimate molecular rates or divergence times, because the individual contribution of each one to molecular evolution cannot be separated [15]–[17].

Models that take into account rate variation across lineages have been proposed in order to obtain better estimates of divergence time: the so called 'relaxed molecular clock models'. These models represent an intermediate position between the 'strict' molecular clock hypothesis and the unconstrained models (that do not distinguish times from rates). They include local clocks [18], and nonparametric approaches such as penalized likelihood [19], Bayesian parametric models [15], [20] and a maximum likelihood approach with discrete rate variation (MLA) recently developed by Paradis [21]. Among these, the Bayesian methods allow the use of prior distributions, which quantify the uncertainty in the values of the unknown model parameters before the data are calculated and offer the opportunity of exploring a wide diversity of alternative models, each of which corresponds to specific assumptions concerning the shape of the tree and the way the rate of substitution changes with time [15].

Although prokaryotes represent the majority of living organisms, and dominated the first 80% of the history of life, the macroevolutionary models established for eukaryotes have been scarcely applied to them [22], and the origin of a bacterial lineage or the way in which it has diversified remains largely unexplored. There are only a few references in the literature about bacterial diversification [22]–[24], and in no case has the reported analysis been as complete as those published on higher organisms.

Among the challenges associated with the study of macroevolutionary patterns in microorganisms, one of the most significant is to determine if the diversification rate is constant or varies over time. The limited studies on bacterial macroevolution have been mainly based on pathogenic bacteria, in which diversification rates seem to vary over time [24]. Controversially, the very few studies

Figure 1. *Aeromonas* species maximum likelihood phylogenetic tree. *E.coli* and *S. enterica* were used as the outgroup. Nodes supported by bootstrap values ≥70% are indicated. The scale bar represents 20% sequence divergence.

on free-living or symbiotic bacteria suggest a constant rate of diversification [22], [23].

The genus *Aeromonas* Stanier 1943 [25] is a γ-*Proteobacteria* (*Proteobacteria*, Bacteria) that comprises a group of Gram-negative, rod-shaped bacteria, which are autochthonous to aquatic environments worldwide and are usual microbiota (as well as primary or secondary pathogens) of fish, amphibians and other animals. Some species, mainly *A. caviae*, *A. hydrophila* and *A. veronii* bv. Sobria, are opportunistic pathogens of humans [26]. Hence, *Aeromonas* constitutes a perfect scenario to study the diversification processes in bacteria due to the huge variety of habitats from which its species can be isolated and its combination of free-living bacteria and host-associated strains.

At present a combination of phenotypic, population genetics and phylogenetic studies constitute the best theoretical and practical approach to delineate bacterial and archaeal species [27], which are defined on the basis of phenotypic properties and whole-genome DNA-DNA hybridization (DDH). Each species must have unique phenotypic properties and exhibit more than 70% DDH among strains and 5°C or lower difference of the thermal denaturation midpoint of DNA-DNA heteroduplexes (ΔTm). Studies using both DDH and 16S rRNA gene sequence data illustrate that if two strains show less than 97% of 16S rRNA gene sequence similarity, they are considered separate species [28], [29]. More recent studies have shown that the 70% cut-off point corresponds to 95% of average nucleotide identity (ANI) of the whole genome and 69% of the conserved DNA between strains. With the analysis restricted to the protein-coding portion of the genome, 70% DDH corresponds to 85% ANI or 79% conserved genes [30], [31]. Traditionally, the *Aeromonas* taxonomy has been based on a phenotypical characterization, although some uncertainties have persisted, even after the analysis of a large number of

Table 1. Age estimates (Ma) of all strains and the major clades of *Aeromonas*.

		Bayesian			MLA		
Clade[a]	N	Age	Lower 95% HPD	Upper 95% HPD	Age	Lower 95% CI	Upper 95% CI
All	37	265.7	126.3	425.3	247.8	223.7	270.5
1	34	208.9	98.9	327.7	185.0	167.0	202.3
3	24	181.9	82.8	281.9	159.1	143.7	174.0
4	20	163.6	79.3	256.8	146.6	132.5	160.4
6	11	143.1	68.7	228.6	132.5	119.7	145.0
2	10	169.7	78.4	271.6	150.3	135.6	164.3
5	9	131.6	61.2	206.7	118.0	106.4	129.2
7	4	64.4	21.4	113.1	62.4	56.4	68.1
8	3	134.0	56.7	225.4	122.2	110.3	133.3

[a]clade numbers appear in the MLA chronogram in Figure 2.
Abbreviations: N, clade size; HPD, the highest posterior density interval; CI, confidence interval.

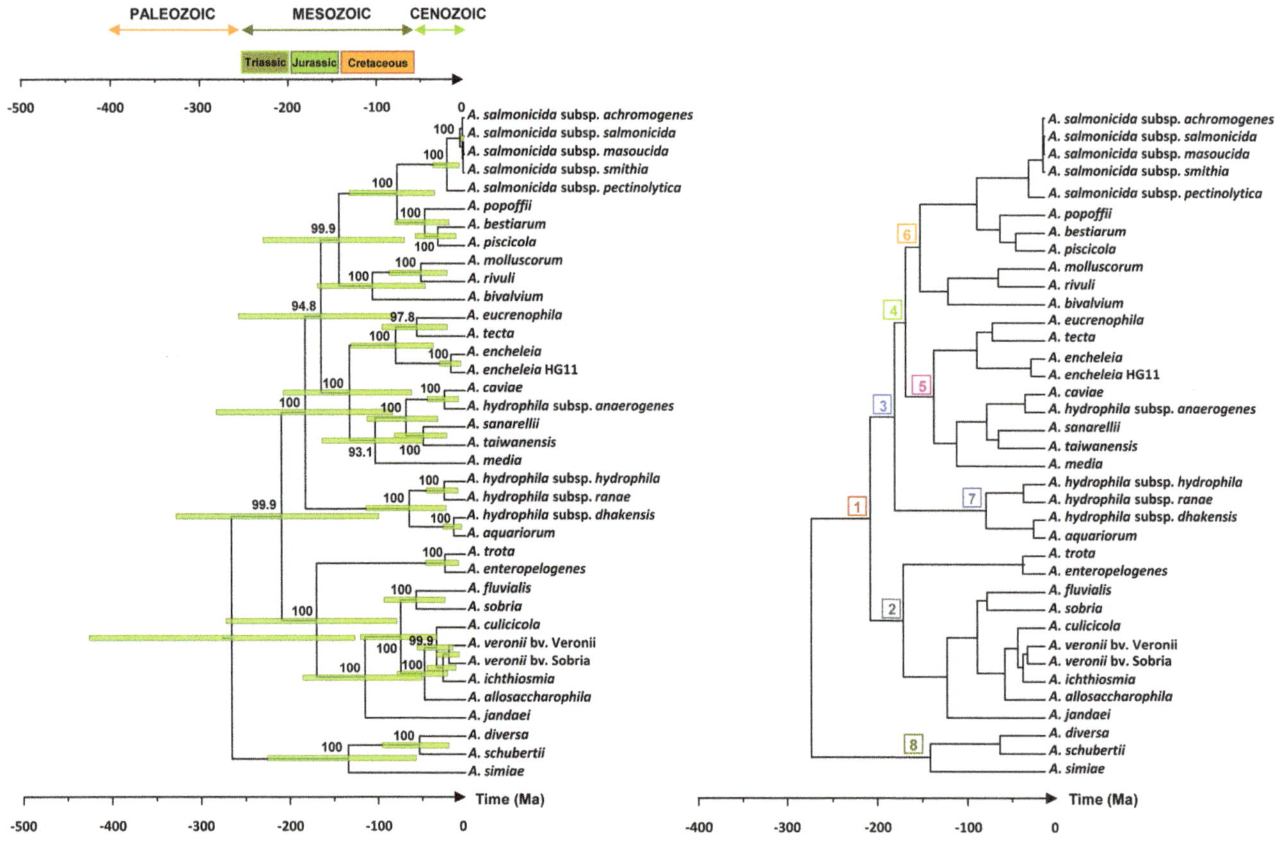

Figure 2. Molecular chronograms of *Aeromonas*. Chronograms were estimated using Bayesian (left) and MLA (right) methods. Bars at the node intersections in the Bayesian chronogram indicate 95% Highest Posterior Density (HPD). Bayesian posterior probability values are shown at the nodes. Time scale is indicated in Mega Annum (Ma). Major *Aeromonas* species clades are indicated by numbers within squares in the MLA chronogram.

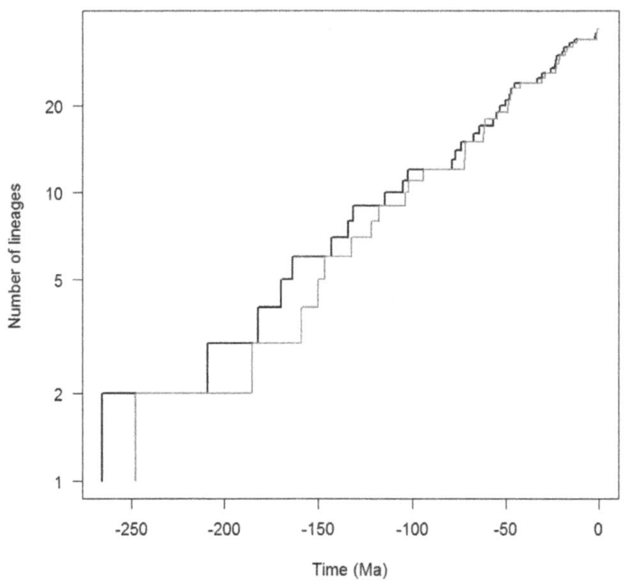

Figure 3. LTT plots for the genus *Aeromonas*. Log-lineage-through-time (LTT) plot for the genus *Aeromonas* based on Bayes (dark line) and MLA (grey line) approaches.

characteristics. To date, most *Aeromonas* species have been taxonomically resolved by phenotypical, molecular and phylogenetic studies. This approach is being widely used in microbial molecular systematics as well as in the phylogenetic analyses of eukaryotic organisms.

We performed a phylogenetic analysis of the genus *Aeromonas* based on the sequences of five housekeeping genes applying Maximum Likelihood (ML) and Bayesian reconstructions and calculated the absolute divergence time by means of Bayesian and Maximum Likelihood Approach (MLA) methods, using the divergence time of *Escherichia coli* and *Salmonella enterica* serovar Typhimurium estimated by Ochman and Wilson [32], [33] as the calibration point. Molecular dating and macro evolutionary birth-death models were used to determine the temporal pattern of lineage diversification and significant changes in diversification rates were estimated using models with constant and variable diversification rates [11], [34]. We evaluated the significance of the gamma statistic, the tree shape and the degree of imbalance as well as the recently developed hypothesis-testing framework that accounts for the possibility that some lineages have not been sampled [35]–[37]. Finally, we discuss our results in light of data reported for macro- and microorganisms, the specific biological characteristics of prokaryotes and current knowledge of macro-diversity through geological time.

Materials and Methods

Data Set

A collection of 37 strains belonging to the genus *Aeromonas* was analyzed, including all species and subspecies recognized to date, and several strains considered synonymous or that have been reclassified. We used only one sequence for each species because the inclusion of more strains of the same species would artificially inflate the number of branching events toward the tip of the trees, producing misleading results [38]. For the analysis that needed outgroup rooting, *Escherichia coli* K12 (GenBank accession number NC000913) and *Salmonella enterica* serovar Typhimurium LT2 (GenBank accession number AE006468) were chosen because, despite belonging to another taxonomic group, they are closely related with *Aeromonas*. Five genes under stabilizing selection for encoded metabolic functions (housekeeping genes), widely used in the phylogeny of *Aeromonas* (*cpn60*, *dnaJ*, *gyrB*, *mdh* and *rpoD*), were selected for the analysis [39–43]. The nucleotide sequences of these genes were determined in our laboratory according to methods previously described or obtained from the GenBank database. All taxa and GenBank accession numbers of the sequences included in this study are listed in the Online Table S1.

Alignment and Phylogenetic Analysis

Sequence data were translated aligned using Clustal X according to the system default parameters and translated back to obtain the nucleotide alignments. The sequences were concatenated with the DAMBE program [44] to be used in posterior analysis. DnaSP software (v.5.10, [45]) was applied to determine the DNA polymorphism data. The best fit models of sequence evolution were implemented according to the Akaike Information Criterion (AIC) scores for substitution models evaluated using jModeltest (v.0.1.1, [46], http://darwin.uvigo.es/software/jmodeltest.html).

Phylogenetic relationships were assessed using Maximum Likelihood (ML) and Bayesian inference. ML analysis was performed with PhyML (v.3.0, [47]). ML tree support was evaluated with 500 bootstrap replicates. The tree from PhyML output was obtained from the website http://www.atgc-montpellier.fr/phyml/and visualized using MEGA (v.5, [48]). The aligned matrix and the ML tree generated in this study are available in TreeBASE (www.treebase.org) Study Accession URL: http://purl.org/phylo/treebase/phylows/study/TB2:S13056.

Both Bayesian reconstruction of phylogeny and molecular dating were determined using BEAST (v.1.6.2, [49]). We performed three independent separate Markov Chain Monte Carlo analyses (MCMC) of 10 million generations each, sampling every thousandth generation. In each case, we used an uncorrelated lognormal relaxed-clock model, with a Yule prior on the tree, a GTR+I+G as a substitution model, the default priors for the relaxed clock parameter and a randomly generated starting tree. The resulting log files were monitored for convergence with the CODA package [50]. Traceplots and effective sampling sizes (EESs) were determined using Tracer (v.1.5, [51]). ESS greater than 200 suggests that MCMC chains were run long enough to obtain a valid estimate of the parameters.

The three BEAST runs were combined using Tracer after a burn-in of 10% of generations and used to estimate the posterior distribution of topologies, the divergence times and other parameter values. Node ages and lower bounds of the 95% highest posterior density intervals for divergence times were calculated using TreeAnnotator (v.1.5.4, http://beast.bio.ed.ac.uk/TreeAnnotator) and visualized using FigTree (v.1.3.1, http://beast.bio.ed.ac.uk/FigTree).

In order to determine the maximum credibility tree, we used the 10,000 posterior trees obtained in each run. After discarding the first 5,000 in each case, we sampled 3,000 trees at random. These were combined and from the 9,000 trees obtained we determined the maximum clade credibility tree with a posterior probability limit of 0.5. The outgroup (*E. coli* and *S. enterica*) was pruned.

Divergent Time Estimations

To test if the sequences evolved in a clock-like manner, we used a clock and a non clock model analysis implemented in BASEML (part of the PAML4 package, [52]). The likelihood values obtained for both models were then compared by a likelihood ratio test

Table 2. Fit of alternative diversity models to LTT plots derived from Bayes and MLA chronograms.

	AIC[a]							ΔAIC$_{RC}$ test[b]			
Method	**Yule**	**Birth-Death**	**Best constant model**	**DDL**	**DDX**	**Yule 2 rates**	**Best variable model**	**ΔAIC$_{RC}$**	***P* value[e]**	**Best model**	**λ (ML) ± se[f]**
Bayes	190.64	192.64	Yule	193.06	192.60	193.28[c]	DDX	−1.2668	0.6359	Yule	0.0119±0.0020
MLA	185.06	187.07	Yule	187.15	186.98	187.17[d]	DDX	−1.9119	0.8928	Yule	0.0129±0.0012

[a]Akaike Information Criterium.
[b]ΔAIC$_{RC}$ test. See text for details.
[c]breakpoint at 13 Ma ago.
[d]breakpoint at 1.3 Ma ago.
[e]*P* value obtained by simulation (5,000 iterations). See text for additional explanation of simulations.
[f]standard error.

(LRT), with LRT = 2 ($L_{clock} - L_{no\ clock}$) and assuming that this statistic was distributed as a χ^2 with n−2 degrees of freedom, where n is the number of taxa in our data set. As the LRT test rejected the strict clock model, relative branching times were estimated using two different approaches: a maximum likelihood with discrete rate variation (MLA) implemented in the R package *ape* v. 3.0–7 [10] and the Bayesian uncorrelated relaxed-clock analysis (Bayes) implemented in BEAST. For the Bayesian analysis, the absolute divergence times were calculated indirectly using the estimated divergence time between *E. coli* and *S. enterica* (140±20 Ma, [32]). As we used an indirect approach to determine the calibration point, we applied a normal distribution as the prior with a mean of 140 Ma and a standard deviation of 10 Ma, providing a prior range of 116.7–163.2 Ma (99% CI). For the maximum likelihood approach (MLA), the ML tree obtained with PhyML was converted to ultrametric using the *chronos* function [21] of the R package *ape* and dated according to the divergence time estimated from *E. coli* and *S. enterica*. This method assumes a discrete variation in rates, so it is possible to categorize branches according to the different rates (we used 10 branch categories). The method calculates the contribution of each branch to the maximum likelihood function by summing the contribution of each rate category weighted by its frequency [21]. We also performed a simulation to find the optimal smoothing parameter (lambda) corresponding to our data, with smoothing values ranging from 10^{-6} to 10^6 with increments of 10, using the same *chronos* function.

Aeromonas Diversification Rates

To visualize the temporal pattern of lineage diversification in *Aeromonas* we performed a semilogarithm lineage-through-time plot (LTT plot) with the R packages *ape* and LASER (v.2.3; [11], [53]

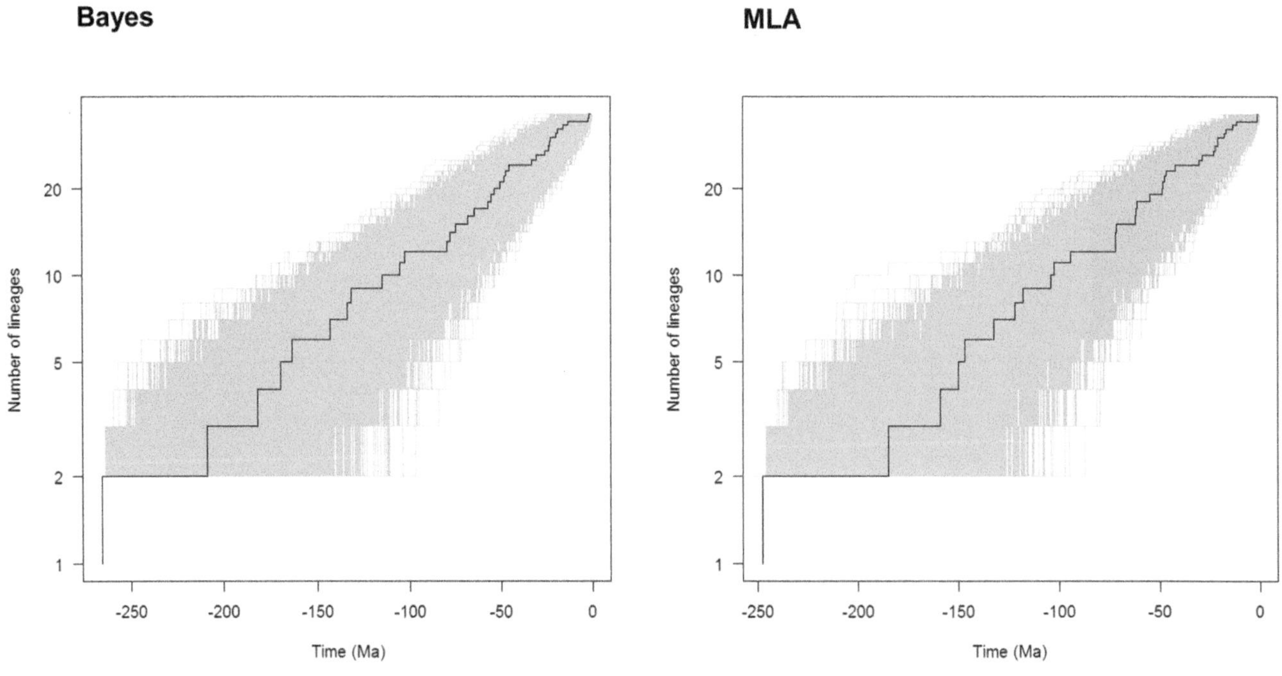

Figure 4. Testing departure of the empirical chronograms of *Aeromonas* from a constant rate diversification model. Dark lines represent the LTT plots obtained for empirical Bayesian (left) and MLA (right) *Aeromonas* phylogenies, while grey lines correspond to the LTT plots of 5,000 simulated phylogenies. In both cases, the root was rescaled to the time to the most recent common ancestor.

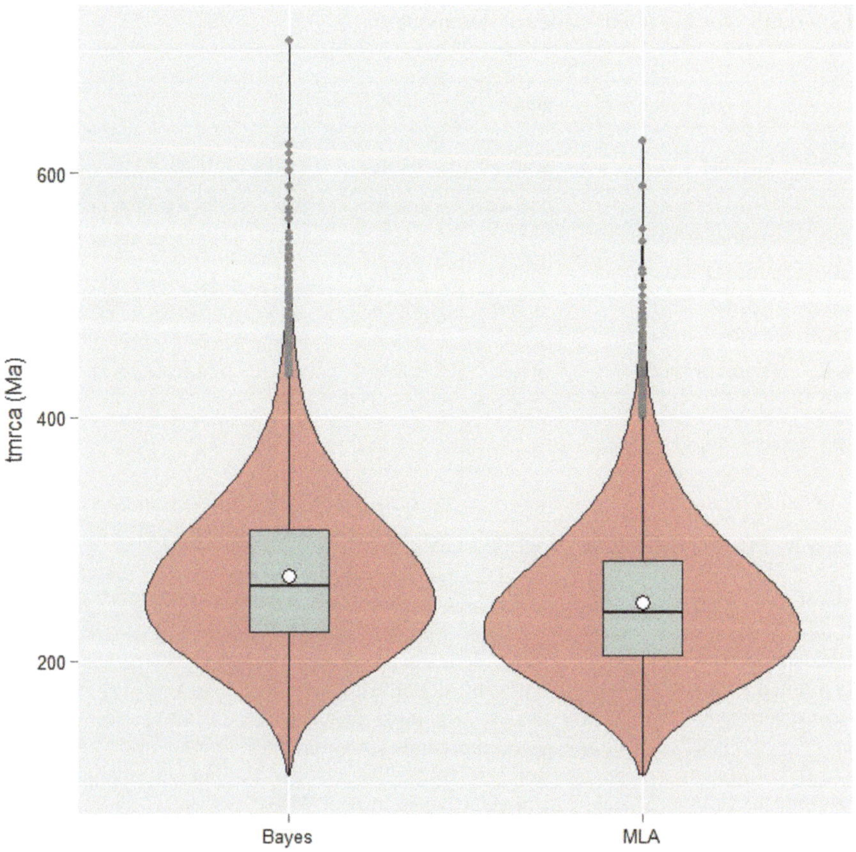

Figure 5. Violin plots of the tmrca distribution for Bayes and MLA simulated trees. The plots show the kernel density estimation of the data (mirrored curves) with a box and whiskers plot overlaid. The plots are scaled so each one has the same total area. The internal box plots indicate the range (whiskers), interquartile range (boxes), median (horizontal black lines) and mean (circles) of the data. The grey points on the plot tails indicate the outliers. See text for more details.

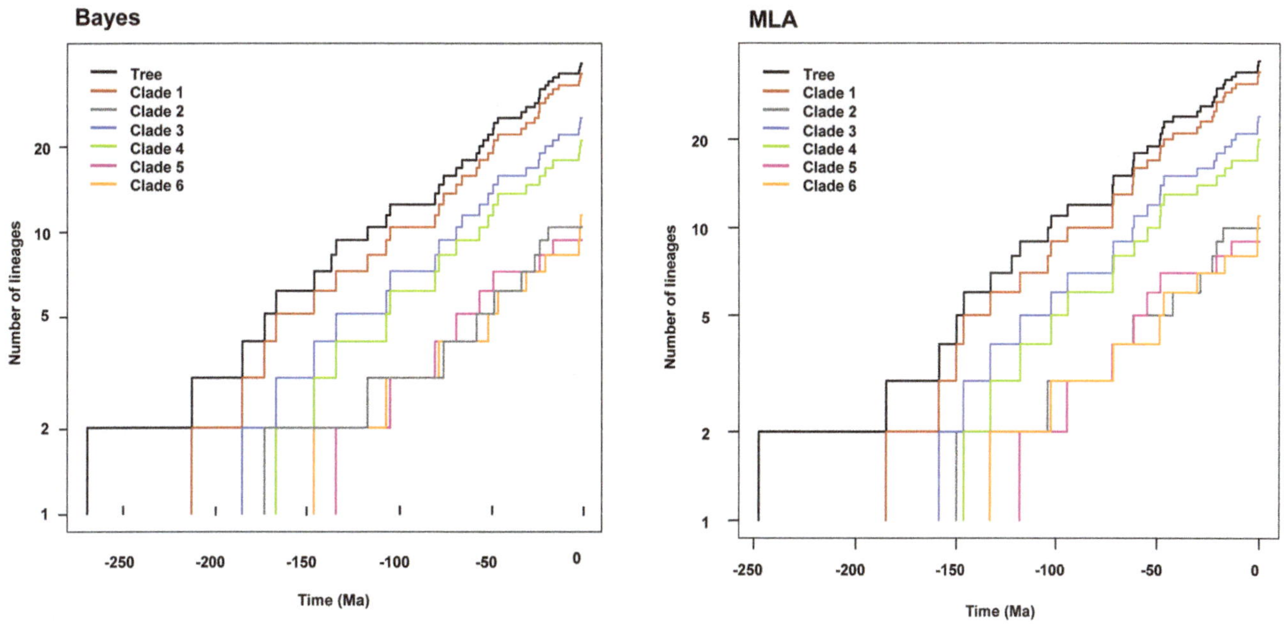

Figure 6. LTT plots for the major clades of *Aeromonas*. Log-lineage-through-time (LTT) plot for the six major clades inferred by Bayesian (left) and MLA (right) analysis.

Table 3. Diversification rates and model of speciation for the major clades of *Aeromonas*.

		Bayesian				MLA			
Clade[a]	N	ΔAIC_{RC}	*P* value	Best model	λ (ML) ± se	ΔAIC_{RC}	*P* value	Best model	λ (ML) ± se
1	34	−1.1976	0.6126	Yule	0.0132±0.0023	−1.2177	0.6031	Yule	0.0143±0.0018
2	10	−1.6910	0.6538	Yule	0.0109±0.0038	−1.5032	0.5467	Yule	0.0115±0.0088
3	24	−1.7612	0.7849	Yule	0.0135±0.0029	−1.1437	0.5377	Yule	0.0147±0.0022
4	20	−1.4868	0.6241	Yule	0.0136±0.0032	−0.5088	0.3563	Yule	0.0146±0.0025
5	9	0.5684	0.2242	Yule	0.0107±0.0040	1.1205	0.2543	Yule	0.0116±0.0031
6	11	0.3735	0.2253	Yule	0.0145±0.0048	1.6440	0.1270	Yule	0.0154±0.0053

[a]clade numbers appear in the MLA chronogram in Figure 2.
Clades 7 and 8 were not analysed due to their low number of species (Fig. 2).
Abbreviations: N, clade size; se, standard error.

using the ultrametric trees obtained from our data by MLA and Bayesian methods.

Three different approaches were used to test significant changes in the diversification rates. First, we used LASER to perform a ML analysis to test whether diversification rates have changed over time, contrasting the likelihood values of the data (branching times derived from the trees) under models with constant diversification rates with those obtained under models where rates varied through time, to detect temporal shifts in diversification in the phylogeny. In this analysis, we included two models with a constant speciation rate (λ), a pure birth model with a $\lambda > 0$ and extinction rate $\mu = 0$ (the Yule process), and a birth-death model with $\lambda > 0$ and $\mu > 0$. Three rate-variable models were also considered: two multi-rate variants of the Yule model and Yule 2-rates that assume the existence of two breakpoints in time, in which λ reaches different values before and after the breakpoints; a logistic density-dependent speciation rate model (DDL) under which λ at time t is modelled as $\lambda(t) = \lambda_0 (1 - N_t/K)$ where λ_0 is the initial speciation rate, N_t the number of lineages at time t and K is a constant analogous to the carrying capacity parameter of population ecology; and an exponential density-dependent model (DDX), in which $\lambda(t) = \lambda_0 N_t^{-x}$, and x controls the value of the rate change in the number of lineages at any point in time. The ΔAIC_{RC} test was used to statistically evaluate the fit of the temporal pattern of lineage diversification in *Aeromonas* to this set of rate-constant and rate-variable models. It was computed as $\Delta AIC_{RC} = AIC_{RC} - AIC_{RV}$ where AIC_{RC} is the AIC score for the best-fit rate-constant model of diversification, and AIC_{RV} is the AIC for the best-fit rate-variable model under consideration. A negative ΔAIC_{RC} value suggests that data are best approximated by a rate-constant model of diversification [11], [34].

Secondly, we determined the gamma (γ) statistic [36] as implemented in LASER from the Bayesian and MLA chronograms. This statistic compares the relative position of nodes in a phylogeny to those expected under a constant diversification rate model, in which the statistic follows a standard normal distribution. The significance of γ was also determined by calculating its value in 5,000 simulated phylogenies obtained under the Yule model of speciation with the same size and diversification rate as those obtained from our data. Phylogenies were simulated using the R package TreeSim [54]. Broadly, positive values of γ signify that nodes are closer to the tips than what is expected under the constant rate model, while negative values might indicate an apparent deceleration [37]. The analysis of diversification shifts using the γ statistic may produce results biased toward negative values if all taxa of the group are not included in the phylogeny. To overcome this problem, Pybus and Harvey [36] developed the Monte Carlo Constant Rates test (MCCR test), which conducts γ statistic analysis for incompletely sampled phylogenies and estimates the significance of negative values of γ, taking into account a possible undersampling in the phylogeny [37]. Although in this analysis we have included all the known species and subspecies of the genus *Aeromonas*, new species are likely to be described in the future. Therefore, to compute the significance of our γ estimates (Bayesian and MLA), we have conducted the MCCR test implemented in LASER. 5,000 phylogenies were simulated with various clade sizes under the Yule model diversification process. Taxa were randomly pruned from the tree to mimic incomplete sampling. The null distribution of the γ statistic was then calculated from these phylogenies and compared with the observed empirical γ.

The shape of a phylogenetic tree contains useful information about the process of cladogenesis. Measuring the degree of imbalance or asymmetry of a tree topology may provide support for the hypothesis that species have the same or different potential for speciation. Under the Yule model, each extant species is equally likely to split into two daughter-species. Several statistics have been introduced for assessing the level of asymmetry of a tree. These statistics are often used to test whether the tree topology differs significantly from a null model with a constant rate of speciation, commonly the Yule model, in which each external branch on a rooted tree has an equal probability of splitting [55]. We have applied two tests that measure the balance of the tree: Colless' index (I_c) [56] and the number of cherries (C_n) [35].

Among the various alternative statistics that measure the balance of phylogenetic trees, I_c is simple, intuitive, and powerful [55], [57]. It computes the sum of absolute values $|L-R|$ at each node of the tree, where L and R are the size of the left and right daughter clades, respectively. This sum is often renormalized by dividing it by its maximum possible value: $(n-1)(n-2)/2$, n being the number of leaves of a tree. Therefore, this statistic varies between 0 and 1: for a completely balanced tree, I_c equals zero, while a value of one indicates that the tree is completely imbalanced.

The mean and standard deviation of Colless' index under the null hypothesis of Yule trees have been computed by Blum et al. [58]. We applied a Colless test based on a Monte Carlo estimate of the *P* value from quantiles of replicate trees generated under the Yule model. We used the R package apTreeshape (v.1, [59]) to

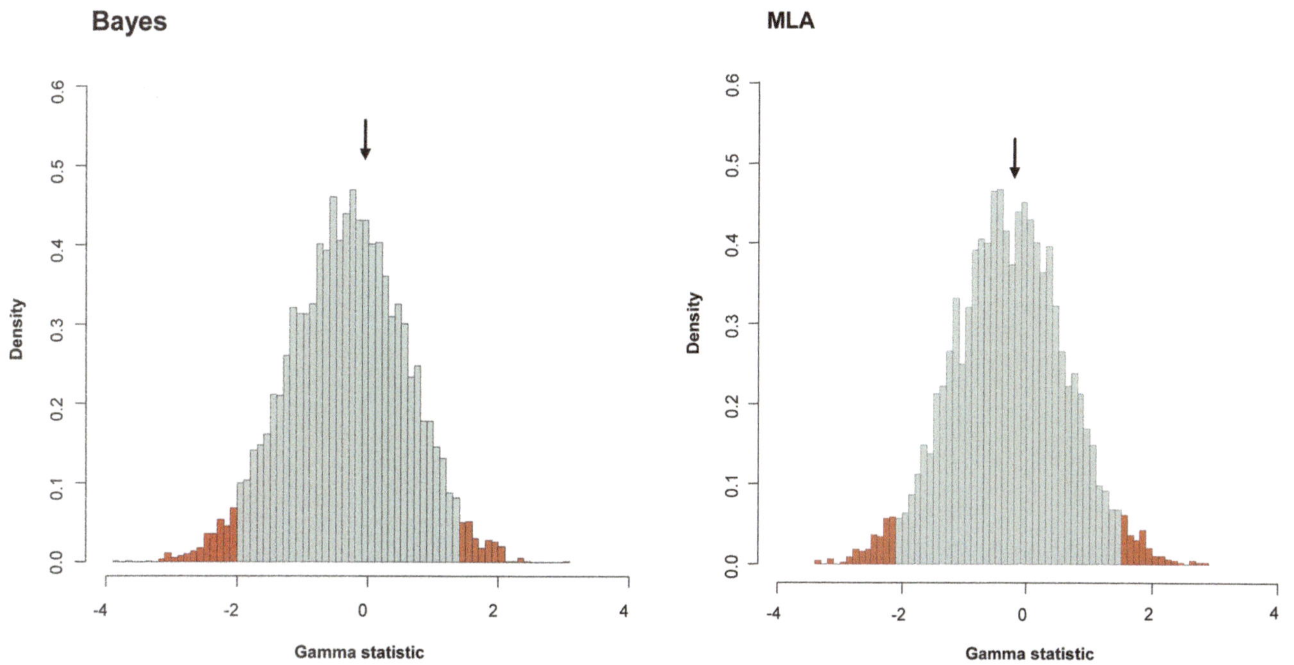

Figure 7. Gamma statistic distribution. Gamma statistic distribution was obtained by simulating 5,000 phylogenies under a Yule model using Bayes (left) and MLA (right) approaches. The arrows indicate the gamma value obtained from our sequences. Red bars indicate the 95% limits of the distribution.

compute the normalized I_c in order to check the balance of our phylogenies and compare them with the Yule model.

McKenzie and Steel [35] considered a simple and easily computed statistic for evaluating the tree shape: the number of cherries of a tree (C_n, where n is the number of tips in a tree). They defined a cherry as a pair of leaves that are adjacent to a common ancestor node. The authors analyzed the distribution of this statistic under the Yule model, and calculated the mean ($E[C_n] = n/3$) and variance ($Var[C_n] = 2n/45$) of the number of cherries. We used the R package *ape* to determine the number of cherries in our phylogenies and then compared the values with those calculated theoretically and by simulation, in order to test the rate of homogeneity across clades.

The remaining analyses and graphs not specified in this section were done in the R environment (R Development Core Team, [60], http://www.r-project.org/) using the packages base, *ape*, LASER and TreeSim.

Table 4. Monte Carlo Constant Rates (MCCR) test.

	Bayesian chronogram[a]			MLA chronogram[b]		
N[c]	Prob.[d]	Critical value	*P* value	Prob.[d]	Critical value	*P* value
50	0.8838	−1.9198	0.6151	0.8762	−1.8759	0.5409
60	0.9245	−2.2675	0.7345	0.9185	−2.2560	0.6788
70	0.9509	−2.4945	0.8314	0.9463	−2.5104	0.7904

[a]$\lambda = 0.0119$ and $t = 265.7$ Ma.
[b]$\lambda = 0.0129$ and $t = 247.7$ Ma.
[c]hypothetical clade size.
[d]probability that *Aeromonas* has N or less species according to the Yule model.

Results

Data Set and Phylogenetic Analysis

The analysis involved 37 *Aeromonas* strains in which we determined the gene sequence of five housekeeping genes (*cpn*60, *dnaJ*, *gyrB*, *mdh* and *rpoD*). The number of total positions in the concatenated sequences was 3,774 bp, with a proportion of 2,140 invariable sites and 1,634 polymorphic sites, 1,423 of which were parsimony informative. The average identity among the concatenated sequences of the *Aeromonas* species was 90.0% (ranging from 86.0 to 91.2%), while the average identity between the *Aeromonas* species and the outgroups *E. coli* and *S. enterica* was 73.9% and 73.2%, respectively. All positions containing gaps and missing data were eliminated in the construction of the different trees. The best model selected for the concatenated sequences was the General Time Reversible (GTR) using a discrete Gamma distribution and a fraction of invariable sites (GTR+G+I). The mean distance between species was 10.0% ±2.6. The uncorrected divergence with the same concatenated sequence between *E. coli* and *Salmonella* was 11.2%. The mean distance between *Aeromonas* species and *E. coli* was 26.1% ±0.5, and 26.8% ±0.5 between *Aeromonas* species and *Salmonella*. Figure 1 shows the *Aeromonas* ML phylogeny, in which the bootstrap support was higher than 70% for the majority of clades. We also performed other phylogenetic reconstructions (Bayesian, Neighbor-Joining, Minimum-Evolution), which gave identical topologies (data not shown).

Divergence Time Estimations

To determine if our sequences evolved in a clock-like manner we applied two models, a clock and a non-clock model analysis implemented in BASEML. The results obtained shown that our data do not support an assumption of a strict molecular clock model ($\chi^2 = 405.3$; d.f. = 35; $P << 0.001$). As the LRT test rejected the strict clock model, we used the Bayesian and MLA approaches to estimate the relative branching times.

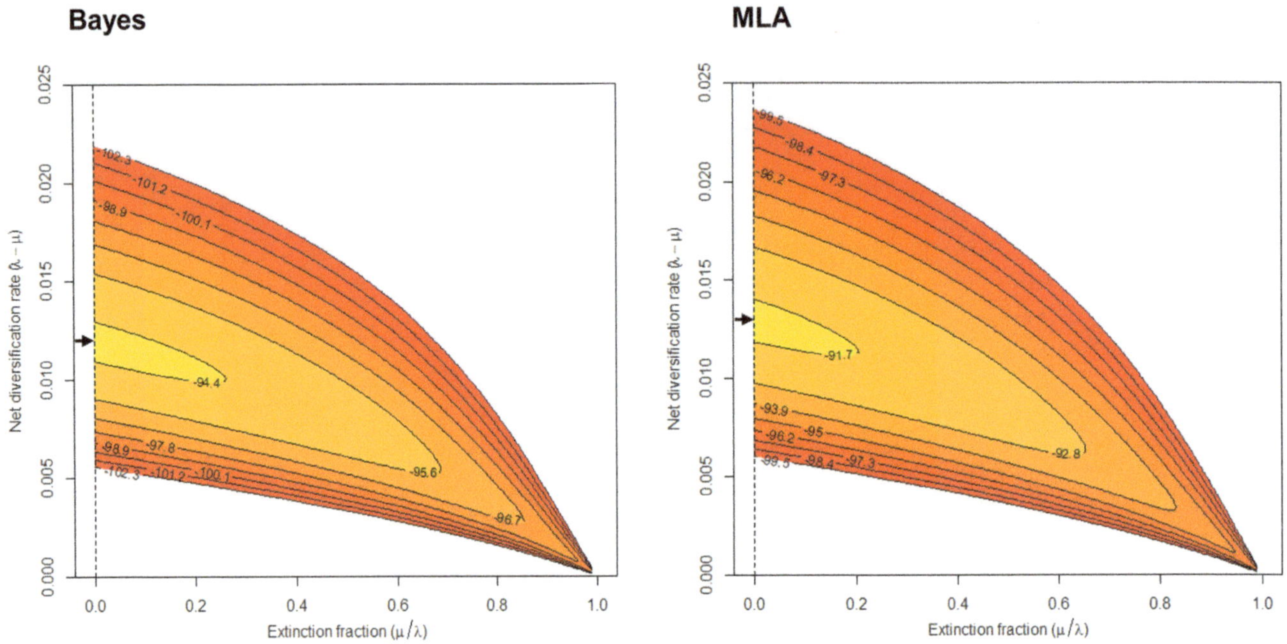

Figure 8. Contour plots of the likelihood surfaces. Contour plots of the likelihood surface were inferred from the relation between the net diversification rate (λ – μ) and extinction fraction (μ/λ) for the Bayes (left) and MLA (right) chronograms. Likelihoods were calculated using the R package LASER. The maximum likelihood estimates (i.e., the peak of the surface) are marked with an arrow.

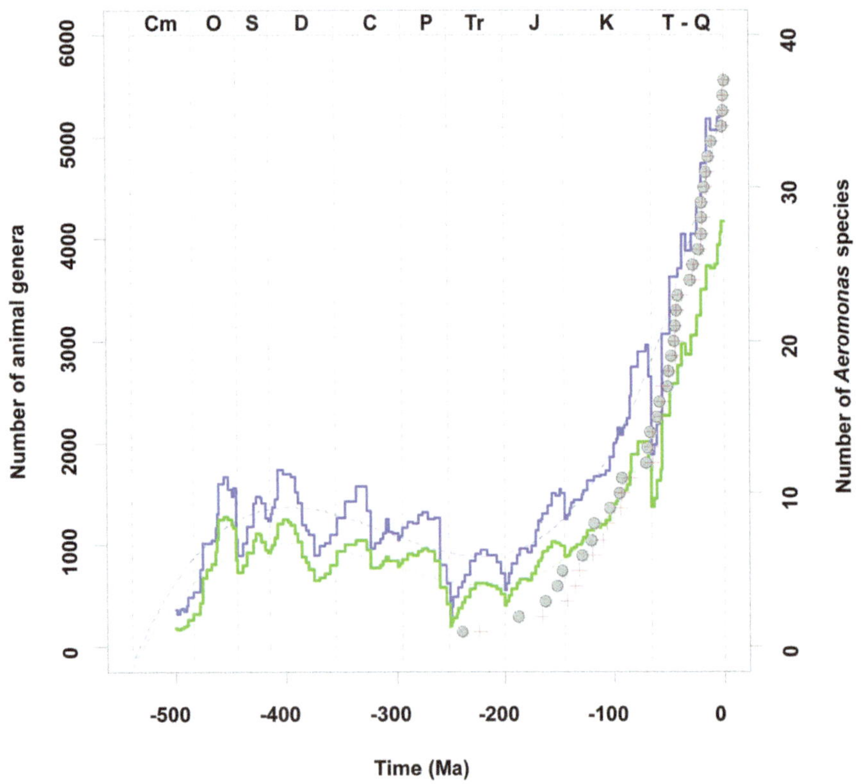

Figure 9. Diversification of *Aeromonas* compared with the number of animal genera versus time. Diversification of *Aeromonas* determined by Bayesian (circles) and MLA (crosses) approaches, compared with the number of animal genera versus time obtained from Sepkoski's compendium, converted to the 2004 Geologic Time Scale [95], [96]. The blue line represents the total number of animal genera while the green line shows the same data with single occurrence and poorly dated genera removed. Dashed lines indicate polynomial fits to data. Abbreviations: Cm; Cambrian, O; Ordovician, S; Silurian, D; Devonian, C; Carboniferous, P; Permian, Tr; Triassic, J; Jurassic, K; Cretaceous, and T-Q; Tertiary-Quaternary.

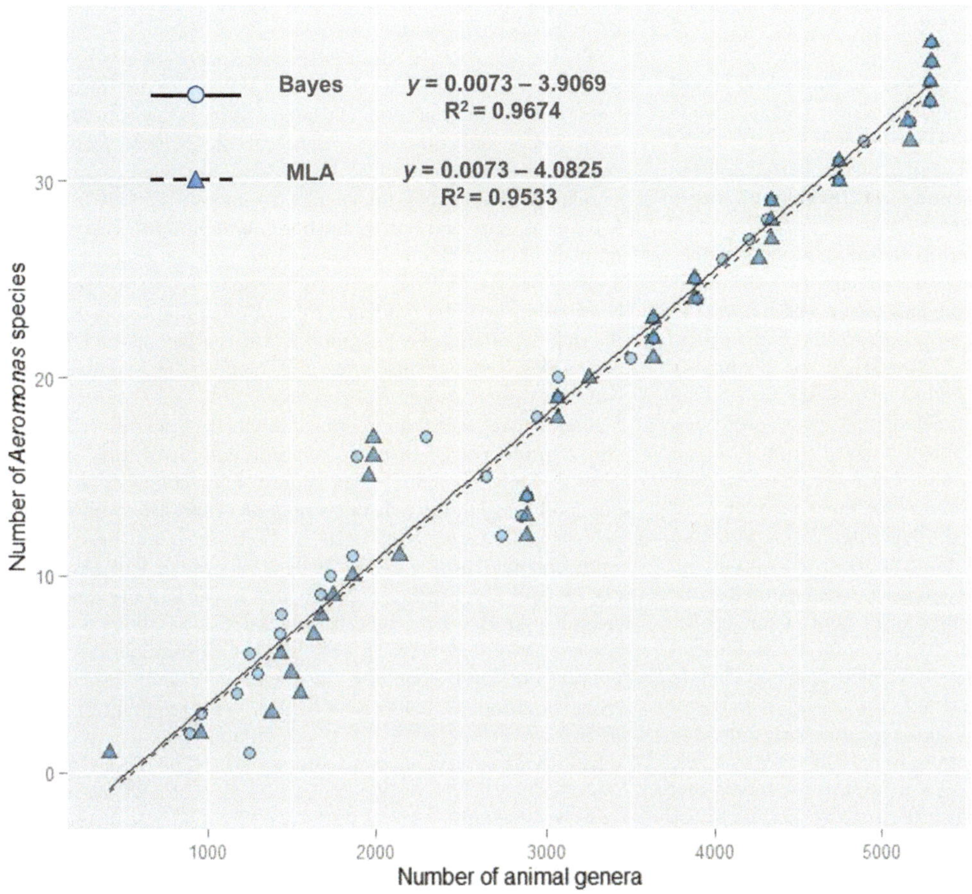

Figure 10. Regression plot between the number of animal genera and the number of *Aeromonas* species. The plot shows the number of animal genera obtained from Sepkoski's data and the number of *Aeromonas* species in the last 250 Ma, applying both the Bayesian and the MLA approaches.

Figure 2 shows the chronograms obtained by Bayesian and MLA analyses. In both trees, all clades were coincident and well supported. Age estimates from Bayesian and MLA chronograms were similar. Table 1 show the divergence times for the major clades obtained in the chronograms. Our estimates for the origin of the genus *Aeromonas* ranged from about 248 to 266 Ma ago, depending on the chronogram construction method. Our molecular data suggest that *Aeromonas* diversification began approximately 250 Ma ago and was completed during the last 50 Ma (Fig. 2).

Aeromonas Diversification Rates

The expected number of lineages versus time (LTT plots) is widely used to characterize clade diversification as a function of time [4], [61]. The semi-logarithmic LTT plots derived from the Bayesian and MLA chronograms are shown in Figure 3. A simple inspection of these plots reveals that the *Aeromonas* lineage accumulation through time appears as a straight line with stochastic fluctuations, which suggests a constant diversification rate. Moreover, the plots do not exhibit any abrupt changes that would suggest the existence of a clear "push of the past" or a clear "pull of the present", which would be expected if there had been a relatively high extinction rate [61].

To confirm if the diversification rate is really constant or has changed over time, we used maximum likelihood to fit the branching times derived from our chronograms to a variety of diversification models [34], [53]. As suggested by Rabosky, we calculated the significance of ΔAIC_{RC} for the set of analyzed models by simulating 5,000 phylogenies of the same size and diversification rate as those obtained from our data under the Yule model and calculating the *P* value from the resulting distributions. As can be seen in Table 2, in both analyses (Bayesian and MLA) the null hypothesis of a Yule model cannot be rejected to a level of significance of $\alpha = 0.05$. In concordance with the differences in divergence times obtained in the Bayesian (265.7 Ma) and MLA (247.7 Ma) chronograms, we also observed a slight difference in the diversification rates obtained with the Bayesian ($\lambda_{Bayes} = 0.0119$) and MLA analysis ($\lambda_{MLA} = 0.0129$). In conclusion, these results suggest that a Yule model of diversification provides the best fit for our data (Table 2).

To corroborate this conclusion, we compared our LTT plots with those obtained from 5,000 simulated trees under a Yule process with the same size and diversification rate, rescaling the root to the time to the most recent common ancestor. Figure 4 shows that the *Aeromonas* LTT plot (dark line) lays within the range of the simulated phylogenies (grey lines).

To verify that there is not an increase in the diversification rate toward the present in our LTT plots (pull of the present), we fit a Yule model to temporal windows that include the last 200, 100, 50 and 25 million years using the R package LASER. The results obtained supplied respective diversification rates of 0.0122,

0.0116, 0.0113 and 0.0123 for the Bayes chronogram and 0.0134, 0.0124, 0.0129 and 0.0137 for the MLA chronogram. Additionally, to corroborate the absence of a decrease in the diversification rate deep in the phylogeny (push of the past), we fit the Yule model to the first 100 and 150 million years, the results obtained being 0.0119 and 0.0127 for the Bayes and 0.0118 and 0.0139 for the MLA chronograms. In this case, we were unable to fit the model to a smaller window due to the low number of speciation events in this period of time.

To study the distribution of time to the most recent ancestor (tmrca), we simulated 9,000 phylogenies under the Yule model assuming that *Aeromonas* is a monophyletic group and follows this diversification pattern. The only constraints were the diversification rates (0.0119 for Bayes and 0.0129 for MLA) and the number of species (37) of our phylogenies. Figure 5 shows the distribution of tmrca for Bayes and MLA simulated phylogenies (BSP and MLASP, respectively). The tmrca mean obtained from BSP was 268.3 Ma (95% CI: 164.1–422.5) and 246.7 Ma from MLASP (95% CI: 148.9–385.8 Ma). The divergence times calculated for the genus *Aeromonas* (265.7 Ma from Bayes and 247.7 Ma from MLA chronograms) fall within the limits of these simulations.

Another important question to consider when verifying the suitability of a diversification model for our data is the determination of the diversification rate variation across lineages in the phylogeny. Figure 6 shows the LTT plots corresponding to the six major clades, which exhibit the same pattern of diversification as the entire tree, and each one fits well to a straight line parallel to that of the entire tree. In all cases the Yule model was selected by Rabosky's ΔAIC_{RC} test as the model that best describes the data (Table 3). The range of diversification rates obtained for this analysis was 0.0107–0.0145 for the Bayes chronogram, and 0.0115–0.0154 for the MLA chronogram. These values are in good agreement with the rates obtained when considering all the analyzed species (0.0119 for Bayes and 0.0129 for MLA; Table 2). Thus, we can conclude that there is little or no among-lineage variation in diversification rates in our *Aeromonas* phylogeny.

The gamma statistic is a powerful tool to test the constancy of diversification rates, and is principally used for comparing models of decreasing speciation rate through time and constant-rate diversification [36], [37]. We thus obtained an estimated γ in both the Bayes (γ=-0.071) and MLA (γ=-0.225) chronograms. Although both γ values were negative, suggesting a possible deceleration of diversification rates through time, they were greater than those corresponding to critical values obtained by simulating 5,000 trees under a constant rate model (Bayes: $\gamma = -2.053$, and MLA: $\gamma = -2.131$) at a level of $\alpha = 0.05$. Thus, a constant diversification rate had to be accepted for our phylogenies. In the case of the Bayesian chronogram, we were also able to compare the gamma statistic calculated from the posterior distribution of trees (mean = 0.046; 95% CI: −0.680–0.837), the values being within the limits of the simulated data for the Bayes chronogram (95% CI: −2.053–1.350) and MLA chronogram (95% CI: −2.131–1.394) (Fig. 7).

It is well known that LTT plot results are sensitive to incomplete taxon sampling [4], [61]. In order to discard the influence of incomplete sampling due to the likely existence of undescribed species, we performed a Monte Carlo Constant Rate (MCCR) test, assuming a possible total number of 50, 60 and 70 *Aeromonas* species (Table 4). The results obtained did not allow us to reject the hypothesis of a constant rate of diversification in neither the Bayes nor the MLA chronograms at a level of $\alpha = 0.05$. These results indicate that although we obtained negative γ values in both cases, they are not significantly negative when compared with the null model of the constant diversification rate, irrespective of the effect of taxon sampling.

The standardized Colless' index value for both the Bayes and MLA chronograms was identical, $I_c = 0.1562$, an unsurprising result considering that both chronograms have the same topology. This value did not differ significantly (P>0.3) from those calculated through the I_c distribution obtained from 5,000 simulated phylogenies under a Yule model (95% CI ranging between 0.0781 and 0.2282 for both chronograms). Likewise, the number of cherries in both chronograms was 13. Both the normal approximation (P>0.6) and the comparison with the cherry distribution in 5,000 simulated trees (P>0.7) allowed us to accept the null hypothesis of a Yule model. The results obtained from the aforementioned statistical tests also support the hypothesis of a constant diversification in *Aeromonas*.

The mean substitution rate obtained from the Bayesian chronogram was 9.80×10^{-4} substitutions per site per Ma (se. = 2.53×10^{-5}). This value almost fully coincides with that obtained through the MLA chronogram: 1.06×10^{-3} substitutions per site per Ma (se. = 2.53×10^{-5}). Both mean substitution rates were obtained by regression analysis of branch lengths of both chronograms and those of the original maximum likelihood tree. As shown in the supplementary material (Fig. S1), there is a robust linear relationship between synonymous substitutions per synonymous site and sequence divergence in our concatenated sequences (slope = 2.8486; $R^2 = 0.9947$). This relationship allowed us to convert our estimates of substitution rates to synonymous substitutions per synonymous site per Ma. The values obtained were 2.79×10^{-3} for Bayesian and 3.02×10^{-3} for MLA, implying a silent substitution rate of 0.3% per Ma, which is slightly lower than the average silent substitution rate estimated for *E.coli* and *Salmonella* (0.45%), assuming that universally distributed proteins evolve at the same rate in enteric bacteria as in mammals [62], [63], and is similar to that observed in other protein coding genes in *Salmonella* and *Escherichia* [64].

In summary, the analysis of the LTT plots obtained with our phylogeny, the fit of the best model of diversification through maximum likelihood, the comparison with null models obtained by simulated trees, the gamma statistic of Pybus and Harvey and the tree imbalance tests all confirm that our phylogenetic trees are best explained by assuming a Yule model of constant diversification. The diversification rate of the genus *Aeromonas* ranges from 0.0119 to 0.0129 per Ma, depending on the dating method used (Bayesian or MLA). This constant rate remains virtually unchanged through time and across the different major clades of the phylogeny.

Discussion

The explosion of molecular data in recent years has culminated in a vast accumulation of prokaryote genomic information. However, this huge amount of information has not been used to unveil the speciation mechanisms of prokaryotes nor to clarify the conflicting hypotheses on the prokaryote species concept [28], [65]–[68]. As a consequence, prokaryotes are still subject to far more controversy than their eukaryotic counterparts. Thus, understanding the evolution of biological diversity of prokaryotes remains a great challenge for biologists [22], [23], [69]. The issue is far from trivial because many problems of extreme importance to human society hinge on understanding prokaryotic diversity and how it will respond to change [69]. Although few studies of this type have been carried out [22]–[24], [69], in our opinion, knowledge of the diversification rate and pattern of a bacterial genus may be useful for understanding prokaryotic evolution [70].

In this work, we studied the phylogeny and diversification rates of *Aeromonas* by applying methods previously used with eukaryotic taxa. Assuming that the cohesion of major phylogenetic groups within the prokaryotes is due to vertical transmission and common ancestry rather than to preferential horizontal gene transfer (HGT), it is possible to construct robust phylogenies reflecting the evolutionary history of bacteria using a sufficient number of orthologous housekeeping (core) genes. In these phylogenies most bacterial species are delineable as discrete evolutionary lineages [71]–[73].

The foregoing does not exclude the existence of HGT, which in fact occurs and has important evolutionary consequences, but it is doubtful that HGT is the essence of modern genome phylogeny [72]. Besides ecological isolation, mechanisms of sexual isolation, such as the obstacles to DNA entry in bacterial cells or restriction endonuclease activity, can significantly reduce the effectivity of HGT [66]. Moreover, as demonstrated in *Salmonella*, *Streptococcus* and *Bacillus*, homologous recombination decays exponentially with sequence divergence, that is, a sequence divergence between two strains of 10% suppresses the recombination rate between them by a factor of about 100 [74], [75]. This suggests that most genes acquired by HGT were probably introduced only rarely and very early in the evolutionary history of these bacterial species [67].

Phylogenetic Relationships of *Aeromonas* Species

Our phylogenetic analysis of the genus *Aeromonas* corroborates the monophyletic origin of this group of bacteria. The chronogram topology obtained from the Bayesian approach coincided fully with the MLA chronogram, confirming the robustness of our phylogeny (Fig. 2). The distribution of the main clades in our trees is in complete agreement with previous similar studies [76]–[78], although divergences occasionally appear in comparisons with phylogenies constructed with single genes [39]–[41]. Our study also provides further evidence for the existence of subspecies at the limit of being considered separate species, for example, *A. hydrophila* subsp. *dhakensis* [79], or for the relocation of what have been considered as new species, such as *A. culicicola* and *A. aquariorum* [26], [79], [80].

In the phylogeny we obtained, the different lineages that lead to the present species showed a mean pairwise divergence of 10% (ranging from 8.8 to 14.1%), a value that is enough to make recombination highly improbable, even if there was a relatively high recombination rate in *Aeromonas* [68]. Further evidence for this assumption was provided by an independent approach, a split decomposition analysis using SplitsTree4 software (Fig. S2). This analysis showed the absence of reticulated phylogenetic structures suggesting no evidence of detectable recombination in *Aeromonas*.

Aeromonas Diversification Rates

Both MLA and Bayesian chronograms suggest that the divergence of the genus *Aeromonas* began at an indeterminate point between the Permian and Triassic periods and has continued exponentially until today (Fig. 2). Slight differences between the Bayesian (265.7 Ma) and MLA (247.7 Ma) ancestry estimates may be due to the relaxed clock method used to infer dates and the prior distribution for the divergence time of the calibration point in the Bayesian versus the minimum and maximum age constraints in the MLA method. Moreover, we are aware that the use of a single calibration point can be a source of uncertainty, which is very difficult to minimize in the absence of more reliable calibration data and the impossibility of accepting the hypothesis of a molecular clock evolution for our sequences. However, if a reliable phylogeny is obtained, as in our *Aeromonas* study, it is possible even with a single calibration point to make useful statements about bacterial divergence times [81]. The calibration point we have used, 140 Ma (120–160 Ma), for the divergence between *Escherichia* and *Salmonella* was proposed by Ochman and Wilson based on calibrated rates of ribosomal RNA divergence. This date roughly coincides with the appearance of the principal niche of *E. coli*, the mammalian intestine. A similar date of divergence for *E. coli* and *Salmonella* (100–130 Ma) was obtained when assuming that universally distributed proteins evolve at the same rate in enteric bacteria as in mammals [62] and a somewhat broader range of divergence times (57–176 Ma) was estimated based on biogeochemical evidence of cyanobacterial divergence [82].

The data obtained from both analyses (Bayes and MLA) were in good agreement although the estimations were obtained from two completely different approaches. Moreover, the MLA method, recently described by Paradis, requires far less computing than the Bayesian approach, yet gives more accurate results than the Penalized Likelihood method [21].

LTT plots, diversification tests and comparisons between simulated and empirical phylogenies give support to the hypothesis of a constant rate of cladogenesis in *Aeromonas* during all the Phanerozoic with no or an undetectable extinction rate (Tables 2 and 3; Fig. 2). The rate of diversification varies between 0.0119 and 0.0129 according to the method used for the analysis (Bayesian and MLA, respectively). Unfortunately, we can not compare our results with those of other authors, since, to our knowledge, no studies have been previously published on the diversification of an entire bacterial genus. Martin et al. [22] used sequences from ribosomal genes of a wide variety of prokaryotes obtained from alpine soils or databases to determine their diversification pattern, which in all cases proved to be constant over time, but without a quantitative estimation of the diversification rates. These results were remarkably homogeneous regardless of the bacterial group analyzed or the method used for constructing chronograms. More recently, Morlon et al. [24] used multilocus and genomic sequence data to determine the diversification rate of *Borrelia burgdorferi* sensu lato, a pathogenic intracellular bacterium. In this case, the pattern of diversification was not constant, with explosive radiations followed by rapid decreases in diversification rates. This different pattern of cladogenesis could be explained considering that *B. burgdorferi* is obligatorily associated with vertebrate and arthropod hosts, which may limit the gene flow between isolated populations and result in a type of diversification similar to that of eukaryotes.

Given the paucity of prokaryotic data, comparison with other bacterial taxa is impossible but comparison with available estimates of diversification rates for eukaryotic taxa may be insightful. *Aeromonas* appears to have a lower rate of diversification than other existing taxa. Our values are close to the minimal diversification rates (0.0162, 0.0092 and 0.0143, respectively) but far from the mean values (0.0753, 0.1859 and 0.0750, respectively) estimated for fish, birds and mammals [83]. This low rate of speciation does not seem related to the rate of substitution calculated from our sequences (0.3% per Ma), which is similar to that obtained for many species of both prokaryotes and eukaryotes [62]–[64].

Allopatric isolation is controversial in prokaryotes. As with many bacterial species, the majority of *Aeromonas* strains are isolated regularly in very different locations, virtually anywhere on the planet. In bacteria, genetic isolation would be achieved only, but not totally, by genetic (DNA) divergence. In more complex organisms (multicellular eukaryotes) the number of mechanisms leading to sympatric reproductive isolation increases considerably (ploidy, hybridization, reproductive behavior) and the number of genes in which one or several mutations can lead to a reproductive isolation is large.

The fact that our data fits a Yule model, a constant rate birth-death process with death rate $\mu = 0$, in which each species evolves independently and produces new species at a constant rate λ, raises some questions. First of all, the parameters of this diversification model, the net diversification rate and the extinction fraction, have been determined by a maximum likelihood method. Generally accepted interpretations of this statistical method tell us that, for a fixed set of data (our sequences) and an underlying statistical model (the Yule model), the method of maximum likelihood selects the values of the model parameters that produce a distribution, giving the observed data the greatest probability. Figure 8 shows the contour plots of the log likelihood surface for our data. These plots reveal that although the maximum likelihood estimate of the extinction fraction (μ / λ) is zero, we cannot completely exclude the possibility that the extinction rate (μ) has a small but appreciable value.

To address this question, we first estimated a 95% confidence interval of the expected relative extinction fraction ($r = \mu$ / λ) by simulation of 5,000 phylogenies under the Yule model and obtained identical results from both Bayes and MLA approaches (0–0.62). On the basis of the divergence data from the basal node of the *Aeromonas* clade (247.7 Ma for Bayes and 265.7 Ma for MLA) and the observed diversity of 37 lineages, we used the estimates (0–0.62) to calculate the net diversification rate ($a = \lambda - \mu$) for $r = 0$ (0.0135 for Bayes and 0.0146 for MLA) and $r = 0.62$ (0.0101 for Bayes and 0.0108 for MLA), using the Magallón and Sanderson approach [84] as implemented in the package LASER. These results suggest that even when assuming a relative extinction rate as high as 0.62, the net speciation rate remains reasonably close to the values obtained in our analysis (0.012 for Bayes and 0.013 for MLA).

Huge populations of prokaryotes are relatively immune to the extinction and founder effects experienced by larger, less abundant, organisms [85], [86]. Bacterial species may be considered as metapopulations (i.e. sets of connected subpopulations that are maximally inclusive and whose boundaries are set by evolutionary cohesive forces) that extend over time and that evolved separately from other species [65]–[67], [87]. Metapopulation models predict that the metapopulation will go extinct only if the ratio between the within-subpopulation extinction and colonization rates is greater than or equal to the availability of habitats for this species [88]. Although geographical barriers to microbial dispersal can be relatively common and physical isolation can play a certain role in microbial evolution, it is reasonable to assume that bacteria have essentially unlimited capacity for dispersal [89]. Bacterial size is on average in the micrometer range and passive dispersal can easily occur via a variety of mechanisms, including transport in the atmosphere, water currents, or transport on or within larger plants and animals and are more likely to be transported long distances [90]. Moreover, bacterial populations may be very large and have high growth rates under favorable environmental conditions, and adopt physiologically inactive states for extended periods of time and survive during unfavorable ones [91].

If we combine this high probability of dispersal with the fact that a free-living heterotrophic, facultatively anaerobic bacteria such as *Aeromonas* can be isolated from virtually every environmental niche, including aquatic habitats, soils, fresh and marine waters, plant surfaces, invertebrates, fish, reptiles, birds, and food [26], and environmental conditions (pH, temperature), the availability of potential habitats for such prokaryotes is astronomically large and therefore, the probability of extinction of the *Aeromonas* metapopulation is null or very low.

As can be seen in Figure 9, since the Permian-Triassic, the diversification of the genus *Aeromonas* runs parallel to the increase of animal genera. This diversification seems to have begun after the Permian-Triassic extinction of approximately 251 million year ago, when more than 90% of marine and terrestrial life became extinct [92], possibly causing a return to an ancient world dominated by microorganisms [86].

Analysis of the fossil record of microbes, fungi, plants and animals shows that the diversity of both marine and continental life, although interrupted by mass extinctions, has increased exponentially since the end of the Precambrian [93]. Fossil records also suggest that after the end-Permian extinction, eukaryotic life, primarily multicellular plants and animals, diversified at an exponential rate through most of the Mesozoic and Cenozoic [92]–[94]. This explosive proliferation of multicellular organisms and their decisive influence on the structure and function of modern ecosystems in the Phanerozoic provided a new universe of potential ecological niches and the corresponding evolutionary opportunities for the bacterial lineages of the Paleozoic and Precambrian [86], [92]. Moreover, phanerozoic plants and animals have changed the ancient biosphere over evolutionary time, modifying biogeochemical cycles that are now intimately linked to the capacity of multicellular organisms to translocate nutrients across mixing boundaries, forcing the diversification of microorganisms to the new trophic structures [86]. The existence of a good correlation between the number of animal genera (according to Sepkoski's data) and the *Aeromonas* diversification in the last 250 Ma (Fig. 10) corroborates this idea.

In conclusion, our results suggest that the diversification of the genus *Aeromonas* began 248–266 Ma ago, remaining constant through time and across the different major clades of the phylogeny, and runs parallel to the exponential increase of animal genera after the Permian-Triassic extinction. Their speciation rate is significantly lower than that found for many eukaryotic taxa, although the absence of quantitative prokaryote data makes comparison almost impossible. Two particular features distinguish our work from previous studies: firstly, it is based on a robust phylogeny of all the species and subspecies of a bacterial genus, and secondly, it has used two proven methods to estimate the absolute speciation rate and the approximate date of origin of *Aeromonas*. Our results appear to confirm those of Martin et al. [22] regarding the constancy of the diversification rate in prokaryotes. Nevertheless, considerable more research is required on other bacterial genera to test if our results are comparable with complete phylogenies of other bacterial taxa, including pathogens and free-living bacteria.

Supporting Information

Figure S1 Regression plot of sequence divergence versus synonymous substitutions per site.

Figure S2 Split decomposition analysis. We used the split decomposition method to infer the 37 *Aeromonas* strains relatedness based on the concatenated sequence of five genes. Node labels refer to strain names (listed in). The split was generated by SplitsTree4 (v 4.13.1; www-ab.informatik.uni-tuebingen.de/software/splitstree4; Huson DH and Bryant D (2006) Mol Biol Evol 23:254–267).

Table S1 *Aeromonas* strains and GenBank accession numbers of gene sequences used in this study.

Acknowledgments

We thank José Castresana (Institut de Biologia Evolutica, CSIC-UPF, Barcelona, Spain) for helping to clarify our doubts about using the BEAST program. We thank Vicenta Albarral and Ariadna Sanglas for their valuable assistance in the experimental work.

Author Contributions

Conceived and designed the experiments: JGL MCF. Performed the experiments: MF. Analyzed the data: JGL. Contributed reagents/materials/analysis tools: MF. Wrote the paper: JGL MCF MF.

References

1. Barraclough TG, Nee S (2001) Phylogenetics and speciation. Trends Ecol Evol 16: 391–399.
2. Nee S (2001) Inferring speciation rates from phylogenies. Evolution 55: 661–668.
3. The Marie Curie SPECIATION Network. Butlin R, Debelle A, Kerth C, Snook RR, Beukeboom LW et al. (2012) What do we need to know about speciation? Trends Ecol Evol 27: 27–39.
4. Nee S, May RM, Harvey PH (1994) The reconstructed evolutionary process. Phil Trans R Soc Lond B 344: 305–311.
5. Sanderson MJ, Donoghue MJ (1996) Reconstructing shifts in diversification rates on phylogenetic trees. Trends Ecol Evol 11: 15–20.
6. Aldous DJ (2001) Stochastic models and descriptive statistics for phylogenetic trees, from Yule to today. Statist Sci 16: 23–34.
7. Nee S (2006) Birth-death models in macroevolution. Annu Rev Ecol Evol Syst 37: 1–17.
8. Rabosky DL, Lovette IJ (2008) Explosive evolutionary radiations: decreasing speciation or increasing extinction through time? Evolution 62: 1866–1875.
9. Morlon H, Potts MD, Plotkin JB (2010) Inferring the dynamics of diversification: a coalescent approach. PLoS Biol 8: e1000493.
10. Paradis E, Claude J, Strimmer K (2004) APE: Analyses of Phylogenetics and Evolution in R language. Bioinformatics 20: 289–290.
11. Rabosky D (2009) LASER: Likelihood Analysis of Speciation/Extinction Rates from Phylogenies. R package version 2.3. Available : http://CRAN.R-project.org/package = laser.
12. Zuckerkandl E, Pauling L (1965) Evolutionary divergence and convergence in proteins. In: Bryson V, Vogel HJ, editors. Evolving Genes and Proteins. New York: Academic Press. 97–16.
13. Kimura M (1968) Evolutionary rate at the molecular level. Nature 217: 624–626.
14. Lemey P, Posada D. 2009. Molecular clock analysis. In: Vandamme AM, Salemi M, Lemey P, editors. The Phylogenetic Handbook. Cambridge: Cambridge University Press. 362–380.
15. Drummond AJ, Ho SY, Phillips MJ, Rambaut A (2006) Relaxed phylogenetics and dating with confidence. PLoS Biol 4: e88.
16. Felsenstein J (1981) Evolutionary trees from DNA sequences: a maximum likelihood approach. J Mol Evol 17: 368–376.
17. Lepage T, Bryant D, Philippe H, Lartillot N (2007) A general comparison of relaxed molecular clock models. Mol Biol Evol 24: 2669–2680.
18. Yoder AD, Yang Z (2000) Estimation of primate speciation dates using local molecular clocks. Mol Biol Evol 17: 1081–1090.
19. Sanderson MJ (2002) Estimating absolute rates of molecular evolution and divergence times: a penalized likelihood approach. Mol Biol Evol 19: 101–109.
20. Huelsenbeck JP, Larget B, Swofford D (2000) A compound Poisson process for relaxing the molecular clock. Genetics 154: 1879–1892.
21. Paradis E (2013) Molecular dating of phylogenies by likelihood methods: a comparison of models and a new information criterion. Mol Phylogenet Evol 67: 436–444.
22. Martin P, Costello EK, Meyer AF, Nemergut DR, Schmidt SK (2004) The rate and pattern of cladogenesis in microbes. Evolution 58: 946–955.
23. Vinuesa P, Silva C, Werner D, Martínez-Romero E (2005) Population genetics and phylogenetic inference in bacterial molecular systematics: the roles of migration and recombination in *Bradyrhizobium* species cohesion and delineation. Mol Phylogenet Evol 34: 29–54.
24. Morlon H, Kemps BD, Plotkin JB, Brisson D (2012) Explosive radiation of a bacterial species group. Evolution 66: 2577–2586.
25. Martin-Carnahan A, Joseph SW. 2005. Genus I. *Aeromonas* Stanier 1943, 213[AL]. In: Garrity GM, Brenner DJ, Krieg NR, Staley JT, editors. Bergey's Manual of Systematic Bacteriology, vol. 2, part B. New York: Springer. 557–578.
26. Janda JM, Abbott SL (2010) The genus *Aeromonas*: taxonomy, pathogenicity and infection. Clin Microbiol Rev 23: 35–73.
27. Lan R, Reeves PR (2001) When does a clone deserve a name? A perspective on bacterial species based on population genetics. Trends Microbiol 9: 419–424.
28. Rosselló-Mora R, Amann R (2001) The species concept for prokaryotes. FEMS Microbiol Rev 25: 39–67.
29. Stackebrandt E, Frederiksen W, Garrity GM, Grimont PA, Kämpfer P et al. (2002) Report of the *ad hoc* committee for the re-evaluation of the species definition in bacteriology. Int J Syst Evol Microbiol 52: 1043–1047.
30. Goris J, Konstantinidis KT, Klappenbach JA, Coenye T, Vandamme P et al. (2007) DNA-DNA hybridization values and their relationship to whole-genome sequence similarities. Int J Syst Evol Microbiol 57: 81–91.
31. Konstantinidis KT, Tiedje JM (2007) Prokaryotic taxonomy and phylogeny in the genomic era: advancements and challenges ahead. Curr Opin Microbiol 10: 504–509.
32. Ochman H, Wilson AC (1987) Evolution in bacteria: evidence for a universal substitution rate in cellular genomes. J Molec Evol 26: 74–86.
33. Ochman H, Wilson AC (1987) Evolutionary history of enteric bacteria. In: Neidhardt FC et al., editors. *Escherichia coli* and *Salmonella typhimurium*: Molecular and Cellular Aspects. Washington: ASM Publications. 1649–1654.
34. Rabosky DL (2006) Likelihood methods for detecting temporal shifts in diversification rates. Evolution 60: 1152–1164.
35. McKenzie A, Steel M (2000) Distributions of cherries for two models of trees. Math Biosci 164: 81–92.
36. Pybus OG, Harvey PH (2000) Testing macro-evolutionary models using incomplete molecular phylogenies. Proc Biol Sci 267: 2267–2272.
37. Fordyce JA (2010) Interpreting the gamma statistic in phylogenetic diversification rate studies: a rate decrease does not necessarily indicate an early burst. PLoS One 5: e11781.
38. Fontaneto D, Tang CQ, Obertegger U, Leasi F, Barraclough TG (2012) Different diversification rates between sexual and asexual organisms. Evol Biol 39: 262–270.
39. Soler L, Yáñez MA, Chacón MR, Aguilera-Arreola MG, Catalán V et al. (2004) Phylogenetic analysis of the genus *Aeromonas* based on two housekeeping genes. Int J Syst Evol Microbiol 54: 1511–1519.
40. Küpfer M, Kuhnert P, Korczak BM, Peduzzi R, Demarta A (2006) Genetic relationships of *Aeromonas* strains inferred from 16S rRNA, *gyrB* and *rpoD* gene sequences. Int J Syst Evol Microbiol 56: 2741–2751.
41. Nhung PH, Hata H, Ohkusu K, Noda M, Shah MM et al. (2007) Use of the novel phylogenetic marker *dnaJ* and DNA-DNA hybridization to clarify interrelationships within the genus *Aeromonas*. Int J Syst Evol Microbiol 57: 1232–1237.
42. Miñana-Galbis D, Urbizu-Serrano A, Farfán M, Fusté MC, Lorén JG (2009) Phylogenetic analysis and identification of *Aeromonas* species based on sequencing of the *cpn60* universal target. Int J Syst Evol Microbiol 59: 1976–1983.
43. Farfán M, Miñana-Galbis D, Garreta A, Lorén JG, Fusté MC (2010) Malate dehydrogenase: a useful phylogenetic marker for the genus *Aeromonas*. Syst Appl Microbiol 33: 427–435.
44. Xia X, Xie Z (2001) DAMBE: software package for data analysis in molecular biology and evolution. J Hered 92: 371–373.
45. Librado P, Rozas J (2009) DnaSP v5: a software for comprehensive analysis of DNA polymorphism data. Bioinformatics 25: 1451–1452.
46. Posada D (2008) JModelTest: Phylogenetic Model Averaging. Mol Biol Evol 25: 1253–1256.
47. Guindon S, Dufayard JF, Lefort V, Anisimova M, Hordijk W et al. (2010) New algorithms and methods to estimate maximum-likelihood phylogenies: assessing the performance of PhyML 3.0. Syst Biol 59: 307–321.
48. Tamura K, Peterson D, Peterson N, Stecher G, Nei M et al. (2011) MEGA5: Molecular evolutionary genetics analysis using maximum likelihood, evolutionary distance, and maximum parsimony methods. Mol Biol Evol 28: 2731–2739.
49. Drummond AJ, Rambaut A (2007) BEAST: Bayesian evolutionary analysis by sampling trees. BMC Evol Biol 7: 214.
50. Plummer M, Best N, Cowles K, Vines K (2006) CODA: Convergence diagnosis and output analysis for MCMC. R News 6: 7–11.
51. Rambaut A, Drummond AJ (2007) Tracer v1.4. Available: http://beast.bio.ed.ac.uk/Tracer.
52. Yang Z (2007) PAML 4: Phylogenetic Analysis by Maximum Likelihood. Mol Biol Evol 24: 1586–1591.
53. Rabosky DL (2006) LASER. A maximum likelihood toolkit for detecting temporal shifts in diversification rates from molecular phylogenies. Evol Bioinform Online 2: 247–250.
54. Stadler T (2010) TreeSim in R-simulating trees under the birth-death model. Available: http://cran.r-project.org/web/packages/TreeSim/index.html.
55. Blum MGB, François O (2005) On statistical tests of phylogeny imbalance: The Sackin and other indices revisited. Math Biosci 195: 141–153.
56. Colless DH (1982) Review of phylogenetics: The theory and practice of phylogenetic systematics. Syst Zool 31: 100–104.
57. Agapow PM, Purvis A (2002) Power of eight tree shape statistics to detect non-random diversification: A comparison by simulation of two models of cladogenesis. Syst Biol 51: 866–872.
58. Blum MGB, François O, Janson S (2006) The mean, variance and limiting distribution of two statistics sensitive to phylogenetic tree balance. Ann Appl Probab 16: 2195–2214.

59. Bortolussi N, Durand E, Blum M, François O (2011) apTreeshape: Analyses of Phylogenetic Treeshape. R package v.1.4–4. Available: http://CRAN.R-project.org/package = apTreeshape.
60. R Development Core Team. R: a language and environment for statistical computing. R Foundation for Statistical Computing, Vienna, Austria. ISBN 3-900051-07-0. Available: http://www.r-project.org/.
61. Ricklefs RE (2007) Estimating diversification rates from phylogenetic information. Trends Ecol Evol 22: 601–610.
62. Doolittle RF, Feng DF, Tsang S, Cho G, Little E (1996) Determining divergence times of the major kingdoms of living organism with a protein clock. Science 271: 470–477.
63. Ochman H (2003) Neutral mutations and neutral substitutions in bacterial genomes. Mol Biol Evol 20: 2091–2096.
64. Degnan PH, Lazarus AB, Brock CD, Wernegreen JJ (2004) Host-symbiont stability and fast evolutionary rates in an ant-bacterium association: cospeciation of *Camponotus* species and their endosymbionts, *Candidatus Blochmannia*. Syst Biol 53: 95–110.
65. Cohan FM (2002) What are bacterial species? Annu Rev Microbiol 56: 457–487.
66. Cohan FM (2002) Sexual isolation and speciation in bacteria. Genetica 116: 359–370.
67. Achtman M, Wagner M (2008) Microbial diversity and the genetic nature of microbial species. Nat Rev Microbiol 6: 431–440.
68. Fraser C, Alm EJ, Polz MF, Spratt BG, Hanage WP (2009) The bacterial species challenge: making sense of genetic and ecological diversity. Science 323: 741–746.
69. Barraclough TG, Hughes M, Ashford-Hodges N, Fujisawa T (2009) Inferring evolutionarily significant units of bacterial diversity from broad environmental surveys of single-locus data. Biol Lett 5: 425–428.
70. Horner-Devine MC, Carney KM, Bohannan BJ (2004) An ecological perspective on bacterial biodiversity. Proc Biol Sci 271: 113–122.
71. Daubin V, Gouy M, Perrière G (2002) A phylogenomic approach to bacterial phylogeny: evidence of a core of genes sharing a common history. Genome Res 12: 1080–1090.
72. Kurland CG, Canback B, Berg OG (2003) Horizontal gene transfer: a critical view. Proc Natl Acad Sci USA 100: 9658–9662.
73. Lerat E, Daubin V, Moran NA (2003) From gene trees to organismal phylogeny in prokaryotes: The case of the *γ-Proteobacteria*. PLoS Biol 1: 101–109.
74. Majewski J, Zawadzki P, Pickerill P, Cohan FM, Dowson CG (2000) Barriers to genetic exchange between bacterial species: *Streptococcus pneumoniae* transformation. J Bacteriol 182: 1016–1023.
75. Fraser C, Hanage WP, Spratt BG (2007) Recombination and the nature of bacterial speciation. Science 315: 476–480.
76. Martínez-Murcia AJ, Monera A, Saavedra MJ, Oncina R, López-Alvarez M et al. (2011) Multilocus phylogenetic analysis of the genus *Aeromonas*. Syst Appl Microbiol 34: 189–199.
77. Martino MA, Fasolato L, Montemurro F, Rosteghin M, Manfrin A et al. (2011) Determination of microbial diversity of *Aeromonas* strains on the basis of multilocus sequence typing, phenotype, and presence of putative virulence genes. Appl Environ Microbiol 77: 4986–5000.
78. Roger F, Marchandin H, Jumas-Bilak E, Kodjo A, BVH C et al. (2012) Multilocus genetics to reconstruct aeromonad evolution. BMC Microbiol 12: 62. doi: 10.1186/1471-2180-12-62.
79. Fusté MC, Farfán M, Miñana-Galbis D, Albarral V, Sanglas A et al. (2012) Population Genetics of the "*Aeromonas hydrophila* Species Complex". In: Fusté MC, editor. Studies in Population Genetics. Croatia: InTech. 39–54.
80. Euzéby JP (2013) LPSN: List of Prokaryotic names with Standing in Nomenclature. Available: http://www.bacterio.cict.fr/.
81. Kuo CH, Ochman H (2009) Inferring clocks when lacking rocks: the variable rates of molecular evolution in bacteria. Biol Direct 4: 35.
82. Battistuzzi FU, Feijao A, Hedges SB (2004) A genomic timescale of prokaryote evolution: insights into the origin of methanogenesis, phototrophy, and the colonization of land. BMC Evol Biol 4: 44.
83. McPeek MA, Brown JM (2007) Clade age and not diversification rate explains species richness among animal taxa. Am Nat 169: E97–106.
84. Magallón S, Sanderson MJ (2001) Absolute diversification rates in angiosperm clades. Evolution 55: 1762–1780.
85. Lynch M, Conery JS (2003) The origins of genome complexity. Science 302: 1401–1404.
86. Butterfield NJ (2011) Animals and the invention of the Phanerozoic Earth system. Trends Ecol Evol 26: 81–87.
87. de Queiroz K (2005) Ernst Mayr and the modern concept of species. Proc Natl Acad Sci USA 102: 6600–6607.
88. Ovaskainen O, Hanski I (2003) Extinction threshold in metapopulation models. Ann Zool Fennici 40: 81–97.
89. Finlay BJ, Clarke KJ (1999) Ubiquitous dispersal of microbial species. Nature 400: 828–828.
90. Fierer N (2008) Microbial biogeography: patterns in microbial diversity across space and time. In: Zengler K, editor. Accessing Uncultivated Microorganisms: from the Environment to Organisms and Genomes and Back. Whashington DC: ASM Press. 95–115.
91. Papke RT, Ward DM (2004) The importance of physical isolation to microbial diversification. FEMS Microbiol Ecol 48: 293–303.
92. Sahney S, Benton MJ (2008) Recovery from the most profound mass extinction of all time. Proc Biol Sci 275: 759–765.
93. Benton MJ (1995) Diversification and extinction in the history of life. Science 268: 52–58.
94. Sahney S, Benton MJ, Ferry PA (2010) Links between global taxonomic diversity, ecological diversity and the expansion of vertebrates on land. Biol Lett 6: 544–547.
95. Sepkoski JJ Jr (2002) A compendium of fossil marine animal genera. Bull Am Paleontol 363: 1–560.
96. Rohde RA, Muller RA (2005) Cycles in fossil diversity. Nature 434: 208–210. Supplemental Table E3.

8

Threatened Bird Valuation in Australia

Kerstin K. Zander[1]*, Gillian B. Ainsworth[2], Jürgen Meyerhoff[3], Stephen T. Garnett[2]

1 The Northern Institute, Charles Darwin University, Darwin NT, Australia, **2** Research Institute for the Environment and Livelihoods, Charles Darwin University, Darwin NT, Australia, **3** Institute for Landscape Architecture and Environmental Planning, Technische Universität Berlin, Berlin, Germany

Abstract

Threatened species programs need a social license to justify public funding. A contingent valuation survey of a broadly representative sample of the Australian public found that almost two thirds (63%) supported funding of threatened bird conservation. These included 45% of a sample of 645 respondents willing to pay into a fund for threatened bird conservation, 3% who already supported bird conservation in another form, and 15% who could not afford to pay into a conservation fund but who nevertheless thought that humans have a moral obligation to protect threatened birds. Only 6% explicitly opposed such payments. Respondents were willing to pay about AUD 11 annually into a conservation fund (median value), including those who would pay nothing. Highest values were offered by young or middle aged men, and those with knowledge of birds and those with an emotional response to encountering an endangered bird. However, the prospect of a bird going extinct alarmed almost everybody, even most of those inclined to put the interests of people ahead of birds and those who resent the way threatened species sometimes hold up development. The results suggest that funding for threatened birds has widespread popular support among the Australian population. Conservatively they would be willing to pay about AUD 14 million per year, and realistically about AUD 70 million, which is substantially more than the AUD 10 million currently thought to be required to prevent Australian bird extinctions.

Editor: David L. Roberts, University of Kent, United Kingdom

Funding: The research was undertaken under the aegis of Australian Research Council Linkage Grant LP0990395 and also benefited from Australian Research Council Discovery Grant DP09875528. The survey was funded by the BLA Stuart Leslie Bird Research Award 2010. The funders had no role in study design, data collection and analysis, decision to publish, or preparation of the manuscript.

Competing Interests: The authors have declared that no competing interests exist.

* Email: kerstin.zander@cdu.edu.au

Introduction

Wild birds have been the subject of economic valuation studies for decades [1,2]. Direct economic values to humans [3] include being hunted for food or sport [4,5], pest control [6] or as objects of tourism [7,8]. However, birds also play less tangible roles that increase the well-being of those who encounter them [9]. Like many ecosystem services [10,11], birds provide utility to humans in ways that are not traded in the market and so their value cannot be obtained from observations of market transactions. Put another way, many people would feel poorer should wild birds no longer enrich their daily lives but there is no market from which their presence can be purchased. Thus quantifying the economic value of wild birds, including values not directly related to use, requires non-market valuation techniques.

Stated preference methods, of which contingent valuation (CV) is one, allow for the estimation of non-market goods for which there is no corroborating market behaviour that would provide reliable measurements [12]. Stated preference methods have been used to value non-market environmental goods for more than 50 years [13], with few alternative methods [14]. The core of a survey-based CV is the creation of a hypothetical market where respondents are asked to state directly their willingness-to-pay (WTP) for the good in question based on information provided to them. Studies using CV have provided a range of meaningful quantitative estimates of the anthropocentric benefits derived from threatened species conservation [15]. Most WTP bird studies have evaluated single, often threatened, bird species [16–20], some considered a category of birds (e.g. migratory birds) [21], some investigated multiple threatened and non-threatened species of which one assessed category was birds [22–24] and some studies investigated particular qualities of birds (e.g. rarity) [25]. Some of the values identified in these studies have been substantial. A meta-analysis of 12 studies found a mean WTP of USD44 per threatened species per year [24]. More recently a CV study found an average WTP for the nationally threatened corncrake (*Crex crex*) of between €7 and €11 among Irish farmers [20].

In Australia 211 bird taxa have been assessed as threatened or Near Threatened using the IUCN Red List criteria [26]. Although investment in conservation of these species has already been substantial [27], preventing their extinction will still cost millions of dollars to pay for actions ranging from direct interventions for individual species through to opportunity costs incurred by retaining habitat that might otherwise be developed [28,29]. Against this are the benefits from birds received by the Australian public who, for the most part, pay for conservation through their taxes. Comparing the costs and benefits (the value) of threatened bird conservation can lead to optimal conservation investment.

In this study we provide a monetary estimate of some of the benefits bird conservation would bring to Australians. Using the CV method, we compared the stated WTP across respondents from different socioeconomic backgrounds, with different attitudes and beliefs about birds and bird conservation, and with different levels of knowledge about birds. We know of no other study in which the diversity of perceptions about birds and the value of all threatened birds, rather than a narrow selection of species, has been estimated for a whole country.

Methods

Willingness-to-pay determinants

Many studies that have evaluated the value of threatened species and peoples' WTP for their conservation, have investigated and found variation in WTP across respondents. Most studies test for age, education, gender and income effects on the WTP, respondents' location (e.g. urban vs. rural) and the distance of the respondent to the species in question, respondents' knowledge of the species and their attitudes towards environmental issues and conservation more generally. The choice of factors thought likely to influence the WTP for threatened bird conservation in Australia, and so tested in this research, arose partly from literature review and partly from qualitative interviews prior to the survey. Apart from commonly used economic and demographic determinants such as income, gender and age, we were particularly interested in the impact of respondents' knowledge about bird identification and their attitudes towards threatened birds on their WTP. Table 1 outlines the variables that we test in this study and the expected impact on peoples' WTP for threatened bird conservation in Australia.

We expected that people with high incomes would be more likely to pay as well as to pay more for threatened bird conservation in Australia, as found for other threatened wildlife (e.g. for the conservation of black-faced spoonbills (*Platalea minor*) [19], for peregrine falcons (*Falco peregrinus*) and shortnose sturgeon (*Acipenser brevirostrum*) [17]). Age has been found to be a consistent predictor which is negatively related to WTP for environmental amenities in general [30] and, for example, for the recovery of the guillemot (*Uria aalge*) population in Spain [31] specifically. Being female is often positively associated with higher WTP for environmental amenities ([30]; and, for example, for biodiversity protection in Germany [32] and the conservation of Mediterranean monk seals (*Monachus monachus*) [33]), and so we hypothesise that women have a higher WTP for wild birds than men.

Table 1. Potential willingness-to-pay (WTP) determinants and their expected impact (positive [+]/negative [−]).

Determinant	Expected impact on WTP
Income	+
Being female	+
Age	−
Interest in birds in general	+
Attitudes towards threatened birds	
Aesthetic value	+
Humanistic value	+
Spiritual value	+
Scientific value	+
Experiential value	+
Existence value	+
Utilitarian value	−
Knowledge of birds, measured by peoples' self-rated ability to identify common birds	+

The attitudinal questions in our survey are based on the categorization of attitudes to wildlife developed by Kellert [34]. For each of Kellert's eight categories of wildlife value, we posed one statement question to each respondent. Research on environmental attitudes and WTP often find a positive relationship between the two [17,23,35].

Finally we integrated a knowledge variable as an indicator for peoples' WTP. This variable is a measure of respondents' self-rated ability to identify common birds. The possible answers were: cannot identify any, can identify some, can identify most and can identify all common birds. The expectation was that those people who have good or expert knowledge about birds, i.e. those who said they can identify most or all common birds, would be willing to pay more for the conservation of threatened species than those with less knowledge. Apart from examples in wildlife valuation where this was evident [16,18,36], this phenomenon was also found for the conservation of threatened livestock breeds [37].

The sample

The survey was delivered online between 16th and 21st of February 2011. We opted for a cost-efficient online survey because other valuation studies have shown that WTP values do not vary significantly across different survey modes [38–40]. The survey was commissioned by a survey company, MyOpinions Pty Ltd, and respondents were paid AUD 3 on completion (at the time of the survey the AUD equalled the USD). The survey was voluntary and anonymous, and ethics approval was obtained from Charles Darwin University Human Research Ethics Committee (H11059). MyOpinions Pty Ltd is accredited to ISO 20252 and ISO 26362, adheres to the "research only" policy governed by industry bodies including the European Society for Opinion and Marketing Research, the Australian Market and Social Research Society and the Association of Market and Social Research Organisations and has an active panel of 300,000 verified respondents drawn from the general public (1.2% of Australian population) who registered (without having received any payment) with the company after recruitment via television, radio, newspaper, and online. Approximately half of the MyOpinions panel has been recruited from offline sources. The sample was selected using a quota random sampling whereby quotas were set to match the national population for gender, age and geographic location. The survey company randomly selected 5,800 members within these quotas and invited them to participate. Of these, 1,229 people agreed to undertake the survey before the topic was revealed. From these, 70 people dropped out before they started the survey. From the remaining 1,159 people, nearly 56% (645 people) completed the survey. The overall response rate of 11% (645/5,800) is consistent with other online surveys [41] where the invited sample tends to be very large to start with to ensure that all survey categories reach their quota quickly.

Questionnaire

The questionnaire had four sections: 1) questions aiming to elicit respondents' attitudes towards birds and bird knowledge, 2) the CV question and a follow-up on the motivations for those not willing to pay, and 3) questions on socioeconomic characteristics (income, education, current employment situation and country of birth). For the CV we used a single-bounded dichotomous yes/no choice question on whether respondents were willing to pay for a stated amount (referred to as a bid) that varied between questionnaires [42,43]. Additional to these bids we offered respondents the opportunity to state their WTP openly, which could be lower or higher than the bid. The number of bids and bid amounts were finalised after a pilot study with 30 respondents.

The chosen bid amounts offered to respondents were (in AUD): 10, 20, 50, 100, 200. As a payment vehicle a contribution to a conservation fund was chosen and the payment was said to be yearly. The bid amounts were randomly rotated during survey distribution so that equal numbers of respondents within the sample group were shown each bid. The wording of the CV question was as follows:

Of the money you might donate to any kind of cause (charitable, conservation etc.), would you be willing to pay the amount shown below per year into a conservation fund for threatened birds?(if you would not donate the amount shown below, please write your preferred amount into the box)
AUD [Bid] per year
Or, please specify your preferred amount AUD........

All attitudinal questions were assessed using a Likert scale [44] format with five potential answers: 'Strongly disagree', 'Disagree', 'Neither agree nor disagree', 'Strongly agree', 'Agree'. To rate respondents' overall interest in birds we asked them to rate their agreement to the statement 'Seeing a new bird fills me with excitement'. We then asked respondents the extent to which they agreed with a set of value-related statements based on the following question: 'Thinking about how you would feel if you knew you had seen an endangered bird, how much do you agree or disagree with these statements?'.

Analysis of CV responses

A respondent's preference for the environmental good in question, and thereby the probability of accepting the bid over declining it, is based on the random utility framework, stating that utility is composed of two parts, a deterministic part which we can observe (v) and a random part (ε) that we cannot measure [45]. The utility (U) derived from environmental good (q) can be written as (y = income):

$$U_j = v(y_j, q) + \varepsilon_j \tag{1}$$

Following Haab and McConnell [46] we assumed a linear utility function, so that the deterministic part of utility can be written as:

$$v_j = \alpha Z_j + \beta(y_i) \tag{2}$$

Where Z is a range of respondents' socio-economic characteristics and y is income for respondent j. The probability that a respondent will accept the offered bid (BID) is then:

$$\Pr(\text{yes}) = [(1 + \exp(\text{-}\alpha Z\text{-}\beta BID)] / 1 \tag{3}$$

Using a probit model we estimated the α's and β's and subsequently estimated both the mean and the median. Assuming a linear functional form both are given by

$$\text{mean/median WTP} = \left(\frac{-\alpha Z}{\beta}\right) \tag{4}$$

and assuming an exponential functional form the mean is given by

$$\text{mean WTP} = \exp\left(\frac{-\alpha Z}{\beta} + 0.5\sigma^2\right) \tag{5}$$

and the median is given by

$$\text{median WTP} = \exp\left(\frac{-\alpha Z}{\beta}\right) \tag{6}$$

.To calculate the mean and median WTP values as well as the 95% confidence intervals (CI), we employed the Stata command 'wtpcikr' [47]. The CIs were calculated using the Krinsky-Robb approach [48], i.e., the standard errors are derived via simulation. For the simulation we used 10,000 draws. In addition to a bid-only model, which includes solely the constant and the bid parameters, we also estimated a model investigating whether the hypothesised WTP determinants are statistically significantly related to the responses toward the offered bids, i.e. accepting or declining them (Yes/No-response). This model is called the 'Covariates model'.

Results

In general willingness-to-pay and protest responses

More than half the respondents (353; 55% of 645) did not want to contribute to a threatened bird fund in general. These people rejected the bid and also did not state an alternative amount that they might be willing to pay. A follow-up question after the CV asking respondents why they did not want to pay (Table 2) was used to separate those people who had zero value for threatened birds (valid 'no' responses) and those who opposed the CV question even though they might value threatened birds (protest responses). Most of the non-contributors said that they could not afford to pay (37%) or that they already donate money to another cause (32%). Some (11%) would not donate to any fund in general, while a few said that their taxes already support the protection of threatened birds (6%) or that they support bird conservation in other ways already (6%).

Respondents who have been identified as protesting against the payment vehicle are usually deleted from the sample [49]. However if those who are categorised as protesters actually have a WTP, then assuming a zero WTP for them would underestimate the economic value of the good in question [50]. On the other hand, the economic value could be overestimated if respondents categorised as protesters but having a zero WTP are ascribed some average value. Following Jakobsson and Dragun [51], we treated positive responses to two of the reasons as protest responses and deleted them from the data set. These were the 39 respondents who would not donate to any fund in general and 21 respondents who believed that their taxes already pay for the protection of threatened birds. This reduced the data set from 645 to 585 respondents. A further 17 responses could not be used because people did not answer most of the questionnaire properly. The final dataset contained 568 responses.

Sample characteristics

With 61% of the 568 respondents being female, the sample constitutes a slight gender bias. Forty percent of the respondents had an income of up to AUD 40,000 per year, 27% of AUD 41,000-80,000 and 7%>AUD 80,000; the remaining 26% of respondents did not reveal their income category. In accordance with the predetermined sample request, respondents were distributed relatively evenly across all age categories (18–24:10%, 25–34:13%, 35–44:17%, 45–54: 21%, 55–64:17%, 65+ :22%). Also by request, the geographical distribution of respondents matched the demographic variation among Australian states (New South Wales 31%, Victoria 25%, Queensland 20%, Western Australia 10%, South Australia 7%, Tasmania 4%, the Australian Capital Territory 2% and the Northern Territory 1%).

Table 2. Stated reasons for not contributing to a threatened bird conservation fund (in %).

Reason	N	% of those not contributing (N = 353)	% of whole sample (N = 645)	Response type
I already donate to a bird conservation fund	1	<1	<1	True zero value
I already donate to another cause	112	32	17	True zero value
I cannot afford to donate any money to a bird conservation fund	131	37	20	True zero value
I support bird conservation in other ways	20	6	3	True zero value
I would not donate to any fund like this in general	39	11	6	Protest
My taxes already support protection of endangered birds	21	6	3	Protest
No answer	21	6	3	True zero value
Other	8	2	1	True zero value

Attitudes towards threatened birds and their conservation

The highest Likert sample means, over four, were found for the three statements 'I would regret that humans had caused the bird to become endangered', 'I think there's a moral obligation to protect the bird' and 'I would feel upset if the bird became extinct' (Table 3), indicating a strong dislike of endangered birds becoming extinct across the whole sample, including many who had a zero WTP. Of the 13% who actively rejected payments, 35% still agreed or strongly agreed that they would both 'not like to see extinction' and 'feel a moral obligation to protect the bird' with 43% regretting that human activities were making it threatened.

To reduce the individual items to underlying latent factors (Table 4) we used factor analysis. Factor analysis is a family of approaches that aim to reduce a number of observed, correlated variables, such as responses to attitudinal questions, by describing linear combinations of the variables that contain most of the information. This information can subsequently be used to reduce the set of variables to a lower number of unobserved latent variables, also called factors [52,53]. Prior to the factor analysis we calculated the Kaiser-Meyer-Olkin (KMO) Measure of Sampling Adequacy. With a value of 0.825 it indicates that the data are suitable for a factor analysis. Subsequently, the Varimax rotated results singled out one factor with an Eigenvalue of 2.2 which explains 63.6% of the variance. The Eigenvalue of a factor measures the variance in all the variables which is accounted for by this particular factor. The higher the Eigenvalue the more the factor contributes to the explanation of the variances in the variables [53]. Among the three remaining factors, only one had an Eigenvalue above one. However this value was only slightly above one (1.13) so we did not use it in the subsequent analysis. To calculate an attitudinal factor score for incorporation into the bid function, we used the items with factor loadings above 0.4. We called this score 'Avicentric'. The higher this score, the more positive the attitude of a respondent towards threatened birds and their protection. The score had a mean value of 16.6 and ranged from a minimum value of 8 to a maximum value of 21.

Responses to offered bids

In total, respondents accepted the offered bid in 25% of the CV questions while they rejected it in 75% of the questions (Table 5). The percentage of respondents accepting the bid diminished as its cost increased. Almost half of the respondents (45%) accepted the lowest bid offered (AUD 10) while only four respondents (3%) answered 'yes' to the highest bid of AUD 200.

From those 428 respondents who did not accept the offered bid, 133 (31%) suggested a maximum WTP that was lower than offered in the CV question. A few respondents (17) had a higher maximum WTP than indicated by the bid they had accepted. The mean annual WTP of the former group was AUD 26.40. The respondents who professed a WTP that was higher than the bid had a mean annual WTP of AUD 115.30. The mean of the bids offered to those respondents was AUD 15 and thus was in the lower range of the offered bid vector.

Bid function estimation

Various model specifications were tested incorporating different WTP determinants. The bid values were log-transformed. Based on the log-likelihood values, information criteria (AIC and BIC) and the pseudo R^2 measure [54] we opted for the model presented in Table 6. This model showed the best performance among the model specifications tested. Overall, with a pseudo R^2 of 0.20, the model performs well [54]. As expected, the bid amount was negative and statistically significant at the 1% level (Table 6). The negative sign denotes that the higher the bid amount the respondent was asked to donate, the lower the probability that the respondent would accept it. The extent to which people were excited about seeing a new bird and had positive attitudes towards bird conservation (Factor 1: 'Avicentric') significantly and positively affected the likelihood of accepting the bid. The higher the respondents' score for this attitudinal variable, the higher was their WTP. These positive impacts on peoples' WTP agreed with our expectations (see Table 1). The fact that people who could not identify any common birds were less likely to accept the offered bid was also consistent with our expectation, because people with poor knowledge of birds probably value them less than those who have made the effort to learn to identify most or all common birds. However, some determinants did not have the positive impact we had expected, e.g. income and younger age groups were insignificant. The only significant age group was 'older than 65', which had a negative effect on WTP. The gender effect was the opposite of what we expected with male respondents being more likely to accept the bids.

Willingness-to-pay estimation and aggregation of estimates

The median WTP estimates for threatened birds in Australia were computed as between AUD 11.30 and AUD 11.55, for the bid-only model and the model including covariates, respectively (Table 7). These figures were aggregated for the population of

Table 3. Responses (in %) to a series of statement questions asking: 'Thinking about how you would feel if you knew you had seen an endangered bird, how much do you agree or disagree with these statements?'

Statement	Strongly disagree (1)	Disagree (2)	Neither Agree nor Disagree (3)	Agree (4)	Strongly agree (5)	Sample Mean
Seeing a new bird fills me with excitement	4	16	32	33	15	3.4
I want to learn more about the bird	1	7	32	48	12	3.6
I want to add it to my bird watching list	14	24	40	18	4	2.8
I would regret that humans had caused the bird to become endangered	<1	2	12	47	39	4.2
I think there's a moral obligation to protect the bird	<1	1	17	49	33	4.1
I feel it's a nuisance when an endangered bird stops development	34	27	23	11	5	2.3
I think the bird has a right to live only if it's beautiful or unusual	45	28	13	8	6	2.0
I feel the needs of people come before those of endangered birds	26	32	32	8	2	2.3
I think government is responsible for the bird's survival, not me	17	38	35	8	2	2.4
I would feel upset if the bird became extinct	2	3	15	47	33	4.1
I would feel privileged or spiritually uplifted	1	4	31	43	21	3.8

working adults (rounded about 11 million [55]). The most conservative estimate (AUD 13.7–14.0 million) assumed that all those who did not accept the invitation to participate in the survey (89% of those requested) had a zero WTP, even though the subject of the questionnaire was not revealed in the invitation. A more realistic, but still conservative, estimate (AUD 69.6–71.1 million) assumed that all of those who failed to complete the questionnaire after accepting the invitation to participate had a zero WTP (44%; Table 7).

Discussion

Based on data collected in the 1990s, about AUD 5 million per year for the next 80 years could reduce Australian bird extinctions to almost zero and reduce the total number of threatened species by 15% [28]. Even assuming this figure has doubled to about AUD 10 million a year [56], this is still less than one dollar a year for the 11 million Australian adults of working age. Respondents to our survey were willing to pay over ten times that amount, around AUD 11 per year (median value), for threatened bird conservation in Australia, even including the non-contributors. This suggests that, even if we assume that all 89% of those who did not respond to the invitation to participate in the survey would be unwilling to pay, an adequate allocation of public funds to threatened bird conservation would be consistent with the benefits gained by the Australian public.

That more than half of the sample were not prepared to pay may also be deceptive. Three percent said they already supported bird conservation in other ways. Three quarters of the 20% unwilling to pay because they could not afford to (see Table 2), nevertheless agreed that humans have a moral obligation to

Table 4. Results of factor analysis extracting four common factors explaining the correlations amongst responses to Likert-type statement questions.

Statement question	Factor 1	Factor 2	Factor 3	Factor 4
I think there's a moral obligation to protect the bird	0.75	−0.16	0.19	−0.02
I would regret that humans had caused the bird to become endangered	0.72	−0.17	0.14	−0.04
I would feel upset if the bird became extinct	0.58	−0.08	0.22	0.19
I would feel privileged or spiritually uplifted	0.54	−0.09	0.30	0.19
I want to learn more about the bird	0.47	−0.07	0.58	0.04
I might tick the bird off my bird watching list	0.25	0.13	0.57	−0.01
I think government is responsible for the bird's survival, not me	0.15	−0.43	0.14	0.05
I think the bird has a right to live only if it's beautiful or unusual	−0.14	0.59	0.05	0.01
I feel it's a nuisance when an endangered bird stops development	−0.19	0.55	0.05	−0.05
I feel the needs of people come before those of endangered birds	−0.34	0.44	−0.06	−0.03
Eigenvalue	2.19	1.13	0.87	0.08

Note: Responses to the first five statements with a loading higher than 0.4 were grouped into one variable which we called 'Avicentric'.

Table 5. Distribution of responses to the WTP bids (in %); N = number of respondents offered the bid.

	Bid (in AUD)					
Response	**10**	**25**	**50**	**100**	**200**	**Total**
Rejected bid	55	57	78	87	97	75
Accepted bid	45	43	22	13	3	25
N	101	111	138	100	118	568

conserve birds. Of the protest votes, 3% felt that government already covered their responsibility towards birds and just 6% actively rejected payment into a conservation fund. Curiously nearly half of the protest group (47%) said they would still pay for threatened birds in general and the majority (83%) would be upset if a bird went extinct. This survey therefore demonstrates that there is substantial support for the conservation of threatened birds across society. It also shows that threatened birds are valued as a group, not just particular species with which people might have a strong affinity.

This strong desire among respondents for wild birds not to go extinct is perhaps surprising given the relatively low rate of membership of bird clubs in Australia. BirdLife Australia has about 25,000 members, supporters and volunteers and around 12,000 Australians participate in BirdLife Australia's Birds in Backyards citizen science program [57,58] (0.3% of the Australian population) with others interested in birds belonging to natural history and avicultural societies (which collectively are likely to have many more members than BirdLife Australia). This is far lower than, for example, the United Kingdom where the Royal Society for the Protection of Birds has over a million members (1.7% of the population, about one in every 60 people of all ages). In the survey 2.5% were or had been a member of a bird club of any type, although 19% said they sometimes went birdwatching. International comparisons of WTP for threatened birds could provide meaningful comparisons of conservation culture, although the comparison would also need to take account of national differences in the probability of joining a society of any type.

Given that this WTP was elicited from people who had not been primed for the questionnaire, one might expect that it would have been higher had people had greater knowledge than provided in the survey introduction. The result that the stated WTP in this study was higher among those more knowledgeable about birds, and that increasing a person's knowledge about a species increases

Table 6. Bid-only and covariates model (probit), depended variable = Yes/No response to offered bid.

	Bid-only model			Covariates model		
Variable	**Coef.**	**SE**$^+$	**p-value**	**Coef.**	**SE**$^+$	**p-value**
Constant	1.296	0.06	0.001	0.54	0.58	0.356
Bid (log)	−0.534	0.24	0.001	−0.60***	0.07	0.001
Age 25–34				−0.17	0.27	0.527
Age 35–44				−0.34	0.26	0.193
Age 45–54				−0.34	0.25	0.170
Age 55–65				−0.32	0.26	0.215
Age>65				−0.50**	0.25	0.049
Female				−0.27**	0.14	0.049
Medium income				0.14	0.15	0.366
High income				0.10	0.26	0.700
Can identify some common birds				0.11	0.18	0.539
Cannot identify any common birds				−0.41*	0.22	0.0610
Attitudinal score 'Avicentric'				0.08**	0.03	0.0110
Excited to see birds				0.37**	0.15	0.0140
Log-likelihood $_{null}$			−317.19		−317.19	
Log-likelihood $_{model}$			−278.25		−254.10	
Pseudo R^2			0.12		0.20	
AIC			560.50		539.67	
BIC			569.18		604.81	
Observations			568		568	

$^+$SE = Standard Error.
*** = 1% significance level; ** = 5% significance level; * = 10% significance level.

Table 7. WTP estimates (in AUD) for Australian threatened bird conservation and aggregation of these estimates.

Variable	Bid-only model	Covariates model
Mean WTP	65.10 [42.27–166.00]	46.61 [33.46–90.35]
[95% CI]		
Median WTP	11.30 [7.16–15.21]	11.55 [7.70–15.23]
[95% CI]		
Aggregation based on median WTP		
Conservative scenario: 11% of adult Australians+ would pay the average median WTP[1)]	13,673,000	13,975,500
Realistic scenario: 56% of adult Australians would pay the average median WTP[2)]	69,608,000	71,148,000

+There are about 11 million adult Australians (rounded; [55]).
[1)]This assumes that all of those people who did not respond to the survey when invited by the survey company (89%) have a zero WTP for threated bird conservation in Australia.
[2)]This assumes that all of the 44% who did not complete the survey have a zero WTP and with the other 56% having a WTP corresponding to the sample.

their stated WTP [18], suggests that there is considerable extra support that could be generated if more people knew about the species that could be lost. While stated and actual WTP differ, there was, no suggestion that interest in conservation peaked among those moderately well-informed (unlike [59]). Similarly there was no suggestion to confirm our hypothesis, in line with other similar studies [33], that women are more likely to support conservation than men, even though men tend to relate differently to birds than women [60]. There was, however, an effect of age, with respondents over 65 being less likely to pay, which is in line with the attitudes of young people towards taking environmental responsibility [61]. This confirmed our hypothesis of a negative relationship between age and WTP [30] with older people (beyond work force age) potentially not able to contribute as much money as younger people [31]. While people in Australia, as in some other wealthy countries like the USA and Japan, may be less directly involved with wildlife than they were historically [62], there still appears to be a strong wish among young people to prevent species extinctions, which is consistent with a substantial interest in retaining wildlife even without seeing it [18,63].

The attitudinal values of those willing to pay for bird conservation also affected their WTP. Thus, as with Spash [23], who found a positive correlation between respondents who held the view that endangered birds have the right to protection and their WTP, there was a positive correlation between WTP and being 'avicentric'. The results are also in line with those presented by Kotchen and Reiling [17] who found that a positive correlation between environmental concern, measured by the NEP (New Ecological Paradigm) scale, and WTP for species protection. However not all results were as expected: 55% of respondents who considered people more important than birds and 42% of those who resented threatened birds impeding development were still willing to contribute to retain them in general. Likewise, 58% of those thinking that a threatened bird has a right to live only if it is beautiful or unusual would still, in principle, be prepared to pay into a conservation fund. Thus, even among those who put the interests of people first, there was a willingness to contribute to bird conservation at sites where there were fewer trade-offs, and an unwillingness to condone extinction. Thus these apparently negative attitudes do not necessarily exclude people from wanting to conserve threatened birds. Generally, however, people expressing a strong emotional response to birds were willing to pay more than those interested simply in seeing them.

Caution is always required with WTP estimates. In this case the major caution is that, while the survey with the dichotomous choice format had an incentive compatible question format, the payment vehicle is not among those that support consequentiality and thus truth telling [43]. Hence we opted for a voluntary payment into a conservation fund as the payment vehicle, rather than a tax increase, which would be compulsory across the entire population. However, in Australia the link between tax and expenditure by government is nearly always indirect, and suggesting a tax increase may have confounded considerations of the value of birds with views about taxation increases in principle. We therefore felt that the link between a conservation fund and conservation action was more explicit and self-evidently voluntary. Also, while the choice of payment vehicle may have led to over-estimation of the WTP, the aggregation is based on the more conservative median WTP, which is lower than the mean. Moreover, some respondents stated that they were willing to pay higher amounts than they were requested to pay based on the bids and some respondents who rejected the offered bid were willing to pay lower amounts, which were on average higher than the estimated median WTP. Overall, we therefore think that the figures presented are sufficiently accurate as an estimate of the Australian population's valuation of threatened birds that they can be used in conservation policy decisions.

Conclusions

There was strong support for the conservation of threatened birds among the Australian public as demonstrated by their willingness-to-pay for their conservation. Nearly half of the respondents said they were willing to pay into a bird conservation fund or did so already. Many of the remainder said they could not afford to pay rather than they would rather not do so. While support was strongest among those with a passion for birds and those who knew most about them, it was by no means confined to this sector of society. Even many of those who would favour development over birds would still be willing to pay to prevent extinctions. The study is notable for valuing a threatened fauna in its entirety rather than any specific bird. It also suggests that funding of threatened species conservation has broad backing from the Australian population.

Acknowledgments

We are grateful for advice on the wider questionnaire, of which this was part, from Heather Aslin and Mike Weston, and for comments on the manuscript from Stuart Butchart.

Author Contributions

Conceived and designed the experiments: KKZ GBA. Performed the experiments: GBA. Analyzed the data: KKZ JM. Contributed reagents/materials/analysis tools: KKZ JM STG. Wrote the paper: KKZ STG.

References

1. Bowker JM, Stoll JR (1988) Use of dichotomous choice nonmarket methods to value the whooping crane resource. Am J Agr Econ 70: 327–381.
2. Rubin J, Helfand G, Loomis J (1991) A benefit-cost analysis of the northern spotted owl. J Forest 89: 25–30.
3. Wenny DG, Devault TL, Johnson MD, Kelly D, Sekercioglu CH, et al. (2011) The need to quantify ecosystem services provided by birds. Auk 128: 1–14.
4. Gascoigne WR, Hoag D, Koontz L, Tangen BA, Shaffer TL, et al. (2011) Valuing ecosystem and economic services across land-use scenarios in the Prairie Pothole Region of the Dakotas, USA. Ecol Econ 70: 1715–1725.
5. Heffelfinger JR, Geist V, Wishart W (2013) The role of hunting in North American wildlife conservation. Int J Environ Stud 70: 399–413.
6. Johnson MD, Kellermann JL, Stercho AM (2010) Pest reduction services by birds in shade and sun coffee in Jamaica. Anim Conserv 13: 140–147.
7. Edwards PE, Parsons GR, Myers KH (2011) The economic value of viewing migratory shorebirds on the Delaware Bay: an application of the single site travel cost model using on-site data. Hum Dimens Wildl 16: 435–444.
8. Şekercioğlu ÇH (2002) Impacts of birdwatching on human and avian communities. Environ Conserv 29: 282–289.
9. Prescott-Allen C, Prescott-Allen R (1986) The First Resource: wild species in the North America economy. Newhaven: Yale University Press. 529 p.
10. Daily GC, Söderqvist T, Aniyar S, Arrow K, Dasgupta P, et al. (2000) The value of nature and the nature of value. Science 289: 395–396.
11. Millennium Ecosystem Assessment (MA) (2005) Ecosystems and human well-being: the assessment series (four volumes and summary). Washington: Island Press.
12. Mitchell RC, Carson RT (1989) Using survey to value public goods. The contingent valuation method. Washington: Resources for the Future. 484 p.
13. Smith VK (2006) Fifty years of contingent valuation. In: Alberini A, Kahn JR, editors. Handbook on contingent valuation. Cheltenham: Edward Elgar. pp. 7–65.
14. Baker R, Ruting B (2014) Environmental policy analysis: a guide to non-market valuation. Productivity Commission Staff Working Paper, Canberra.
15. Loomis JB, White DS (1996) Economic benefits of rare and endangered species: summary and meta-analysis. Ecol Econ 18:197–206.
16. Loomis J, Ekstrand E (1998) Alternative approaches for incorporating respondent uncertainty when estimating willingness to pay: the case of the Mexican spotted owl. Ecol Econ 27: 29–41.
17. Kotchen MJ, Reiling SD (2000) Environmental attitudes, motivations, and contingent valuation of nonuse values: a case study involving endangered species. Ecol Econ 32: 93–107.
18. Wilson C, Tisdell C (2007) How knowledge affects payment to conserve an endangered bird. Contemp Econ Policy 25: 226–237.
19. Jin J, Wang Z, Liu X (2008) Valuing black-faced spoonbill conservation in Macao: a policy and contingent valuation study. Ecol Econ 68: 328–335.
20. Hynes S, Hanley N (2009) The "Crexcrex" lament: estimating landowners' willingness to pay for corncrake conservation on Irish farmland. Biol Conserv 142: 180–188.
21. Brouwer R, van Beukering P, Sultanian E (2008) The impact of the bird flu on public willingness to pay for the protection of migratory birds. Ecol Econ 64: 575–585.
22. Bulte E, Kooten CV (1999) Marginal valuation of charismatic species: implications for conservation. Environ Resour Econ 14: 119–130.
23. Spash CL (2000) Multiple value expression in contingent valuation: economics and ethics. Environ Sci Technol 34: 1433–1438.
24. Martín-López B, Montes C, Benyas J (2008) Economic valuation of biodiversity conservation: the meaning of numbers. Conserv Biol 22: 624–635.
25. Booth JE, Gaston KJ, Evans KL, Armsworth PR (2011) The value of species rarity in biodiversity recreation: a birdwatching example. Biol Conserv 144: 2728–2732.
26. Garnett ST, Szabo JK, Dutson G (2011) The Action Plan for Australian Birds 2010. Melbourne: CSIRO. 456 p.
27. Garnett ST, Crowley GM, Balmford A (2003) The costs and effectiveness of funding the conservation of Australian threatened birds. BioScience 53: 658–665.
28. McCarthy MA, Thompson CJ, Garnett ST (2008) Optimal investment in conservation of species. J Appl Ecol 45: 1428–1435.
29. McCarthy DP, Donald PF, Scharlemann JPW, Buchanan GM, Balmford A, et al. (2012) Financial costs of meeting two global biodiversity conservation targets: current spending and unmet needs. Science 338: 946–949.
30. Carson RT, Wilks L, Imber D (1994) Valuing the preservation of Australia's Kakadu conservation zone. Oxford Econ Pap 46: 727–749.
31. Loureiro ML, Ojea E (2008) Valuing local endangered species: the role of intra-species substitutes. Ecol Econ 68: 362–369.
32. Meyerhoff J, Angeli D, Hartje V (2012) Valuing the benefits of implementing a national strategy on biological diversity – the case of Germany. Environ Sci Policy 23: 109–119.
33. Langford I, Kontogianni A, Skourtos M, Georgiou S, Bateman I (1998) Multivariate mixed models for open-ended contingent valuation data: willingness to pay for conservation of monk seals. Environ Resour Econ 12: 443–456.
34. Kellert S (1986) Social and perceptual factors in the preservation of animal species. In: Norton BG, editor. The preservation of species. Princeton: Princeton University Press. pp. 50–73.
35. Ojea E, Loureiro ML (20070 Altruistic, egoistic and biospheric values in willingness to pay (WTP) for wildlife. Ecol Econ 63: 807–814.
36. Wilson C, Tisdell C (2005) Knowledge of birds and willingness to support their conservation: an Australian case study. Bird Conserv Int 15: 225–235.
37. Zander KK, Signorello G, De Salvo M, Gandini G, Drucker AG (2013) Assessing the total economic value of threatened livestock breeds in Italy. Ecol Econ 93: 219–229.
38. Fleming CM, Bowden M (2009) Web-based surveys as an alternative to traditional mail methods. J Environ Manage 90: 284–292.
39. Olsen SB (2009) Choosing between Internet and mail survey modes for choice experiment surveys considering non-market goods. Environ Resour Econ 44: 591–610.
40. Lindhjem H, Navrud S (2011) Are Internet surveys an alternative to face-to-face interviews in contingent valuation? Ecol Econ 70: 1628–1637.
41. Sax LJ, Gilmartin SK, Bryant AN (2003) Assessing response rates and non-response bias in web and paper surveys. Res High Educ 44: 409–432.
42. Hanemann M (1984) Welfare evaluations in contingent valuation experiments with discrete responses. Am J Agr Econ 67: 332–341.
43. Carson RT, Hanemann WM (2005) Contingent Valuation. In: Mäler K-G, Vincent JR, editors. Handbook of Environmental Economics. Amsterdam: North-Holland. pp. 821–936.
44. Likert R (1932) A technique for the measurement of attitudes. Arch Psychol 140: 1–55.
45. Bateman IJ, Carson RT, Day B, Hanemann WM, Hanley N, et al. (2002) Economic valuation with stated preference techniques - a manual. Cheltenham: Edward Elgar. 480 p.
46. Haab TC, McConnell KE (2002) Valuing environmental and natural resources: the econometrics of non-market valuation. Cheltenham: Edward Elgar. 352 p.
47. Jeanty PW (2007) wtpcikr: Constructing Krinsky and Robb confidence interval for mean and median willingness to pay (WTP) using Stata. North American Stata Users' Group Meetings 2007, 8.
48. Krinsky I, Robb AL (1986) On approximating the statistical properties of elasticities. Rev Econ Stat 68: 715–719.
49. Morrison MD, Blamey RK, Bennett JW (2000) Minimising payment vehicle bias in contingent valuation studies. Environ Resour Econ 16: 407–422.
50. Meyerhoff J, Liebe U (2006) Protest beliefs in contingent valuation: explaining their motivation. Ecol Econ 57: 583–594.
51. Jakobsson KM, Dragun AK (2001) The worth of a possum: valuing species with the contingent valuation method. Environ Resour Econ 19: 211–227.
52. Loehlin JC (2004) Latent variable models. An introduction to factor, path, and structural equation analysis. Mahwah: Lawrence Erlbaum. 332 p.
53. Child D (2006) The essentials of factor analysis. New York: Continuum. 180 p.
54. Hanemann M, Kanninen B (1999) The statistical analysis of discrete-response CV data. In: Bateman IJ, Willis KG, editors. Valuing environmental preferences: theory and practice of the contingent valuation method in the US, EU, and developing countries. Oxford: Oxford University Press. pp. 302–441.
55. Australian Bureau of Statistics (ABS) (2012) Census Population and Housing 2011. Canberra.
56. Possingham HP, McCarthy MA (2013) Submission to the Senate Committee on the Effectiveness of Threatened Species and Ecological Communities' Protection in Australia. https://senate.aph.gov.au/submissions/comittees/viewdocument.aspx?id = 57e87a9d-39af-4770-8de2-2eb4d7e6268f Accessed 2014 March.
57. BirdLife Australia (2011) Annual Report 2011. Melbourne: BirdLife Australia.
58. BirdLife Australia (2013) Volunteer. Melbourne: BirdLife Australia. http://www.birdlife.org.au/support-us/volunteer/ Accessed 2014 March.
59. McFarlane BL (1994) Specialization and motivations of birdwatchers. Wildlife Soc B 22: 361–370.

60. Hvenegaard GT (2002) Birder specialization differences in conservation involvement, demographics, and motivations. Hum Dimens Wildl 7: 21–36.
61. Wray-Lake L, Flanagan CA, Osgood DW (2010) Examining trends in adolescent environmental attitudes, beliefs, and behaviors across three decades. Environ Behav 42: 61–85.
62. Balmford A, Beresford J, Gree J, Naidoo R, Walpole M, et al. (2009) A global perspective on trends in nature-based tourism. PLoS Biol 7: e1000144.
63. Zander KK, Sing Tyan P, Jinam C, Tuen AA, Garnett ST (2014) Wild and valuable? Tourists' preferences for improvements in orang-utan conservation. Conservat Soc 12: 27–42.

9

Prioritizing Populations for Conservation Using Phylogenetic Networks

Logan Volkmann[1], Iain Martyn[2], Vincent Moulton[3], Andreas Spillner[4], Arne O. Mooers[1,2]*

1 Department of Biological Sciences, Simon Fraser University, Burnaby, British Columbia, Canada, **2** Interdisciplinary Research in the Mathematical and Computational Sciences (IRMACS) Centre, Simon Fraser University, Burnaby, British Columbia, Canada, **3** School of Computing Sciences, University of East Anglia, Norwich, United Kingdom, **4** Department of Mathematics and Computer Science, University of Greifswald, Greifswald, Germany

Abstract

In the face of inevitable future losses to biodiversity, ranking species by conservation priority seems more than prudent. Setting conservation priorities within species (i.e., at the population level) may be critical as species ranges become fragmented and connectivity declines. However, existing approaches to prioritization (e.g., scoring organisms by their expected genetic contribution) are based on phylogenetic trees, which may be poor representations of differentiation below the species level. In this paper we extend evolutionary isolation indices used in conservation planning from phylogenetic trees to phylogenetic networks. Such networks better represent population differentiation, and our extension allows populations to be ranked in order of their expected contribution to the set. We illustrate the approach using data from two imperiled species: the spotted owl *Strix occidentalis* in North America and the mountain pygmy-possum *Burramys parvus* in Australia. Using previously published mitochondrial and microsatellite data, we construct phylogenetic networks and score each population by its relative genetic distinctiveness. In both cases, our phylogenetic networks capture the geographic structure of each species: geographically peripheral populations harbor less-redundant genetic information, increasing their conservation rankings. We note that our approach can be used with all conservation-relevant distances (e.g., those based on whole-genome, ecological, or adaptive variation) and suggest it be added to the assortment of tools available to wildlife managers for allocating effort among threatened populations.

Editor: Tom Gilbert, Natural History Museum of Denmark, Denmark

Funding: Funded by the Natural Sciences and Engineering Research Council of Canada (Discovery grant to AOM), http://www.nserc-crsng.gc.ca/, Simon Fraser University (fellowship to LAV and IM), and Yale University (YIBS fellowship, AOM). The funders had no role in study design, data collection and analysis, decision to publish or preparation of the manuscript.

Competing Interests: The authors have declared that no competing interests exist.

* E-mail: amooers@sfu.ca

Introduction

Extinctions due to human impacts are now unavoidable: even optimistic scenarios predict significant changes in biodiversity by the year 2100 [1,2], with most extinction starting with the loss of isolated populations [3,4].

One prime conservation goal is to preserve genetic variation [5,6], both as a representation of past evolution and raw material for future evolution [7] and, potentially, as a surrogate for improved ecosystem function [8]. However, not all genetic lineages are equally important, with more isolated lineages warranting additional interest because of their expected contribution to total variation [5,8,9]. Indices of evolutionary isolation have been developed to rank species on a phylogenetic tree based on unique and shared evolutionary history (*e.g.*, [10–13]). These metrics use rooted phylogenetic trees with edge lengths as input (Figure 1), and rank tips with less shared history as requiring more urgent conservation attention. For example, the Zoological Society of London has made this approach operational in their "Edge of Existence" programme (www.edgeofexistence.org). In the United States, taxonomic distinctiveness is one of several explicit criteria for prioritizing conservation attention [14]. The extension to populations within species would seem to be straightforward.

Below the species level, Ryder [15] advocated the use of *evolutionarily significant units* (ESUs) to identify populations with genetic variation in need of long-term conservation; this was expanded by Moritz [16] and Waples [17] to the concepts of *management units* (MUs) and *distinct population segments* (DPS), respectively, for species that had undergone more recent range fragmentation. All these population-based approaches have enjoyed wide usage in population genetic studies (*e.g.*, [18–21]), and are the basis for identifying populations worthy of protection in law. Importantly, ESUs assume that the relationships among populations can be represented by a bifurcating tree. However, bifurcating trees often fail to capture the relationships among populations [22]. DPSs and MUs can deal with populations that have more complex interrelationships (Figure 1), but neither of these designations is designed to prioritize among populations. This would seem a major shortcoming if populations do need to be prioritized for conservation attention.

Previous authors have shown that the logic of measuring and maximizing phylogenetic diversity [6], which forms the basis for tree-based prioritization schemes, can be generalized to phylogenetic networks [23–25]. Here we show that the prioritization approaches for trees can also be adapted for populations within species by extending evolutionary isolation indices from trees to networks. We develop efficient algorithms to compute these

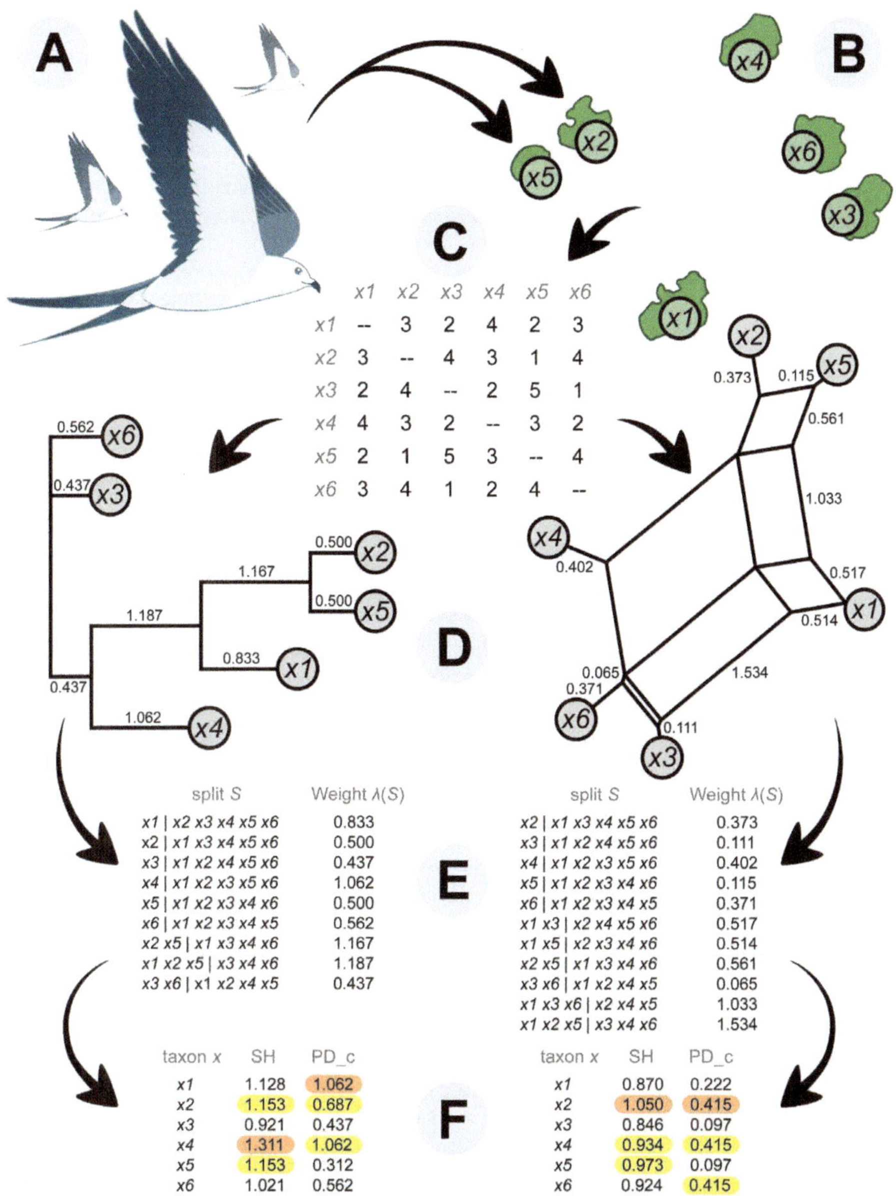

Figure 1. Using pairwise distances to rank species or populations. Consider a hypothetical group of taxa (A)—a set of closely-related species or populations of a single species—that is distributed across several islands in an archipelago (B). Differences among the taxa, labeled *x1* through *x6*, can be organized into a pairwise distance matrix (C). We can represent this matrix either as a phylogenetic tree or as a phylogenetic network (D), where a set of weighted *splits* describes the relationships among the taxa (E). Altogether, these splits represent the group's phylogenetic diversity (PD). By selecting subsets of splits that exclude a given taxon, we can calculate each taxon's contribution to the total PD of the tree or network (F). The *Shapley metric* (SH) and *expected PD complementarity* (PD_c) are different approaches for ranking taxa based on split data. Note that the highest-scoring taxa (highlighted values) can differ considerably depending on the type of metric used and whether the splits come from a tree or network. We discuss the reasons for these differences and methods for ranking taxa in Section (*ii*) of the main text.

indices for NeighborNet networks [26,27], and illustrate their use with heuristic data from two imperiled species, the spotted owl (*Strix occidentalis* Xantus de Vesey 1860) and the mountain pygmy-possum (*Burramys parvus* Broom 1896). The new approach to assessing population differentiation might be of immediate practical use to those tasked with managing discrete populations of a threatened species, and may allow for new policy associated with conservation triage [28].

Methods

We present our approach for prioritizing populations in three steps. First, we briefly review the various approaches for measuring diversity and evolutionary isolation on bifurcating trees of taxa. We then review the properties of NeighborNet networks as a representation of pairwise evolutionary distances and describe how to prioritize taxa by their expected contribution to biodiversity. In File S1, we outline efficient algorithms for estimating evolutionary isolation on NeighborNet networks. Finally, we illustrate the new method of population prioritization using two small published datasets.

(i) Diversity measures on trees and networks

The concept of evolutionary isolation can be understood in terms of a species' biological distinctiveness, which we might measure by comparing its adaptive or non-adaptive traits to those of related species. More generally, our goal is to measure a taxon's contribution to the current and/or future "diversity" in a set of taxa. Several different approaches for quantifying such diversity have been proposed. One of the earliest, described by Weitzman [23], is *expected diversity*. Rather than score taxa individually, this approach seeks to identify the set of taxa that will retain the most diversity on a future tree, given some measure of diversity and a probability of persistence for each potential combination of taxa. Although Weitzman's original diversity metric was rather general, he did consider an example of biological character-state differences that could be represented on a phylogenetic tree.

On such a tree, every taxon contributes an amount of unique evolutionary information denoted by the length of the branch (or *edge*) linking it to all other taxa (Figure 1) [6,23]. This length may be calibrated in units of time (*e.g.*, millions of years) or in raw or inferred genetic distances. Looking specifically at biological systems, Witting and Loeschcke [29] and Faith and Walker [30] combined Weitzman's [23] expected diversity framework with Faith's [6] concept of *phylogenetic diversity* (PD), the latter which specifically calculates the sum of all branch lengths on a tree (see next section). Like Weitzman [23], this *expected PD* approach can be used to identify a set of taxa that maximizes the amount of total tree length retained, given a set of extinction probabilities for the tips.

The related *k of n problem* [6] seeks to identify the most diverse subset of k taxa (*i.e.*, the one that maximizes PD) on a tree of size n. Faith [31] and Weitzman [32] explored the special case where $k=1$, which Faith [33] refers to as the *PD complementarity* of a given taxon.

An independently-derived approach based on Game Theory ([10], first published 2005) explicitly considers the individual contribution of each taxon to future diversity. Like Weitzman's [23] expected diversity framework, all possible subsets of taxa on a tree may persist. By calculating the amount of unique information each taxon contributes to future subsets (*i.e.*, the average length of the edge linking the taxon to all possible future trees), one can rank taxa in order of their relative impact on future diversity. This *Shapley metric* (SH) is almost identical to the ad-hoc *evolutionary distinctness* (ED) metric used by the Zoological Society of London in their Edge of Existence programme (www.edgeofexistence.org). The major difference between the two is that the ED metric is explicitly measured on a rooted tree, as opposed to the more general undirected graph that SH takes as input [34].

The Shapley metric was further refined by Steel *et al.* [35] and named HED (for *heightened evolutionary distinctiveness*). HED is the expected contribution of a given taxon to future subsets of taxa where the subsets are weighted by their probability of persistence. In this case, the focal taxon is assumed to persist (*i.e.*, its probability of extinction does not affect its HED score). On trees, HED is formally equivalent to a form of PD complementarity where the contribution of a taxon is measured with respect to all possible subsets, each weighted by their probability of persistence [33]. Weitzman [32] also arrived at this formulation ten years earlier, which he termed the "distinctiveness" of a taxon, in the context of his "Noah's Ark Problem" of biodiversity preservation. Using Faith's [33] terminology, HED, which combines the concepts of expected PD with PD complementarity, might be considered *expected PD complementarity*.

As a final antecedent, Minh *et al.* [24,25,36] extended PD to phylogenetic networks and presented algorithms for solving the k of n problem to maximize diversity for a given subset size. They referred to this metric as *split diversity* (SD).

In this context it should be possible to measure the PD contribution of individual taxa on a phylogenetic network. Critically for our purposes, the two metrics we use here (SH and HED) do not require a rooted phylogenetic tree, and so can be adapted to networks in the same way that PD indices can [24,25,27,36]. SH and HED are formally defined in File S1 and discussed further below. In short, if we do not have probabilities of extinction for taxa, we assume all future subsets of taxa are equally likely, and calculate SH. If we can estimate (even broadly) the probabilities of persistence of all taxa, we can weight future subsets by their probability, and use HED.

(ii) Interpreting phylogenetic networks, Shapley values, and HED

NeighborNet [26] is a method that permits the representation of pairwise distances between taxa in the form of a network. An important property of NeighborNet networks is that they permit the representation of relationships among the underlying taxa that cannot be depicted on any phylogenetic tree. For example, to the extent that populations exchange migrants, the between-population genetic distance data (F_{ST}) may yield many alternative trees, none of which accurately reflect the actual relationships among these populations (*e.g.*, [37]). The NeighborNet framework, by contrast, accommodates for such phylogenetic uncertainty and will always yield a single network with positive edge lengths, permitting calculation of SH and HED. If a pairwise distance matrix is tree-like (*i.e.*, yields only one possible phylogeny) the resulting NeighborNet output will resemble a phylogenetic tree. Where there is no tree-like history, a network representation should be more informative. Indeed, for many distance matrices (including Example A below, results not shown), the assumptions necessary to produce a tree are not met, and a neighbor-joining tree, for example, produces negative edge lengths. Here, a network representation would definitely be preferred [26].

An example of a very simple matrix of pairwise distances and the resulting network is depicted in Figure 1. Each edge or set of parallel edges in the network corresponds to a partition of the underlying set of taxa into two non-overlapping subsets, called a *split* (S). The *edge length* reflects the *weight* of the split ($\lambda(S)$)—in other words, a component of the pairwise distance (F_{ST}, for example) separating any two taxa. Thus, just as a phylogenetic tree represents a collection of weighted splits (Σ) [38], where each branch of the tree denotes a split, a NeighborNet network represents a weighted collection of splits of the underlying set of taxa. As Figure 1 illustrates, the distance between two tips on a network (*i.e.*, the shortest path between two taxa) represents the observed distance in the distance matrix.

Whether represented on a tree or a network, every split system contains information on the overall diversity of its constituent taxa [5,23]. The conservation planning metric *phylogenetic diversity* (PD) [6] can be calculated for split systems as

$$PD_{(\Sigma,\lambda)}(Y)=\sum_{\substack{A|B\in\Sigma \\ A\cap Y\neq 0, B\cap Y\neq 0}} \lambda(A|B)$$

where Y is a subset of taxa on the tree or network and $\lambda(A|B)$ is the weight of the split between two non-overlapping groups A and

B of taxa. Note that the overall PD for both trees [6,39] and networks [24,25,36] is simply the sum of all split weights (Figure 1).

A very simple approach for measuring an individual taxon's PD contribution, illustrated in Figure 1, is to consider the change in PD when this taxon is removed from the tree or network [40]. This *PD complementarity* (PD_c) metric can be expressed as

$$PD\_c(x) = PD_{(\Sigma,\lambda)}(X) - PD_{(\Sigma,\lambda)}(\bar{X})$$

where X is the set of all taxa in the tree or network and $\bar{X}$ is the subset where a given taxon x has been removed from the underlying distance matrix.

We can also extend the metrics SH and HED from trees to NeighborNet networks using similar ideas for extending PD calculations from trees (*e.g.*, [6,27,29]) to networks (*e.g.*, [24,25,36,41]). On a tree, the Shapley value (ψ_x^{sh}) for taxon x can be defined as the mean split weight of the set of splits defining $x|\bar{X}$, where $\bar{X}$ represents all unique possible subsets of the taxon set X that do not contain x. Importantly, Haake *et al.* [10] present a formal proof that the Shapley value for x can also be calculated as a weighted sum of all the edge lengths on a tree, with the weights determined by the sizes of the sets containing x. This can be presented compactly using split notation as

$$\psi_x^{sh}(\Sigma,\lambda) = \sum_{S\in\Sigma} \frac{|\bar{S}(x)|}{|X||S(x)|}\lambda(S)$$

where (Σ,λ) is the set of splits defined by the network and their weights, $|X|$ is the total number of taxa, $|S(x)|$ is the size of a split set containing the taxon x, $|\bar{S}(x)|$ is the size of the complementary set that does not contain x, and $\lambda(S)$ (following the notation from Minh *et al.* [24,25]) is the split weight, equal to the edge length separating $S(x)$ from $\bar{S}(x)$. To calculate the Shapley value for taxon $x1$ in the network in Figure 1, we take the first split $x2|x1,x3,x4,x5,x6$ to be composed of $|S(x1)|=5$ and $|\bar{S}(x1)|=1$ and $\lambda(S)=0.373$, the second split $x3|x1,x2,x4,x5,x6$ to be composed of $|S(x1)|=5$ and $|\bar{S}(x1)|=1$ and $\lambda(S)=0.111$ and so on. With a taxon set containing six elements, $|X|=6$ and the Shapley value for taxon $x1$ is 0.870 (Figure 1).

As with a phylogenetic tree, the sum of Shapley values will always equal the sum of all parallel split weights in the network. Because the shape of a network reflects the relative distances among its taxa, we should expect outlying taxa (*i.e.*, those connected to the rest of the network by long edges, like taxon $x4$) to show higher values for ψ_x^{sh}. Thus, the Shapley values calculated for a network can reflect the relative degree of isolation of each taxon based on molecular, morphological, or any other relevant distance measure.

Though conceptually similar, the calculation of HED (ψ_x^{hed}) is somewhat more complex, as it accounts for differences in the probability of extinction $p(y)$ for each taxon:

$$\psi_x^{head}(\Sigma,\lambda) =$$

$$\sum_{S\in\Sigma}\left(\prod_{y\in(S(s)-\{x\})} p(y)\right)\cdot\left(1-\left(\prod_{y\in(\bar{S}(x))} p(y)\right)\right)\cdot\lambda(S)$$

Here, the first product operator considers $p(y)$ for every taxon in $S(x)$ but excludes $p(y)$ for taxon x itself [27,35]. The second product operator considers $p(y)$ for every taxon in $\bar{S}(x)$. Unlike SH, the sum of HED scores will not equal the sum of split weights in the split system. We also note that $\lambda(S)$ will influence HED more strongly than $p(y)$ for outlying taxa. Thus, the ranking order for highly isolated populations should be similar for SH and HED, regardless of which populations have a higher extinction probability.

A more detailed mathematical treatment of the SH and HED metrics and efficient algorithms for their computation are given in File S1. For the datasets in this paper, we used the implementation of NeighborNet in the SplitsTree software package [41] to compute networks. For a given matrix of pairwise distances, this yields the network together with the corresponding collection of weighted splits. We also developed custom R scripts (available in File S1) [42] to compute SH and HED on the outputs from SplitsTree.

(iii) Application

We present SH and HED ranking for two datasets based on putatively neutral genetic markers. In the first example (*A*), the size of each population (and hence the probability of extinction for each population) is not known. In the second example (*B*), population sizes are known, allowing us to estimate separate probabilities of extinction for each population.

We selected our two examples based on the following criteria: (1) The species as a whole is of conservation interest (*i.e.*, vulnerable, endangered, or critically endangered), (2) its distribution is fragmented (*i.e.*, we can define multiple populations), (3) sampling efforts have covered its entire range, and (4) genetic analyses have been published or the raw sequence data made publicly available.

Readers should note that the primary goals of this article are to introduce and illustrate our network ranking approach, not to advocate new management decisions for the taxa described below.

Example A.. Spotted owls (*Strix occidentalis*) are distributed throughout late-succession conifer forests in western North America [43]. Four subspecies are currently recognized (Figure 2a): *S. o. caurina* from southern British Columbia to northwest California, *S. o. occidentalis* in California and Nevada, *S. o. lucida* in Utah, Colorado, Arizona, New Mexico, and northern Mexico, and *S. o. juanaphillipsae* in central Mexico [44,45]. Populations in the United States continue to decline due largely to poor timber harvesting practices, but also as a result of climate change and the westward expansion of barred owls (*S. varia* Barton 1799) [46]. *S. o. caurina* (the northern spotted owl) and *S. o. lucida* (the Mexican spotted owl) are threatened subspecies under the United States' Endangered Species Act, and *S. o. occidentalis* (the California spotted owl) is a subspecies of special concern in the state of California [47]. Spotted owls in the American Southwest "sky islands" (mostly *S. o. lucida*) are particularly fragmented and perhaps most suitable for population-level conservation [48]. Although genetic data for the Mexican subspecies remain poor, we can construct a reasonably complete representative phylogenetic network for subspecies in the United States.

Spotted owl mitochondrial sequences were obtained from Genbank (accession numbers AY833608–AY833644, AY836774–AY836776, DQ230843–DQ230888) and aligned in Mega v. 5 [49] using MUSCLE [50]. These sequences comprise about 1105 bp of the control (D-loop) region and represent 86 haplotypes from 32 populations in the United States and Mexico (Figure 2b; Table 1) [48,51]. We ran a standard analysis of molecular variance (AMOVA) [52] on all 298 aligned sequences in Arlequin v. 3.5 [53] using the Kimura 2-Parameter model [54] to compute distances among haplotypes (Φ_{ST}). This procedure

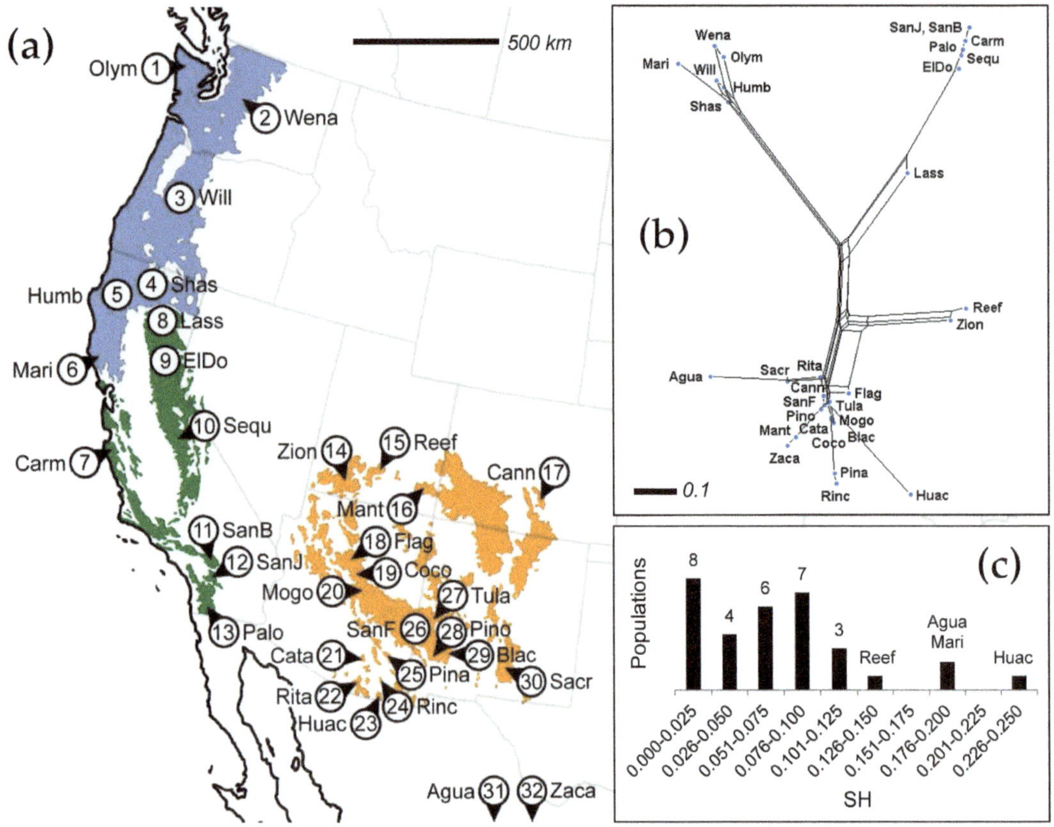

Figure 2. Conservation prioritization of spotted owl (*Strix occidentalis*) populations. (a) Distribution of spotted owls in the United States and the populations sampled by Barrowclough *et al.* [48,51]. Shaded areas denote suitable habitat based on forest cover data [73]. Colors denote the subspecies *S. o. caurina* (blue), *S. o. occidentalis* (green), and *S. o. lucida* (orange). Populations 31 and 32 represent the *S. o. juanaphillipsae* subspecies in Mexico (range not shown). (b) NeighborNet of sampled populations based on mtDNA differentiation (pairwise Φ_{ST} values). (c) Histogram of SH values, highlighting the populations with the highest scores. See Table 1 for an explanation of abbreviations used.

generated a pairwise differentiation matrix for the 32 populations (Table S1). A NeighborNet based on this matrix (Figure 2b) [26] was then constructed in SplitsTree v. 4.11 [41] under default assumptions. Negative Φ_{ST} values were treated as being equal to zero. Because the size of each population is not known, for the purposes of illustration, we gave each population an extinction probability $p_i = 0.5$ when calculating HED—an approach similar to the "PD50" metric used by FISHBASE (www.fishbase.org) [55].

Example B. Mountain pygmy-possums (*Burramys parvus*) are alpine specialists restricted to three small regions of the Australian Alps (Figure 3a). The species depends on block streams and block fields found above 1,400 meters—habitats less than 10 km^2 in total extent [56]. The areas where mountain pygmy-possums still occur are particularly sensitive to destruction and fragmentation. Surveys conducted in the 1990s estimated the adult population size to be 2,600 [57]. A decade later this number had decreased to below 2,000 [56], with signs of continued decline [58]. At present, the IUCN lists mountain pygmy-possums as critically endangered [59].

Because of its restricted distribution and high extinction risk, the species has been subject to extensive population genetic research [58,60–62]. Unlike our example with spotted owls, direct estimates of population sizes are available, within-population sample sizes are uniformly large, and genetic data are available across the mountain pygmy-possums' entire range. This provides us with an opportunity to compare SH to HED and assess the effect of variable population sizes on conservation ranking.

We used a published matrix of genetic differentiation (F_{ST}) based on data from 8 microsatellite loci [58] to construct a phylogenetic network for 13 mountain pygmy-possum populations (Figure 3b). Our methods for generating NeighborNet outputs, and for computing SH and HED, were the same as above.

We modeled the probabilities of extinction for individual populations (p_i) of a given size (n_i) as a negative exponential

$$p_i \propto e^{-cn_i}$$

where the constant of proportionality c is $-\ln(P)/N$, with P being the probability that the entire species goes extinct and N being the total census size of the species (the sum of n_i). We used a conservative 100-year extinction probability for the entire species, $P = 0.4$, to derive HED (see [63]).

Results

Example A

As expected for a set of lineages with a recent history of gene flow, the network for spotted owls is quite non tree-like (Figure 2b). However, populations with the greatest degrees of genetic differentiation, relative to all other populations, occupy nodes subtending the longest edges. Populations at relatively isolated

Table 1. Spotted owl populations sampled by Barrowclough *et al.* [48,51] and ranked by Shapley value (SH) and heightened evolutionary distinctiveness (HED).

Pop.	Code	Subspecies	State	Sampling Locality	*n* ind.	*n* hap.	SH	HED
1	Huac	*lucida*	AZ	Huachuca Mountains	5	2	0.242	7.431E-03
2	Agua	*juanaphillipsae*	—	Aguascalientes, Sierra Fria, Mexico	1	1	0.191	3.983E-03
3	Mari	*caurina*	CA	Marin County	8	3	0.177	4.067E-03
4	Reef	*lucida*	UT	Capitol Reef National Park	9	4	0.133	3.683E-03
5	Wena	*caurina*	WA	Cascade Range	10	8	0.111	1.777E-03
6	Olym	*caurina*	WA	Olympic Peninsula	10	4	0.106	1.729E-03
7	Zion	*lucida*	UT	Zion National Park	7	4	0.105	2.851E-03
8	SanB	*occidentalis*	CA	San Bernardino Mountains	15	1	0.091	9.418E-04
9	SanJ	*occidentalis*	CA	Mount San Jacinto	15	1	0.091	9.418E-04
10	Rinc	*lucida*	AZ	Rincon Mountains	8	4	0.089	2.542E-03
11	Will	*caurina*	OR	Willamette National Forest	15	8	0.081	7.621E-04
12	Carm	*occidentalis*	CA	Carmel Valley	10	1	0.079	5.187E-04
13	ElDo	*occidentalis*	CA	El Dorado National Forest	15	4	0.077	6.214E-04
14	Zaca	*juanaphillipsae*	—	Zacatecas, Sierra de Urica, Mexico	1	1	0.076	2.079E-03
15	Palo	*occidentalis*	CA	Mount Palomar	8	1	0.074	3.994E-04
17	Humb	*caurina*	CA	Humboldt and Siskiyou Counties	30	11	0.072	4.528E-04
16	Sequ	*occidentalis*	CA	Sierra National Forest	15	6	0.071	3.422E-04
18	Pina	*lucida*	AZ	Pinaleno Mountains Graham County	4	2	0.071	1.605E-03
19	Shas	*caurina*	CA	Klamath and Shasta National Forests	16	8	0.067	3.638E-04
20	Mant	*lucida*	UT	Manti-La Sal National Forest	2	2	0.054	1.276E-03
21	Sacr	*lucida*	NM	Sacramento Mountains	8	6	0.047	1.284E-03
22	Lass	*occidentalis*	CA	Lassen National Forest	11	6	0.041	1.006E-04
23	Flag	*lucida*	AZ	San Fransisco Peaks	4	4	0.027	2.144E-04
24	Blac	*lucida*	NM	Black Range	8	6	0.026	3.461E-04
25	Coco	*lucida*	AZ	Coconino Plateau	15	9	0.022	1.039E-04
26	Cata	*lucida*	AZ	Santa Catalina Mountains	5	3	0.021	6.359E-05
27	Pino	*lucida*	NM	Pinos Altos Mountains	5	4	0.020	2.832E-05
28	SanF	*lucida*	NM	San Fransisco Mountains	7	4	0.017	1.020E-06
29	Cann	*lucida*	CO	Near Canon City	4	4	0.017	8.505E-07
30	Rita	*lucida*	AZ	Santa Rita Mountains	4	4	0.017	1.374E-05
31	Mogo	*lucida*	AZ	Mogollon Mesa	8	6	0.017	8.578E-07
32	Tula	*lucida*	NM	Tularosa Mountains	15	12	0.017	2.034E-06

Number of individuals (*n* ind.), number of haplotypes (*n* hap.), SH, and HED scores from the present study are reported.

nodes, such as those from Mount San Jacinto and the Huachuca Mountains, share few mutations with neighboring populations and subsequently exhibit higher pairwise Φ_{ST} values (Table 2; Table S1). Conversely, the (uncorrected) pairwise Φ_{ST} values for closely-related populations are either negative (as great as -1) or close to zero, indicating higher levels of genetic differentiation within these populations than among them [52].

We observe strong geographic structure across the United States consistent with current subspecific designations (Figure 2b). Populations of *S. o. lucida* exhibit a more star-like phylogenetic network that may reflect historical isolation in the "sky islands" of the American Southwest [48]. The intermediate position of the Lassen National Forest population, in contrast, may be due to its location near the point of contact between southern *S. o. caurina* and northern *S. o. occidentalis* [51].

The results of our SH and HED ranking are shown in Table 1. As expected, populations at relatively isolated tips score higher than those closer to the interior of the network (Figure 2b, c). The rankings are highly consistent between the two metrics (Spearman rank correlation =0.91), and the same populations receive top ranking for both SH and HED.

Example B

As with spotted owls, the most genetically differentiated populations of mountain pygmy-possums, namely those in the northern and southern areas of their range, occupy nodes that are separated from most other populations by long edges (Figure 3b). Overall the structure of our network is in good agreement with the species' present distribution (Figure 3a). Given the habitat requirements and limited dispersal ability of mountain pygmy-possums, it is not likely that Mount Buller and Kosciusko National Park still exchange migrants with the Bogong High Plains [58]. In contrast, the close grouping of central populations in our

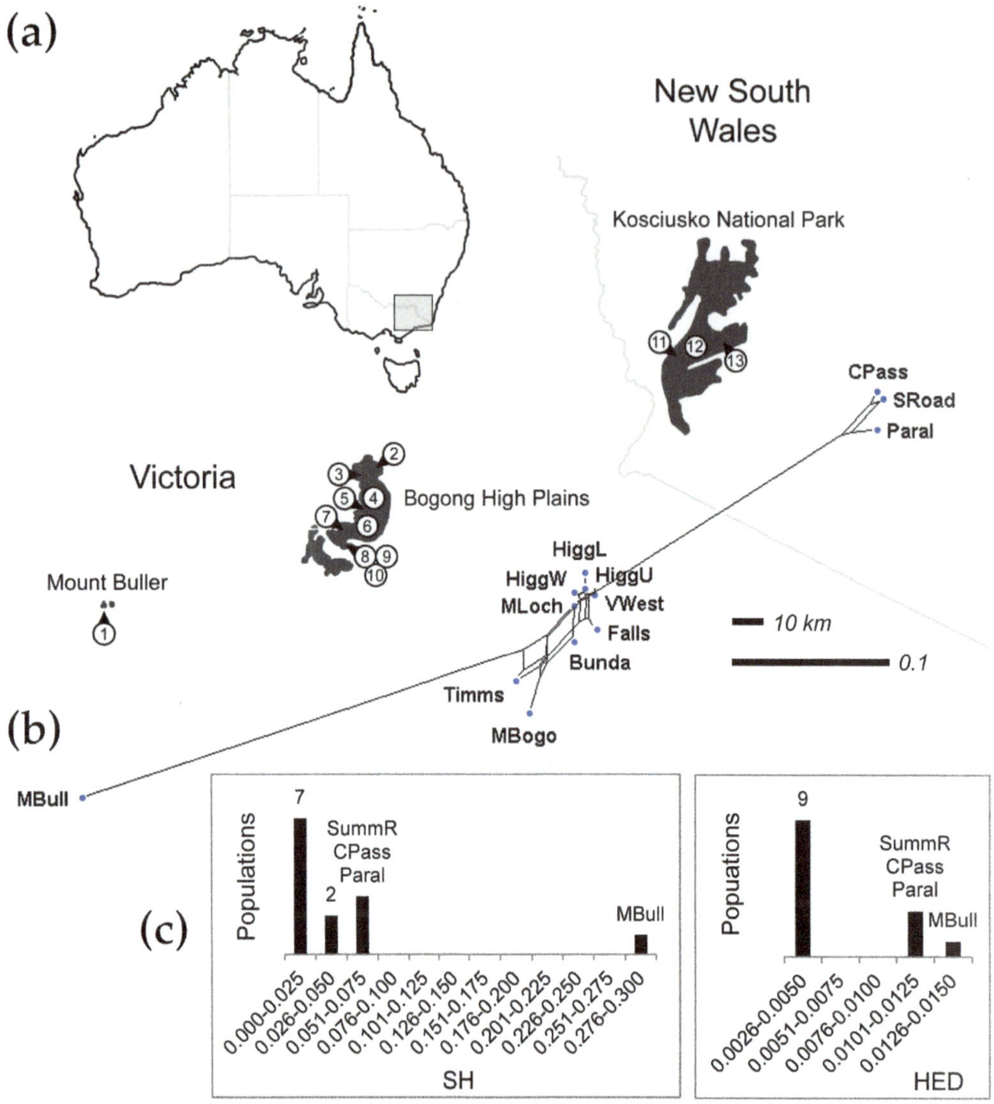

Figure 3. Conservation prioritization of mountain pygmy-possum (*Burramys parvus*) populations. (a) Distribution of mountain pygmy-possums in Australia (gray inset), showing populations sampled by Mitrovski *et al.* [58]. Shaded areas denote suitable habitat above 1,400 m. (b) NeighborNet of sampled populations based on microsatellite differentiation (pairwise F_{ST} values). (c) Histograms of SH and HED values, highlighting the populations with the highest scores. See Table 2 for an explanation of abbreviations used.

phylogenetic network, and subsequently their low SH and HED, is consistent with a shared history and/or recent gene flow.

The ranking results are shown in Table 2. Again, the phylogenetic network for mountain pygmy-possums reflects geographic distribution. Although we did not make *a priori* group assignments based on sampling location, the 13 populations still partition into northern, central, and southern regions. Again, outlying populations on the network tend to receive higher SH and HED scores. Unsurprisingly, the small and isolated Mount Buller population consistently ranks highest. For HED, no bias towards small or large populations is apparent; populations with high extinction probabilities do not necessarily receive high scores [35]. Again, although ranking order changes slightly between SH and HED, the two methods provide roughly equivalent rankings (Spearman rank correlation = 0.97, Figure 3c). High-ranking populations are similar in both cases.

We note that SH and HED calculations on a network consider a taxon's distance from *all other* taxa. Thus, although the three northern populations are closely related to each other, they still receive high SH and HED scores because of the long branches separating them from the central and southern populations (Table 2; Figure 3c).

Discussion

The premise of conservation below the species level is not novel. Faith's original [6] discussion of prioritizing taxa also considered populations on a tree. Several economically-important taxa have received population-level management since the late 1980s, *e.g.*, Atlantic salmon (*Salmo salar* Linnaeus 1758) [64], brown trout (*Salmo trutta* Linnaeus 1758) [65] and yellowfin tuna (*Thunnus albacares* Bonnaterre 1788) [66]. Managing species at the population level implies at least an informal ranking scheme, one which would rely, for example, on estimates of habitat patch size or effective population size [67]. Habitat degradation, climate change, and the demands of a growing human population have

Table 2. Mountain pygmy-possum populations sampled by Mitrovski *et al.* [58] and ranked by Shapley value (SH) and heightened evolutionary distinctiveness (HED).

Pop.	Code	Region	Sampling Locality	*n* ind.	*n* all.	*r*	*N*	*pi*	SH	HED
1	MBull	Southern	Mount Buller	66	3.38	2.29	150	0.9072	0.292	1.497E-02
2	CPass	Northern	Charlottes Pass	44	6.00	5.21	45	0.9712	0.072	1.153E-02
3	Paral	Northern	Paralyser	40	6.63	5.77	22	0.9858	0.071	1.089E-02
4	SummR	Northern	Summit Road	43	6.13	5.16	25	0.9839	0.068	1.138E-02
5	MBogo	Central	Mount Bogong	42	6.50	5.77	100	0.9372	0.045	4.497E-03
6	Timms	Central	Timm Spur	120	6.88	5.32	120	0.9251	0.028	3.859E-03
7	Falls	Central	Falls Creek	35	6.63	5.73	30	0.9807	0.020	3.051E-03
8	HiggL	Central	Mount Higginbotham L	17	5.25	5.25	50	0.9681	0.019	2.850E-03
9	Bunda	Central	Bundara	78	7.25	5.45	120	0.9251	0.018	2.858E-03
10	HiggW	Central	Mount Higginbotham W	59	7.63	6.01	250	0.8502	0.013	2.659E-03
11	VWest	Central	Pretty Valley West	69	7.00	5.95	50	0.9681	0.010	2.541E-03
12	HiggU	Central	Mount Higginbotham U	56	7.25	5.76	50	0.9681	0.010	2.420E-03
13	Mloch	Central	Mount Loch	93	7.00	5.80	400	0.7714	0.010	2.646E-03

Number of individuals (*n* ind.), number of alleles (*n* all.), allelic richness (*r*), and adult population sizes (*N*) are reported from previously-published data. Probabilities of extinction (*pi*, with $P = 0.4$), SH, and HED scores from the present study are also shown.

ensured the continued fragmentation of species' ranges over the next century (see, *e.g.*, pikas (*Ochotona princeps* Richardson 1828) [68]). In the midst of such rapid change, managing an imperiled species over its entire range may no longer be feasible, such that population rankings may be necessary.

Phylogenetic diversity measures have previously been adapted for non-treelike population genetic data (*e.g.*, [24,25,36]). However, the PD complementarity scores that can be obtained from these methods are contingent, *i.e.*, subject to change if extinction alters the shape of the network. Ours is the first ranking scheme to consider a taxon's contribution to all possible future networks (*sensu* Weitzman [23]), a potentially relevant framework for preserving future biodiversity. Given the stochastic nature of extinction, the general ranking systems offered by SH and HED may be more useful to wildlife managers than those that only consider the present structure of a phylogenetic network. Unlike previous approaches based on PD (*e.g.*, [24,25,36]), SH and HED rankings allow one to lengthen or shorten the list of taxa to conserve in the event that resources become more or less available.

Molecular techniques are now inexpensive and robust enough to make population genetic sampling a standard component of conservation planning, and we argue that a phylogenetic network approach offers insight into a species' population structure complementary to the current statistical assessments of differentiation employed by MUs and DPSs [16,17]. We encourage researchers to employ such networks in future population genetic studies to provide conservation agencies with more informative analysis of datasets. Genotyping at multiple loci will provide more accurate estimates of population differentiation and allow for more sophisticated analyses of conservation-relevant processes such as recent demographic history and gene flow [69].

We acknowledge that the mathematical shortcomings of Φ_{ST} and F_{ST} estimators [70] may influence the magnitude and ranking of SH and HED scores, depending on the number of loci measured and the distribution of genetic diversity in a set of taxa. Our intention here is not to solve these theoretical problems but to demonstrate our network-based prioritization method with existing data. Newly-developed metrics such as Jost's D can be used to calculate SH and HED just as readily as traditional Φ_{ST} and F_{ST} distances, and we encourage the use of such unbiased estimators whenever such data are available. Indeed, any conservation metric of difference (*e.g.*, ecological, genomic, adaptability) can be used.

Several properties of the networks described here invite further investigation. In both our heuristic datasets, geographically peripheral populations are more genetically isolated, meaning they would rank highly on SH and HED. However, this was based on only very few putatively neutral markers. Two related questions concern how processes such as demographic history and current patterns of gene flow map onto genetic isolation as we measure it here, and also how phylogenetic networks map onto networks produced from ecological data (*e.g.*, niche use differences among populations).

We do not advocate relying solely on genetic isolation when deciding where and how resources should be allocated at the population level. Total population genetic diversity (*i.e.*, number of haplotypes) might also be considered. We note that in our examples, low-ranking populations tend to be geographically close to one another, meaning that their haplotypes are often shared. Important differences in ecology and adaptability [71] and current and future connectivity [72], must also be considered. However, our network approach and ranking system based on genetic differentiation can supplement existing systems of MUs and DPSs to improve the conservation of evolutionarily distinct populations in a world of increasing pressures and limited resources.

Supporting Information

File S1 Mathematical treatment of SH and HED and annotated R code for calculating both metrics.

Table S1 Pairwise genetic distances (Φ_{ST}) for spotted owl (*Strix occidentalis*) populations based on data from Barrowclough *et al.* [48,51], with negative values set to zero.

Acknowledgments

We thank the Isaac Newton Center (Cambridge University) for facilitating our collaboration, and M Hart, JB Joy, M Steel, three anonymous reviewers and the FAB Lab at Simon Fraser University for discussion and comments on the manuscript. All R scripts used in this paper are available in File S1.

Author Contributions

Conceived and designed the experiments: AOM VM AS. Performed the experiments: LV IM AS. Analyzed the data: LV AOM. Contributed reagents/materials/analysis tools: IM AS. Wrote the paper: LV AOM.

References

1. Sala OE, Chapin FS, Armesto JJ, Berlow E, Bloomfield J, et al. (2000) Global biodiversity scenarios for the year 2100. Science 287: 1770–1774.
2. Jenkins M (2003) Prospects for biodiversity. Science 302: 1175–1177.
3. Hughes JB, Daily GC, Ehrlich PR (1997) Population diversity: its extent and extinction. Science 278: 689–692.
4. Ceballos G, Ehrlich PR (2002) Mammal population losses and the extinction crisis. Science 296: 904–907.
5. Vane-Wright RI, Humphries CJ, Williams PH (1991) What to protect?—Systematics and the agony of choice. Biol Conserv 55: 235–254.
6. Faith DP (1992) Conservation evaluation and phylogenetic diversity. Biol Conserv 61: 1–10.
7. Bowen BW, Roman J (2005) Gaia's handmaidens: the Orlog model for conservation biology. Conserv Biol 19: 1037–1043.
8. Redding DW, Mooers AØ (2010) Can systematists help decide the relative worth of bits of biodiversity? Systematist 32: 4–8.
9. May RM (1990) Taxonomy as destiny. Nature 347: 129–130.
10. Haake CJ, Kashiwada A, Su FE (2008) The Shapley value of phylogenetic trees. J Math Biol 56: 479–497.
11. Pavoine S, Ollier S, Dufour AB (2005) Is the originality of a species measurable? Ecol Lett 8: 579–586.
12. Redding DR, Mooers AØ (2006) Incorporating evolutionary measures into conservation prioritization. Conserv Biol 20: 1670–1678.
13. Isaac NBJ, Turvey ST, Collen B, Waterman C, Baillie JEM (2007) Mammals on the EDGE: conservation priorities based on threat and phylogeny. PLoS One 2: e296.
14. Fay JJ, Thomas WL (1983) Endangered species listing and recovery priority guidelines. Fed Regist 48: 43098–43105.
15. Ryder OA (1986) Species conservation and systematics: the dilemma of subspecies. Trends Ecol Evol 1: 9–10.
16. Moritz C (1994) Defining 'evolutionarily significant units' for conservation. Trends Ecol Evol 9: 373–375.
17. Waples RS (1991) Pacific salmon, *Oncorhynchus* spp., and the definition of 'species' under the endangered species act. Mar Fish Rev 53: 11–22.
18. Firestone KB, Elphinstone MS, Sherwin WB, Houlden BA (1999) Phylogeographical population structure of tiger quolls *Dasyurus maculatus* (Dasyuridae: Marsupialia), an endangered carnivorous marsupial. Mol Ecol 8: 1613–1625.
19. Gorbics CS, Bodkin JL (2001) Stock structure of sea otters (*Enhydra lutris kenyoni*) in Alaska. Mar Mammal Sci 17: 632–647.
20. Haig SM, Wagner RS, Forsman ED, Mullins TD (2001) Geographic variation and genetic structure in spotted owls. Conserv Genet 2: 25–40.
21. Jensen EL, Govindarajula P, Russello MA (2013) When the shoe doesn't fit: applying conservation unit concepts to western painted turtles at their northern periphery. Conserv Genet. doi: 10.1007/s10592-013-0535-2.
22. Fitch WM (1997) Networks and viral evolution. J Mol Evol 44: S65–S75.
23. Weitzman ML (1992) On diversity. Q J Econ 107: 363–405.
24. Minh BQ, Pardi F, Klaere S, von Haeseler A (2009) Budgeted phylogenetic diversity on circular split systems. IEEE/ACM Trans Comput Biol Bioinform 6: 22–29.
25. Minh BQ, Klaere S, von Haeseler A (2009) Taxon selection under split diversity. Syst Biol 57: 586–594.
26. Bryant D, Moulton V (2004) Neighbor-Net: an agglomerative method for the construction of phylogenetic networks. Mol Biol Evol 21: 255–265.
27. Martyn I, Kuhn TS, Mooers AØ, Moulton V, Spillner A (2012) Computing evolutionary distinctiveness indices in large scale analysis. Algorithms Mol Biol 7: 6.
28. Bottrill MC, Joseph LN, Carwardine J, Bode M, Cook C, et al. (2008) Is conservation triage just smart decision making? Trends Ecol Evol 23: 649–654.
29. Witting L, Loeschcke V (1995) The optimization of biodiversity conservation. Biol Conserv 71: 205–207.
30. Faith DP, Walker PA (1996) DIVERSITY – TD. In: Faith DP, Nicholls AO, editors. BioRap: rapid assessment of biodiversity. Volume 3: tools for assessing biodiversity priority areas. Canberra: The Australian BioRap Consortium. pp. 63–74.
31. Faith DP (1996) Conservation priorities and phylogenetic pattern. Conserv Biol 10: 1286–1289.
32. Weitzman ML (1998) The Noah's Ark Problem. Econometrica 66: 1279–1298.
33. Faith DP (2008) Threatened species and the potential loss of phylogenetic diversity: conservation scenarios based on estimated extinction probabilities and phylogenetic risk analysis. Conserv Biol 22: 1461–1470.
34. Hartmann K (2013) The equivalence of two phylogenetic biodiversity measures: the Shapley value and the Fair Proportion index. J Math Biol 67: 1163–1170.
35. Steel M, Mimoto A, Mooers AØ (2007) Hedging our bets: the expected contribution of species to future phylogenetic diversity. Evol Bioinform Online 3: 237–244.
36. Minh BQ, Klaere S, von Haeseler A (2007) Phylogenetic diversity on split networks. Technical Report NI07090-PLG. Cambridge: Isaac Newton Institute. 19 p.
37. Poczai P, Mátyás KK, Szabó I, Varga I, Hyvönen J, et al. (2011) Genetic variability of thermal *Nymphaea* (Nymphaeaceae) populations based on ISSR markers: implications on relationships, hybridization, and conservation. Plant Mol Biol Rep 29: 906–918.
38. Buneman P (1971) The recovery of trees from measures of dissimilarity. In: Hodson FR, Kendall DG, Tautu P, editors. Mathematics in the archaeological and historical sciences. Edinburgh: Edinburgh University Press. pp. 387–395.
39. Steel M (2005) Phylogenetic diversity and the greedy algorithm. Syst Biol 54: 527–529.
40. Faith DP (1994) Phylogenetic pattern and the quantification of organismal biodiversity. Phil Trans R Soc Lond B 345: 45–58.
41. Huson DH, Bryant D (2006) Application of phylogenetic networks in evolutionary studies. Molec Biol Evol 23: 254–267.
42. R-Development Core Team (2009) R: a language and environment for statistical computing. Vienna: R Foundation for Statistical Computing. 1706 p.
43. Gutiérrez RJ, Franklin AB, Lahaye WS (1995) Spotted owl (*Strix occidentalis*). In: Poole A, editor. The birds of North America online. Ithaca: Cornell Lab of Ornithology. Available: http://bna.birds.cornell.edu/bna/species/179. Accessed 28 November 2011.
44. Dickerman RW (1997) Geographic variation in southwestern US and Mexican spotted owls, with the description of a new subspecies. In: Dickerman RW, editor. The era of Allen R. Phillips: a festschrift. Albuquerque: Horizon Communications. pp. 45–48.
45. Barrowclough GF, Gutiérrez RJ, Groth JG (1999) Phylogeography of spotted owl (*Strix occidentalis*) populations based on mitochondrial DNA sequences: gene flow, genetic structure, and a novel biogeographic pattern. Evolution 53: 919–931.
46. US Fish and Wildlife Service (2011) Revised recovery plan for the northern spotted owl (*Strix occidentalis caurina*). Portland, OR: US Fish and Wildlife Service. 277 p.
47. Gould GI (1985) Management of spotted owls by the California Department of Fish and Game. In: Gutiérrez RJ, Carey AB, editors. Ecology and management of the spotted owl in the Pacific Northwest. Portland, OR: US Department of Agriculture, Forest Service. pp. 21–26.
48. Barrowclough GF, Groth JG, Mertz LA, Gutiérrez RJ (2006) Genetic structure of Mexican spotted owl (*Strix occidentalis lucida*) populations in a fragmented landscape. Auk 123: 1090–1102.
49. Tamura K, Peterson D, Peterson N, Stecher G, Nei M, et al. (2011) MEGA5: molecular evolutionary genetics analysis using maximum likelihood, evolutionary distance, and maximum parsimony methods. Mol Biol Evol 28: 2731–2739.
50. Edgar RC (2004) MUSCLE: multiple sequence alignment with high accuracy and high throughput. Nucl Acids Res 32: 1792–1797.
51. Barrowclough GF, Groth JG, Mertz LA, Gutiérrez RJ (2005) Genetic structure, introgression, and a narrow hybrid zone between northern and California spotted owls (*Strix occidentalis*). Mol Ecol 14: 1109–1120.
52. Excoffier L, Smouse PE, Quattro JM (1992) Analysis of molecular variance inferred from metric distances among DNA haplotypes: application to human mitochondrial DNA restriction data. Genetics 131: 479–491.
53. Excoffier L, Laval G, Schneider S (2005) Arlequin (version 3.0): An integrated software package for population genetics data analysis. Evol Bioinform Online 1: 47–50.
54. Kimura M (1980) A simple method for estimating evolutionary rates of base substitutions through comparative studies of nucleotide sequences. J Mol Evol 16: 111–120.
55. Faith DP, Reid CAM, Hunter J (2004) Integrating phylogenetic diversity, complementarity, and endemism for conservation assessment. Conserv Biol 18: 255–261.
56. Heinze D, Broome LS, Mansergh I (2004) A review of the ecology and conservation of the mountain pygmy-possum *Burramys parvus* (Broom 1896). In: Hillis DM, Moritz C, Mable BK, editors. The biology of the Australian possums and gliding possums. Sunderland, MA: Sinauer Associates. pp. 254–267.
57. Mansergh I, Broome L (1994) The mountain pygmy-possum of the Australian Alps, Australia. Sydney: New South Wales University Press. 114 p.
58. Mitrovski P, Heinze DA, Broome L, Hoffmann AA, Weeks AR (2007) High levels of variation despite genetic fragmentation in populations of the endangered mountain pygmy-possum, *Burramys parvus*, in alpine Australia. Mol Ecol 16: 75–87.

59. Menkhorst P, Broome L, Driessen M (2008) Burramys parvus. IUCN Red List of Threatened Species. Available: http://www.iucnredlist.org/details/full/3339/0. Accessed 28 November 2011.
60. Osborne MJ, Norman JA, Christidis L, Murray ND (2000) Genetic distinctness of isolated populations of an endangered marsupial, the mountain pygmy-possum, *Burramys parvus*. Mol Ecol 9: 609–613.
61. Mitrovski P, Heinze DA, Guthridge K, Weeks AR (2005) Isolation and characterization of microsatellite loci from the Australian endemic mountain pygmy-possum, *Burramys parvus* Broom. Mol Ecol Notes 5: 395–397.
62. Mitrovski P, Hoffmann AA, Heinze DA, Weeks AR (2008) Rapid loss of genetic variation in an endangered possum. Biol Lett 4: 134–138.
63. Mooers AØ, Faith DP, Maddison WP (2008) Converting endangered species categories to probabilities of extinction for phylogenetic conservation prioritization. PLoS One 3: e3700.
64. Fontaine PM, Dodson JJ, Bernatchez L, Slettan A (1997) A genetic test of metapopulation structure in Atlantic salmon (*Salmo salar*) using microsatellites. Can J Fish Aquat Sci 54: 2434–2442.
65. Ferguson A (1989) Genetic differences among brown trout, *Salmo trutta*, stocks and their importance for the conservation and management of the species. Freshw Biol 21: 35–46.
66. Ward RD (1995) Population genetics of tunas. J Fish Biol 47: 259–280.
67. Lande R, Barrowclough GF (1987) Effective population size, genetic variation, and their use in population management. In: Soulé ME, editor. Viable populations for conservation. Cambridge: Cambridge University Press. pp. 87–123.
68. Galbreath KE, Hafner DJ, Zamudio KR (2009) When cold is better: climate-driven elevation shifts yield complex patterns of diversification and demography in an alpine specialist (American pika, *Ochotona princeps*). Evolution 63: 2848–2863.
69. Marko PB, Hart MW (2011) The complex analytical landscape of gene flow inference. Trends Ecol and Evol 26: 448–456.
70. Jost L (2008) GST and its relatives do not measure genetic differentiation. Mol Ecol 17: 4015–4026.
71. Bonin A, Nicole F, Pompanon F, Miaud C, Taberlet P (2007) Population adaptive index: a new method to help measure intraspecific genetic diversity and prioritize populations for conservation. Conserv Biol 21: 697–708.
72. Geffen E, Waidyaratne S, Dalén L, Angerbjörn A, Vila C, et al. (2007) Sea ice occurrence predicts genetic isolation in the artic fox. Mol Ecol 16: 4241–4255.
73. US Geological Service (2013) The national atlas of the United States of America. Washington, DC: US Department of Interior, Geological Service. Available: http://www.nationalatlas.gov/biology.html. Accessed 28 November 2011.

10

Body Size Reductions in Nonmammalian Eutheriodont Therapsids (Synapsida) during the End-Permian Mass Extinction

Adam K. Huttenlocker*¤

Department of Biology and Burke Museum of Natural History and Culture, University of Washington, Seattle, Washington, United States of America

Abstract

The extent to which mass extinctions influence body size evolution in major tetrapod clades is inadequately understood. For example, the 'Lilliput effect,' a common feature of mass extinctions, describes a temporary decrease in body sizes of survivor taxa in post-extinction faunas. However, its signature on existing patterns of body size evolution in tetrapods and the persistence of its impacts during post-extinction recoveries are virtually unknown, and rarely compared in both geologic and phylogenetic contexts. Here, I evaluate temporal and phylogenetic distributions of body size in Permo-Triassic therocephalian and cynodont therapsids (eutheriodonts) using a museum collections-based approach and time series model fitting on a regional stratigraphic sequence from the Karoo Basin, South Africa. I further employed rank order correlation tests on global age and clade rank data from an expanded phylogenetic dataset, and performed evolutionary model testing using Brownian (passive diffusion) models. Results support significant size reductions in the immediate aftermath of the end-Permian mass extinction (ca. 252.3 Ma) consistent with some definitions of Lilliput effects. However, this temporal succession reflects a pattern that was underscored largely by Brownian processes and constructive selectivity. Results also support two recent contentions about body size evolution and mass extinctions: 1) active, directional evolution in size traits is rare over macroevolutionary time scales and 2) geologically brief size reductions may be accomplished by the ecological removal of large-bodied species without rapid originations of new small-bodied clades or shifts from long-term evolutionary patterns.

Editor: Richard J. Butler, University of Birmingham, United Kingdom

Funding: Funding was provided by the National Science Foundation Doctoral Dissertation Improvement Grant Program (1209018), Evolving Earth Foundation, University of Washington Department of Biology WRF-Hall Fellowship, Burke Museum of Natural History and Culture Vertebrate Paleontology Fellowship, and the Society of Vertebrate Paleontology Richard Estes Memorial Grant. The funders had no role in study design, data collection and analysis, decision to publish, or preparation of the manuscript.

Competing Interests: The author has declared that no competing interests exist.

* E-mail: ahuttenlocker@gmail.com

¤ Current address: Department of Biology, University of Utah, and Natural History Museum of Utah, Salt Lake City, Utah, United States of America

Introduction

Body size is an important biological trait that affects organismal fitness and imparts a strong influence on ecological, physiological, and life history attributes of species [1–4]. As such, size and size-related traits have received considerable attention from paleobiologists interested in characterizing the tempo and mode of morphological evolution in extinct groups. Famous examples include 'Cope's rule,' an apparent tendency for animal lineages to achieve larger sizes over time [5–7], and more recently the 'Lilliput effect,' short-term size reductions in animal taxa following mass extinction events [8–10]. In the case of Cope's rule, workers historically attributed patterns of body size increase to non-random selection within lineages over geologic time, an intuitive conclusion when surveying some of the best examples of macroevolutionary transitions in the fossil record [11–14]. However, several studies suggest that many of these important paleontological transitions (e.g., expansions in dinosaur body sizes) were in fact influenced by Brownian processes of trait evolution, random tendencies for lineages to diffuse through size space with increasing disparity [7,15–18]. Considerably less attention has been paid to the underlying influences of Lilliput-type patterns and mass extinctions. Background patterns of origination/extinction during the Neogene indicate lower turnover rates (i.e., long species durations) in smaller-bodied mammals, possibly due to their higher reproductive rates, large population sizes, and frequent sleep-or-hide burrowing behavior compared to large mammals [19,20]. Nevertheless, size-related extinction risk remains controversial [21–23], and the supposed chaotic and indiscriminate nature of mass extinctions, as in the end-Cretaceous and end-Permian mass extinctions, calls into question whether these dynamics are applicable to rapid, global-scale catastrophes [24].

Though originally reserved for within-lineage size decreases in survivor taxa [8], the term 'Lilliput' has been repurposed as a pattern that may encompass a number of other underlying processes, such as differential extinction of large-bodied taxa and rapid diversifications of new small-bodied clades following mass extinctions [9,10], or a combination of these [25]. These patterns have been particularly well documented in Triassic marine invertebrate communities following the end-Permian mass extinction [9,25–28] and anecdotally in terrestrial vertebrates of the earliest Triassic *Lystrosaurus* Assemblage Zone, Karoo Basin of

South Africa [29–31]. However, the extent to which the end-Permian mass extinction influenced body size reductions and possible departures from background patterns of body size evolution are imprecisely known. This is in part because the temporal and phylogenetic components of body size evolution before and after the extinction have never been compared. Here, I use time series analysis and a reference phylogeny of Permian through Triassic eutheriodont therapsids (therocephalians and cynodonts) to better understand the tempo and mode of body size evolution in the therapsid forerunners of mammals during a mass extinction interval. Notably, therocephalian eutheriodonts originated at relatively large sizes during the Middle Permian (estimated maximum skull length, ~40 centimeters), while many of their Triassic relatives, including some cynodonts and baurioid therocephalians, achieved comparatively diminutive sizes (estimated skull length ~2.5–3.0 centimeters in brasilodontids and early Mammaliaformes). Therocephalians and cynodonts provided a rare opportunity to study post-extinction body size shifts in terrestrial vertebrates due to their range of sizes, relative generic richness in earliest Triassic faunas, and differential survivorship of the end-Permian extinction [32,33]. I identify short-term reductions in maximum and mean body sizes during the earliest Triassic, and suggest that geologic patterns of size reductions in lowermost Triassic rocks do not reflect rapid diversifications of new small-bodied clades or changes in evolutionary patterns following the extinction. Rather, observed short-term body size reductions arose from interplay between a backdrop of increasing size disparity and a temporary culling of large-bodied therapsid predators during the end-Permian extinction. Institutional abbreviations are listed in Appendix S1.

Approach

Stratigraphic (Non-phylogenetic) Patterns

Non-random extinction patterns may be discernible with regard to some organismal traits, especially when studied in the context of existing patterns of trait evolution during background intervals. Non-phylogenetic investigations of selectivity in Cretaceous-Paleogene marine invertebrates have suggested that mass extinction may bring about changes in selectivity patterns that reset the stage for biotic recoveries. This departure from normal (background) selectivity patterns has been termed "non-constructive selectivity" [24]. Patterns of ecomorphological change have been well studied in stratigraphic series, particularly in Permo-Triassic and Cretaceous-Paleogene marine faunas [24,34,35], but similar approaches have only been recently applied to body size in terrestrial vertebrates. For example, Wilson [36] identified significant post-extinction body size reductions in a succession of Cretaceous-Paleogene mammals from northeastern Montana, and suggested constructive selectivity against large, dietary specialists possibly due to depressed primary productivity during that time. Similar patterns have been suggested in earliest Triassic terrestrial disaster faunas, including selection for small-bodied generalists with drought-tolerant food sources and fossorial or semi-aquatic habits [29,30,32,37]. I propose that investigations of selectivity during the end-Permian mass extinction should address firstly whether non-random patterns are discernible and, if so, whether they deviate from pre-existing patterns of trait evolution. First, I test whether maximum and mean body sizes decreased significantly in the earliest Triassic *Lystrosaurus* Assemblage Zone (hereafter 'AZ') using non-phylogenetic time series and statistical comparisons of size distributions between temporally successive eutheriodont assemblages. Second, I chose to implement time series modeling and rank correlation tests to evaluate whether reductions, if present, signified transient shifts or longer-term directional size shifts both across the extinction and during pre-extinction intervals. A transient shift following mass extinction (an expectation of Lilliput-type patterns) would be indicated by short-term shifts in stratigraphic series without a significant mean step change across the entire sampled interval (or during pre-extinction intervals).

Geologic controls and sampling. Sampling of non-isotaphonomic intervals and unevenness of sampling may bias the quality and size controls of the present sample. These can include but are not restricted to available outcrop area, collection intensity, and sediment coarseness and sorting, which should be considered given the comparatively small sizes observed in the Triassic *Lystrosaurus* AZ. Recent assessments of geologic controls on paleodiversities have found potential associations between outcrop area and taxonomic richness of pre-extinction assemblages, but the outlying *Lystrosaurus* AZ exhibits unexpectedly modest richness that is unexplainable by outcrop area alone [38–41], in spite of intense collecting effort and large numbers of specimens from this interval. A comparison of the diversities of therocephalians and cynodonts from these assemblages yields comparable results (Appendix S2). Also, differences in grain sizes may produce differential size-sorting in fossil assemblages, where coarse sediments deposited in high-energy environments may preferentially preserve larger, more robust skeletal remains. However, the fine-grained floodplain deposits and paleosols of the Permian Balfour Formation give way to rhythmically-bedded sheet sandstone bodies and thick channel sands and conglomerates of the Triassic Katberg Formation, producing a general coarsening upward in the Permo-Triassic boundary sequence of the Karoo. It is therefore unlikely that this coarsening upward has produced a bias in small-bodied eutheriodonts in the *Lystrosaurus* AZ. While taphonomic and collecting biases are important considerations, I suggest that they would have had limited impacts on diversity and body size shifts observed in the *Lystrosaurus* AZ.

Phylogenetic Patterns

Size shifts in post-extinction faunas, if present, may have been underpinned to some extent by stochastic processes of trait evolution. Long-term negative shifts observed throughout the sampled interval may indicate background patterns of trait evolution (e.g., general tendencies toward body size reductions). Size shifts may alternatively represent rapid diversifications of new clades whose success was associated with the trait in question (in this case, small body size). Rapid diversification of small-bodied taxa following extinction is one explanation for Lilliput-type patterns, although few examples have been demonstrated [25,42]. Nevertheless, an inadequate understanding of phenotypic evolution during background intervals hinders comparisons of post-extinction dynamics. It is therefore informative to compare stratigraphic patterns of trait evolution to those patterns inferred by cladogenetic branching events. This approach allows investigators to identify directionality, characterize the timing and location of important cladogenetic events, and distinguish those features from more general (long-term) features of phenotypic evolution during background intervals. I chose to implement ancestor-descendant tests in order to (1) evaluate the relative frequency at which smaller descendants evolved from larger ancestors in a reconstructed lineage of ancestor-descendant pairs, and to (2) identify the branches (internodes) in which those negative shifts occurred. Diversification rate shifts were also assessed to determine whether post-extinction radiations of small, Triassic taxa during biotic recovery could have contributed to observed shifts in size distributions of eutheriodont faunas. Finally,

I evaluated the fit of phylogenetic models to determine whether trait evolution was best approximated by directional, 'stabilized,' or initially rapid but decelerating ('early burst') models, or a Brownian motion model. These approaches establish a baseline of pre-extinction evolutionary dynamics, and help to determine whether selectivity during the extinction was constructive (i.e., whether it mirrored background patterns) or non-constructive in nature. I argue here that extinctions were largely size selective, but were constructive, dependent upon existing patterns of trait evolution, and did not result in shifts from long-term phylogenetic patterns in the study group.

Methods 1: Stratigraphic Patterns

Database Assembly

Specimen sampling. I assembled a database of specimen measurements and stratigraphic occurrences for 72 species of Permian and Triassic therocephalians and cynodonts. The majority of measurements were taken directly from museum specimens, but data for some non-African taxa that were inaccessible during the study period were taken from the literature. Permission to access collections was obtained from all visited institutions prior to the study period. By incorporating data from museum specimens, I was able to assess variation within species, including maximum sizes, rather than representative samples from the literature. A rigorous, collections-based survey therefore allows previous literature-based assessments of body size evolution in Permo-Triassic therapsids to be improved upon [17,18]. Taxa were assigned to one or more stratigraphic bins, which mainly follow the biozonation of the Karoo Basin, the best-sampled and most complete terrestrial Permo-Triassic boundary succession in the world [43,44]. The Karoo biozonation includes distinct assemblage zones identified by their fossil tetrapod associations: *Eodicynodon*, *Tapinocephalus*, *Pristerognathus*, *Tropidostoma*, *Cistecephalus*, *Dicynodon*, *Lystrosaurus*, and *Cynognathus* AZs. Additional global data, collected to distinguish non-regional geologic and phylogenetic trends in all sampled Permian and Triassic taxa, covered a wider stratigraphic interval so that late Middle Triassic (Ladinian) and Late Triassic (Carnian, Norian, Rhaetian) global standard stages were also included as distinct bins. The sample was limited to Permian and Triassic genera only, and to specimens having well-preserved skulls for complete skull length measurements. Several Early Jurassic nonmammaliaform cynodonts from the Karoo (e.g., *Diarthrognathus*, *Pachygenelus*, *Tritylodon*) and those lacking complete skulls were therefore excluded from the dataset. The resulting database included 347 complete cranial specimens and 72 species of Permian through Triassic therocephalians and cynodonts from the Karoo Basin and contemporaneous basins. Sources for occurrence data, including voucher specimens and absolute dates, are provided in Appendix S3. Stratigraphic nomenclature and dates for Middle-to-Late Triassic basins follow Abdala et al. [45], Abdala and Ribeiro [46], Irmis et al. [47] and references cited therein.

Measurements. From the complete database, 57 measurements were retained for time series analysis of the Karoo assemblages, and 72 measurements were retained for rank order analyses of the global dataset representing the maximum basal skull lengths of the 72 sampled taxa from the stratigraphic interval of their first appearance (therocephalians: $N = 37$; cynodonts: $N = 35$). Basal skull length (BSL), measured from the tip of the snout to the occiput in millimeters (mm), was chosen as a proxy for body size for several reasons. Firstly, many of the fossil taxa examined here are best represented by cranial material, which is more accessible than postcranial elements and is relatively diagnostic for nonmammalian therapsids (contributing to the accuracy of specimen identifications). Moreover, investigation of BSL provides direct comparisons to other published studies using similar proxies. Skull length also correlates well with other linear skeletal measurements in therapsids such as femur length [17,18]. Measurements were recorded to the nearest mm using digital Mitutoyo calipers. For each taxon, the 'max lnBSL' represents the natural log of BSL for the largest measurable individual observed within a given interval. By assessing only the maximum lnBSL of each taxon (rather than taxon averages), these measures permitted both a conservative estimate of size reductions from one interval to the next while also reducing the effects of differences in relative abundance.

Evaluation of Time Series Models

Time series data represent a sequence of observations repeated over elapsed time, and can be conceptualized in a paleontological context as ancestor-descendant relationships (in as much as they assume that traits in daughter populations are a product of evolutionary changes from preceding populations, lineages, or clades) [48,49]. I performed time series analysis to evaluate regional patterns of body size evolution in a stratigraphic succession of therocephalians and cynodonts from the Permian-Triassic Beaufort Group, Karoo Basin, South Africa. First, I evaluated the probability that the body size samples from the Late Permian (*Cistecephalus* and *Dicynodon* AZs) and earliest Triassic (*Lystrosaurus* AZ) were drawn from equivalent distributions using a Kolmogorov-Smirnov (K-S) test. Then, I evaluated the likelihood that observed negative shifts in size distributions (if present) represent a random evolutionary walk (rather than a long-term directional trend) by assessing the record of body size evolution over eight stratigraphic bins of the Karoo succession. Mean lnBSL was estimated for each of eight stratigraphic bins by taking the average of the maximum known lnBSL for each taxon within a given bin, therefore representing a per taxon average (rather than an average of all specimens, which could otherwise be skewed by taxa with unusually high abundances). In contrast to the rank order correlation tests, which examined only size at first appearance (discussed below), taxa whose ranges spanned multiple assemblage zones were included in each bin in which they occurred in order to capture more fully the variance within each bin. Mean sizes were estimated in each bin for all eutheriodonts, as well as separately for therocephalians and cynodonts in order to evaluate the pervasiveness of potential trends among the therapsid subclades.

Mean lnBSL and disparity (as measured by variance) were estimated from the database for each Karoo biozone (average bin duration $\approx$3 Myr) and imported into the 'R' version 2.15.1 statistical environment using the 'paleoTS' package [48,49]. The 'paleoTS' package permited evaluation of temporal patterns of body size evolution across the Karoo succession, using several models that fall into two general categories: non-directional and directional. Non-directional models implemented by the 'paleoTS' package include unbiased random walk (or 'URW') and stasis. URW models evolution as a volatile Brownian process with observable step variance but no mean step change, whereas stasis models evolution with modest deviations that are uncorrelated and normally distributed around the trait mean [49]. Directional evolution is modeled as a generalized random walk ('GRW') having a positive or negative mean step change. The relative fit of non-directional and directional evolutionary models was evaluated using likelihood methods and goodness-of-fit tests. My expectation is that, if short-term (Lilliput-type) size reductions are a unique signature of the Triassic *Lystrosaurus* AZ, then longer-term size

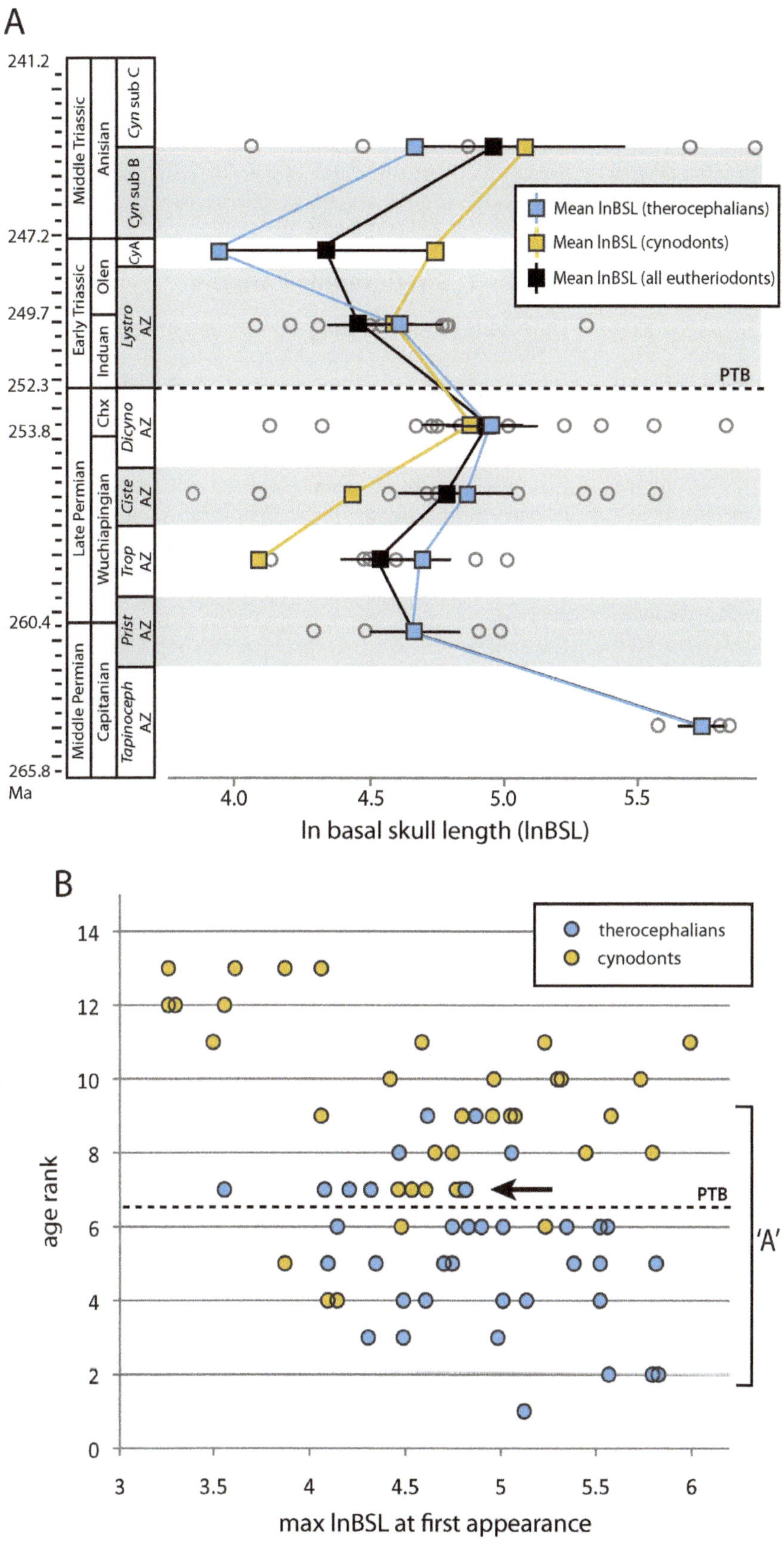

Figure 1. Karoo time series and age rank results for observed geologic distributions of maximum basal skull length (lnBSL). A, Regional patterns of trait evolution in Karoo eutheriodonts (black) (therocephalian subclade = blue; cynodont subclade = gold) fit a non-directional model throughout the Karoo succession (see Table 1). Squares represent clade averages (with 95% confidence intervals), gray circles represent maximum known sizes for each species recorded in a given stratigraphic interval (Ma, millions of years ago). **B,** Spearman's rank correlation tests on

global age rank data (used for the phylogenetic sampling) reveal a weak negative trend in species maximum sizes ($\rho = -0.261$; $p = 0.025$) with a short-term bottleneck in size maxima in the Early Triassic *Lystrosaurus* AZ (arrow). Permian-Triassic boundary (PTB) represented by dashed line.

reductions should not be manifested in the Karoo time series, with non-directional models best fitting the data (i.e., URW and stasis). If, on the other hand, body size evolution was driven by longer-term trends beginning in the Permian, then a directional model with a negative mean step change should best fit the data, thereby refuting the rapid and short-term perturbation predicted by the Lilliput effect.

The relative fit of the three models (URW, stasis, and GRW) was assessed using Akaike's information criterion corrected for small sample sizes (AICc). AICc is an information loss metric that is used to compare models both through their goodness-of-fit and model complexity (i.e., the number of adaptable model parameters). AICc diminishes the problem of overfitting complex models, and is widely used for small paleontological and neontological datasets. Akaike weights ('AW') are estimated by converting the individual AICc scores for each candidate model into a proportional value representing the model's relative support (with the sum of the values adding to 1.0 across all models); the model with the highest weight represents that with the least information loss (and, thus, the most desirable model). A more complete description of AICc and its application to time series and phylogenetic model fitting can be found in Hunt and Carrano [49].

Rank Order Correlations

Rank order statistics provide additional tests for identifying directional trends in ranked paleontological data (in this case age ranks, clade ranks and patristic distances) [16,50–53]. 'Age rank' represents an ordered numerical representation of the stratigraphic bin in which a taxon first appeared, particularly useful where estimated durations of bins and positions of taxa therein are only imprecisely known. Rank order correlations were performed here to assess the relationship between size and age rank, as well as the relationships between size, clade rank and patristic distance (Methods 2). Non-parametric assessments of binned age ranks permitted comparisons to results of the clade rank analyses, which used similar rank order tests. Stratigraphic bins were divided into 13 ranks (average duration ≈5 Ma), increased from eight in the Karoo-only time series to accommodate late-Middle and Late Triassic taxa. Each taxon in the global dataset was assigned an age rank, according to the ranked stratigraphic bin in which it first appeared, and an associated maximum lnBSL within that bin. Correlations between size and age rank were assessed using Spearmans' rho (ρ) and Kendall's tau (τ) rank order tests. These tests allowed potentially positive or negative trends in the global dataset to be evaluated as a monotonic function (the Spearman's rank correlation describing the strength of a monotonic association, and Kendall's tau assessing the probability of observing concordant and discordant pairs among ranked variables). Importantly, unlike their parametric equivalent, Pearson's product moment correlation, the tests are robust even when the association is not linear. Three separate data subsets were analyzed to evaluate directional size trends in eutheriodonts ($N = 72$) and therocephalian ($N = 37$) and cynodont ($N = 35$) subclades from the Middle Permian through Late Triassic. For Eutheriodontia and its subclades, I evaluated ρ and τ coefficients for negative trends in two ways, assessing 1) data subsets that included taxa only having first appearances in the Permian and 2) the total Permian-Triassic dataset. By examining Permian-only and excluding Triassic originations, I was able to assess the extent to which general trends in the total Permian-Triassic dataset were influenced by new size distributions of post-extinction taxa (versus pre-existing or 'background' patterns of trait evolution). Trends were considered only weakly significant if both coefficients were negative but non-significant for one of the tests.

Results 1: Stratigraphic Patterns

Time Series Model Fitting

My evaluation of stratigraphic (i.e., non-phylogenetic) patterns failed to identify long-term negative trends in eutheriodont body size evolution, instead showing a short-term reduction in mean sizes in the Triassic *Lystrosaurus* AZ. Visual inspection of the Karoo time series data suggests a transient decrease in mean size from the Late Permian *Dicynodon* AZ to the *Lystrosaurus* AZ, which rebounded subsequently in the *Cynognathus* AZ (Fig. 1). Kolmogorov-Smirnov tests on size distributions from the *Lystrosaurus* AZ suggest significant decreases when compared against the Late Permian *Cistecephalus* and *Dicynodon* AZs (therocephalians: $D = 0.61$, $p < 0.001$; cynodonts: $D = 0.38$, $p = 0.021$). A binomial test on the Karoo dataset using the *binom.test* function of paleoTS [49] found no systematic tendency for decreases in mean size of eutheriodonts over the Karoo succession, suggesting that this geologically abrupt shift was not part of an observable long-term trend. This may, however, reflect the poor ability of the statistical test to reject the null hypothesis of a random walk, given the small number of stratigraphic bins available. I therefore used the small sample size version of Akaike's Information Criterion (AICc) to evaluate the relative fit of directional (generalized random walk or 'GRW') and non-directional (unbiased random walk or 'URW' and evolutionary stasis) candidate models, but non-directional models found the best support as size differences accrued with noticeable trait variance but with a step mean of zero (Table 1). Observed size decreases in the *Lystrosaurus* AZ therefore cannot be attributed to a long-term directional trend when evaluated as a time series.

Age Ranks and Rank Order Correlations

Rank order correlations between age rank and size at first appearance, on the other hand, may indicate a weak negative trend for some groups such as therocephalians, which originated at their largest sizes in the Permian (Fig. 1B, Table 2). This correlation, however, is likely influenced both by a lack of originations of new large Triassic therocephalians (pulling higher age ranks into the lower end of size space) and the ranking and evaluation of data as a monotonic function. When Triassic taxa are removed from the analysis, correlation coefficients are only weakly negative and non-significant, indicating an inability of the tests to identify long-term negative trends from Permian size distributions. This result may be owed to the somewhat smaller sample size of the Permian-only data set (27 versus 37 therocephalians), but most Permian bins reflect the persistence of large-bodied forms and high disparity (variance) up until the Permo-Triassic boundary. Permian cynodonts, however, were difficult to evaluate due to their low taxonomic richness during the Permian and, consequently, small sample sizes. In summary, Early Triassic eutheriodont faunas were on average smaller-bodied than in preceding Permian assemblages. However, Permian stratigraphic trends alone provide little evidence of long-term declines in mean body sizes.

Table 1. Time series model fitting of body size evolution in Karoo therocephalians and cynodonts.

	GRW			URW			Stasis		
	AICc	AW	μ, σ^2	AICc	AW	σ^2	AICc	AW	θ, ω
Eutheriodontia$_{Karoo}$	19.848	0.025	−0.029, 0.042	*14.601*	*0.349*	*0.045*	**13.435**	**0.625**	**4.794, 0.077**
Therocephalia$_{Karoo}$	*16.669*	*0.159*	*−0.038, 0.000*	**14.540**	**0.462**	**0.031**	*14.942*	*0.378*	*4.818, 0.079*
Cynodontia$_{Karoo}$	16.937	0.033	0.052, 0.000	**11.495**	**0.503**	**0.008**	*11.658*	*0.464*	*4.702, 0.023*

GRW, generalized random walk (directional); URW, unbiased random walk;
AICc, Akaike's Information Criterion corrected for small sample size; AW, Akaike weight;
μ, step mean; σ^2, step variance; θ, trait mean; ω, trait variance;
bold = best model; italics = non-negligible model (AW ≥1/8 best model).

Methods 2: Phylogenetic Patterns

Reference Cladogram, Clade Ranks and Patristic Distances

Cladistic data were also ranked to permit the evaluation of directional trends in size data with respect to relative branching events in a reference cladogram (Fig. 2). I constructed an informal supertree primarily from the literature, although some new taxa and specimens are included here in an expanded analysis of therocephalian therapsid relationships updated from [33,54,55]. Inclusion of these additional taxa facilitated the most comprehensive phylogenetic sample of therocephalian therapsids to date, updated from Sigurdsen et al. [55] (see Dataset S1). I follow several sources for cynodont phylogeny, which has received greater attention in the literature [43,56–60]. 'Clade ranks' (CR) were assigned based on the number of nodes passed along the backbone of the tree (e.g., Cynodontia to Mammaliaformes) without incorporating branching events of clades off the primary axis. This metric has provided a useful tool for assessing long-term evolutionary patterns along the primary axis of the tree with respect to early mammaliaforms [50,61]. 'Patristic distances' (PD) represented the number of nodes passed through from the hypothetical ancestor at the base of the tree to each of the tips [52]. Therocephalians and cynodonts were analyzed separately for correlations with CR because these two main subclades replicate CR values (Fig. 3). PD values bear no such constraint and were estimated directly from the complete reference phylogeny in Figure 2. Rank order correlation coefficients were estimated from Permian-only and combined Permian-Triassic datasets as in the aforementioned age rank analysis. Statistical results are presented in Table 3.

Ancestor-Descendant Tests

Ancestor-descendant tests were performed as an extension of the CR analyses in order to assess tendencies toward negative versus positive size changes in the phylogenetic tree between reconstructed ancestor and descendant nodes [15,16]. Whereas CR analyses allowed interpretations only from observed tip data, this analysis relies on squared-change parsimony to reconstruct the hypothetical traits of ancestors (which cannot be observed, only estimated). Ancestral states were estimated using squared-change parsimony in Mesquite version 2.0 [62]. For each nested clade, I then manually estimated the mean step change, the number of positive and negative changes between internodes (internodes having zero change were not counted), and the skewness, providing measures of tendency that could contribute to directional evolution in the tree. Results are presented in Table 4.

Diversification Rate Shifts

To assess the hypothesis that shifts in mean body size represented post-extinction radiations of small-bodied eutheriodont clades, a phylogeny was constructed using all studied taxa in Figure 2 and tested for variations in diversification rates using the topology-based method of Chan and Moore [63]. The software SymmeTREE version 1.1 was used to evaluate the topological distribution of taxic diversity across branches in the reference phylogeny (i.e., its symmetry) and to identify regions of the tree that contribute to significant imbalance in diversity. This operation was performed by calculating the Δ_1 and Δ_2 statistics, functions of two likelihood ratios that represent the probability of observing a shift in diversification rate along the internal branch of a three-taxon tree (composed of a local outgroup and two basalmost subclades of the ingroup). For each node, the ratios compared the likelihood of yielding the observed distribution of diversity under an equal-rates Markov branching model to that under a two-rate model. Importantly, the inclusion of non-contemporaneous clades

Table 2. Results of rank correlations, Spearman's rho (ρ) and Kendall's tau (τ), for global age ranks.

Age rank *v.* lnBSL		***N* taxa**	**ρ**	***p*(ρ)**	**τ**	***p*(τ)**
Eutheriodontia	(Permian taxa only)	32	−0.089	0.621	−0.054	0.693
Therocephalia	(Permian taxa only)	27	−0.153	0.443	−0.089	0.559
Cynodontia	(Permian taxa only)	5	–	–	–	–
Eutheriodontia	(Permian + Triassic)	72	**−0.261**	**0.025**	**−0.173**	**0.037**
Therocephalia	(Permian + Triassic)	37	**−0.370**	**0.023**	**−0.254**	**0.035**
Cynodontia	(Permian + Triassic)	35	−0.265	0.115	−0.113	0.362

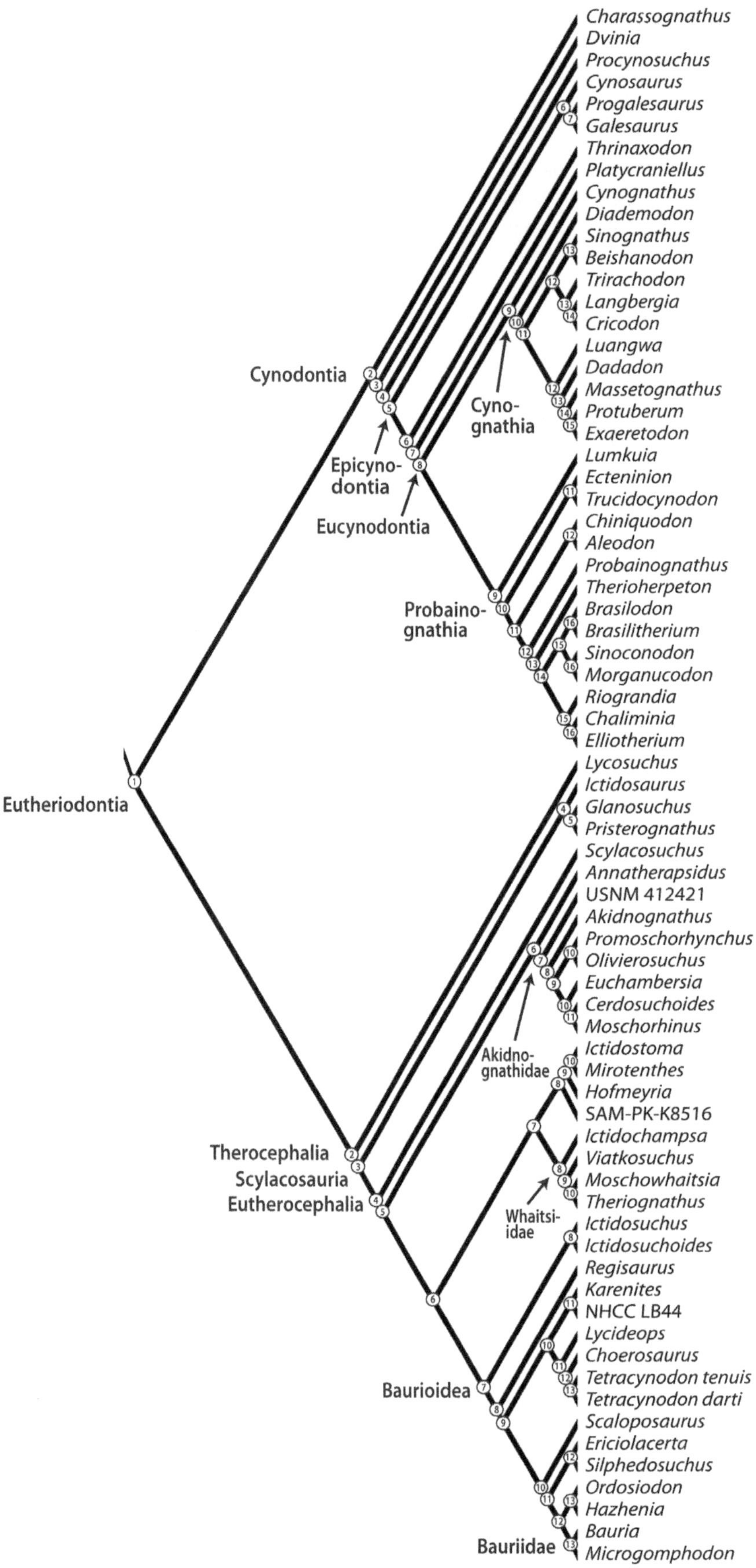
Charassognathus
Dvinia
Procynosuchus
Cynosaurus
Progalesaurus
Galesaurus
Thrinaxodon
Platycraniellus
Cynognathus
Diademodon
Sinognathus
Beishanodon
Trirachodon
Langbergia
Cricodon
Luangwa
Dadadon
Massetognathus
Protuberum
Exaeretodon
Lumkuia
Ecteninion
Trucidocynodon
Chiniquodon
Aleodon
Probainognathus
Therioherpeton
Brasilodon
Brasilitherium
Sinoconodon
Morganucodon
Riograndia
Chaliminia
Elliotherium
Lycosuchus
Ictidosaurus
Glanosuchus
Pristerognathus
Scylacosuchus
Annatherapsidus
USNM 412421
Akidnognathus
Promoschorhynchus
Olivierosuchus
Euchambersia
Cerdosuchoides
Moschorhinus
Ictidostoma
Mirotenthes
Hofmeyria
SAM-PK-K8516
Ictidochampsa
Viatkosuchus
Moschowhaitsia
Theriognathus
Ictidosuchus
Ictidosuchoides
Regisaurus
Karenites
NHCC LB44
Lycideops
Choerosaurus
Tetracynodon tenuis
Tetracynodon darti
Scaloposaurus
Ericiolacerta
Silphedosuchus
Ordosiodon
Hazhenia
Bauria
Microgomphodon
Cynodontia
Epicynodontia
Eucynodontia
Cynognathia
Probainognathia
Eutheriodontia
Akidnognathidae
Therocephalia
Scylacosauria
Eutherocephalia
Whaitsiidae
Baurioidea
Bauriidae

Figure 2. Complete reference cladogram with nodes numbered for estimation of patristic distances. Sources include: therocephalians, Huttenlocker [54], Huttenlocker et al. [33], Sigurdsen et al. [55]; cynodonts, Hopson & Kitching [56], Abdala et al. [45], Sidor & Hancox [57], Ranivoharimanana et al. [60], Oliveira et al. [58], Liu & Olsen [59].

sampled from disjunct stratigraphic bins may violate an assumption of the equal-rates Markov model by heterogeneously sampling diversity through time [64,65]. Therefore, five trees were subsampled from the reference phylogeny, representing a cross-section of diversity for each of the five best sampled assemblage zones spanning the Permian-Triassic transition: *Tropidostoma*, *Cistecephalus*, *Dicynodon*, *Lystrosaurus*, and *Cynognathus* AZs. Following Botha-Brink and Angielczyk [65], trees were pruned such that they only included taxa and ghost lineages present within the given stratigraphic bin. The trees were then evaluated in SymmeTREE to identify diversification rate shifts in post-extinction clades (e.g., bauriids; cynognathians or other eucynodonts) or in clades that diversified before the extinction (e.g., eutherocephalians; basal cynodonts).

Evaluation of Phylogenetic Models

Finally, stochastic models of Brownian evolution were fitted to observed patterns of body size evolution in a comprehensive global phylogeny of eutheriodonts, permitting a marriage of temporal and topology-based methods. Phylogenetic model fitting was performed in R using 'ape' and 'geiger' packages [49]. The global reference phylogeny was exported into Newick notation and zero length branches that constrained ghost lineages of their sister clades were augmented using the smoothing distribution of the *date.phylo()* function (http://www.graemetlloyd.com/methdpf.html) [66]. Skull length data were imported and joined to individual taxon labels (corresponding to 71 tip taxa present on the tree), then assigned to the body size vector. Using 'geiger,' I evaluated the relative fit of four phylogenetic models in order to approximate the mode of body size evolution observed on the tree: Brownian motion (BM), Brownian motion with trend (BM_T), Ornstein-Uhlenbeck (OU), and early burst Brownian motion (EB). BM models evolution as a process of passive diffusion (equivalent to an unbiased random walk); BM_T, models evolution as a generalized random walk having a positive or negative mean step change (equivalent to a 'directional trend'); OU models trait

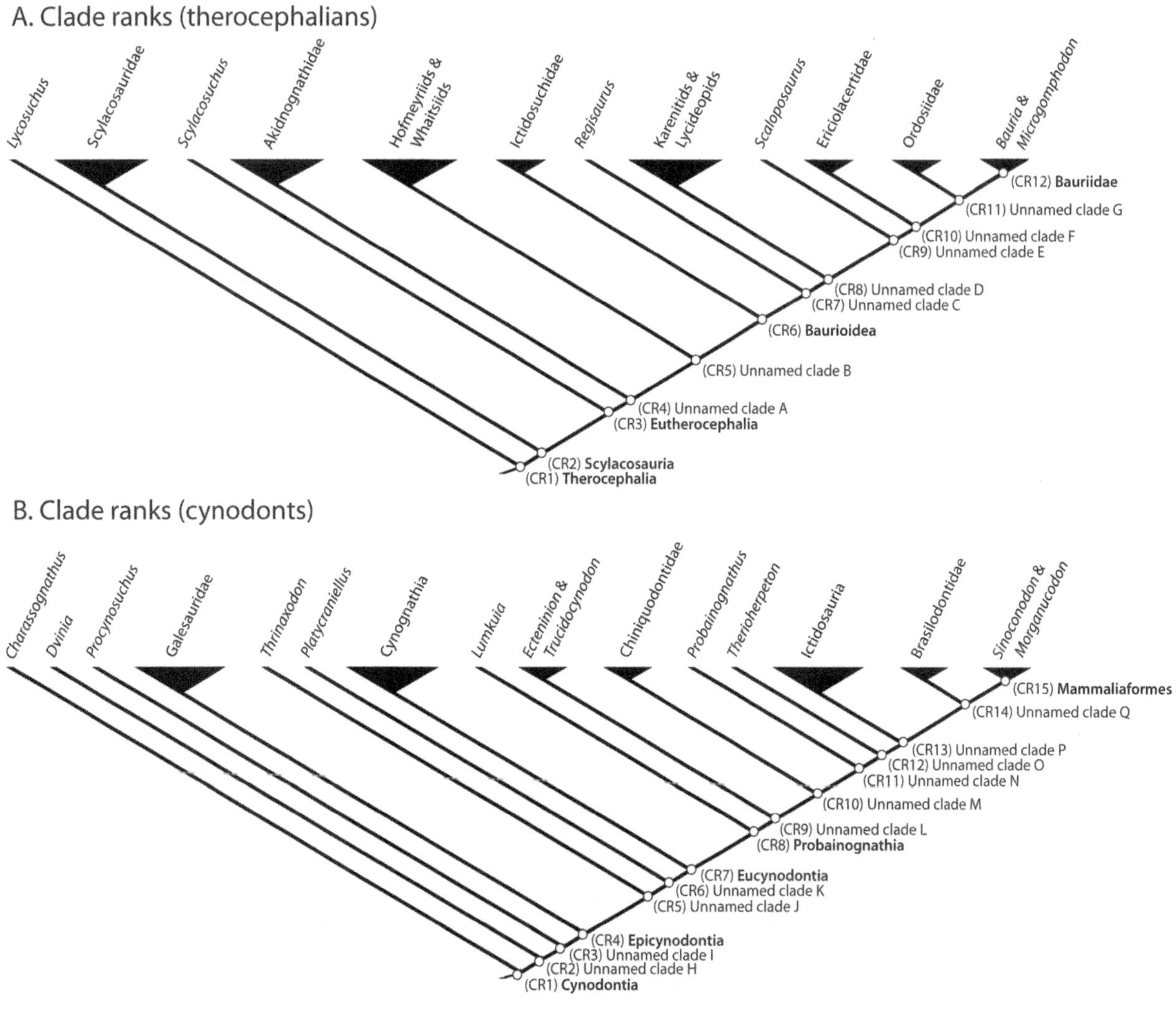

Figure 3. Ranking of side-branches within the two major subgroups: Therocephalia (A) and Cynodontia (B) (simplified from reference cladogram in Fig. 2).

Table 3. Results of rank correlations, Spearman's rho (ρ) and Kendall's tau (τ), for clade ranks and patristic distances.

Clade rank *v.* lnBSL		***N* taxa**	**ρ**	***p*(ρ)**	**τ**	***p*(τ)**
Therocephalia	(Permian taxa only)	27	**−0.608**	**0.001**	**−0.488**	**<0.001**
Cynodontia	(Permian taxa only)	4	–	–	–	–
Therocephalia	(Permian + Triassic)	37	**−0.622**	**<0.001**	**−0.484**	**<0.001**
Cynodontia	(Permian + Triassic)	34	**−0.371**	**0.032**	−0.234	0.064
Patristic distance *v.* lnBSL		***N* taxa**	**ρ**	***p*(ρ)**	**τ**	***p*(τ)**
Eutheriodontia	(Permian taxa only)	31	−0.135	0.467	−0.136	0.306
Therocephalia	(Permian taxa only)	27	**−0.508**	**0.007**	**−0.407**	**0.005**
Cynodontia	(Permian taxa only)	4	–	–	–	–
Eutheriodontia	(Permian + Triassic)	71	**−0.281**	**0.019**	**−0.189**	**0.022**
Therocephalia	(Permian + Triassic)	37	**−0.546**	**<0.001**	**−0.417**	**<0.001**
Cynodontia	(Permian + Triassic)	34	−0.192	0.279	−0.093	0.439

evolution as a constrained process approaching some optimal value (sometimes equated to 'stabilizing selection'); and EB approximates a Brownian process in which the rate of trait evolution exponentially decreases ('decays') over time. Models were fitted to the data using 'geiger,' and likelihood ratio tests were performed to determine whether multiparameter models should

Table 4. Summary of ancestor-descendant step changes in the therocephalian and cynodont trees (ancestral states reconstructed using squared-change parsimony).

Eutheriodontia	**BSL_{anc}**	**$lnBSL_{anc}$**	**Mean Δ**	**+**	**–**	**Skewness**
Therocephalia	170	5.135	−0.006	9	25	1.213
Scylacosauria	170	5.135	−0.006	9	25	1.199
Eutherocephalia	162	5.087	−0.023	8	23	1.314
Unnamed clade A	151	5.017	−0.022	8	22	1.282
Unnamed clade B	140	4.941	−0.032	6	16	1.273
Baurioidea	135	4.905	−0.081	3	11	1.431
Unnamed clade C	128	4.852	−0.095	2	10	1.493
Unnamed clade D	120	4.787	−0.098	2	9	1.474
Unnamed clade E	56	4.025	0.077	2	3	1.379
Unnamed clade F	53	3.970	0.110	2	2	1.180
Unnamed clade G	153	5.030	−0.270	1	1	–
Cynodontia	61	4.110	0.016	15	15	0.173
Unnamed clade H	61	4.110	0.016	15	15	0.165
Unnamed clade I	60	4.094	0.018	15	14	0.153
Epicynodontia	48	3.871	0.026	15	13	0.093
Unnamed clade J	92	4.521	−0.026	13	13	−0.124
Unnamed clade K	88	4.477	−0.025	13	12	−0.129
Eucynodontia	219	5.389	−0.064	12	12	−1.370
Probainognathia	112	4.718	−0.045	7	5	−1.531
Unnamed clade L	123	4.812	−0.058	6	5	−1.402
Unnamed clade M	134	4.897	−0.096	4	5	−1.091
Unnamed clade N	82	4.406	−0.065	3	4	−1.464
Unnamed clade O	33	3.496	0.075	3	3	1.609
Unnamed clade P	30	3.401	0.109	3	2	1.509
Unnamed clade Q (incl. Mammaliaformes)	28	3.332	0.032	1	1	–

Ancestral basal skull length estimates (BSL_{anc}) are in mm; mean change refers to log change between ancestor-descendant pairs; +, number of internodes having a positive step change; –, number of internodes having a negative step change.

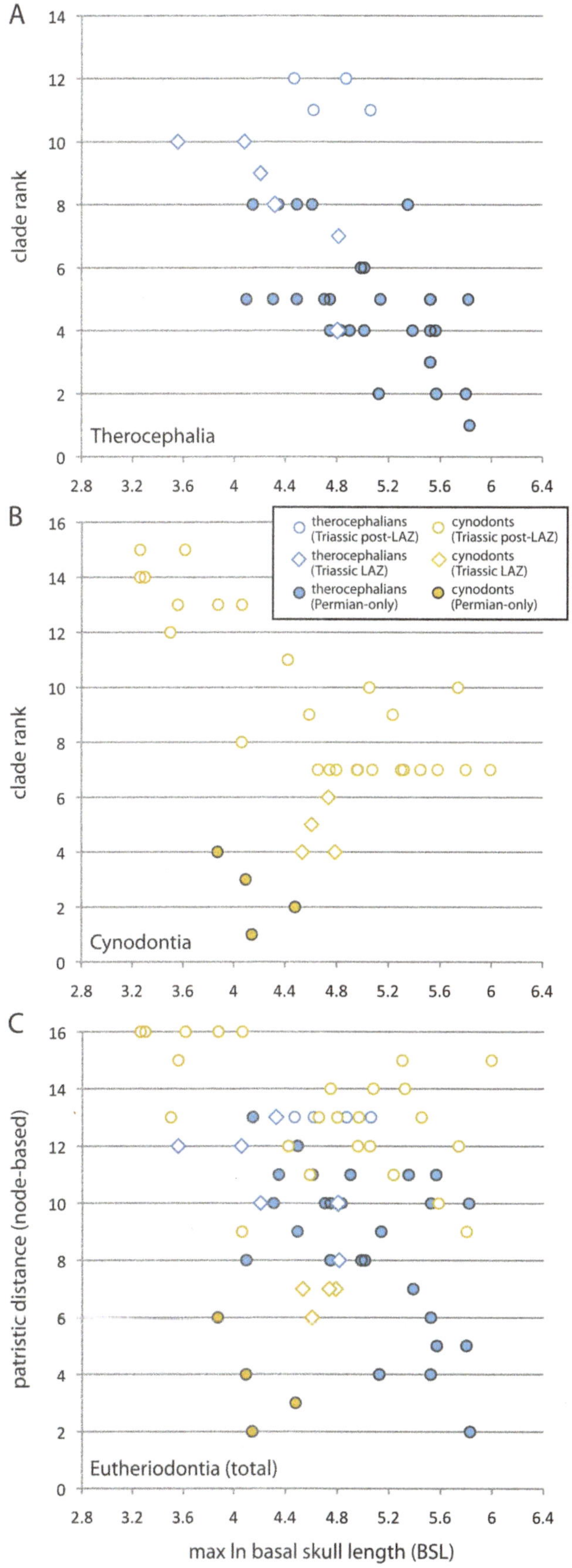

Figure 4. Spearman's rank correlation tests on clade rank (A, B) and patristic distance (C) versus maximum basal skull length for the global dataset of Middle Permian through Late Triassic eutheriodonts. Solid circles represent Permian taxa (therocephalian subclade = blue; cynodont subclade = gold), open diamonds represent Early Triassic *Lystrosaurus* Assemblage Zone (LAZ) taxa, and open circles represent Triassic post-LAZ taxa.

be selected over the null BM model. Goodness-of-fit tests implementing AICc were also used to evaluate the candidate models as in the time series analysis.

Results 2: Phylogenetic Patterns

Cladogenetic Branching

Cladistically-informed analyses were better able to detect long-term body size trends than stratigraphic data alone (Fig. 4; Table 3), with some groups expanding into smaller sizes relatively early in their phylogenetic histories and continuing in more deeply nested clades. Size exhibits a strong negative correlation with clade rank in therocephalians ($\rho = -0.622$, $p < 0.001$; $\tau = -0.484$, $p < 0.001$) and a weakly negative correlation in cynodonts ($\rho = -0.371$, $p = 0.032$; $\tau = -0.234$, $p = 0.064$). The same result is true for therocephalians whether analyzed with or without Triassic taxa, indicating that size reductions began in relatively shallowly-nested nodes in the tree (i.e., early-diverging groups which include Permian taxa) and continued into the Triassic. Correlations with patristic distance have lower statistical support, but are still significant for Permian-only and Permian-Triassic therocephalians indicating the influence of therocephalians on the overall pattern. I carried out ancestor-descendant tests following the method of Carrano [16] to supplement the rank correlation tests and to assess the tendency and relative frequency of negative size changes between reconstructed ancestor-descendant nodes. The results in Table 4 also indicate a strong tendency toward size reductions in therocephalians, with the number of ancestor-descendant decreases outnumbering increases by as much as three to five times. Cynodonts, which originated at relatively small sizes, showed no such tendency with the number of size decreases approximately equaling the number of increases.

Diversification Rate Shifts

Results of rank order correlation and ancestor-descendant tests indicate that body sizes decreased on average in progressively deeper nested therocephalian subclades, supporting the prior expectation that body size shifts were not necessarily influenced by post-extinction diversity shifts (although they may have been associated with earlier diversifications). Accordingly, no significant rate shifts were detected in tree topologies constructed from Triassic *Lystrosaurus* or *Cynognathus* AZ diversity partitions, suggesting that either sampling is too poor to detect a shift or that the impact of the extinction on diversification rates in eutheriodonts was limited. The only significant diversification rate shift occurred at the early-Late Permian divergence of Eutherocephalia (Fig. 5A,B), a result that is consistent with the relatively high rates of per capita origination and extinction in Late Permian eutherocephalians [33,67]. This early diversification rate shift also mirrors previous findings in stratigraphically contemporaneous bidentalian dicynodonts [65]. Alternatively, the high diversity of therapsids in the Late Permian *Tropidostoma* and *Cistecephalus* AZs could be a sampling artifact [39]. Nevertheless, no shift is detected in the well-sampled *Lystrosaurus* AZ following the extinction.

Phylogenetic Model Fitting and Body Size Evolution

Goodness-of-fit tests found that active 'directional' (BM_T) and 'stabilizing' (OU) models fit poorly when compared to other

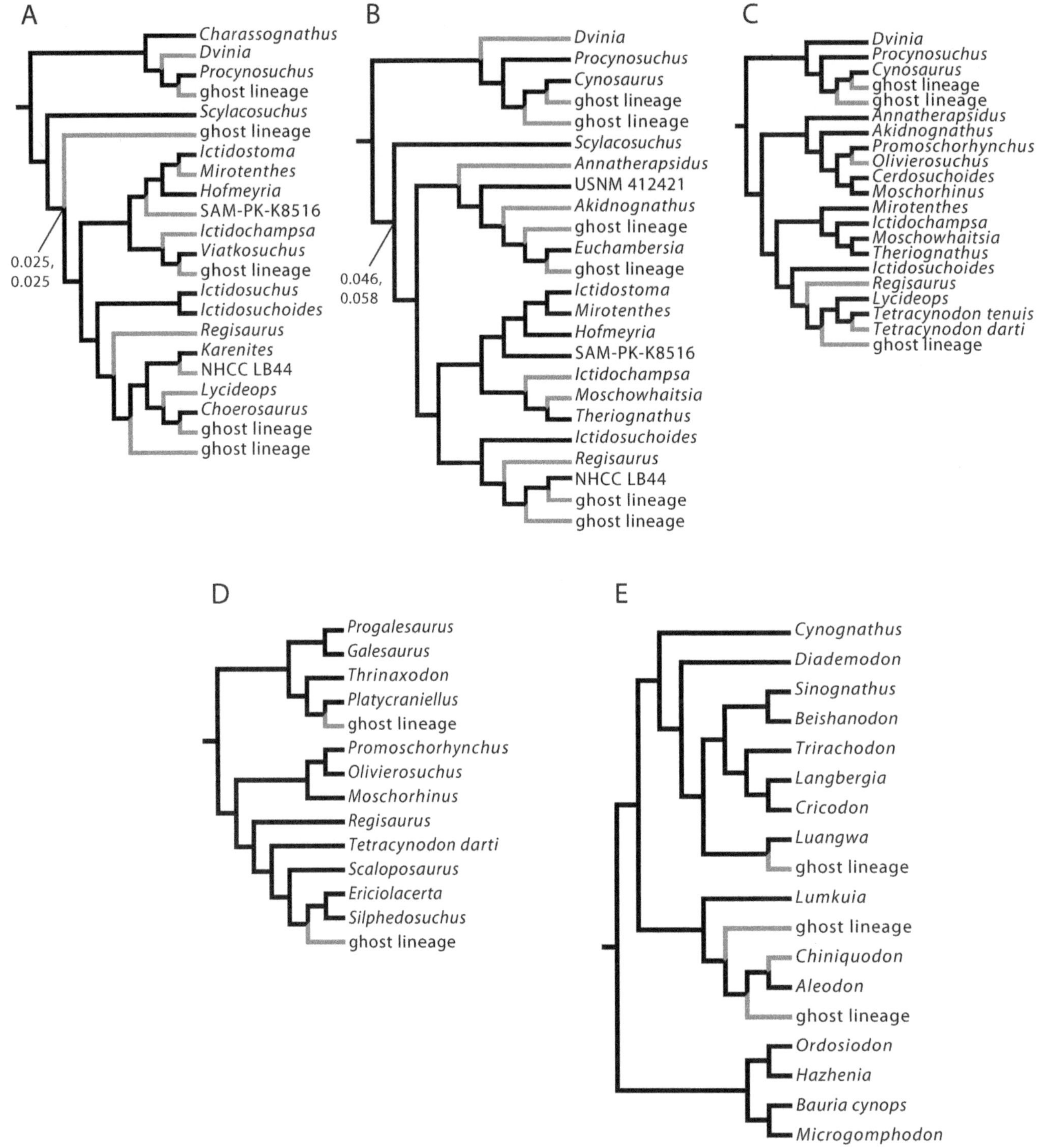

Figure 5. Cladograms subsampled from five stratigraphic bins for analysis of diversification rate shifts. A, *Tropidostoma* Assemblage Zone. **B**, *Cistecephalus* Assemblage Zone. **C**, *Dicynodon* Assemblage Zone. **D**, *Lystrosaurus* Assemblage Zone. **E**, *Cynognathus* Assemblage Zone. Gray branches represent ghost lineages for terminal taxa or clades that appear in a succeeding stratigraphic interval. Numbers at nodes represent *p*-values near or below 0.05 for the Δ_1 (top) and Δ_2 (bottom) statistics. Significant shifts are observed along the branch between the node of Eutherocephalia and non-*Scylacosuchus* eutherocephalians for the *Tropidostoma* AZ subsample (A), and a marginally significant shift is observed along the branch between Eutheriodontia and Eutherocephalia for the *Cistecephalus* AZ (B).

candidate models of passive, Brownian motion evolution (BM and EB). Likelihood ratio tests and comparisons of Akaike weights indicate that EB best approximates the mode of body size evolution across the global dataset of eutheriodonts (Figure 6; Table 5). Body size evolution was better approximated by the null BM model when therocephalians were analyzed separately, indicating that derived cynodont subclades strongly influenced the decay rate parameter for the EB model. In particular, late

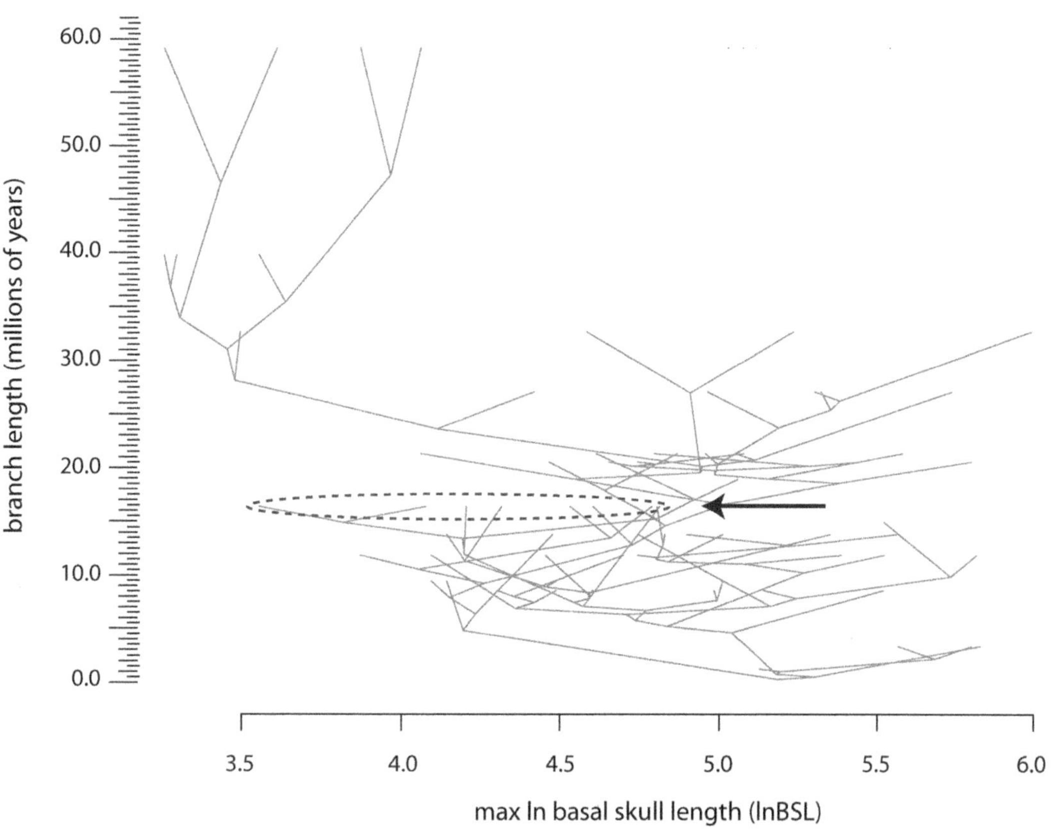

Figure 6. Traitgram of body size evolution in a global sample of Permian through Triassic eutheriodonts, approximating an 'early burst' Brownian process (EB) of trait evolution. Arrow denotes location of *Lystrosaurus* Assemblage Zone taxa along the time-axis. Graphic created using the *traitgram()* function in 'picante.'

Middle and Late Triassic cynodont clades (e.g., stem-mammaliaforms and other probainognathians) that include members from outside of the Karoo succession became more constrained in size space with decreasing variance over time, but not other lineages that lived in the immediate aftermath of the end-Permian extinction in the Karoo. In short, reductions in sizes of *Lystrosaurus* AZ eutheriodonts were likely to have been underpinned by passive processes, rather than long-term directional trends or constraints on evolutionary modes during the end-Permian extinction.

Discussion

Body size evolution before and immediately after the end-Permian extinction cannot demonstrably be classified as active or "driven" sensu McShea [15], because size increases were generally associated with as many or more decreases and with expanding size disparity. Moreover, short-term reductions in mean and maximum sizes in the earliest Triassic *Lystrosaurus* AZ were not associated with changes in these background patterns of Brownian evolution, but were instead a product of the temporary ecological removal of medium to large-bodied predators (in addition to anagenetic size reduction in at least one predator lineage, *Moschorhinus* [68]). In spite of its devastating ecological impacts,

Table 5. Phylogenetic model fitting results.

	BM			BM_T			OU			EB		
	AICc	AW	β	AICc	AW	β,μ,θ	AICc	AW	β,α	AICc	AW	β,r
Eutheriodontia$_{total}$	*113.72*	*0.276*	*0.05*	*115.61*	*0.107*	*0.05,0.01,5.15*	*113.41*	*0.270*	*0.06,0.02*	**112.79**	**0.346**	**0.09,−0.03**
Therocephalia$_{total}$	**51.49**	**0.439**	**0.05**	*53.48*	*0.162*	*0.05,0.01,5.28*	*52.51*	*0.263*	*0.06,0.04*	*53.85*	*0.134*	*0.05,−0.006*
Cynodontia$_{total}$	*58.32*	*0.118*	*0.05*	60.48	0.049	0.05,0.01,4.11	59.00	0.068	0.06,0.03	**54.99**	**0.764**	**0.17,−0.08**

Model selection based on comparison of corrected AICs and Akaike weights (AW) for Brownian motion (BM), Brownian motion with trend (BM_T), Ornstein-Uhlenbeck (OU), and early burst (EB) models;
AICc, Akaike's Information Criterion corrected for small sample size; AW, Akaike weight;
β, BM rate parameter; μ, step mean; θ, trait mean; α, constraint parameter; r, decay rate;
bold = best model; italics = non-negligible model (AW $\geq$1/8 best model).

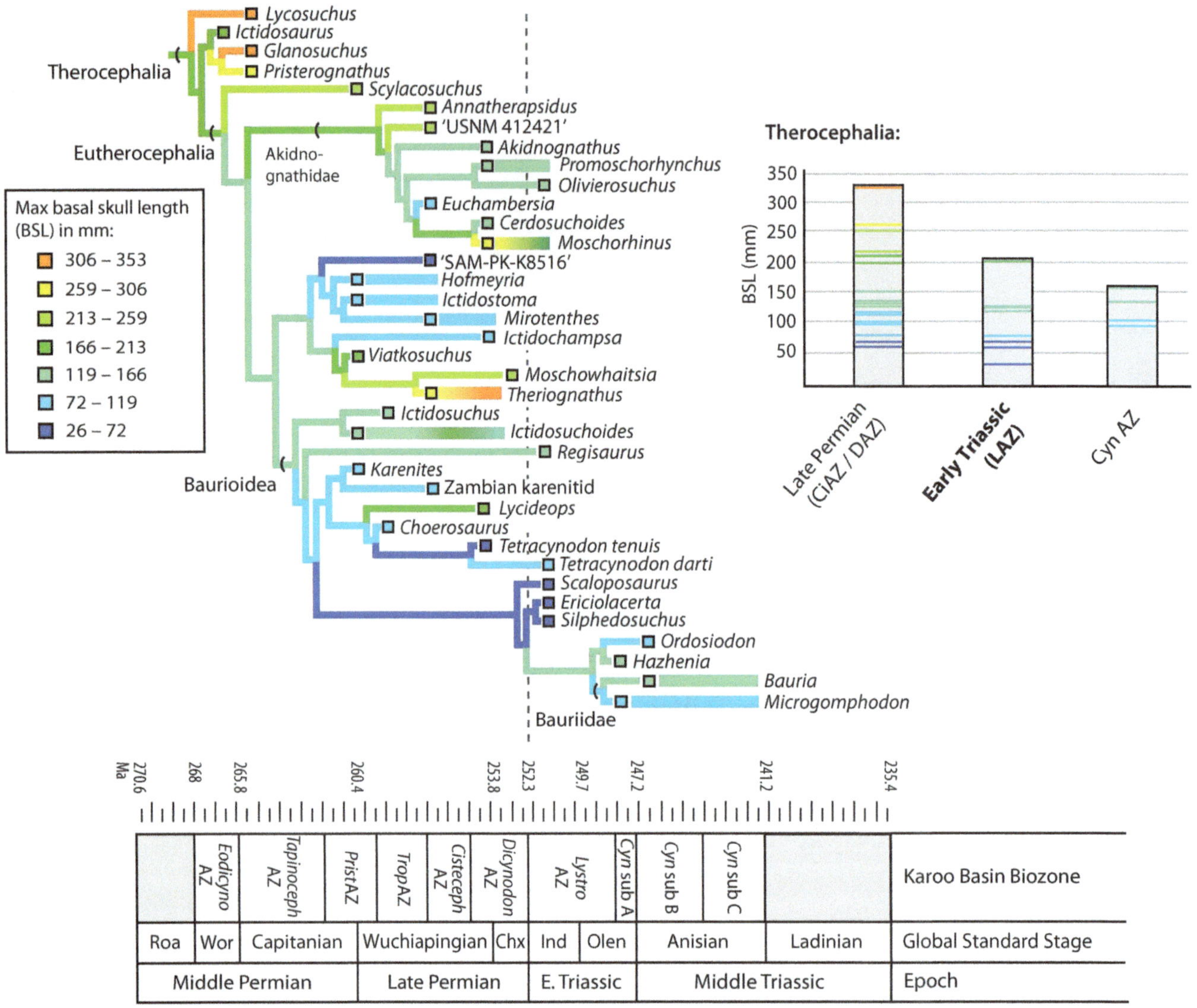

Figure 7. Ancestral state reconstructions for basal skull length in Permian and Triassic therocephalians. Colored horizontal bars indicate ranges of tip taxa spanning more than one stratigraphic bin. Note that significant decreases in mean sizes of major subgroups across the Permian-Triassic boundary (dashed line) were associated with the loss of large-bodied taxa and the survival of pre-existing small-bodied taxa or clades. Abbreviations: CiAZ, *Cistecephalus* Assemblage Zone; CynAZ, *Cynognathus* Assemblage Zone; DAZ, *Dicynodon* Assemblage Zone; LAZ, *Lystrosaurus* Assemblage Zone.

short-term shifts during the end-Permian extinction had negligible long-lasting effects on subsequent patterns of body size evolution in eutheriodont therapsids. Summary diagrams of eutheriodont body size evolution are provided in Figures 7 and 8, showing frequent size reductions beginning in the Permian with occasional increases in some lineages (contributing to increased disparity), and an overall drop in body size distributions in the Triassic *Lystrosaurus* AZ. Though in conflict with traditional views of the fossil record, this finding corroborates more recent suggestions that large-scale patterns of trait evolution rarely yield active, directional trends [17,18,48]. Negative shifts were likely passive and, in the case of therocephalians, may underscore a fundamental asymmetry in size evolution: that size decreases can often occur with higher frequency than increases due to physiological and/or environmental constraints on size [69]. The success of small-bodied lineages may therefore represent a more general feature of size evolution operating during background intervals, thereby representing a case of constructive selectivity as recently noted in Mesozoic mammals during the end-Cretaceous extinction [36,70]. Early cynodonts apparently do not mirror the therocephalian pattern given the present sampling. However, it is noteworthy that they exhibited unusually small sizes relatively early in their evolutionary history, remained small for much of the Late Permian and earliest Triassic, and radiated into a wide range of sizes that included large-bodied forms only after the extinction of other large theriodont predators (e.g., *Theriognathus*, *Moschorhinus*, and gorgonopsians which were excluded in the present analysis). An unexpected result is the protracted and increasingly constrained evolutionary rates in late Middle to Late Triassic cynodonts. The factors influencing their constrained phenotypic evolution are beyond the scope of the present analysis, but may include intrinsic factors (developmental canalization) or extrinsic selective pressures (competition with larger-bodied archosaurs or other emerging clades) that merit future study.

The sudden and short-lived nature of Early Triassic size reductions is compatible with some ecological explanations for

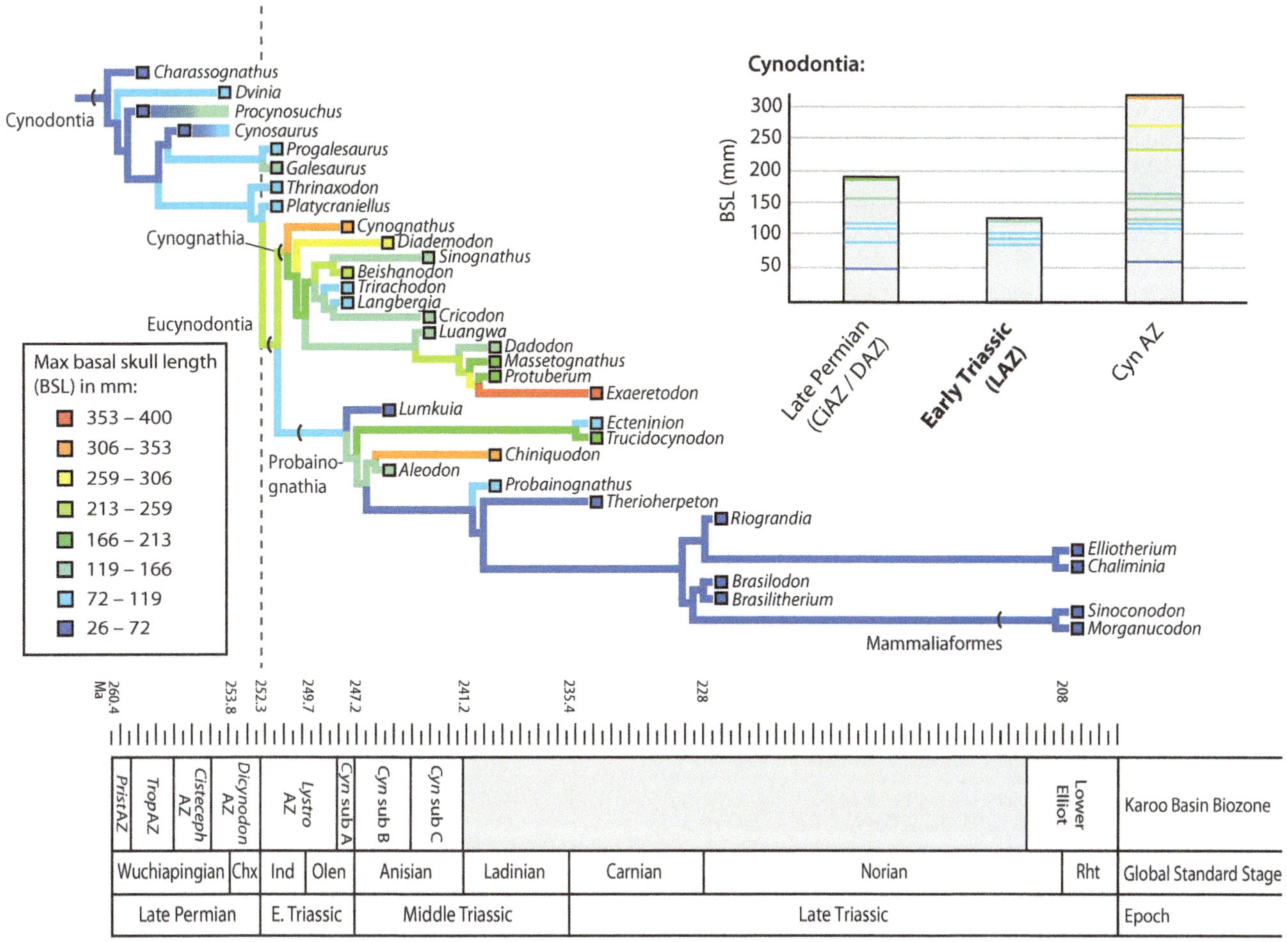

Figure 8. Ancestral state reconstructions for basal skull length in Permian and Triassic cynodonts. Colored horizontal bars indicate ranges of tip taxa spanning more than one stratigraphic bin (e.g., *Procynosuchus*, *Cynosaurus*). Permian-Triassic boundary indicated by dashed line. Abbreviations: CiAZ, *Cistecephalus* Assemblage Zone; CynAZ, *Cynognathus* Assemblage Zone; DAZ, *Dicynodon* Assemblage Zone; LAZ, *Lystrosaurus* Assemblage Zone.

Lilliput effects. Suggested mechanisms driving Lilliput patterns have included 1) within-taxon size reductions in relict lineages, 2) extinction of large taxa, and/or 3) post-extinction radiations of small taxa [9,10]. The phylogenetic models evaluated here do not support a post-extinction radiation of small taxa. In fact, several small-bodied *Lystrosaurus* AZ taxa and clades have Permian ghost lineages (Figs. 7, 8), implying preferential survivorship of these small-bodied lineages. Unfortunately, this latter scenario is difficult to assess because the small sample size of surviving tetrapod genera within the Karoo (only four observed records) is not amenable to the desired statistical tests (e.g., logistic regression of size-specific survivorship). At least one large-bodied eutheriodont, *Moschorhinus*, exhibited within-lineage size reductions across the extinction boundary, implying that the extinction also had the potential to act upon microevolutionary processes in a constructive manner [68]. Similar body size reduction characterizes the Permian and Triassic species of the dicynodont *Lystrosaurus* [65,71].

This interplay of Lilliput mechanisms parallels that of some Permian-Triassic marine invertebrate clades in some respects, but differs in others. Global surveys of marine gastropods suggest strong within-lineage size decreases as well as size-selective extinction [26]. Earliest Triassic foraminifera from South China provide evidence of all three underlying processes [25], although within-lineage effects drive global size distributions of foraminifera most strongly [42]. Notably, within-lineage decreases in lingulid brachiopods underscore hitherto unexplored ecophysiological effects, as they were associated with more frequent interruptions to growth and slower overall growth rates (perhaps in response to ocean acidification and hypercapnia) [28]. Within-lineage size decreases are observed in some nonmarine tetrapods (e.g., *Moschorhinus* and *Lystrosaurus*), but more general shifts toward smaller terrestrial vertebrate taxa were likely accentuated by the ecological removal of large-bodied Permian species (although small size did not necessarily ensure survival). Recent surveys of bone histology in these Permian-Triassic survivor taxa provide little evidence that Triassic therapsids grew more slowly than their Permian predecessors, and in some cases they may have grown more quickly or over shorter durations [65,68]. More data on the life histories of earliest Triassic terrestrial vertebrates will shed further light on the physiological and ecological underpinnings of these large-scale patterns and selectivity during the end-Permian biotic crisis.

Supporting Information

Appendix S1 List of institutional abbreviations.

Appendix S2 Outcrop area and sampled diversity of Karoo Basin therocephalians and cynodonts.

Appendix S3 Age and clade rank data for therocephalian and cynodont subclades. Basal skull length measurements (BSL) are in mm.

Dataset S1 Phylogenetic methods and supporting NEXUS file.

Acknowledgments

I thank C. Sidor and J. Botha-Brink for supervising this project. F. Abdala and B. Rubidge (BP), E. Butler (NMQR), E. De Kock and J. Neveling (CGS), P. Holroyd (UCMP), S. Kaal and R. Smith (SAM), C. Mehling (AMNH), A. Milner (NHMUK), and R. Rubidge (RC) are acknowledged for specimen access. S. Herring, E. Nesbitt, C. Sidor, and G. Wilson provided guidance and helpful suggestions on various aspects of this project. M. Maliska and D. Vilhena provided help with some of the 'R' scripts, and G. Hunt helped with time series analysis and the 'paleoTS' package. F. Abdala and K. Angielczyk offered insightful discussions on body size evolution and therapsid diversity during the end-Permian extinction. I also thank R. Butler for handling the manuscript submission, as well as N. Campione and J. Fröbisch for their constructive reviews.

Author Contributions

Conceived and designed the experiments: AH. Performed the experiments: AH. Analyzed the data: AH. Contributed reagents/materials/analysis tools: AH. Wrote the paper: AH.

References

1. Peters R (1983) The Ecological Implications of Body Size. Cambridge University Press. 344 p.
2. Calder WA III (1984) Size, Function, and Life History. Cambridge: Harvard University Press. 431 p.
3. Stearns SC (1992) The Evolution of Life Histories. Oxford: Oxford University Press. 249 p.
4. Gotthard K (2001) Growth strategies of ectothermic animals in temperate environments. In Atkinson D, Thorndyke M, editors. Environment and Animal Development: Genes, Life Histories, and Plasticity. Oxford: BIOS Scientific Publishers. pp. 287–303.
5. Cope ED (1887) The Origin of the Fittest. New York: Appleton and Company.
6. Cope ED (1896) The Primary Factors of Organic Evolution. Chicago: The Open Court Publishing Company.
7. Stanley SM (1973) An explanation for Cope's rule. Evolution 27:1–26.
8. Urbanek A (1993) Biotic crises in the history of Upper Silurian graptoloids: a paleobiological model. Hist Biol 7:29–50.
9. Twitchett RJ (2007) The Lilliput effect in the aftermath of the end-Permian extinction event. Palaeogeogr Palaeoclimatol Palaeoecol 252:132–144.
10. Harries PJ, Knorr PO (2009) What does the 'Lilliput Effect' mean? Palaeogeogr Palaeoclimatol Palaeoecol 284:4–10.
11. Gould SJ (1997) Cope's rule as psychological artefact. Nature 385:199–200.
12. Laurin M (2004) The evolution of body size, Cope's rule and the origin of amniotes. Syst Biol 53:594–622.
13. Kingsolver JG, Pfennig DW (2004) Individual-level selection as a cause of Cope's Rule of phyletic size increase. Evolution 58:1608–1612.
14. Hone DWE, Benton MJ (2005) The evolution of large size: how does Cope's Rule work? Trends Ecol Evol 20:4–6.
15. McShea DW (1994) Mechanisms of large-scale evolutionary trends. Evolution 48(6):1747–1763.
16. Carrano MT (2006) Body-size evolution in the Dinosauria. In: Carrano MT, Blob RW, Gaudin TJ, Wible JR, editors. Amniote Paleobiology: Perspectives on the Evolution of Mammals, Birds, and Reptiles. Chicago: University of Chicago Press. pp. 225–268.
17. Sookias RB, Butler RJ, Benson RBJ (2012a) Rise of dinosaurs reveals major body-size transitions are driven by passive processes of trait evolution. Proc R Soc Lond B Biol Sci 279(1736):2180–2187.
18. Sookias RB, Benson RBJ, Butler RJ (2012b) Biology, not environment, drives major patterns in maximum tetrapod body size through time. Biol Lett 8(4):674–677.
19. Liow LH, Fortelius M, Bingham E, Lintulaakso K, Mannila H, et al. (2008) Higher origination and extinction rates in larger mammals. Proc Natl Acad Sci USA 105:6097–6102.
20. Liow LH, Fortelius M, Lintulaakso K, Mannila H, Stenseth NC (2009) Lower extinction risk in sleep-or-hide mammals. Am Nat 173:264–272.
21. McKinney ML (1997) Extinction vulnerability and selectivity: combining ecological and paleontological views. Annu Rev Ecol Syst 28:495–516.
22. Cardillo M, Bromham L (2001) Body size and extinction risk in Australian mammals. Conserv Biol 15:1435–1440.
23. Matthews LJ, Arnold C, Machanda Z, Nunn CL (2011) Primate extinction risk and historical patterns of speciation and extinction in relation to body mass. Proc R Soc Lond B Biol Sci 278:1256–1263.
24. Jablonski D (2005) Mass extinctions and macroevolution. In: Vrba ES, Eldredge N, editors. Paleobiology 31(supplement to 2): Macroevolution: Diversity, Disparity, Contingency .pp. 192–210.
25. Song H, Tong J, Chen ZQ (2011) Evolutionary dynamics of the Permian-Triassic foraminifer size: Evidence for Lilliput effect in the end-Permian mass extinction and its aftermath. Palaeogeogr Palaeoclimatol Palaeoecol 308:98–110.
26. Payne JL (2005) Evolutionary dynamics of gastropod size across the end-Permian extinction and through the Triassic recovery interval. Paleobiology 31(2):269–290.
27. Luo G, Lai X, Shi GR, Jiang H, Yin H, et al. (2008) Size variation of conodont elements of the *Hindeodus-Isarcicella* clade during the Permian-Triassic transition in South China and its implications for mass extinction. Palaeogeogr Palaeoclimatol Palaeoecol 264:176–187.
28. Metcalfe B, Twitchett RJ, Price-Lloyd N (2011) Changes in size and growth rate of 'Lilliput' animals in the earliest Triassic. Palaeogeogr Palaeoclimatol Palaeoecol 308:171–180.
29. Smith RMH (1995) Changing fluvial environments across the Permian-Triassic boundary in the Karoo Basin, South Africa and possible causes of tetrapod extinctions. Palaeogeogr Palaeoclimatol Palaeoecol 117:81–104.
30. Sahney S, Benton MJ (2008) Recovery from the most profound mass extinction of all time. Proc R Soc Lond B Biol Sci 275:759–765.
31. Retallack G (2013) Permian and Triassic greenhouse crises. Gondwana Research 24:90–103.
32. Botha J, Smith RMH (2006) Rapid vertebrate recuperation in the Karoo Basin of South Africa following the end-Permian extinction. J Afr Earth Sci 45:502–514.
33. Huttenlocker AK, Sidor CA, Smith RMH (2011) A new specimen of *Promoschorhynchus* (Therapsida: Therocephalia: Akidnognathidae) from the Lower Triassic of South Africa and its implications for theriodont survivorship across the Permo-Triassic boundary. J Vertebr Paleontol 31:405–421.
34. Erwin DH (1998) The end and the beginning: recoveries from mass extinctions. Trends Ecol Evol 13:344–349.
35. Erwin DH, Bowring SA, Yugan J (2002) End-Permian mass extinctions: a review. In: Koeberl C, MacLeod KG, editors. Catastrophic Events and Mass Extinctions: Impacts and Beyond. Geological Society of America Special Paper 356. pp. 363–383.
36. Wilson GP (2013) Mammals across the K/Pg boundary in northeastern Montana, U.S.A.: dental morphology and body-size patterns reveal extinction selectivity and immigrant-fueled ecospace filling. Paleobiology 39:429–469.
37. Smith RMH, Botha J (2005) The recovery of terrestrial vertebrate diversity in the South African Karoo Basin after the end-Permian extinction. Comptes Rendus Palevol 4:623–636.
38. Irmis RB, Whiteside JH, Kammerer CF (2013) Non-biotic controls of observed diversity in the paleontologic record: An example from the Permo-Triassic Karoo Basin of South Africa. Palaeogeogr Palaeoclimatol Palaeoecol 372:50–61.
39. Fröbisch J (2013) Vertebrate diversity across the end-Permian mass extinction – separating biological and geological signals. Palaeogeogr Palaeoclimatol Palaeoecol 372:50–61.
40. Fröbisch J (2014) Synapsid diversity and the rock record in the Permian-Triassic Beaufort Group (Karoo Supergroup), South Africa. In: Kammerer CF, Angielczyk KD, Fröbisch J, editors. Early Evolutionary History of the Synapsida. Dordrecht: Springer. pp. 305–319.
41. Walther M, Fröbisch J (in press) The quality of the fossil record of anomodonts (Synapsida, Therapsida). Comptes Rendus Palevol http://dx.doi.org/10.1016/j.crpv.2013.07.007
42. Rego BL, Wang SC, Altiner D, Payne JL (2012) Within- and among-genus components of foraminiferan size evolution during mass extinction, recovery, and background intervals. Paleobiology 38:627–643.
43. Rubidge BS (1995) Biostratigraphy of the Beaufort Group (Karoo Supergroup). South African Committee for Stratigraphy Biostratigraphic Series 1, Geological Survey of South Africa. 46 p.
44. Rubidge BS (2005) Reuniting lost continents – Fossil reptiles from the ancient Karoo and their wanderlust. South Afr J Geol 108:135–172.
45. Abdala F, Neveling J, Welman J (2006) A new trirachodontid cynodont from the lower levels of the Burgersdorp Formation (Lower Triassic) of the Beaufort

Group, South Africa and the cladistic relationships of Gondwanan gomphodonts. Zool J Linn Soc 147:383–413.
46. Abdala F, Ribeiro AM (2010) Distribution and diversity patterns of Triassic cynodonts (Therapsida, Cynodontia) in Gondwana. Palaeogeogr Palaeoclimatol Palaeoecol 286:202–217.
47. Irmis RB, Martz JW, Parker WG, Nesbitt SJ (2010) Re-evaluating the correlation between Late Triassic terrestrial vertebrate biostratigraphy and the GSSP-defined marine stages. Albertiana 38:40–52.
48. Hunt G (2007) The relative importance of directional change, random walks, and stasis in the evolution of fossil lineages. Proc Natl Acad Sci USA 104:18404–18408.
49. Hunt G, Carrano MT (2010) Models and methods for analyzing phenotypic evolution in lineages and clades. In: Alroy J, Hunt G, editors. Quantitative Methods in Paleobiology: Paleontological Society Papers v. 16. The Paleontological Society. pp. 245–269.
50. Norell MA, Novacek MJ (1992) The fossil record and evolution: comparing cladistic and paleontologic evidence for vertebrate history. Science 255:1690–1693.
51. Carrano MT (2000) Homoplasy and the evolution of dinosaur locomotion. Paleobiology 26:489–512.
52. Sidor CA (2001) Simplification as a trend in synapsid cranial evolution. Evolution 55:1419–1442.
53. Sidor CA (2003) Evolutionary trends and the origin of the mammalian lower jaw. Paleobiology 29: 605–640.
54. Huttenlocker AK (2009) An investigation into the cladistic relationships and monophyly of therocephalian therapsids (Amniota: Synapsida). Zool J Linn Soc 157:865–891.
55. Sigurdsen T, Huttenlocker AK, Modesto SP, Rowe TB, Damiani R (2012) Reassessment of the morphology and paleobiology of the therocephalian *Tetracynodon darti* (Therapsida), and the phylogenetic relationships of Baurioidea. J Vertebr Paleontol 32(5):1113–1134.
56. Hopson JA, Kitching JW (2001) A probainognathian cynodont from South Africa and the phylogeny of nonmammalian cynodonts. Bull Mus Comp Zool 156:5–35.
57. Sidor CA, Hancox PJ (2006) Elliotherium kersteni, a new tritheledontid from the lower Elliot Formation (Upper Triassic) of South Africa. J Paleontol 80:333–342.
58. Oliveira TV, Soares MB, Schultz CL (2010) *Trucidocynodon riograndensis* gen. nov. et sp. nov. (Eucynodontia), a new cynodont from the Brazilian Upper Triassic (Santa Maria Formation). Zootaxa 2382:1–71.
59. Liu J, Olsen P (2010) The phylogenetic relationships of Eucynodontia (Amniota: Synapsida). J Mamm Evol 17:151–176.
60. Ranivoharimanana L, Kammerer CF, Flynn JJ, Wyss AR (2011) New material of *Dadadon isaloi* (Cynodontia, Traversodontidae) from the Triassic of Madagascar. J Vertebr Paleontol 31:1292–1302.
61. Sidor CA, Hopson JA (1998) Ghost lineages and "mammalness": assessing the temporal pattern of character acquisition in the Synapsida. Paleobiology 24:254–273.
62. Maddison WP, Maddison DR (2007) Mesquite: a modular system for evolutionary analysis. Version 2.0. http://mesquiteproject.org
63. Chan KMA, Moore BR (2004) SymmeTREE: whole-tree analysis of differential diversification rates. Bioinformatics 21:1709–1710.
64. Ruta M, Pisani D, Lloyd GT, Benton MJ (2007) A super-tree of Temnospondyli: cladogenetic patterns in the most species-rich group of early tetrapods. Proc R Soc Lond B Biol Sci 274:3087–3095.
65. Botha-Brink J, Angielczyk KD (2010) Do extraordinarily high growth rates in Permo-Triassic dicynodonts (Therapsida, Anomodontia) explain their success before and after the end-Permian extinction? Zool J Linn Soc 160:341–365.
66. Brusatte SL, Benton MJ, Ruta M, Lloyd GT (2008) Superiority, competition, and opportunism in the evolutionary radiation of dinosaurs. Science 321:1485–1488.
67. Huttenlocker AK (2013) The Paleobiology of South African Therocephalian Therapsids (Amniota, Synapsida) and the Effects of the End-Permian Extinction on Size, Growth, and Bone Microstructure. Unpublished PhD dissertation. University of Washington, Seattle. 414 p.
68. Huttenlocker AK, Botha-Brink J (2013) Body size and growth patterns in the therocephalian *Moschorhinus kitchingi* (Therapsida: Eutheriodontia) before and after the end-Permian extinction in South Africa. Paleobiology 39:253–277.
69. Evans AR, Jones D, Boyer AG, Brown JH, Costa DP, et al. (2012) The maximum rate of mammal evolution. Proc Natl Acad Sci USA 109:4187–4190.
70. Wilson GP, Evans AR, Corfe IJ, Smits PD, Fortelius M, et al. (2012) Adaptive radiation of multituberculate mammals before the extinction of dinosaurs. Nature 483:457–460.
71. Botha J, Smith RMH (2007) *Lystrosaurus* species composition across the Permian-Triassic boundary in the Karoo Basin of South Africa. Lethaia 40:125–137.

11

Range-Wide Latitudinal and Elevational Temperature Gradients for the World's Terrestrial Birds: Implications under Global Climate Change

Frank A. La Sorte[1]*, Stuart H. M. Butchart[2], Walter Jetz[3], Katrin Böhning-Gaese[4,5]

1 Cornell Laboratory of Ornithology, Cornell University, Ithaca, New York, United States of America, **2** BirdLife International, Cambridge, United Kingdom, **3** Department of Ecology and Evolutionary Biology, Yale University, New Haven, Connecticut, United States of America, **4** Biodiversity and Climate Research Centre (BiK-F), Senckenberg Gesellschaft für Naturforschung, Frankfurt (Main), Germany, **5** Department of Biological Sciences, Goethe Universität, Frankfurt (Main), Germany

Abstract

Species' geographical distributions are tracking latitudinal and elevational surface temperature gradients under global climate change. To evaluate the opportunities to track these gradients across space, we provide a first baseline assessment of the steepness of these gradients for the world's terrestrial birds. Within the breeding ranges of 9,014 bird species, we characterized the spatial gradients in temperature along latitude and elevation for all and a subset of bird species, respectively. We summarized these temperature gradients globally for threatened and non-threatened species and determined how their steepness varied based on species' geography (range size, shape, and orientation) and projected changes in temperature under climate change. Elevational temperature gradients were steepest for species in Africa, western North and South America, and central Asia and shallowest in Australasia, insular IndoMalaya, and the Neotropical lowlands. Latitudinal temperature gradients were steepest for extratropical species, especially in the Northern Hemisphere. Threatened species had shallower elevational gradients whereas latitudinal gradients differed little between threatened and non-threatened species. The strength of elevational gradients was positively correlated with projected changes in temperature. For latitudinal gradients, this relationship only held for extratropical species. The strength of latitudinal gradients was better predicted by species' geography, but primarily for extratropical species. Our findings suggest threatened species are associated with shallower elevational temperature gradients, whereas steep latitudinal gradients are most prevalent outside the tropics where fewer bird species occur year-round. Future modeling and mitigation efforts would benefit from the development of finer grain distributional data to ascertain how these gradients are structured within species' ranges, how and why these gradients vary among species, and the capacity of species to utilize these gradients under climate change.

Editor: David L. Roberts, University of Kent, United Kingdom

Funding: This work was funded through a postdoctoral fellowship at the Cornell Lab of Ornithology. The funders had no role in study design, data collection and analysis, decision to publish, or preparation of the manuscript.

Competing Interests: The authors have declared that no competing interests exist.

* E-mail: fal42@cornell.edu

Introduction

A variety of responses are available for species as global climate change progresses [1,2] and changing climatic conditions increasingly impact species' performance and fitness. The two primary options, which are not necessarily independent, are for species' populations to shift their geographic ranges to regions containing suitable climatic conditions, or to remain and adapt phenotypically or genetically [3,4]. One of the most important immediate responses is for species to track their geographic climatic associations across space. This may be possible provided that opportunities are available for geographic range-shifts, species have the physiological and behavioral capacity to take advantage of them, and they can keep pace with the velocity of climate change [5]. Geographic niche tracking has been observed under climate change over geological time scales [6] and more recently under modern climate change [7,8].

Two environmental gradients appear particularly relevant when considering geographic niche tracking under past and current climate change: surface latitudinal and elevational temperature gradients. Both gradients represent natural environmental features whose location and general form remain consistent across ecological time-scales, and there is empirical evidence that many taxa including birds are currently tracking these gradients under climate change [7,9]. Latitudinal temperature gradients are strongest outside the tropics, especially in the Northern Hemisphere where the most extensive land masses are located [10]. Elevational temperature gradients are stronger and less variable than latitudinal temperature gradients but are geographically more restricted [11]. Latitudinal and elevational temperature gradients are not equally available for all species nor are they uniformly distributed within species' geographic ranges. Under current climate change projections, there is no evidence to suggest the strength of elevational temperature gradients are expected to change; whereas there is evidence that the strength of latitudinal temperatures gradients may change over broad geographic regions [2]. Due to latitudinal asymmetry in surface warming, latitudinal temperature gradients are projected to weaken in the Northern

Hemisphere [2,12], which is expected to result in diminished seasonality and increased vegetation growth at the high northern latitudes [13,14].

Life history, morphological, and physiological traits are known to vary among species' populations at geographic scales [15] and, in some cases, this variation is correlated with latitudinal or elevational gradients [16,17] and associated temperature gradients [18]. This variation in ecological traits across latitudinal or elevational gradients can in turn affect reproductive potential and rates of population growth, especially at range limits [19,20]. Under climate change, the structure of latitudinal and elevational temperature gradients can be modified, affecting population growth rates, especially at the leading and trailing edge of the gradient within species' ranges [21,22]. Specifically, climate change can result in population growth increasing at the leading edge and declining at the trailing edge of the gradient, with the increased likelihood of range shifts at the leading edge and extirpation of populations at the trailing edge.

The strength of latitudinal and elevational temperature gradients within species current geographic ranges relative to projected changes in temperature under climate change can determine the potential for range-shifts to occur under climate change. For species whose geographic ranges have similar spatial dimensions in relationship to latitudinal or elevational temperature gradients, the steeper the gradient, the broader the species' geographic climatic niche (i.e., species' association with climatic conditions across space) and the greater the likelihood that a larger component of current climatic associations will be retained within the geographic range as temperatures increase (Figure 1A,B,C). In contrast, the shallower the gradient, the narrower the species' geographic climatic niche and the smaller the likelihood that a significant geographic component of current climatic conditions will be retained within the geographic range as temperatures increase (Figure 1D,E,F). One potential consequence of shallow latitudinal or elevational temperature gradients is the formation of range-shift gaps [10]. Here, increasing temperatures can result in the complete loss of current climatic associations within species' current geographic ranges, an outcome that may substantially hinder successful range-shifts. Range-shift gap width is also likely to be negatively correlated with the strength of the temperature gradients (Figure 1), suggesting species' distributions containing very weak gradients, as found with many tropical species, are particularly susceptible to the consequences of climate change [10]. In total, climate change can modify latitudinal or elevational temperature gradients within species' geographic ranges, which in turn can alter population dynamics at range boundaries and range-shift potential based on the strength of those gradients relative to the magnitude of climate change.

Here, we estimate latitudinal and elevational temperature gradients within the geographic ranges of *ca.* 90% of the world's avifauna, numbering 9,014 species. To evaluate whether bird species that are currently threatened with extinction also face particularly high risks under climate change, we structure our analysis to compare species classified as threatened ($n = 878$) vs. non-threatened with extinction ($n = 8,136$) on the IUCN Red List [23]. We additionally examine the ability of species' geography (range size, shape, and orientation relative to the equator) to explain variation in elevational and latitudinal temperature gradients among species. These characteristics of species' ranges tend to be geographically similar across taxa [24] and are determined in part by spatial variation in climate and topography [25]. Lastly, to assess the degree of correspondence between projected warming and the availability of elevational and latitudinal temperature gradients within species' ranges, we consider how gradient strength is related to variation in projected changes in temperature under climate change.

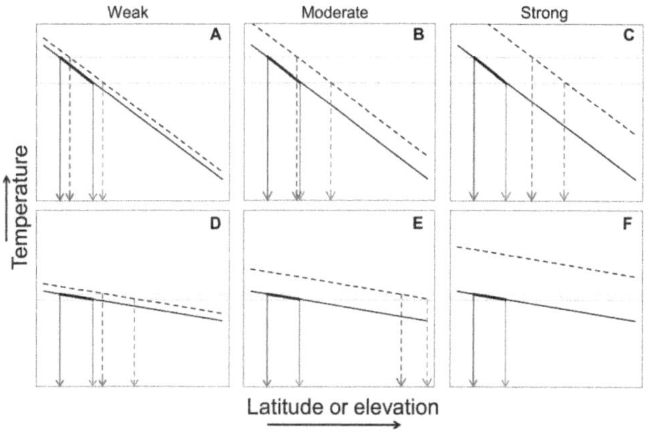

Figure 1. Conceptual figures of six geographic range-shift scenarios for species containing steep (A,B,C) or shallow (D,E,F) latitudinal or elevational temperature gradients within their geographic ranges. The spatial breadth of the latitudinal or elevational gradient within the geographic range is identical in each case. The current temperature gradient (solid line) and the projected temperature gradient under climate change (dashed line) are shown for each scenario. The location of the geographic range of the species along the temperature gradient is depicted with a bold line between the solid blue and red arrows. The species' leading (blue arrow) and trailing (red arrow) range limits are shown based on the current gradient (solid arrows) and the projected gradient under climate change (dashed arrows). Three gradations in projected climate change are shown: weak (A, D), moderate (B, E), and strong climate change (C, F). The horizontal dotted lines identify the boundaries of each species' geographic climatic niche as defined along the temperature gradient. Everything else being equal, species with steeper environmental temperature gradients have broader geographic climatic niches and need to shift their geographic range less in order to track climate change.

Materials and Methods

Data sources and preparation

We acquired range maps for the world's birds from BirdLife International and NatureServe [26] whose data sources are described in Buchanan, Donald and Butchart [27]. We considered breeding/resident ranges only and excluded species that were not well suited for our analysis, specifically those associated with marine environments. This resulted in a total of 9,014 extant species for analysis (see Table S1). These species were classified as either threatened ($n = 878$) or non-threatened with extinction ($n = 8,136$) using the IUCN Red List [23]. Here threatened refers to species identified as vulnerable, endangered, or critically endangered under the IUCN Red List.

Species range data was analyzed using a gridded system having a cylindrical equal-area projection and a cell area of 3,091 km^2. We defined species' geographic breeding ranges as the spatial combination of equal area cells with ≥50% terrestrial surface area (based on the proportion of the cell that contained non-marine terrain) that intersected each species' geographic range polygon. We placed each species into one of six biogeographical realms [28] based on the realm that contained the greatest proportion of each species' gridded range (Nearctic, Palaearctic, IndoMalaya, Neotropics, Afrotropics, and Australasia). Species whose ranges occurred primarily in Antarctica or Oceania were excluded from analysis. From the 9,014 species, we acquired minimum and maximum elevation associations for 4,978 species from BirdLife International [23]. The distribution of these species across the six realms did not show geographic biases, with the proportion of species in each realm not diverging significantly from expectation ($\chi^2 = 0.029$, df = 5, $P = 1.0$).

Temperatures within species' geographic ranges were estimated using the annual average of 1950–2000 average-monthly temperatures from WorldClim [29] gridded and analyzed at a 30-arcsec resolution (*ca.* 1 km at the equator). Elevations within species' ranges were estimated using the USGS global digital elevation model (GTOPO30) gridded and analyzed at a 30-arcsec resolution. Tropospheric lapse rates, the association between temperature and elevation within the troposphere, were estimated globally using a gridded model at a resolution of 2.5° (*ca.* 278 km at the equator) for the period 1948–2001 [11].

Broad-scale, expert-based range maps are only coarsely defined estimates of species occurrence and reliably characterize presences only above 100 km grain [30,31]. Here, we are interested in the form of latitudinal and elevational temperature gradients (i.e., shallow or steep) as defined across the geographic breadth of species' distributions independent of patterns of occurrence. The use of relatively high resolution global estimates of temperature and elevation provided the detail of data necessary to estimate the full structure of these gradients for individual species, especially in situations where geographic ranges were small or temperature gradients were non-linear. For the purpose of this baseline study we assume that low probability of occurrence naturally incurred at the finer spatial grain are not biasing the geographic and species comparisons. We acknowledge that this assumption is not necessarily valid and will require careful follow-up assessments at local scales or at broader scales once more detailed distribution data becomes available.

Projected changes in temperature (temperature anomalies) were estimated using projections averaged over four Atmosphere Ocean General Circulation Models (AOGCMs). The four AOGCMs were CNRMCM3, CSIRO-Mk3.0, ECHam5, and MIROC 3.2. The gridded projections from these models were downscaled to 30 arc-minute resolution using pattern scaling with the MarkSim weather generator [32]. The projections were based on the A2 scenario, which is currently considered to be the most relevant under current greenhouse gas emission rates [33,34]. The anomalies represent the change in temperature between the twentieth century control 30-year normal (1961–1990) and the 30-year period 2081–2100. The temperature anomalies were bilinearly interpolated to match the resolution of the 3,091 km^2 equal-area grid.

Elevational temperature gradients

We estimated latitudinal and elevational temperature gradients separately in order to better represent each gradient's unique characteristics. The presence of elevational temperature gradients is dictated by the presence of mountain systems or orographic features, which show substantial geographic variation (see Figure S1). Mountain systems often contain strong topographic heterogeneity resulting in a mosaic of climatic conditions, which are poorly estimated using interpolated weather station data [35,36]. Here, we use elevation data, which does not rely on similar interpolation methods, to estimate elevational temperature gradients within species' ranges.

Unlike latitudinal temperature gradients, the strength of elevational temperature gradients are broadly consistent from the equator to the poles. Tropospheric lapse-rate averages 6.2°C km^{-1} over the continents, with a range of 4.5 to 6.5°C km^{-1} from the polar latitudes to the equator, respectively [11]. To account for this geographic variation, we calculated the average tropospheric lapse-rate within each species' geographic range. To account for the high variability in the extent and spatial distribution of orographic features and the decline in terrestrial area with increasing elevation resulting in strongly skewed distributions of elevation within species' ranges, we estimated two components of the elevational temperature gradient for each species: (1) the overall magnitude and (2) the evenness of the gradient. We then combined tropospheric lapse-rate with magnitude and evenness to generate one metric designed to represent the overall strength of elevational temperature gradients within each species' range.

More specifically, for species whose minimum and maximum elevation associations had both been estimated ($n = 4{,}978$), we excluded elevations within the geographic range below and above these associations. We then identified the actual minimum and maximum elevations within each species' range based on our gridded elevation data, which in combination with species' documented elevation associations were used to calculate each species' realized vertical range extent. Vertical range extent was used to estimate the magnitude of each species' elevational temperature gradient. To assess how evenly the vertical range extent was distributed across each species' range, we first summed the number of 30 arc-second elevation cells in all 100-m elevational bands found within each species range. We then used the Lorenz curve and associated Gini coefficient [37–39], explained below, to estimate how evenly the cells were distributed across elevation bands. Here, the 100-m bands were first ranked from lowest to highest elevation. The frequency of cells within these bands was then used to plot the Lorenz curve, i.e. the cumulative proportion of the number of cells in each band (x-axis) against the corresponding cumulative proportion of their elevations (y-axis). Gini coefficients (or Gini ratio) were then calculated, which is defined as twice the area contained between the Lorenz curve and the line of perfect equality (or perfect evenness). The Gini coefficient has a range from 0 to 1, high to low evenness, respectively. To ease interpretation, for each species we took the product of the vertical range extent, average tropospheric lapse rate, and the inverse of evenness (1 – Gini coefficient) to generate a single index for the elevational temperature gradient. We interpret this index as the overall strength of the elevational temperature gradient across each species' range: larger values indicate the presence of a strong and more evenly distributed gradient; lower values indicate the presence of a weak gradient due to either a small vertical range extent or an unevenly distributed vertical range extent.

Latitudinal temperature gradients

Outside the tropics, average annual temperature declines on average 0.7°C for each degree of latitude in the Northern Hemisphere and on average 0.5°C for each degree of latitude in the Southern Hemisphere (Figure S2). With one degree of latitude equal to approximately 111 km, this translates to a decline of 1°C for every 150 km in the Northern Hemisphere and a decline of 1°C for every 197 km in the Southern Hemisphere.

We estimated the latitudinal temperature gradient within the breeding ranges of 9,014 species while controlling for variation in elevation. For our approach, we first found the average tropospheric lapse rate within each species' range. We then compiled gridded 30 arc-second average annual temperatures and elevations within each species' range. We chose annual average temperature because matching each species' breeding season distribution with their unique breeding season temperatures, especially for migratory species that only spend a few weeks to months on the breeding grounds, was not feasible. A likely consequence of using annual average temperature is that estimates of latitudinal temperature gradients for species that breed outside the tropics are likely to be steeper than they would be if estimates were based on the breeding season alone. Beyond these qualitative differences, using annual average temperature is unlikely to alter our qualitative conclusions. To control for the effect of elevation on temperature, we applied quantile regression to temperature as a function of elevation for 11 quantiles (0.0, 0.1,...0.9, 1.0). The residuals for the quantile whose slope most closely matched the tropospheric lapse rate estimated for each species was retained for further analysis. This approach allowed us to interpret latitudinal temperature gradients as the gradient along a fixed elevation within each species' range.

The relationship between these residuals and latitude was then assessed for each species using segmented linear regression [40]. If species' ranges occurred in both hemispheres, the range was first split in half at the equator and each portion was analyzed separately. We tested if the relationship between temperature and latitude contained one or two segments using Davie's test [41], which tests for a non-zero difference in the slope parameters of the segmented relationship. Relationships with two segments are more likely to occur if species' distributions are intersected by the Tropics of Cancer or Capricorn (Figure S2). When there was statistical evidence for a segmented relationship, we estimated a break point between segments and slope coefficients for each segment. When there was no statistical evidence for multiple segments, a single slope coefficient was estimated. Coefficients estimated south of the equator were multiplied by -1 to allow for comparison between the two hemispheres (i.e., latitudinal temperature gradients north and south of the equator are both negative; Figure S2). For species whose ranges were estimated by one slope coefficient, this coefficient was used to represent the overall gradient. For species whose ranges contained multiple slope coefficients, we used a weighted average to summarize the gradient. Weights were based on the latitudinal extent of the region where each slope coefficient was estimated. Based on this procedure, larger negative values indicate steeper latitudinal temperature gradients and values approaching zero indicate shallower latitudinal temperature gradients.

Analysis

To summarize how steep elevational and latitudinal temperature gradients are on average within species' ranges globally, we first calculated the average elevational temperature gradient and median latitudinal temperature gradient within each equal-area cell for all species whose geographic ranges intersected that cell. To summarize how gradients differed for threatened and non-threatened species, we quantified how the gradients varied by species' median latitude. This analysis was implemented separately for species in each of the six biogeographical realms. For elevational temperature gradients, we used generalized additive models (GAM; [42]) with a Gaussian error distribution. For latitudinal temperature gradients, we used robust MM-type polynomial regression [43,44]. GAM was selected because it adjusts automatically to the nonlinear associations observed between elevational temperature gradients and species' median latitude. This feature also supported the use of GAM to examine how species richness, range size, elevational extent, and Gini coefficient varied by species' median latitude for threatened and non-threatened species. Robust polynomial regression was selected for latitudinal temperature gradients because of the presence of many extreme outliers that were poorly modeled with GAM. In addition, the general curvilinear form of the relationship was well represented with a polynomial function. The order of the polynomial function was determined using robust Wald-type or deviance-type tests of nested model pairs. Median latitude was defined as the median of the full latitudinal extent of each species' range.

We used robust multiple regression to examine the ability of geographic and climatic predictors to explain the variability observed in species latitudinal and elevational temperature gradients. The four predictors were estimated within each species' geographic range and included average projected temperature anomaly, and geographic range size, shape, and orientation. Range size was $\log_{10}$ transformed before analysis. Range shape and orientation were estimated based on a principle component analysis (PCA) of the latitude and longitude of the grid cells contained within each species geographic range. PCA was conducted using a singular value decomposition of the centered data matrix. The square roots of the two eigenvalues were used to estimate range shape, where the minimum eigenvalue was divided by the maximum eigenvalue. Here, values approaching zero indicating elongated ranges and values approaching one indicating circular ranges. Range orientation was estimated based on the angle of the major axis of the PCA eigenvector from the equator. Values for range orientation were defined from 0° (parallel with the equator) to 90° (perpendicular to the equator). We evaluated our four predictors for multi-collinearity and singularity using variance inflation factors (VIFs) where predictors with VIF>5 indicate cause for concern and VIF>10 indicate the presence of significant collinearity [45]. We retained all four predictors after they were deemed to be statistically independent (VIF≤1.6).

To evaluate differences between threatened and non-threatened species, we included in addition to the four continuous predictors, a species' IUCN Red List status as threatened or non-threatened with extinction. For the analysis of the latitudinal temperature gradient, we also considered if the median latitude of a species' range was located within or outside the tropics. Median latitudes between 23.5°S and 23.5°N latitude were considered tropical. The tropical/extratropical classification was included because it defined a major feature of the gradient (Figure S2). In total, we examined eight robust multiple regression models that consisted of the two gradients, the four continuous predictors, associated categorical predictors, and all pairwise interactions.

All analyses were conducted in R version 3.0.1 [46]. Segmented regression was conducted using the 'segmented' library, robust regression using the 'robust' and 'robustbase' libraries, and GAM using the 'mgcv' library [42]. We used the default optimization procedure to estimate the degree of smoothing in the GAMs [42]. The Lorenz curves were estimated using the 'ineq' library and the Gini coefficients were estimated using the 'reldist' library.

Results

Threatened and non-threatened bird species presented similar latitudinal trends in species richness with peaks for both groups occurring in the tropics (Figure 2A and Figure S3). Range sizes for threatened species were significantly smaller on average except for species located at the most extreme northern and southern latitudes (Figure 2B and Figure S3). For the 4,978 species with documented elevational associations, threatened species had smaller vertical range extents within their geographic ranges (Figure 2C and Figure S3) and threatened species had higher elevational evenness within their geographic ranges in the central tropics and more equivalent elevational evenness elsewhere (Figure 2D and Figure S3). Projected increases in temperature within species' ranges were highest for species located at higher

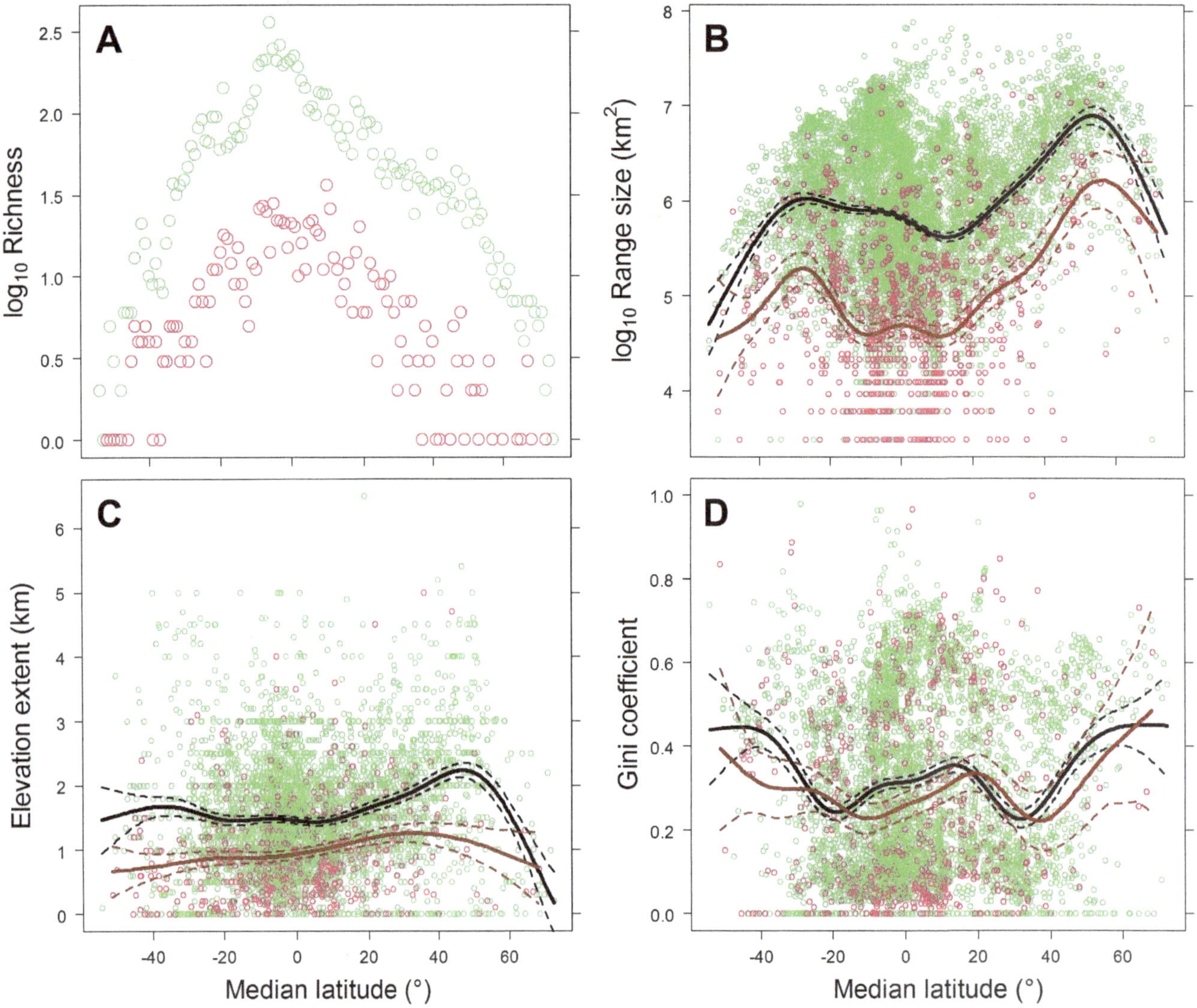

Figure 2. Plots summarizing geographic patterns for species' currently identified as threatened (red points; *n* = 878) and non-threatened (green points; *n* = 8,136) with extinction by the median latitude of each species' range. (A) The number of species' whose median latitudes occur within one degree latitudinal bands; (B) species' range size; (C) the elevational extent within species' ranges for 4,978 species with minimum and maximum elevation associations; and (D) the evenness of the distribution of 100 m elevations bands within species' ranges based on the Gini coefficient (0 = high evenness; 1 = low evenness; see Materials and Methods for details) for 4,978 species with minimum and maximum elevation associations. The trend lines are the fits of generalized additive models for species threatened (red) and non-threatened (black) with extinction.

latitudes in the Northern Hemisphere and lowest for species located at lower latitudes in the Southern Hemisphere (Figure S4).

There was strong geographic variation in the steepness of elevational temperature gradients (Figure 3). When considering the combined effects of elevational extent and elevational evenness (Figure S3) with tropospheric lapse rate, the strongest gradients were found for species located in Africa, the western part of North and South America, and the Tibetan Plateau region. In contrast, species in Australasia, the islands of IndoMalaya and the lowlands of the Neotropics had the weakest elevational temperature gradients (Figure 3). Threatened species were associated with weaker elevational temperature gradients across the six biogeographical realms, with the most significant differences occurring in the Neotropics and Afrotropics (Figure 3).

After controlling for elevation, latitudinal temperature gradients were on average shallowest for species in the tropics and steepest for species outside the tropics (Figure 4). The steepest latitudinal temperature gradients occurred in the Northern Hemisphere. Australasia had the strongest gradients in the Southern Hemisphere, which did not match the strength observed in the Northern Hemisphere. There was no evidence that the latitudinal temperature gradients differed between threatened and non-threatened species across the six biogeographical realms (Figure 4).

Variation in elevational temperature gradients among species was poorly explained by our four predictors (Table 1 and Figure S5). Species with higher projected temperature increases and larger geographic ranges had steeper elevational temperature gradients, and this was the case for both threatened and non-threatened species. Species with geographic ranges that were elongated in shape and oriented perpendicular to the equator had stronger gradients, and this was more pronounced for non-threatened species.

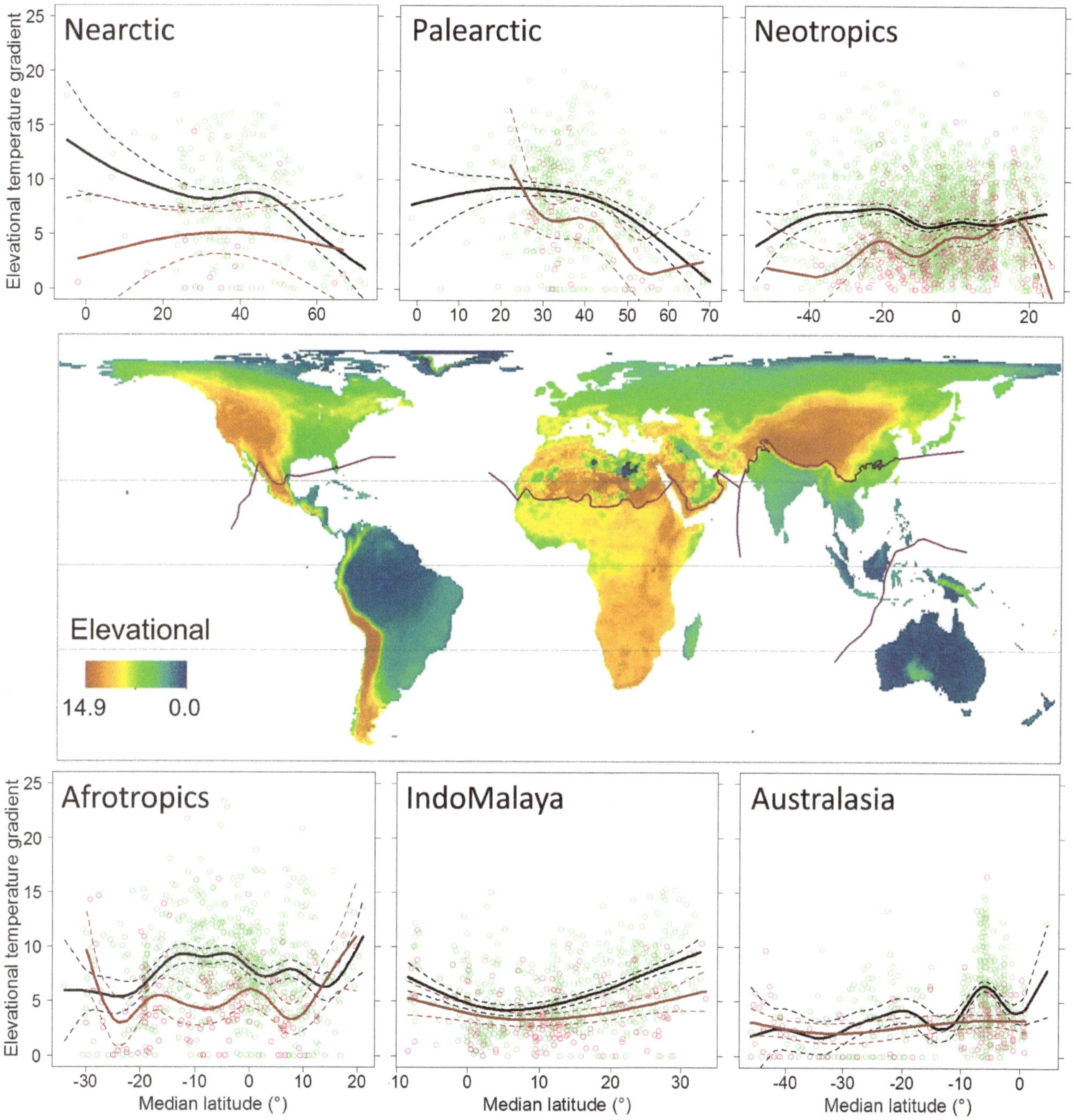

Figure 3. The elevation temperature gradient index summarized across species' geographic ranges (map) and summarized within six biogeographical realms (plots) as a function of the median latitude of each species' range. The temperature gradient was estimated spatially within species' current ranges (see Materials and Methods for details). In the plots, red points are threatened species ($n = 766$) and green points are non-threated species ($n = 4{,}212$). Trend lines are the fits of generalized additive models for threatened (red) and non-threatened species (black). The solid line is the equator and the dashed lines are the Tropics of Cancer and Capricorn (23.5°N and 23.5°S latitude, respectively).

Variation in latitudinal temperature gradients was better explained by our four predictors, with evidence for significant and strong differences between tropical and non-tropical species (Table 2 and Figure S6). Tropical species had weaker relationships for all four predictors with little evidence for differences between threatened and non-threatened species. For threatened and non-threatened extratropical species, species with higher projected temperature increases, larger ranges and ranges that were elongated in shape and oriented parallel with the equator had steeper latitudinal temperature gradients. For threatened extra-tropical species, gradients were even steeper for species with larger ranges.

Discussion

Species face unique challenges under climate change, with the availability and strength of range-shift opportunities along latitudinal or elevational temperature gradients playing a signif-

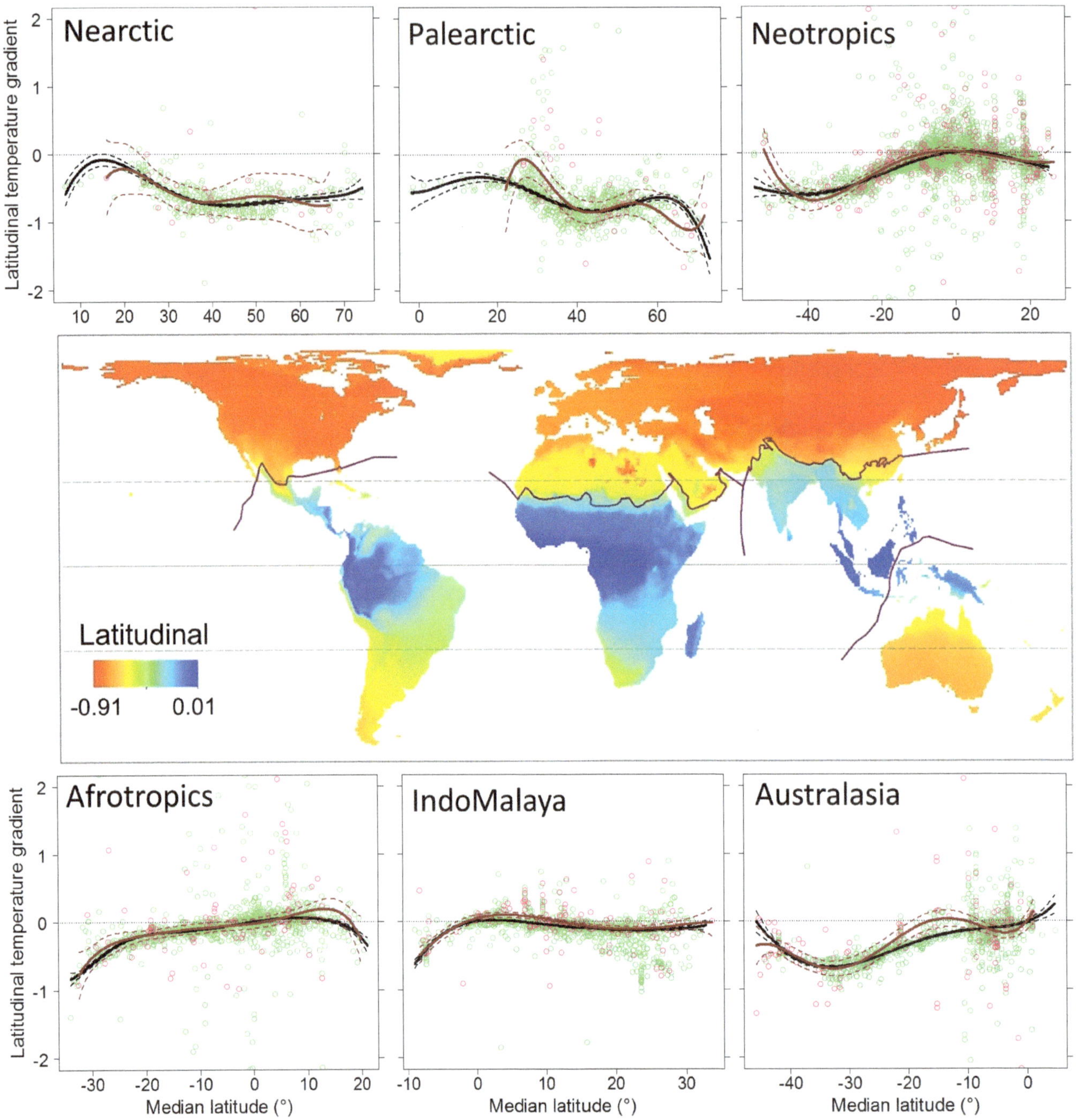

Figure 4. The latitudinal temperature gradient index summarized across species' geographic ranges (map) and summarized within six biogeographical realms (plots) as a function of the median latitude of each species' range. The temperature gradient was estimated spatially within species' current ranges (see Materials and Methods for details). In the plots, red points are threatened species ($n = 878$) and green points species non-threated species ($n = 8{,}136$). Trend lines are the fits of generalized additive models for threatened (red) and non-threatened species (black). The solid line is the equator and the dashed lines are the Tropics of Cancer and Capricorn (23.5°N and 23.5°S latitude, respectively).

icant role in determining species' likelihood of persistence. In this study we provide the first estimation of these gradients for the world's avifauna relative to species' IUCN Red List category, geography, and projected exposure to climate change. Our approach expands upon current methods for estimating species' geographic climatic associations by measuring these two gradients individually using fine-grained environmental data, which more accurately captures their unique spatial structures. The application of this approach has the potential to increase biological realism and predictive quality in current large-scale modeling efforts by removing problematic assumptions and minimizing potential biases. Our findings suggest threatened species are at a distinct disadvantage globally based on their association with shallower elevational temperature gradients. Latitudinal temperature gradients are primarily relevant for species located outside the tropics, a region where few bird species occur year-round, including threatened species whose associations with latitudinal temperature gradients did not differ from non-threatened species. Our results

Table 1. Coefficients and test statistics from robust linear models examining predictors of elevational temperature gradients as estimated within the geographic ranges of 4,978 bird species.

Model	Coef.	SE	*t*	*P*
Intercept	3.149	0.380	8.29	<0.001
Anomaly	0.976	0.108	9.05	<0.001
Threat	−3.117	0.895	−3.48	<0.001
Anomaly × Threat	0.344	0.273	1.26	0.2090
($R^2 = 0.07$)				
Intercept	3.600	0.434	8.30	<0.001
Size	0.511	0.074	6.92	<0.001
Threat	−1.872	1.076	−1.74	0.082
Size × Threat	0.002	0.213	0.01	0.993
($R^2 = 0.06$)				
Intercept	6.988	0.134	52.01	<0.001
Shape	−1.155	0.310	−3.73	<0.001
Threat	−2.940	0.295	−9.96	<0.001
Shape × Threat	1.580	0.707	2.23	0.026
($R^2 = 0.06$)				
Intercept	5.845	0.105	55.85	<0.001
Orientation	0.018	0.002	8.18	<0.001
Threat	−1.824	0.280	−6.52	<0.001
Orientation × Threat	−0.014	0.005	−2.68	0.008
($R^2 = 0.06$)				

Predictors include projected temperature Anomaly, geographic range Size, Shape, and Orientation and if the species is threatened with extinction (Threat).

also indicate a degree of correspondence between range-shift opportunities and climate change, in that species in regions that are projected to experience greater warming are more likely to be associated with steeper temperature gradients.

The strength of latitudinal temperature gradients is determined in part by the location, size, shape and orientation of species' geographic ranges. Not surprisingly, the strongest latitudinal temperature gradients were associated with species whose ranges occurred outside the tropics, especially at the higher latitudes in the North Hemisphere where the largest land masses are located. Species in these regions tend to have larger geographic ranges (see Figure 2B) and ranges that tend to be elongated parallel to the equator. As our results indicate, the strength of the latitudinal temperature gradient is of sufficient size within this region that geographic ranges with narrow latitudinal extents are still able to capture a significant gradient. In contrast, the spatial distribution of elevation temperature gradients reflects the form and extent of orographic features and how species' distributions intersect these features. We found that the strength of these gradients varied substantially across the globe and, unlike latitudinal temperature gradients, differed significantly between threatened and non-threatened species. In agreement with other studies [47,48], species in Australia and the Islands of IndoMalaya stood out as particularly vulnerable, with some of the shallowest elevational gradients, followed by species located in the lowlands of the Neotropics. In contrast to latitudinal gradients, our ability to predict species' elevational temperature gradients was limited, reflecting the greater geographic heterogeneity in the location and structure of these gradients.

The strength of both temperature gradients was positively correlated with projected changes in temperature, but only for species outside the tropics with latitudinal temperature gradients. The greatest increases in temperature under climate change are projected to occur at the high northern latitudes (see Figure S4). Our findings therefore suggest a geographic correspondence with the greatest projected warming occurring in regions with the strongest latitudinal temperature gradients. However, few bird species occur at the high northern latitudes year round. The proportion of migratory species (*ca.* 19% of extant bird species [49]) within breeding communities increasing linearly as you travel north from the equator, approaching 100% at the highest latitudes [50]. Migratory species spend the bulk of their annual cycle in migration or on the non-breeding ranges at lower latitudes. Thus, very few bird species are likely to benefit substantially from this correspondence.

IUCN Red List categories of extinction risk are assigned using the Red List criteria, which have quantitative thresholds relating to population and range size, structure and trends [23]. Only 3.5% of threatened or near threatened bird species (77/2193) are assessed as threatened by climate change so severely that the global population may be declining rapidly or very rapidly over three generations, and hence potentially contributing to their Red List category through the A criterion. Of these, only 26 (1.2%) actually qualify under the A criterion. Our findings suggest this proportion is likely to increase, especially for threatened species associated with shallow elevational temperature gradients. As confirmed in our analysis, threatened species occur primarily in the tropics and their distributions are defined by smaller geographic ranges and elevational extents. Hence, threatened species may be at a geographic disadvantage under climate change with limited associations with elevational or latitudinal temperature gradients, although we found no evidence for the latter. Our findings suggest

Table 2. Coefficients and test statistics from robust linear models examining predictors of latitudinal temperature gradients as estimated within the geographic ranges of 9,014 bird species.

Model	Coef.	SE	t	P
Intercept	−0.288	0.021	−13.61	<0.001
Anomaly	−0.092	0.006	−16.52	<0.001
Tropics	0.199	0.028	7.14	<0.001
Threat	0.027	0.047	0.58	0.564
Tropics × Threat	0.028	0.019	1.48	0.138
Anomaly × Tropics	0.094	0.008	12.20	<0.001
Anomaly × Threat	−0.016	0.013	−1.21	0.225
($R^2 = 0.34$)				
Intercept	0.148	0.045	3.32	0.001
Size	−0.120	0.007	−17.09	<0.001
Tropics	0.014	0.048	0.28	0.776
Threat	0.130	0.061	2.12	0.034
Tropics × Threat	−0.034	0.020	−1.66	0.098
Size × Tropics	0.080	0.008	10.38	<0.001
Size × Threat	−0.026	0.011	−2.47	0.014
($R^2 = 0.34$)				
Intercept	−0.655	0.011	−61.97	<0.001
Shape	0.096	0.025	3.82	<0.001
Tropics	0.591	0.012	47.91	<0.001
Threat	−0.005	0.335	−0.02	0.988
Tropics × Threat	−0.032	0.061	−0.52	0.606
Shape × Tropics	−0.134	0.028	−4.70	<0.001
Shape × Threat	0.129	1.018	0.13	0.900
($R^2 = 0.35$)				
Intercept	−0.687	0.006	−107.55	<0.001
Orientation	0.002	0.000	13.52	<0.001
Tropics	0.613	0.008	77.18	<0.001
Threat	−0.027	0.019	−1.42	0.155
Tropics × Threat	−0.026	0.019	−1.34	0.180
Orientation × Tropics	−0.002	0.000	−12.49	<0.001
Orientation × Threat	0.001	0.000	4.49	<0.001
($R^2 = 0.34$)				

Predictors included projected temperature Anomaly and geographic range Size, Shape, and Orientation and if the species is tropical (Tropics) or threatened with extinction (Threat).

threatened species that occur in the central tropics have higher elevational evenness across their range, which is a likely consequence of their distributions occurring on steep montane slopes outside of lowland areas, which should result in stronger elevational temperature gradients for these species. However, our analysis suggests that the differences in elevational evenness between threatened and non-threatened species was not substantial enough to overcome the smaller elevational extents contained within threatened species' geographic ranges. Thus, our findings suggest the weaker elevational temperature gradients identified for threatened species is due in large part to a consistent association with smaller elevational extents within their ranges.

When considering both gradients in combination, two regions stand out as having both strong elevational and latitudinal temperature gradients: western North America and Central Asia (see Figures 3–4). These results suggest species in these regions will have greater opportunities to track their climatic niches geographically. Depending on how these temperature gradients intersect spatially within species' ranges, some combination of these gradients could be used. Similarly, two regions stand out as having both shallow elevational and latitudinal temperature gradients: the islands of IndoMalaya and the Amazon Basin in South America (see Figures 3–4). Species in these regions are likely to develop substantial range-shift gaps under climate change [10], resulting in limited opportunities to track their climatic niche geographically. Bird species occurring in the Amazon Basin are also considered to have low adaptive capacity, which may increase their vulnerability to climate change [51].

Species in tropical montane regions are more likely to have restricted lateral and vertical distributions [48,52] and are also more likely to be threatened with extinction (see Figure S3). Even though in many cases these species occur in some of the most extensive montane systems in the world, our findings suggest that the overall steepness of elevational gradients is constrained by their limited representation within species' geographic ranges. Local climatic heterogeneity, however, may allow for microclimatic adjustments that could compensate for broader climatic trends [53–56]. The level of climatic heterogeneity is determined by local topographic variation and also by species' habitat associations [57]. For example, species in forested habitats may have greater opportunities for microclimatic adjustments relative to species in open habitats [58]. In general, however, behavior resulting in microclimatic adjustments might not be sufficient to address the full consequences of climate change [59].

Even if latitudinal and elevational temperature gradients are readily available, dispersal along these gradients is not guaranteed. Dispersal abilities are determined in large part by species' morphological, physiological and behavioral characteristics and adaptations [60,61], all of which are likely to be altered under rapid climate change [62]. Dispersal under climate change may be constrained by unsuitable habitat, population or habitat fragmentation [58,63], or interspecific competition [64]. In addition, gene flow that occurs in association with dispersal can both support and hinder range-shift potential [65]. Gene flow among populations at the leading edge of the range has been found to benefit fitness, whereas gene flow from the interior to the leading edge has been identified as maladaptive [66,67]. Dispersal abilities also vary tremendously among species and between taxa [60]. Even highly vagile taxa such as birds do not necessarily have the full complement of behavioral or physiological characteristics needed to support successful dispersal, which is considered particularly relevant for tropical species [68,69]. Although there is evidence tropical montane birds have moved upslope under climate change [70], the low annual variation in temperatures within the tropics relative to temperate regions may result in stronger physiological barriers to dispersal for tropical species [52,71,72]. When tropical habitats are fragmented through human activities – a process that is occurring at an increasing rate – dispersal can become even more constrained [73,74]. If species' responses are hindered by dispersal limitations or interspecific competition, as is expected for many tropical montane bird communities, substantial ecological disruptions are likely to occur [75].

In contrast, species that breed outside the tropics that are also migratory [76] or have larger geographic ranges tend to have stronger natal dispersal abilities [77–79]. Thus migratory species that are broadly distributed and breed outside the tropics might be in a better position to track elevational or latitudinal temperature gradients. However, exceptions exist [80] and factors that are strongly correlated with dispersal abilities might not be similarly correlated with broad-scale distributional responses under climate change [81]. Nevertheless, current evidence indicates distributional responses for birds along latitudinal [7] and elevational temperature gradients [82] are lagging behind warming trends, suggesting stronger responses could develop as climate change progresses and time lags are overcome [7].

We estimated temperature gradients in this study using relatively high resolution environmental data that was finer than the resolution of the species' range data [30]. This grain mismatch resulted in temperature gradients likely being estimated over fine-grain sections of species' ranges where probability of occurrence is low. Our main results focus on the geographic form of the gradient across the spatial breadth of species' distributions, but future work will benefit from further examination of potential biases this grain mismatch may have on comparisons. How patterns of occurrence are structured within species' ranges, based on proportion of occupancy and level of aggregation [83], may affect the chances of successful dispersal along these gradients. For example, sparse or fragmented patterns of occurrence may diminish the chances of successful dispersal, while more continuous patterns of occurrence may improve them. Alternatively, large contiguous regions with low probabilities of occurrence might contain the steepest gradients with very little dispersal value, as when lowland species' distributions border mountain systems. These remaining uncertainties highlight the need for advancing the spatial grain of our global biodiversity knowledge and for the necessary collaborative data mobilization, integration and quality control to realize this [31].

In addition to temperature, other climate factors and conditions are likely to be important in determining the strength and importance of latitudinal and elevation temperature gradients in defining range-shift potential. Two that have received particular attention are changes in precipitation regimes and changes in climatic variability. In contrast to temperature, precipitation has more complex spatial and temporal variability and more complex associations with latitudinal and elevational gradients. Projected changes in precipitation are also more ambiguous [2], which adds uncertainty to current estimates on how species are likely to respond to climate change [84]. A second factor is climatic variability, which is projected to increase through an increased frequency of extreme climate events [2]. Temporal climatic variability is considered an important factor defining species' distributions limits [85] and increasing climatic variability can have varied effects on species' populations [86] including the potential to hinder range shifts [87] and, especially under rapid climate change, increase extinction risk [88].

Additional research is needed to improve our understanding of how latitudinal and especially elevational temperature gradients are structured within species' ranges and the ability of species, especially tropical, to take advantage of these gradients. In particular, studies using higher resolution climate data (e.g., [22]) for species with varying spatial patterns of occurrence and forms of elevational or latitudinal gradients would be informative. Birds are often used as a model biological system due to high data availability, but less vagile taxa also need to be considered such as plants or vertebrate ectotherms that have alternative modes of dispersal and different physiological and behavior capacities. With birds, examining how these gradients are structured within non-breeding ranges, where migratory birds tend to occur during the majority of the annual cycle and are typically less well studied, would be beneficial [9]. Efforts to better understand the role of precipitation and climate variability in defining range-shift potential along latitudinal and elevational temperature gradients would support more comprehensive inferences.

Current efforts to project species' distributions or to estimate movement corridors under climate change often assume perfect and complete dispersal along temperature gradients (e.g., [89]). Temperature gradients are typically estimated using coarse-grained interpolated weather-station data based on current climatic conditions where the contrasting spatial forms of latitudinal and especially elevational temperature gradients are often poorly summarized [35,36]. These methods may exaggerate estimated dispersal distances [90], especially at high northern latitudes where latitudinal temperature gradients are projected to weaken under climate change [12]. These considerations in addition to the correlative nature of these models have promoted the development of more mechanistic or process-based methods intended to improve biological realism and minimize bias in current projections [91]. Our findings suggest the quality of model predictions may be improved by considering the spatial form and strength of species' demographic responses [92] to projected latitudinal and elevational temperature gradients within species' ranges and along possible movement corridors

Lastly, in order to maximize range-shift potential under climate change, conservation efforts are needed to maximize the size of areas of contiguous habitat, minimize the extent of habitat fragmentation and promote the permeability of the matrix of habitats around key sites. In particular, efforts to assess and promote connectivity at broad geographic extents are likely to be the most beneficial (e.g., [93,94]) where existing gradients are represented as continuously as possible across space [95]. The more thoroughly this can be achieved, the greater the potential for species to track changing climatic conditions based on their unique dispersal abilities and ecological and environmental requirements.

Conclusions

The availability of opportunities for geographic responses under climate change reflects the interplay between each species' geographic distribution and their associated latitudinal and elevational temperature gradients (with the extent to which these opportunities will be utilized depending on each species' dispersal abilities and the consequences of changing gene flow). The distribution of these gradients across the world follows clearly defined geographic patterns, but their availability depends on how they intersect with species' distributions. The location, size, and shape of species' geographic range and projected changes in temperature all contain a degree of relevance in determining the strength of these temperature gradients. However, the strength of latitudinal and elevational temperature gradients varied considerably among species, constraining our ability to make robust predictions for all regions globally. Nevertheless, we found that species that are currently at risk of extinction are consistently at the greatest disadvantage, especially based on the availability of elevational temperature gradients.

Supporting Information

Figure S1 Global patterns of elevation based on the USGS global digital elevation model (GTOPO30) gridded at a 30 arc-second resolution (*ca.* 1 km at the equator). The solid grey line is the equator and the dashed grey lines are the Tropics of Cancer and Capricorn (23.5°N and 23.5°S latitude, respectively).

Figure S2 Global terrestrial annual mean temperature for 1950–2000 (WorldClim) averaged within 0.1 degree latitudinal bands. The dotted lines indicate the Tropics of Cancer and Capricorn (23.5°N and 23.5°S latitude, respectively).

Figure S3 Maps summarizing features of the global distribution of terrestrial avifauna across an equal area grid (cell area: 3,091 km^2). (A) The richness of all species (n = 9,014) and (B) the richness of species threatened with extinction (n = 878). (C) Geographic range size for all species (n = 9,014) and (D) elevational extent for 4,978 species with minimum and maximum elevation associations. (E) The evenness of the distribution of elevation within each species' range based on the Gini coefficient and Lorenz curve (0 = high evenness; 1 = low evenness) for 4,978 species with minimum and maximum elevation associations. Maps are displayed using the Behrmann equal-area cylindrical projection and the color ramps use Jenk's natural break classification. The solid grey line is the equator and the dashed grey lines are the Tropics of Cancer and Capricorn (23.5°N and 23.5°S latitude, respectively). Red lines distinguish boundaries between six biogeographical realms.

Figure S4 Projected temperature anomalies summarized across species' geographic ranges (map) and summarized within six biogeographical realms (plots) as a function of the median latitude of each species' range. In the plots, red points are threatened species (n = 878) and green points species non-threatened species (n = 8,136). Trend lines are the fits of generalized additive models for threatened (red) and non-threatened species (black). The solid line is the equator and the dashed lines are the Tropics of Cancer and Capricorn (23.5°N and 23.5°S latitude, respectively).

Figure S5 Fit of robust linear regression models to four predictors of elevational temperature gradients estimated within the geographic ranges of 4,978 bird species. Red points and lines are threatened species (n = 766) and green points and black line are non-threatened species (n = 4,212).

Figure S6 Fit of robust linear regression models to four predictors of latitudinal temperature gradients estimated within the geographic ranges of 9,014 bird species. Green points and lines are for tropical species (n = 6,720), brown points and lines are for extratropical species (n = 2,294). The solid lines are the fits for non-threatened species (n = 4,214) and the dashed lines are the fits for threatened species (n = 766).

Table S1 The 9,014 species considered in the analysis and their estimated latitudinal and elevational temperature gradients, designated biogeographical realm and threat status (1 = threatened and 0 = non-threatened with extinction), projected temperature anomaly, and geographic range size, shape and orientation.

Acknowledgments

We thank I. I. Mokhov for providing global lapse-rate values and three anonymous reviewers for their constructive suggestions. We thank the large number of experts and organizations who contribute information incorporated into BirdLife International's extinction risk assessments for all species on the IUCN Red List, including spatial information incorporated into the distribution maps.

Author Contributions

Conceived and designed the experiments: FALS KBG WJ. Analyzed the data: FALS. Wrote the paper: FALS KBG WJ SHMB. Collected the data: SHMB.

References

1. Karl TR, Trenberth KE (2003) Modern global climate change. Science 302: 1719–1723.
2. IPCC (2013) Climate change 2013: The physical science basis. In: T. Stocker, Q. Dahe, G.-K. Plattner, editors. Fifth assessment report of the intergovernmental panel on climate change. Cambridge and New York: Intergovernmental Panel on Climate Change. 127 p.
3. Skelly DK, Joseph LN, Possingham HP, Freidenburg LK, Farrugia TJ, et al. (2007) Evolutionary responses to climate change. Conserv Biol 21: 1353–1355.

4. Parmesan C (2006) Ecological and evolutionary responses to recent climate change. Annu Rev Ecol Evol Syst 37: 637–669.
5. Loarie SR, Duffy PB, Hamilton H, Asner GP, Field CB, et al. (2009) The velocity of climate change. Nature 462: 1052–1055.
6. Martínez-Meyer E, Peterson AT, Hargrove WW (2004) Ecological niches as stable distributional constraints on mammal species, with implications for Pleistocene extinctions and climate change projections for biodiversity. Glob Ecol Biogeogr 13: 305–314.
7. La Sorte FA, Jetz W (2012) Tracking of climatic niche boundaries under recent climate change. J Anim Ecol 81: 914–925.
8. Tingley MW, Monahan WB, Beissinger SR, Moritz C (2009) Birds track their Grinnellian niche through a century of climate change. Proc Natl Acad Sci U S A 106: 19637–19643.
9. La Sorte FA, Jetz W (2010) Avian distributions under climate change: Towards improved projections. J Exp Biol 213: 862–869.
10. Colwell RK, Brehm G, Cardelús CL, Gilman AC, Longino JT (2008) Global warming, elevational range shifts, and lowland biotic attrition in the wet tropics. Science 322: 258–261.
11. Mokhov II, Akperov MG (2006) Tropospheric lapse rate and its relation to surface temperature from reanalysis data. Izvestiya, Atmospheric and Oceanic Physics 42: 430–438.
12. Xu YY, Ramanathan V (2012) Latitudinally asymmetric response of global surface temperature: Implications for regional climate change. Geophys Res Lett 39: L13706.
13. Xu L, Myneni RB, Chapin III FS, Callaghan TV, Pinzon JE, et al. (2013) Temperature and vegetation seasonality diminishment over northern lands. Nature Clim Change 3: 581–586.
14. Mao J, Shi X, Thornton P, Hoffman F, Zhu Z, et al. (2013) Global latitudinal-asymmetric vegetation growth trends and their driving mechanisms: 1982–2009. Remote Sens 5: 1484–1497.
15. Jetz W, Ashton KG, La Sorte FA (2009) Phenotypic population divergence in terrestrial vertebrates at macro scales. Ecol Lett 12: 1137–1146.
16. Jonas CS, Geber MA (1999) Variation among populations of *Clarkia unguiculata* (Onagraceae) along altitudinal and latitudinal gradients. Am J Bot 86: 333–343.
17. Naya DE, Spangenberg L, Naya H, Bozinovic F (2012) Latitudinal patterns in rodent metabolic flexibility. Am Nat 179: E172–E179.
18. Munch SB, Salinas S (2009) Latitudinal variation in lifespan within species is explained by the metabolic theory of ecology. Proc Natl Acad Sci U S A 106: 13860–13864.
19. Stanton-Geddes J, Tiffin P, Shaw RG (2012) Role of climate and competitors in limiting fitness across range edges of an annual plant. Ecology 93: 1604–1613.
20. Hargrove L, Rotenberry JT (2011) Breeding success at the range margin of a desert species: Implications for a climate-induced elevational shift. Oikos 120: 1568–1576.
21. Webb T, Bartlein PJ (1992) Global change during the last 3 million years - climatic controls and biotic responses. Annu Rev Ecol Syst 23: 141–173.
22. Bennie J, Hodgson JA, Lawson CR, Holloway CTR, Roy DB, et al. (2013) Range expansion through fragmented landscapes under a variable climaBirdLife Internationalte. Ecol Lett 16: 921–929.
23. BirdLife International (2012) The BirdLife checklist of the birds of the world, with conservation status and taxonomic sources. BirdLife International, Cambridge.
24. Baselga A, Lobo JM, Svenning J-C, Araújo MB (2012) Global patterns in the shape of species geographical ranges reveal range determinants. J Biogeogr 39: 760–771.
25. Pigot AL, Owens IPF, Orme CDL (2010) The environmental limits to geographic range expansion in birds. Ecol Lett 13: 705–715.
26. BirdLife International and NatureServe (2012) Bird species distribution maps of the world. BirdLife International, Cambridge, and NatureServe, Arlington.
27. Buchanan GM, Donald PF, Butchart SHM (2011) Identifying priority areas for conservation: A global assessment for forest-dependent birds. PLoS ONE 6: e29080.
28. Udvardy MDF (1975) A classification of the biogeographical provinces of the world. IUCN Occasional Paper. Morgues: IUCN. 48 p.
29. Hijmans RJ, Cameron SE, Parra JL, Jones PG, Jarvis A (2005) Very high resolution interpolated climate surfaces for global land areas. Int J Climatol 25: 1965–1978.
30. Hurlbert AH, Jetz W (2007) Species richness, hotspots, and the scale dependence of range maps in ecology and conservation. Proc Natl Acad Sci U S A 104: 13384–13389.
31. Jetz W, McPherson JM, Guralnick RP (2012) Integrating biodiversity distribution knowledge: Toward a global map of life. Trends Ecol Evol 27: 151–159.
32. Jones PG, Thornton PK, Heinke J (2009) Generating characteristic daily weather data using downscaled climate model data from the IPCC's fourth assessment. CGIAR Research Program on Climate Change, Agriculture and Food Security. 23 p.
33. Raupach MR, Marland G, Ciais P, Le Quéré C, Canadell JG, et al. (2007) Global and regional drivers of accelerating CO_2 emissions. Proc Natl Acad Sci U S A 104: 10288–10293.
34. Beaumont LJ, Hughes L, Pitman AJ (2008) Why is the choice of future climate scenarios for species distribution modelling important? Ecol Lett 11: 1135–1146.
35. Scherrer D, Schmid S, Körner C (2011) Elevational species shifts in a warmer climate are overestimated when based on weather station data. Int J Biometeorol 55: 645–654.
36. Graae BJ, De Frenne P, Kolb A, Brunet J, Chabrerie O, et al. (2012) On the use of weather data in ecological studies along altitudinal and latitudinal gradients. Oikos 121: 3–19.
37. Nijssen D, Rousseau R, Cleland EE (1998) The Lorenz curve: A graphical representation of evenness. Coenoses 13: 33–38.
38. Lorenz MO (1905) Methods of measuring concentrations of wealth. J Am Statist Assoc 9: 209–219.
39. Dagum C (1980) The generation and distribution of income, the Lorenz curve and the Gini ratio. Econ Appl 33: 327–367.
40. Muggeo VMR (2003) Estimating regression models with unkown break-points. Stat Med 22: 3055–3071.
41. Davies RB (1987) Hypothesis testing when a nuisance parameter is present only under the alternative. Biometrika 74: 33–43.
42. Wood SN (2006) Generalized additive models: An introduction with R. Boca Raton: Chapman & Hall/CRC. 391 p.
43. Yohai VJ (1987) High breakdown-point and high efficiency estimates for regression. Ann Stat 15: 642–665.
44. Koller M, Stahel WA (2011) Sharpening wald-type inferences in robust regression for small samples. Comput Stat Data Anal 55: 2504–2515.
45. Menard S (1995) Applied logistic regression analysis. Thousand Oaks: Sage Publications. 98 p.
46. R Development Core Team (2014) R: A language and environment for statistical computing. Vienna: R Foundation for Statistical Computing.
47. Anderson AS, Reside AE, Vanderwal JJ, Shoo LP, Pearson RG, et al. (2012) Immigrants and refugees: The importance of dispersal in mediating biotic attrition under climate change. Glob Change Biol 18: 2126–2134.
48. La Sorte FA, Jetz W (2010) Projected range contractions of montane biodiversity under global warming. Proc R Soc Lond B Biol Sci 277: 3401–3410.
49. Kirby JS, Stattersfield AJ, Butchart SHM, Evans MI, Grimmett RFA, et al. (2008) Key conservation issues for migratory land- and waterbird species on the world's major flyways. Bird Conserv Int 18: S49–S73.
50. Somveille M, Manica A, Butchart SHM, Rodrigues ASL (2013) Mapping global diversity patterns for migratory birds. PLoS ONE 8: e70907.
51. Foden WB, Butchart SHM, Stuart SN, Vié J-C, Akçakaya HR, et al. (2013) Identifying the world's most climate change vulnerable species: A systematic trait-based assessment of all birds, amphibians and corals. PLoS ONE 8: e65427.
52. McCain CM (2009) Vertebrate range sizes indicate that mountains may be 'higher' in the tropics. Ecol Lett 12: 550–560.
53. Luoto M, Heikkinen RK (2008) Disregarding topographic heterogeneity biases species turnover assessments based on bioclimatic models. Glob Change Biol 14: 1–12.
54. Randin CF, Engler R, Normand S, Zappa M, Zimmermann NE, et al. (2009) Climate change and plant distribution: Local models predict high-elevation persistence. Glob Change Biol 15: 1557–1569.
55. Engler R, Randin CF, Thuiller W, Dullinger S, Zimmermann NE, et al. (2011) 21st century climate change threatens mountain flora unequally across Europe. Glob Change Biol 17: 2330–2341.
56. Gillingham PK, Huntley B, Kunin WE, Thomas CD (2012) The effect of spatial resolution on projected responses to climate warming. Divers Distrib 18: 990–1000.
57. Suggitt AJ, Gillingham PK, Hill JK, Huntley B, Kunin WE, et al. (2011) Habitat microclimates drive fine-scale variation in extreme temperatures. Oikos 120: 1–8.
58. Reif J, Flousek J (2012) The role of species' ecological traits in climatically driven altitudinal range shifts of central European birds. Oikos 121: 1053–1060.
59. Suggitt AJ, Stefanescu C, Paramo F, Oliver T, Anderson BJ, et al. (2012) Habitat associations of species show consistent but weak responses to climate. Biol Lett 8: 590–593.
60. Clobert J, Danchin E, Dhondt AA, Nichols JD (2001) Dispersal. Oxford: Oxford University Press. 452 p.
61. Watkinson AR, Gill JA (2001) Climate change and dispersal. In: J. M. . Bullock, R. E. . Kenward, R. S. . Hails, editors. Dispersal ecology. Oxford: Blackwell Publishing. pp. 410–428.
62. Travis JMJ, Delgado M, Bocedi G, Baguette M, Bartoń K, et al. (2013) Dispersal and species' responses to climate change. Oikos 122: 1532–1540.
63. Forero-Medina G, Joppa L, Pimm SL (2010) Constraints to species' elevational range shifts as climate changes. Conserv Biol 25: 163–171.
64. Jankowski JE, Robinson SK, Levey DJ (2010) Squeezed at the top: Interspecific aggression may constrain elevational ranges in tropical birds. Ecology 91: 1877–1884.
65. Møller AP, Flensted-Jensen E, Mardal W (2006) Dispersal and climate change: A case study of the arctic tern *Sterna paradisaea*. Glob Change Biol 12: 2005–2013.
66. Kremer A, Ronce O, Robledo-Arnuncio JJ, Guillaume F, Bohrer G, et al. (2012) Long-distance gene flow and adaptation of forest trees to rapid climate change. Ecol Lett 15: 378–392.
67. Sexton JP, Strauss SY, Rice KJ (2011) Gene flow increases fitness at the warm edge of a species' range. Proc Natl Acad Sci U S A 108: 11704–11709.
68. Moore RP, Robinson WD, Lovette IJ, Robinson TR (2008) Experimental evidence for extreme dispersal limitation in tropical forest birds. Ecol Lett 11: 960–968.

69. Dexter KG (2010) The influence of dispersal on macroecological patterns of lesser Antillean birds. J Biogeogr 37: 2137–2147.
70. Freeman BG, Class Freeman AM (in press) Rapid upslope shifts in new guinean birds illustrate strong distributional responses of tropical montane species to global warming. Proc Natl Acad Sci U S A.
71. Janzen DH (1967) Why mountain passes are higher in the tropics. Am Nat 101: 233–249.
72. Ghalambor CK, Huey RB, Martin PR, Tewksbury JJ, Wang G (2006) Are mountain passes higher in the tropics? Janzen's hypothesis revisited. Integr Comp Biol 46: 5–17.
73. Gillies CS, Beyer HL, St Clair CC (2011) Fine-scale movement decisions of tropical forest birds in a fragmented landscape. Ecol Appl 21: 944–954.
74. Ibarra-Macias A, Robinson WD, Gaines MS (2011) Experimental evaluation of bird movements in a fragmented neotropical landscape. Biol Conserv 144: 703–712.
75. Urban MC, Tewksbury JJ, Sheldon KS (2012) On a collision course: Competition and dispersal differences create no-analogue communities and cause extinctions during climate change. Proc R Soc Lond B Biol Sci 279: 2072–2080.
76. Dawideit BA, Phillimore AB, Laube I, Leisler B, Böhning-Gaese K (2009) Ecomorphological predictors of natal dispersal distances in birds. J Anim Ecol 78: 388–395.
77. Böhning-Gaese K, Caprano T, van Ewijk K, Veith M (2006) Range size: Disentangling current traits and phylogenetic and biogeographic factors. Am Nat 167: 555–567.
78. Laube I, Graham CH, Böhning-Gaese K (2013) Intra-generic species richness and dispersal ability interact to determine geographic ranges of birds. Glob Ecol Biogeogr 22: 223–232.
79. Laube I, Korntheuer H, Schwager M, Trautmann S, Rahbek C, et al. (2013) Towards a more mechanistic understanding of traits and range sizes. Glob Ecol Biogeogr 22: 233–241.
80. Winkler DW, Wrege PH, Allen PE, Kast TL, Senesac P, et al. (2005) The natal dispersal of tree swallows in a continuous mainland environment. J Anim Ecol 74: 1080–1090.
81. Angert AL, Crozier LG, Rissler LJ, Gilman SE, Tewksbury JJ, et al. (2011) Do species' traits predict recent shifts at expanding range edges? Ecol Lett 14: 677–689.
82. Forero-Medina G, Terborgh J, Socolar SJ, Pimm SL (2011) Elevational ranges of birds on a tropical montane gradient lag behind warming temperatures. PLoS ONE 6: e28535.
83. La Sorte FA, Hawkins BA (2007) Range maps and species richness patterns: Errors of commission and estimates of uncertainty. Ecography 30: 649–662.
84. McCain CM, Colwell RK (2011) Assessing the threat to montane biodiversity from discordant shifts in temperature and precipitation in a changing climate. Ecol Lett 14: 1236–1245.
85. Mac Arthur RH (1972) Geographical ecology. New York: Harper & Row. 269 p.
86. Drake JM (2005) Population effects of increased climate variation. Proc R Soc Lond B Biol Sci 272: 1823–1827.
87. Early R, Sax DF (2011) Analysis of climate paths reveals potential limitations on species range shifts. Ecol Lett 14: 1125–1133.
88. Mustin K, Dytham C, Benton TG, Travis JMJ (2013) Red noise increases extinction risk during rapid climate change. Divers Distrib 19: 815–824.
89. Lawler JJ, Ruesch AS, Olden JD, McRae BH (2013) Projected climate-driven faunal movement routes. Ecol Lett 16: 1014–1022.
90. Bocedi G, Pe'er G, Heikkinen RK, Matsinos Y, Travis JMJ (2012) Projecting species' range expansion dynamics: Sources of systematic biases when scaling up patterns and processes. Methods Ecol Evol 3: 1008–1018.
91. Dormann CF, Schymanski SJ, Cabral J, Chuine I, Graham C, et al. (2012) Correlation and process in species distribution models: Bridging a dichotomy. J Biogeogr 39: 2119–2131.
92. Schurr FM, Pagel J, Cabral JS, Groeneveld J, Bykova O, et al. (2012) How to understand species' niches and range dynamics: A demographic research agenda for biogeography. J Biogeogr 39: 2146–2162.
93. Hole DG, Willis SG, Pain DJ, Fishpool LD, Butchart SHM, et al. (2009) Projected impacts of climate change on a continent-wide protected area network. Ecol Lett 12: 420–431.
94. Hole DG, Huntley B, Arinaitwe J, Butchart SHM, Collingham YC, et al. (2011) Toward a management framework for networks of protected areas in the face of climate change. Conserv Biol 25: 305–315.
95. Kubisch A, Degen T, Hovestadt T, Poethke HJ (2013) Predicting range shifts under global change: The balance between local adaptation and dispersal. Ecography 36: 873–882.

Automatic Detection of Key Innovations, Rate Shifts, and Diversity-Dependence on Phylogenetic Trees

Daniel L. Rabosky*

Department of Ecology and Evolutionary Biology and Museum of Zoology, University of Michigan, Ann Arbor, Michigan, United States of America

Abstract

A number of methods have been developed to infer differential rates of species diversification through time and among clades using time-calibrated phylogenetic trees. However, we lack a general framework that can delineate and quantify heterogeneous mixtures of dynamic processes within single phylogenies. I developed a method that can identify arbitrary numbers of time-varying diversification processes on phylogenies without specifying their locations in advance. The method uses reversible-jump Markov Chain Monte Carlo to move between model subspaces that vary in the number of distinct diversification regimes. The model assumes that changes in evolutionary regimes occur across the branches of phylogenetic trees under a compound Poisson process and explicitly accounts for rate variation through time and among lineages. Using simulated datasets, I demonstrate that the method can be used to quantify complex mixtures of time-dependent, diversity-dependent, and constant-rate diversification processes. I compared the performance of the method to the MEDUSA model of rate variation among lineages. As an empirical example, I analyzed the history of speciation and extinction during the radiation of modern whales. The method described here will greatly facilitate the exploration of macroevolutionary dynamics across large phylogenetic trees, which may have been shaped by heterogeneous mixtures of distinct evolutionary processes.

Editor: Sergios-Orestis Kolokotronis, Fordham University, United States of America

Funding: Financial support was provided by the Miller Institute for Basic Research in Science at the University of California, Berkeley, by the University of Michigan, and by NSF- DEB-1256330. The funders had no role in study design, data collection and analysis, decision to publish, or preparation of the manuscript.

Competing Interests: The author has declared that no competing interests exist.

* E-mail: drabosky@umich.edu

Introduction

Perhaps the most general feature of biological diversity on Earth is the extent to which it varies - either through space, through time, or among different kinds of organisms. Biologists have long been fascinated by the observation that some groups of organisms contain far more species than other groups. For example, within vertebrates, lineages such as tetrapods (22000+ species), therian mammals (5000+ species), and teleosts (30000+ species) are several orders of magnitude more diverse than their respective sister clades (lungfishes, 6 species; monotremes, 5 species; holostean fishes, <10 species). This phylogenetic variation in species richness is mirrored by analogous variation in diversity through time. Paleontological evidence indicates that species richness has undergone dramatic changes during the past 550 million years [1,2]. Finally, contemporary species richness varies dramatically among geographic and climatic regions [3,4]. At least in part, the causes of phylogenetic, temporal, and spatial variation in species richness are thought to reside in the evolutionary processes of speciation and extinction. Consequently, there has been great interest in studying historical patterns of species diversification through time, towards understanding how and why speciation and extinction rates might vary through time, through space, and among clades [5,6,7,8,9].

The fossil record has provided insight into the temporal dynamics of species diversification [10,11], but analyses have generally been restricted to groups with exceptional fossil records and/or to relatively coarse temporal and phylogenetic scales. Because of the difficulties in applying paleontological approaches to many groups of organisms that lack adequate fossil records, there is great interest in extracting information about macroevolutionary dynamics from time-calibrated phylogenetic trees of extant species only [8,12]. The increase in the availability of such phylogenies has helped catalyze a surge of methodological [13,14,15] and meta-analyses [16,17,18,19] on the temporal dynamics of speciation and extinction through time. At the same time, a range of new approaches have been developed to assess the extent to which rates of species diversification vary among lineages [20,21] or in association with character states [22,23,24,25].

To date, few macroevolutionary studies have simultaneously accounted for rate variation through time and among lineages [6,13,26]. Increasing evidence suggests that failing to accommodate rate variation through time and among lineages can lead to profoundly biased parameter estimation [27] and conceptually flawed interpretations of the factors that regulate species richness within clades or regions [26,28].

In this article, I introduce a new framework for studying patterns of rate variation through time and among lineages using time-calibrated phylogenies of extant species. The approach is premised on the idea that phylogenetic trees are frequently shaped by heterogeneous mixtures of distinct processes. For example, some phylogenies may reflect mixtures of both diversity-dependent and constant-rate diversification processes (Figure 1). There is already considerable evidence that many empirical phylogenies have been shaped by multiple distinct evolutionary processes

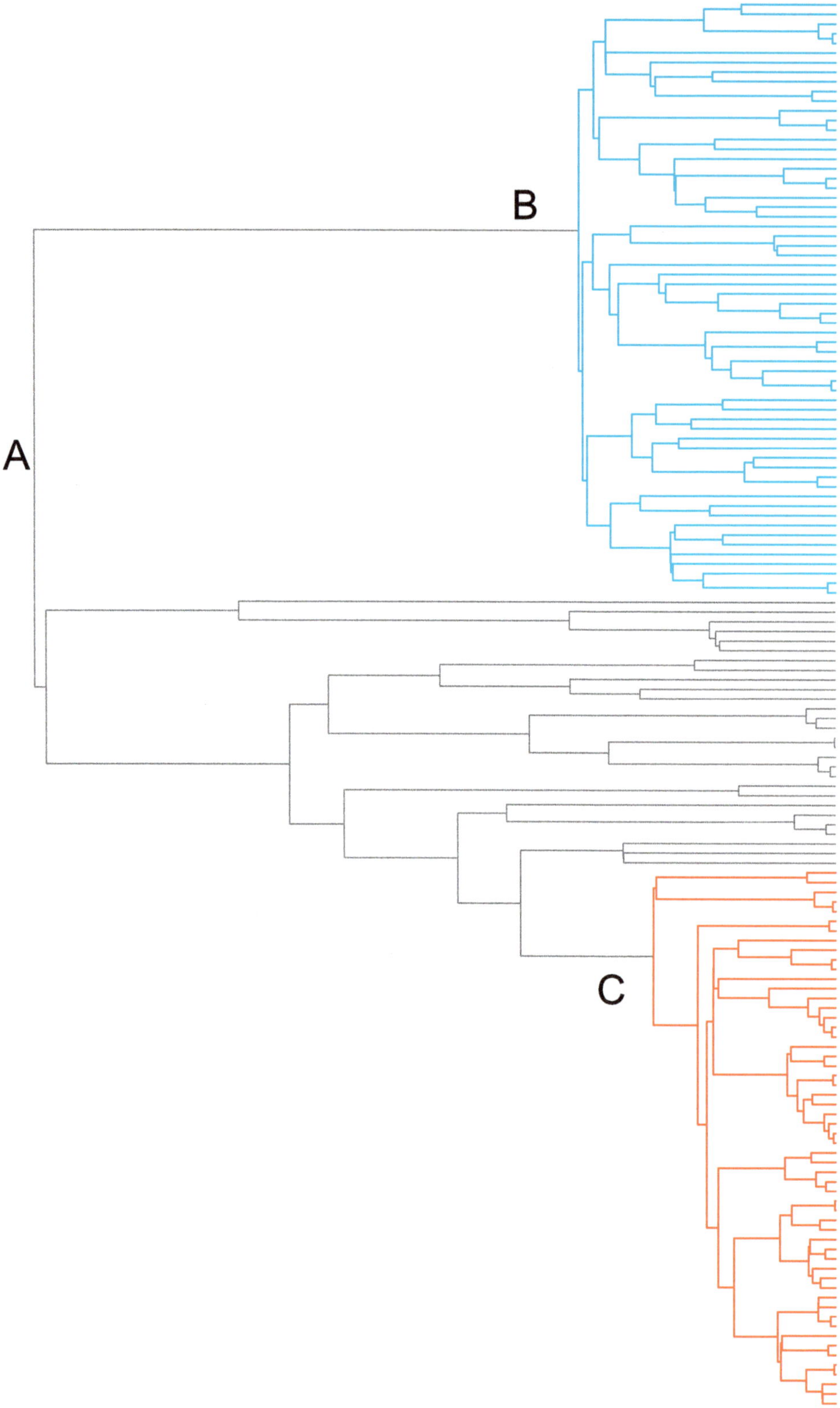

Figure 1. Example of tree simulated under mixture of three distinct evolutionary processes. (A) Clade diversification under constant-rate "background" diversification process with $\lambda = 0.032$ and $\mu = 0$. (B) Shift to new adaptive zone with subsequent diversity-dependent regulation of speciation and diversity-independent extinction (blue branches; $\lambda_0 = 0.395$; $K = 66$; $\mu = 0.041$). (C) Another lineage shifts to diversity-dependent speciation regime (red branches; $\lambda_0 = 0.21$; $K = 97$; $\mu = 0.012$). Total tree depth is 100 time units. Despite undergoing two distinct diversity-dependent slowdowns in the rate of speciation, the overall gamma statistic [63] for the tree is positive ($\gamma = 2.51$) and provides no evidence for changes in the rate of speciation through time. Note that a tree with three distinct processes contains two distinct transitions between processes.

[13,29], and challenges of modeling such data are expected to increase with phylogenetic tree size.

My general approach assumes that shifts between macroevolutionary regimes occur across the branches of a phylogenetic tree under a compound Poisson process. This framework has been used previously to model among-lineage variation in rates of molecular evolution [30]. The number of such transitions between distinct processes is assumed to follow a Poisson distribution. Rather than assume a fixed number of distinct processes on a given phylogenetic tree, I use reversible jump Markov Chain Monte Carlo [31] (hereafter, rjMCMC) to automatically explore the universe of models that differ in the number of distinct evolutionary regimes. The method thus enables exploration of a vast state space of possible models to explain a given phylogenetic diversification pattern.

The method described here differs from previous methods in several key respects. First, the method does not assume that rates of speciation and extinction are constant through time within clades, thus relaxing the assumption of time-homogeneous diversification used in most previous multi-model approaches [20,32]. Second, the location and number of distinct evolutionary processes ("regimes") represent random quantities that are themselves estimated from the data. In addition, by adopting a Bayesian approach, we can algorithmically explore a greater number of candidate models than is possible with incremental (e.g., stepwise) information-theoretic approaches [20]. Because rjMCMC samples diversification models in proportion to their posterior probability [31,33], model selection emerges automatically from the analysis. Finally, the method provides marginal distributions of speciation and extinction rates for every branch in a phylogenetic tree.

Materials and Methods

Compound Poisson Process Model of Diversification Rate Variation

The model assumes that phylogenetic trees are shaped by a countable set of distinct and potentially dynamic evolutionary processes of speciation and extinction. Transitions between processes, or "events", are assumed to occur across the branches of the phylogeny under a compound Poisson process [30]. Let ξ_i denote the mapping of the i'th transition to a specific location on the tree; thus, ξ denotes a unique location on a specific branch of the tree. Nodes and branches descended from a mapped transition ξ_i inherit the evolutionary process, denoted by Φ_i, that begins at point ξ_i. The process Φ_i terminates at terminal branches, or at the next downstream transition (Figure 1). Thus, the occurrence of a transition defines a connected subgraph of adjacent nodes, but does not necessarily include all of the descendent nodes downstream of a particular transition.

Any tree is necessarily governed by at least one process that begins at the root node and the number of additional transitions is a Poisson-distributed random variable with rate parameter Λ. In the MCMC implementation of the model described below, new transitions can be added to the tree, and existing transitions can be moved or deleted from the tree. The addition of a transition results in a new evolutionary process that is decoupled from the parent process. For example, consider a phylogeny with dynamics governed by just a single process, Φ_R. If a transition occurs at position ξ_i, then all lineages downstream from point ξ_i are governed by a new evolutionary process, Φ_i. Formally, each possible count of transitions defines a diversification model, and we denote a model with k distinct transitions by M_k. The addition of a process to a tree with k transitions thus entails a jump from model M_k to M_{k+1}. There is no upper bound on the number of transitions, as multiple transitions can occur on a given branch. The minimum model, with a single process, corresponds to $k = 0$ and contains zero transitions.

I assume that each process Φ represents a distinct time-varying process of speciation (λ) and a constant background rate of extinction (μ). We used an exponential change function to model variation in speciation rates through time within a particular process, such that

$$\lambda(t_i) = \lambda_{0,i} \exp(z_i t_i) \quad (1)$$

where λ is the rate of speciation for a process at time t_i relative to the start of the process and where $\lambda_{0,i}$ and z_i represent the initial speciation rate and rate change parameter for the i'th transition. For notational clarity, parameters associated with the root process are denoted with zero subscripts: for example, $\lambda_{0,0}$ and z_0 correspond to the initial speciation rate and rate-change parameter for the process associated with the root of the tree. This model is equivalent to the SPVAR model from Rabosky and Lovette [14]. The exponential change function is a natural choice for modeling both time-varying and diversity-dependent speciation, because an exponential change in speciation with respect to time closely approximates a linear change in speciation with respect to diversity [34]. The full model thus includes the possibility that a single time-varying diversification process describes the entire phylogeny [14] as well as the possibility that many independent time-varying processes govern evolutionary dynamics across the tree.

I implemented the model in a Bayesian framework. Bayesian approaches have already been used effectively to model single processes on phylogenetic trees [35,36], but in this case, the number of distinct processes is itself a random quantity. I constructed a transdimensional Markov chain that could move between models containing different numbers of processes. This is known as "reversible jump" Markov Chain Monte Carlo (rjMCMC), as it involves probabilistic "jumps" between model subspaces of different dimensionality [31]. An attractive feature of this approach is that the Markov chain samples diversification models in proportion to their posterior probability. Thus, the relative probabilities of diversification models with (0, 1, 2, 3.... k) distinct processes can be computed immediately by tabulating the relative frequencies of those models in the MCMC output. Several recent studies have used rjMCMC to study variation in rates of phenotypic evolution across phylogenetic trees [37,38].

Bayesian Implementation

The full model contains parameters for the overall rate at which transitions occur (Λ), as well as location (ξ) and diversification parameters (λ, z, μ) for each transition. I simulated a Markov process that (i) permitted incremental transitions to new diversification models ($M_k \rightarrow M_{k+1}$ or $M_k \rightarrow M_{k-1}$), and (ii) updates to parameters of the current models. Note that my usage of the word model in this context can refer to either the overall compound Poisson process model, or to submodels with distinct numbers of processes (e.g., M_1, M_2, ... M_N). The Markov chain is updated using the following moves: (1) a transition is added to the tree, (2) a transition is deleted from the tree, (3) the position of an existing transition (ξ_i) is updated, (4) the rate at which transitions occur is updated (Λ), (5) the initial speciation rate for the i'th transition is changed ($\lambda_{0,i}$), (6) the rate-change parameter for the i'th transition (z_i) is updated, and (7) the extinction rate for the i'th transition is changed (μ_i).

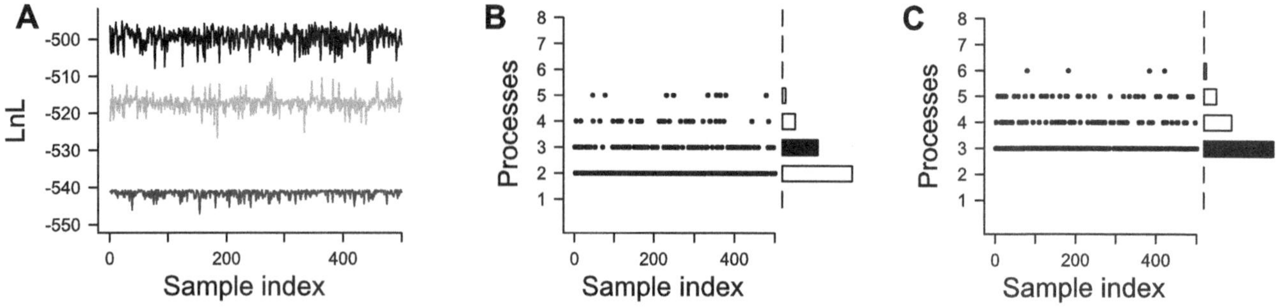

Figure 2. BAMM analysis of example tree (Figure 1). Example tree was simulated under three distinct processes (one constant rate and two diversity dependent processes; two transitions in total). The tree was analyzed under (i) the full multi-process BAMM model with time-variable speciation; (ii) a constrained multi-process BAMM with time-constant speciation; and (iii) a fully constrained 1-process constant-rate birth-death model. (A) Log-likelihoods for thinned MCMC chains for the constant rate birth-death process (bottom), the time-constant multi-process model (middle), and the full BAMM model with time-varying speciation (top). (B) Numbers of transitions during rjMCMC sampling when model is constrained to time-constant speciation rates; sidebar gives frequency distribution of sampled states. (C) Numbers of transitions under full BAMM model with time-variable speciation processes. Black sidebar denotes true number of transitions in generating model. The true number of transitions was estimated correctly only when the assumption of time-constant rates was relaxed.

For within-model moves that do not involve changes in the dimensionality of the full model, acceptance probabilities follow the standard Metropolis-Hastings formulation [39,40], or

$$\min\left\{1,\frac{f(\theta')}{f(\theta)}\frac{\pi(\theta')}{\pi(\theta)}\frac{q(\theta|\theta')}{q(\theta'|\theta)}\right\} \tag{2}$$

where θ and θ' are parameter vectors corresponding to current and proposed states, $f(\bullet)$ and $\pi(\bullet)$ are the corresponding likelihood and prior densities, and $q(\theta' \mid \theta)$ is the relative probability of proposing a move to parameter vector θ' given that the current state is θ.

The acceptance probability for moves that transition between models requires a more general formulation [31,41], of which standard Metropolis-Hastings is a special case. In the present framework, we propose to jump from some model M_k with parameter vector θ to a new model M_{k+1} with parameter vectors θ' and ψ, where θ denotes parameters that are common to both models and ψ denotes parameters that occur in the proposed model but not the current model. To move between models, we generate a random vector v from some known density, $q(v)$. We

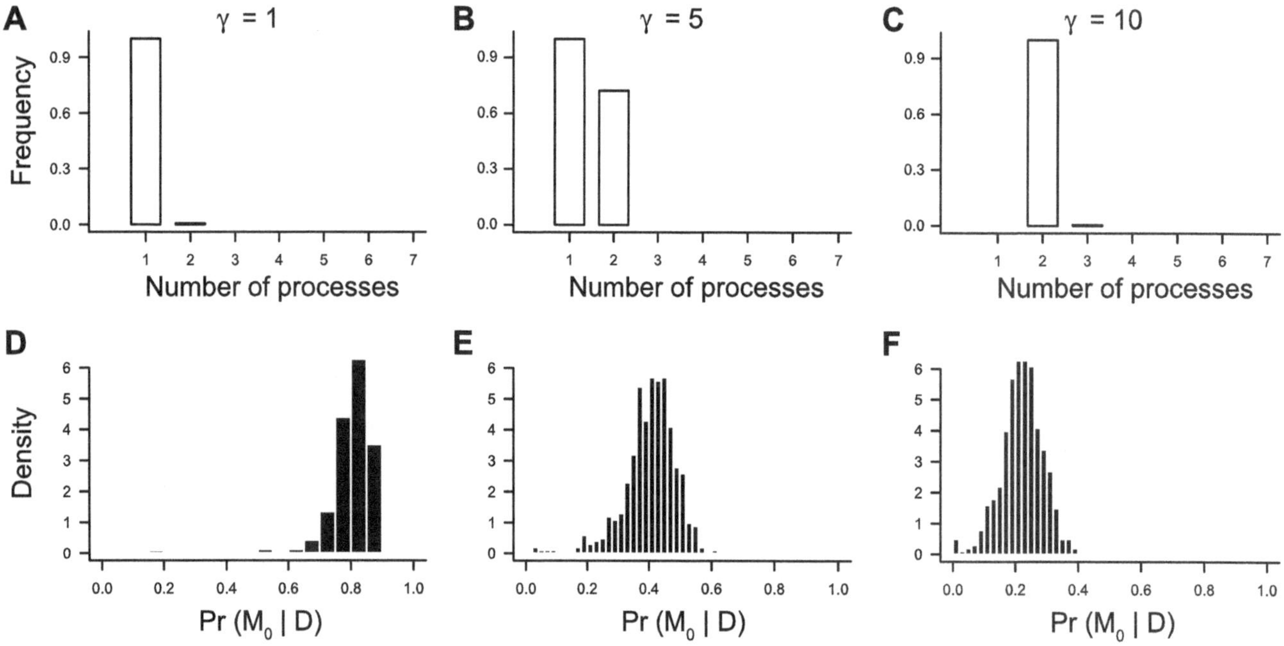

Figure 3. BAMM analyses of constant-rate phylogenies and prior on Poisson rate parameter (γ). Histograms in (A–C) display the frequency distribution of the estimated number of processes in the model with the maximum a posteriori (MAP) probability as a function of three different priors on the Poisson rate parameter Λ ($\gamma = 1$; $\gamma = 5$; $\gamma = 10$). This "best-fit" model was simply the model that was visited most often during the MCMC simulation of the posterior. (D–F) show the distribution of posterior probabilities for the true model (M_0). With a relatively flat prior on models ($\gamma = 10$), the MAP model is biased towards a model with 2 processes (= 1 transition). However, the posterior probability of the true model M_0 remains substantial (F), and M_0 nonetheless had a posterior probability greater than 0.10 for the vast majority of simulations. Results are based on 500 simulated phylogenies per γ scenario.

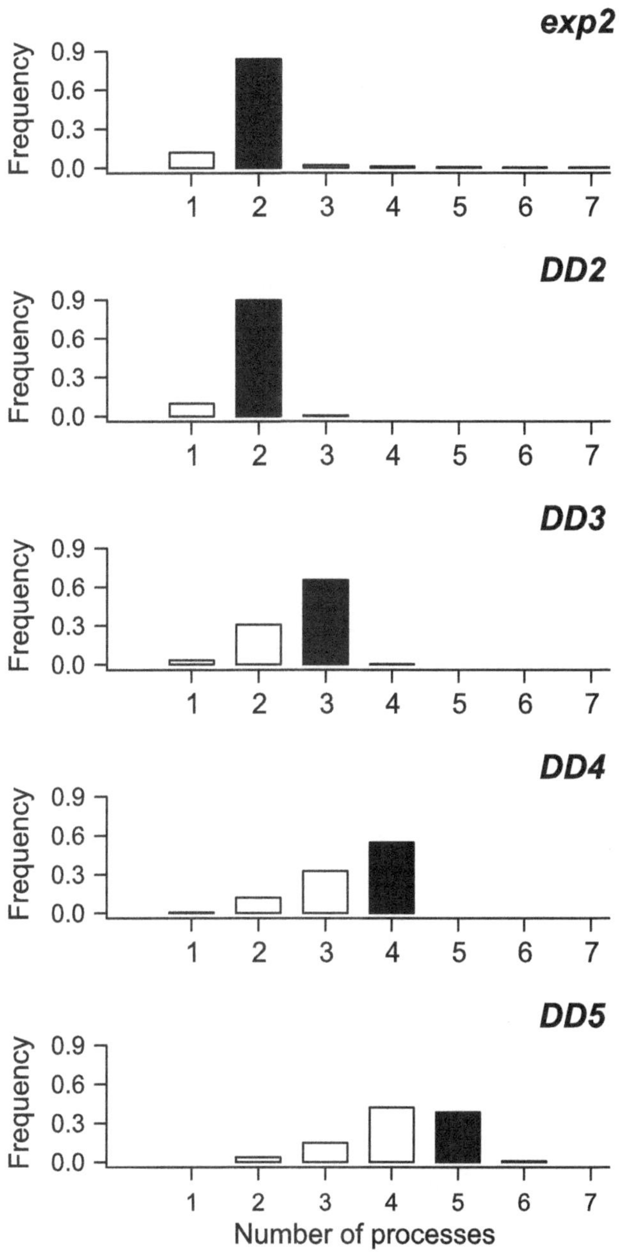

Figure 4. Frequency distribution of evolutionary rate regimes estimated using BAMM, compared with true number of processes. For each simulation, the estimated number of processes was simply the model that was most frequently sampled during MCMC simulation of the posterior distribution. Black bars denote true number of processes in generating model. For example, 84% of trees simulated under a single-shift exponential change model (exp2; top panel; two processes in the generating model) were correctly inferred to have been generated under a two-process model. For a diversity-dependent model with five processes (DD5), power to detect the true number of processes is lower, although though most analyses (80.4%) recovered either 4 or 5 process models as the MAP model. Results for each model are based on 500 simulated phylogenies and used a conservative $\gamma = 1$ prior on the expected number of non-root processes (see Figure 3).

then map the current state and the random vector to the new state (θ', ψ) through use of a mapping function $g(\theta, v)$. The random vector v has a number of elements equal to the number of parameters in ψ, thus satisfying the dimension-matching requirement for transdimensional moves. The acceptance probability for this move is given by

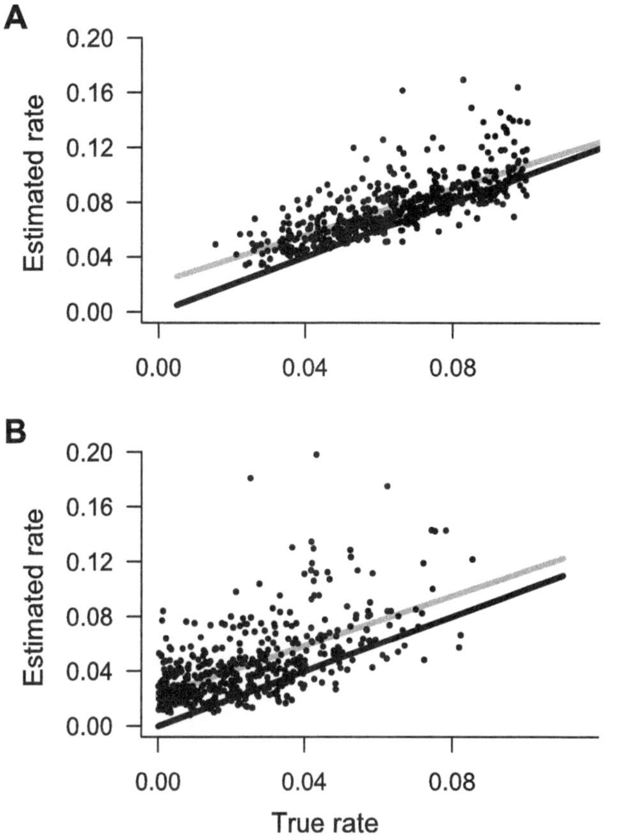

Figure 5. BAMM estimates of speciation and extinction rates for phylogenies simulated under constant-rate birth-death process. (A) Relationship between speciation rate in generating model and reconstructed mean rate across the tree under BAMM. Solid black line: identity line, expected if $\lambda_{TRUE} = \lambda_{ESTIMATED}$. Solid gray line: fitted OLS regression to estimates (black points) obtained using full BAMM model (multiple processes with time-variable speciation rates). (B) Corresponding extinction rate estimates for same set of trees.

$$\min\left\{1, \frac{f(\theta',\psi)}{f(\theta)} \frac{\pi(\theta',\psi)}{\pi(\theta)} \frac{\pi(M_{k+1})}{\pi(M_k)} \frac{q(M_k|M_{k+1})q(\theta|\theta')}{q(M_{k+1}|M_k)q(\theta'|\theta)q(v)} \left|\frac{\partial g(\theta,v)}{\partial(\theta,v)}\right|\right\} \quad (3)$$

where $q(M_k \mid M_{k+1})$ denotes the probability of proposing a move from model M_{k+1} to model M_k, $\pi(M_k)$ is the prior probability of model M_k, and the last term is the determinant of the Jacobian matrix for the transition from the vector (θ, v) to (θ', ψ) via the mapping function $g(\bullet)$. The corresponding reverse move is deterministic and the acceptance probability is given by the inverse of the numerator in equation 3, with the exception of the case where k equals 0 or 1 (discussed below).

In the model described here, an increase from M_k to M_{k+1} involves the addition of four new parameters to the process: $\psi = (\xi_{k+1}, \lambda_{0,k+1}, z_{k+1}, \mu_{k+1})$. During model-jumping proposals, all parameters θ are mapped to θ' via the identity function, such that $\theta' = \theta$. The mapping from v to ψ, or $g(v)$, was also defined using simple identity relationships:

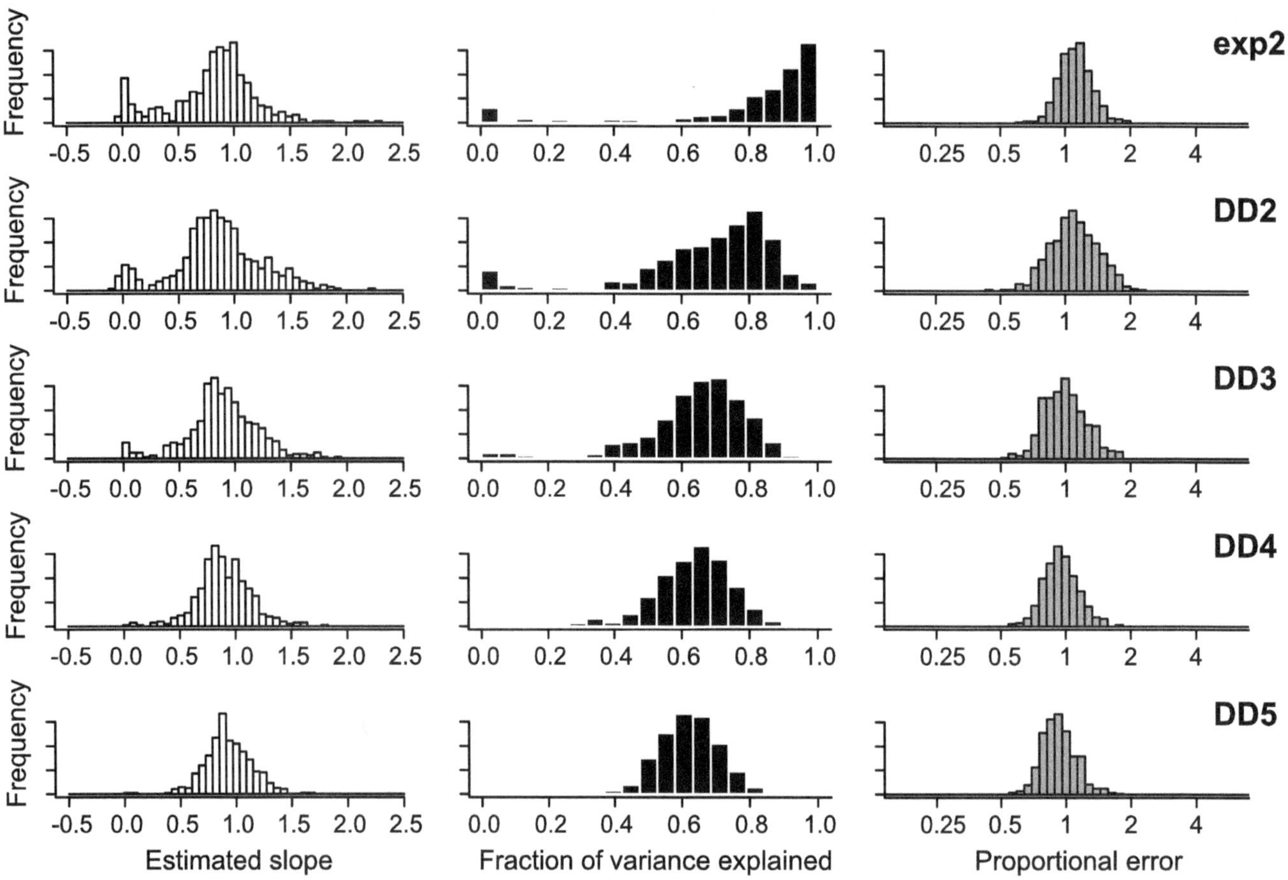

Figure 6. Precision and bias of BAMM in the estimation of branch-specific rates of speciation. Phylogenies were simulated under 5 distinct evolutionary scenarios. For each simulated phylogeny, I reconstructed branch-specific speciation rates using BAMM and modeled these as a function of the true branch rates from the generating model. Frequency distributions of the estimated slope of this relationship are shown in the left column for each simulation scenario. Center column denotes corresponding r^2 values from the same OLS regressions. Right column is distribution of mean relative rate differences (RRD) for each scenario. A value of 1 implies that, on average, branch-specific speciation estimates are unbiased; a value of 0.5 would imply that branch-specific estimates are, on average, equal to 50% of the true value. Results for each simulation scenario are based on 500 simulated phylogenies (thus giving 500 slopes, r^2 values, and RRD values for each simulation scenario).

$$\begin{aligned} \xi_{k+1} &= v_1 \\ \lambda_{0,k+1} &= v_2 \\ z_{k+1} &= v_3 \\ \mu_{k+1} &= v_4 \end{aligned} \qquad (4)$$

Variables v_1, v_2, v_2, and v_3 were sampled from the corresponding prior distributions for ξ, λ_0, z, and μ. The Jacobian term reduces to the identity matrix and has a determinant of 1.

Under the compound Poisson process, the overall (whole-tree) rate at which transitions occur under the model is Λ. The prior ratio for models M_{k+1} to M_k, given Λ, is simply the ratio of Poisson densities with k+1 and k transitions, or

$$\frac{\pi(M_{k+1}|\Lambda)}{\pi(M_k|\Lambda)} = \frac{(\Lambda)^{k+2}e^{-\Lambda}/(k+1)!}{(\Lambda)^{k+1}e^{-\Lambda}/k!} = \frac{\Lambda}{k+1} \qquad (5)$$

Table 1. Relationship between branch-specific BAMM estimates of extinction and true rates in the simulation model.

Model	processes	slope [a]	r^2	PE [b]
Exponential change (exp2)	2	0.76	0.59	1.27
Diversity-dependent (DD2)	2	0.81	0.16	1.85
Diversity-dependent (DD3)	3	0.75	0.11	1.68
Diversity-dependent (DD4)	4	0.81	0.13	1.67
Diversity-dependent (DD5)	5	0.82	0.10	1.66

[a]Slope and r^2 denote the estimated slope and variance explained by the relationship between true and estimated extinction rates for 500 trees simulated under each model.
[b]PE is the mean proportional error across all simulations under a given model.

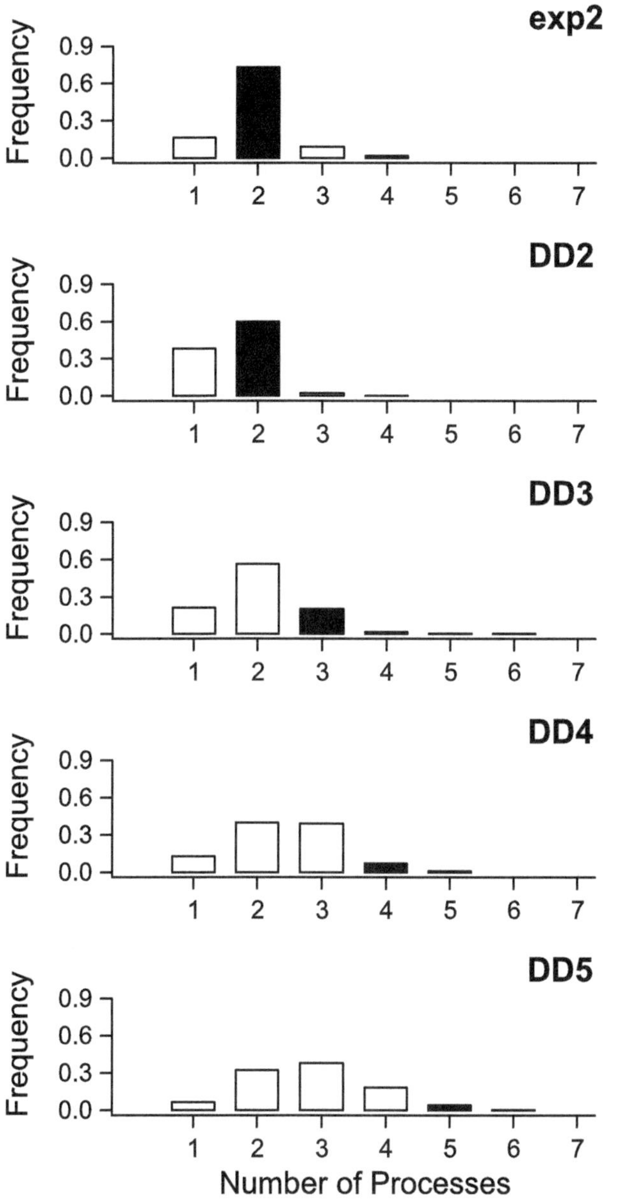

Figure 7. Frequency distribution of evolutionary rate regimes estimated using MEDUSA, compared with true number of processes. Phylogenies were simulated under 5 distinct evolutionary scenarios. For each simulation, the number of distinct rate partitions was estimated using the stepwise AICc algorithm as implemented in MEDUSA. Black bars denote true number of processes in generating model. MEDUSA consistently underestimates the true number of processes in simulated datasets when rates of speciation vary through time. Comparable results for BAMM using the same set of simulated datasets are shown in Figure 4. A total of 500 simulated datasets were analyzed per diversification scenario.

To compute the proposal probability q(M_k | M_{k+1}), let d_{k+1} represent the probability of making a move that deletes one of $k+1$ transitions on the tree, and let b_k represent the probability of adding a transition when there are currently k transitions on the tree. The probabilities d_k and b_k are typically equal to 0.5, since addition and deletion moves are equiprobable for most values of k. However, when the tree includes just a single process and no transitions ($k = 0$), the relative probability of adding a transition is equal to 1.0, as the root process cannot be deleted. In this case, $d_1 = 0$ and $b_1 = 1$, because only additions of transitions (and not deletions) can be proposed. This leads to an asymmetrical proposal ratios for adding transitions (when $k = 0$) and for deleting transitions (when $k = 1$). When $k = 0$, the ratio d_2/b_1 is equal to 0.5. When there are two processes on the tree ($k = 1$), the proposal ratio is asymmetrical and b_{k-1}/d_k is equal to 2.0. This compensates for the excess of gain proposals that occur when there is just a single process on the tree. Otherwise, q(M_{k+1} | M_k) = q(M_k | M_{k+1}) = 1.

Because elements of v are sampled from prior distributions for ξ, λ_0, z, and μ, the prior on ψ in the numerator of equation 3 is equal to the density q(v) in the denominator, as in [30]. The acceptance probability for the addition of a transition is thus a function of the likelihood ratio, the prior ratio for models M_{k+1} and M_k, and the proposal ratio for the models, or

$$\min\left(1, L\frac{\Lambda T d_{k+1}}{(k+1)b_k}\right) \tag{6}$$

where L is the likelihood ratio of current and proposed states.

The acceptance probability for a move that deletes one of k transitions from the tree involves inverting the ratio term from equation (6) and modifying subscripts to reflect the fact that we are proposing a move to a state with k–1 transitions. The proposal ratio becomes b_{k-1}/d_k, leading to an overall acceptance probability of

$$\min\left(1, L\frac{k b_{k-1}}{\Lambda T d_k}\right) \tag{7}$$

The positions of transitions were updated using global and local moves. A global move entailed sampling a new map location ξ from tree and allowed transitions to shift to any point on the tree with uniform probability. A local move involved shifting the position of a transition by a small random quantity that was sampled from a uniform distribution. I fixed the ratio of global:local proposals at 1:10 for all analyses described here. The acceptance probability for a move that changes the position of a transition is equal to min(1, L).

To update any of the zero-bounded rate parameters in the model (Λ, λ_i, μ_i), I used a proportional shrinking-expanding proposal [42], such that

$$r' = re^{\eta(U-0.5)} \tag{8}$$

where r is the current value of the rate parameter, U is a random variable sampled from a uniform (0, 1) distribution, and η is a tuning parameter. The acceptance probability of a move that updates such a rate parameter is

$$\min\left(1, Le^{\eta(U-0.5)}\right) \tag{9}$$

Finally, I used a sliding window proposal to update the value of the rate change parameter z_i. Here, a random variable is sampled from a uniform ($-\delta$, δ) distribution and added to the current value of the parameter; δ is a tuning parameter that can be modified to increase the efficiency of the MCMC sampling. The proposal ratio for the sliding window proposal is 1.0, and the acceptance probability is min(1, L).

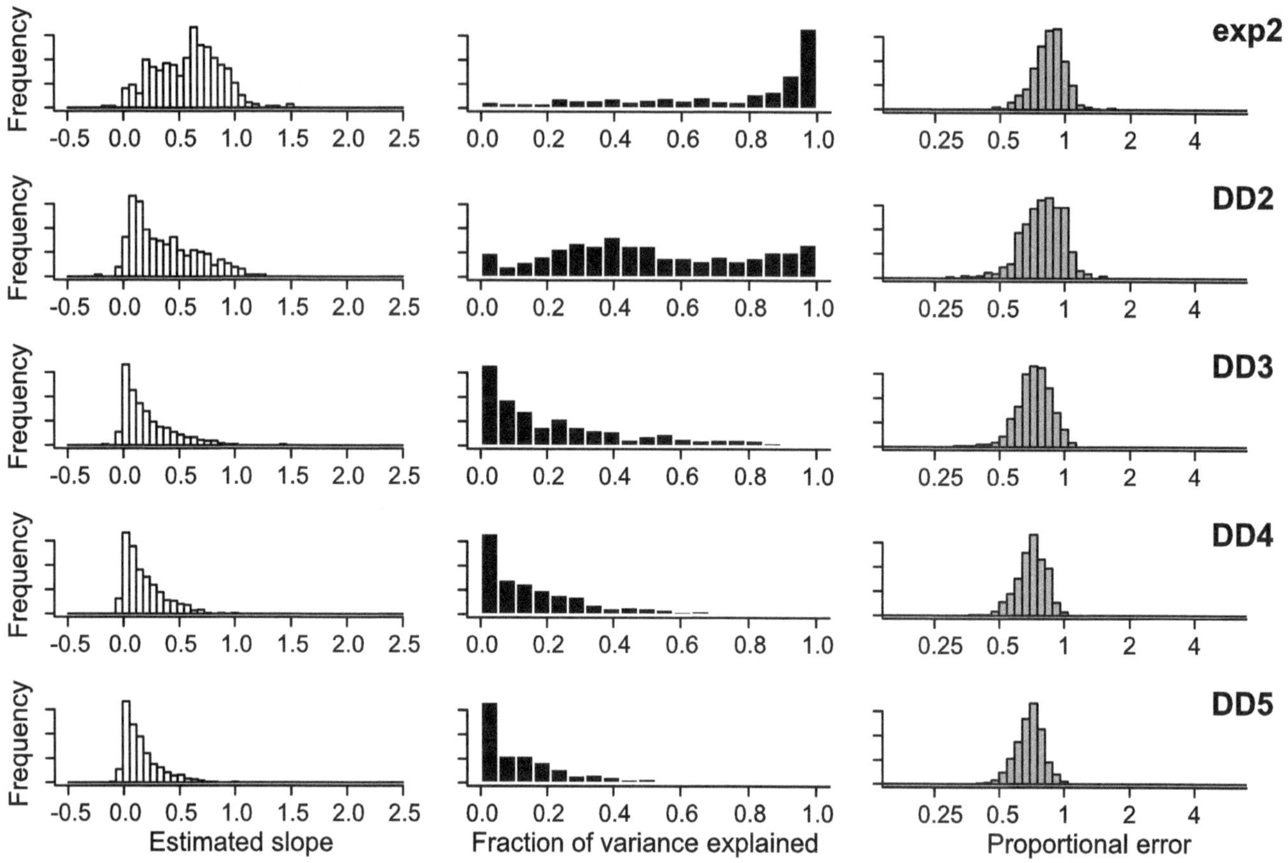

Figure 8. Precision and bias of MEDUSA in the estimation of branch-specific rates of speciation. For each simulated phylogeny, MEDUSA was used to estimate the number, location, and parameters of diversification rate shifts. The resulting branch-specific rates of speciation were compared with the true branch rates from the generating model. Results are based on the same simulated datasets analyzed with BAMM and can be directly compared to those shown in Figure 6. Branch-specific speciation rates estimated with MEDUSA show little correspondence with true rates when rates vary through time, at least in comparison to rates estimated with BAMM.

I placed a uniform (0, *T*) prior density on the location of transitions, assuming simply that all positions on the tree are equiprobable. Thus, during the addition of a new transition, we sample a new map location at random from (0, *T*). I placed relatively flat exponential priors on λ and μ and a normal (mean = 0; variance = 0.05) prior on z; the latter choice was motivated by the fact that $z = 0$ corresponds to a constant-speciation diversification process (equation 1). I placed an exponential prior on Λ, the parameter of the Poisson distribution that serves as a prior on the number of transitions on the tree. Larger values for the rate parameter of this exponential distribution imply a greater number of transitions on the tree. I denote this exponential prior on the number of transitions by γ.

Likelihoods were computed on branches using a discretization of the constant-rate birth death model that enabled us to approximate time-dependent and diversity-dependent rate variation. Following the notation from Maddison et al. [23], let $D(t)$ represent the probability that some lineage at time t evolves into a clade identical to the observed descendant clade, and let $E(t)$ represent the probability that the lineage goes extinct before the present. Following [43], let t_N be the initial time for such an interval, in units of time before the present, and let t be some earlier time (closer to the root), such that $t > t_N > 0$. It is straightforward to write down the change in D and E as a function of time, such that

$$\frac{dD}{dt} = -(\lambda+\mu)D(t) - 2\lambda D(t)E(t) \tag{10}$$

and

$$\frac{dE}{dt} = \mu - (\lambda+\mu)E(t) + \lambda E(t)^2 \tag{11}$$

Let E_0 and D_0 denote the initial values of the speciation and extinction probability for a given interval Δt over which $E(t)$ and $D(t)$ must be computed. The analytical solution to equation (11), given E_0, is

$$E(t) = 1 - \frac{(1-E_0)(\lambda-\mu)}{(1-E_0)\lambda - e^{-(\lambda-\mu)t}(\mu-\lambda E_0)} \tag{12}$$

which is identical to equation (12a) from FitzJohn et al. (2009) under the substitution $E_0 = 1 - f$, where f is the sampling fraction of the phylogeny. FitzJohn et al. [43] demonstrated how the birth-death model could be extended to account for incomplete taxon sampling (via the sampling fraction f), but their results allow the calculation of probabilities along any segment of a branch of a phylogenetic tree, provided that E_0 and D_0 are available for the

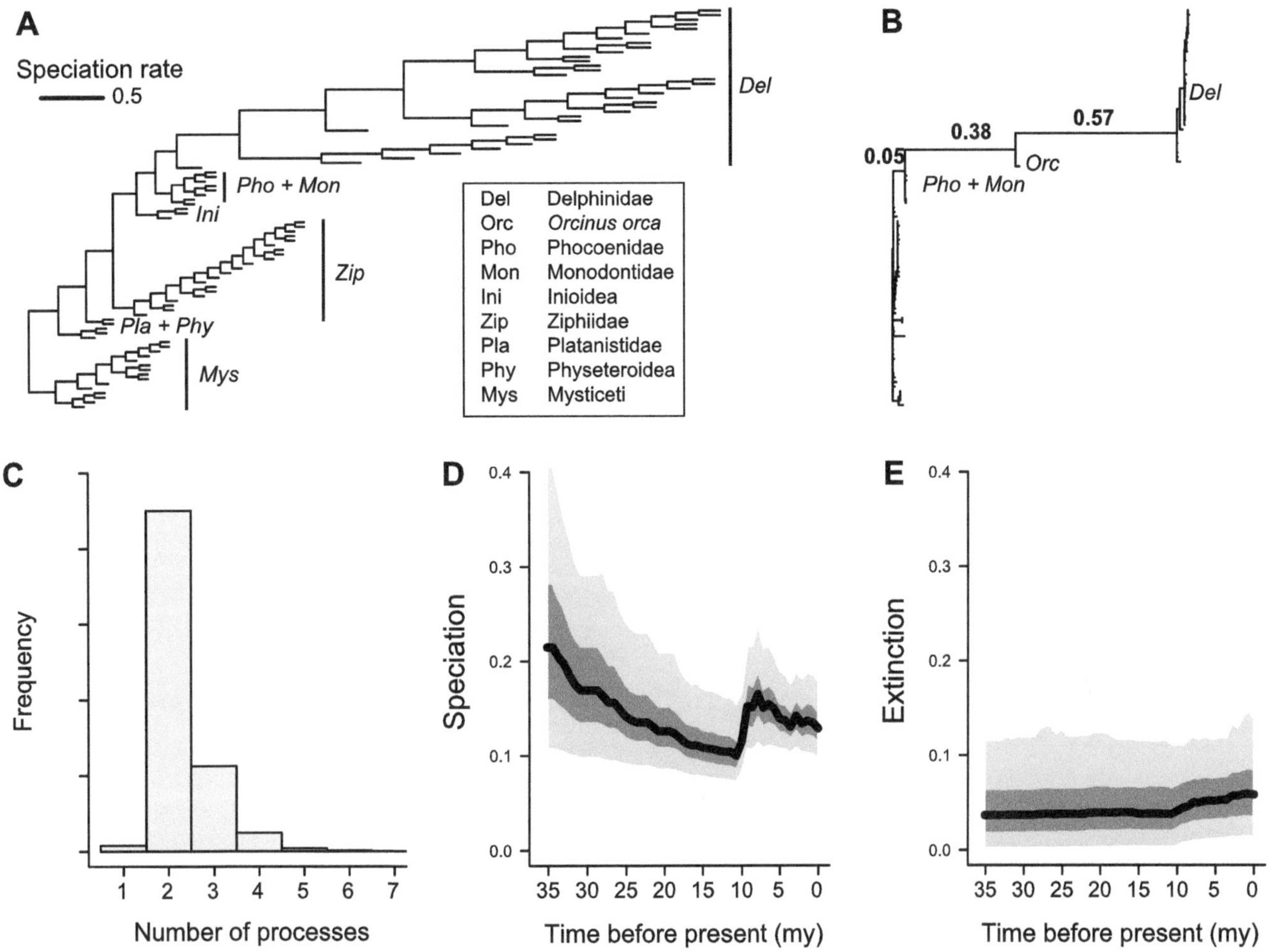

Figure 9. Dynamics of cetacean diversification through time as revealed by BAMM analysis. (A) Phylogeny of cetaceans [51], with branch lengths drawn proportional to their marginal speciation rate as estimated using BAMM. A large increase in the rate of speciation (>6-fold) occurred in one of the ancestral branches leading to the Delphinidae (including or excluding the killer whale, *Orcinus orca*). Despite this increase, the overall trend is towards decelerating rates through time. (B) Cetacean phylogeny with branch lengths scaled by the posterior probability that they contain a rate shift. Numbers above branches denote branch-specific shift probabilities. The probability that a rate shift occurred on at least one of these three branches was 0.975. No other branches had shift probabilities exceeding 0.02. (C) Posterior distribution of the number of distinct processes (including the root process) on the cetacean phylogeny. A two-process model vastly outperforms a one-process model. (D) Speciation rates through time during the extant cetacean radiation; distinct shaded regions denote (from bottom) 0.05, 0.25, 0.50, 0.75, and 0.95 quantiles on the posterior distribution of rates at a given point in time. Massive spike in mean speciation rates at 7.5 Ma corresponds to the early radiation of the Delphinidae clade. (E) Corresponding extinction through time curve.

beginning of the interval over which the probabilities are to be computed. Equation (12) can be substituted into equation (10), and the resulting expression simplifies to

$$D(t) = \frac{e^{(\mu-\lambda)\Delta t} D_0 (\lambda - \mu)^2}{\left[\lambda - \lambda E_0 + e^{(\mu-\lambda)\Delta t}(E_0 \lambda - \mu)\right]^2} \tag{13}$$

where Δt is the duration of the focal interval, between time tN and time t. As we begin with known conditions D0 and E0 at the beginning of the focal segment, we can set tN = 0 for the purposes of our calculations. This immediately simplifies equation (13) from [43] and leads to equation (13) above. As demonstrated by FitzJohn et al. [43], these calculations reduce to the speciation-extinction model for phylogenetic trees developed earlier [12].

To discretize the rate calculations, I broke each branch of the phylogenetic tree into segments and computed the mean speciation rate under the exponential change model (equation 1) for the corresponding process. I then assumed constant rate diversification within each branch segment. For each branch segment, the initial speciation and extinction rates D_0 and E_0 are equal to the terminal values for the preceding segment. I made this design choice to facilitate rapid likelihood calculations on large phylogenetic trees and, as demonstrated below, this discretization performs well across a range of simulated datasets. I used a step size of 1.0 time units for all calculations. If a branch was particularly short, such that this step size exceeded the length of the branch, the entire branch was assumed to have a single rate equal to the mean rate along its length.

To compute the likelihood of the full tree under a given set of parameters, we perform the calculations described above on each terminal branch of the phylogeny. Initial values of D_0 and E_0 at the tips of the tree were set to 1.0 and 0.0 respectively. It is straightforward to modify these values to account for incomplete taxon sampling if only a fraction f of the total species in a clade have been included in a phylogenetic tree.

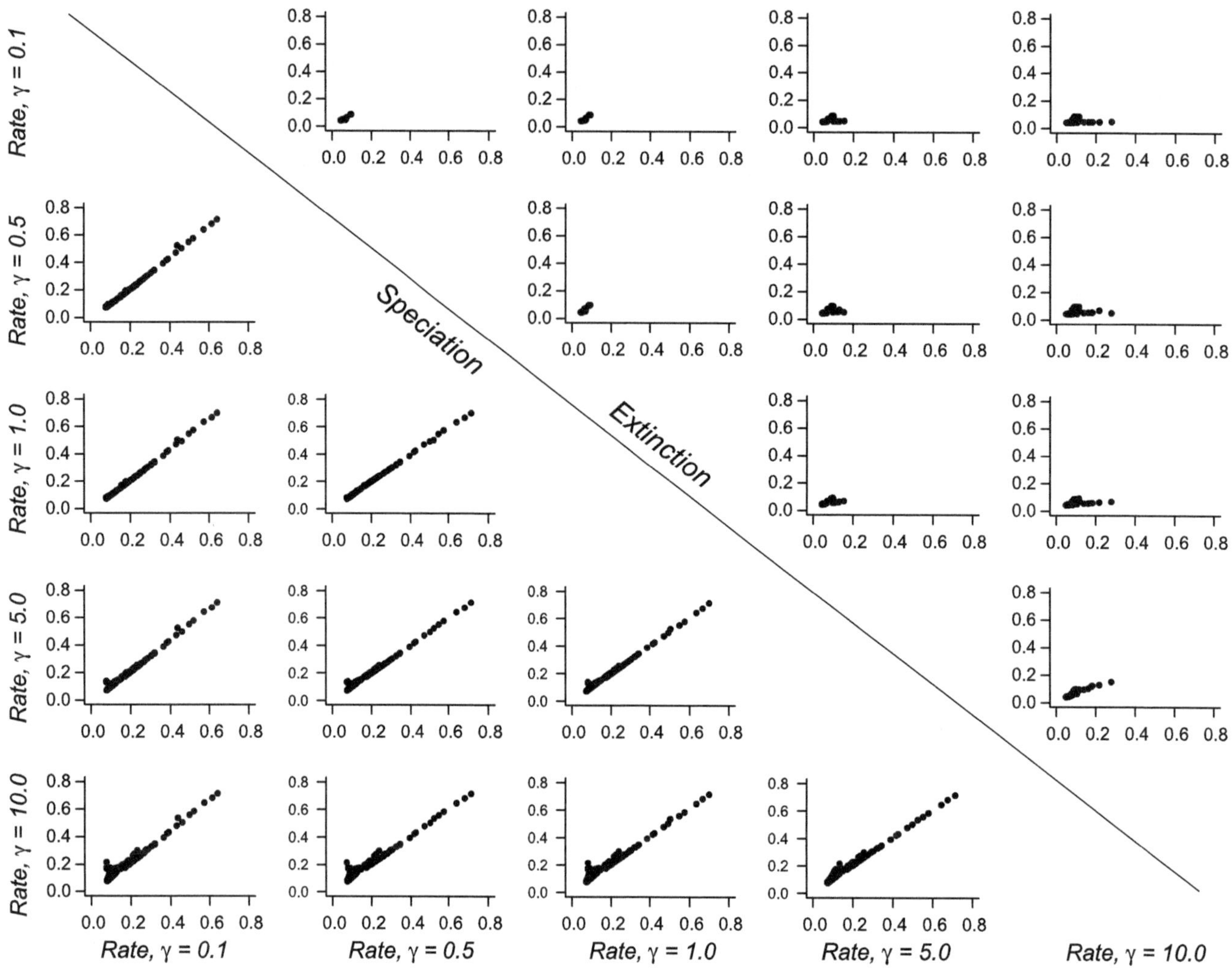

Figure 10. Sensitivity of marginal rate estimates to prior on Poisson rate parameter. Each panel shows a pairwise plot comparing branch-specific (marginal) diversification rate estimates for two values of γ for the Cetacean dataset, with results for speciation and extinction separated by the diagonal. Speciation rate estimates for the cetaceans are remarkably robust to choice of prior: even $\gamma = 10$ and $\gamma = 0.1$ yield strikingly similar marginal distributions for branch-specific speciation rates. This is generally not true for extinction, where mean marginal rates for each branch were more sensitive to prior formulation. However, extinction was nonetheless estimated to be low overall regardless of the prior γ.

When $f > 0$, we can set $D_0 = f$ and $E_0 = 1 - f$ for the initial calculations at the tips of the tree [43]. This correction assumes that species are missing at random from the phylogeny, which may not be valid for many datasets [44,45,46,47]. As in the BiSSE calculations [23], these calculations flow "rootwards" from these terminal branches towards the root. When terminal probabilities have been computed for both descendant branches from a given internal node, the left (D_L) and right (D_R) branch probabilities were combined as $\lambda\ D_R(t)D_L(t)$, where λ is the speciation rate at the focal node [23]. The calculations then continue down the branch subtending this node. The likelihood of the full tree is the value of D after combining these probabilities at the root node. Likelihoods were conditioned on the occurrence of a root node and on the survival of both descendent branches from the root speciation event [12,23].

I implemented the compound Poisson process model of rate variation described above in a C++ program, which I refer to as BAMM. BAMM (Bayesian Analysis of Macroevolutionary Mixtures) can estimate the number of distinct evolutionary regimes across phylogenetic trees and estimates marginal distributions of speciation and extinction rates for each branch in a phylogenetic tree. The model allows extinction rates to exceed speciation rates. BAMM and associated documentation is available from the BAMM project website (www.bamm-project.org). The program operates on fully bifurcating phylogenetic trees of extant species. The implementation allows users to analytically account for incomplete taxon sampling under the assumption of random taxon sampling [43].

Analysis of Simulated Datasets

To evaluate performance of the compound Poisson process model of diversification rate variation, I simulated phylogenetic trees under six general diversification models. I first considered a simple constant-rate birth death process (model CR; 1 process), to evaluate parameter bias and the frequency of overfitting when the generating model does not include a heterogeneous mixture of processes. Given the widespread interest in identifying well-supported rate shifts and key innovations on phylogenetic trees, we are particularly interested in the frequency with which the model described here will incorrectly identify a multi-process model as having the maximum *a posteriori* probability, when the true generating model is a single process model. To assess whether my results were sensitive to choice of prior on the Poisson rate parameter Λ, I analyzed constant-rate phylogenies under three different prior parameterizations, corresponding to $\gamma = 1$, $\gamma = 5$, and $\gamma = 10$. All other analyses used a prior of $\gamma = 1.0$, which is conservative in the context of these analyses (see results).

I also considered a model where a pure-birth diversification process shifts to an exponential change process at some point in time (model exp2; 2 processes). Finally, I considered four variants of diversity-dependent multi-process models. In each case, I assumed that a pure-birth process at the root of the tree underwent multiple (1, 2, 3, or 4) shifts to independent and decoupled diversity-dependent speciation-extinction processes (models DD2, DD3, DD4, DD5). I conducted 500 simulations per scenario.

Each multiprocess simulation was conducted by first simulating a pure-birth phylogeny for 100 time units with $\lambda = 0.032$. I then randomly chose a time T_s on the interval (40, 95) for the occurrence of a rate shift. A shift was then assigned randomly to one of the lineages that existed at time T_s. I then sampled parameters for the new process (see below). The tree was then broken at the shift point, and a new subtree was simulated forward in time from the shift point under the new process parameters. For trees with more than two processes, this procedure was repeated until the target number of processes had been added. For the exp2 model, this consisted of sampling λ, z, and μ for the shift process uniformly on the following intervals: λ, (0.05, 0.50); z, (−0.10, 0.05); μ, (0.0, 0.45). Thus, the addition of an exponential change process could have resulted in either an increase in rates through time (if $z > 0$) or a decrease (if $z < 0$). For all simulations, I required that subtrees contained at least 25 and fewer than 1000 terminal taxa; any simulations failing to meet this criterion were automatically rejected.

For the diversity-dependent models, diversification dynamics followed a linear diversity-dependent model [48]. The rate of speciation was thus a function of the number of coeval lineages in the subclade, or

$$\lambda(n) = \lambda_0 \left(1 - {}^{n_t}\!/_{K}\right) \quad (14)$$

where K is the clade-specific carrying capacity, and n_t is the number of lineages in the subclade at time t. Note that the occurrence of a shift event results in a decoupling of dynamics from the parent process. To parameterize the diversity-dependent processes, I sampled λ_0 from a uniform (0.05, 0.40) distribution, K from a uniform (25, 250) distribution, and μ from a uniform (0, 0.05) distribution. For the constant-rate birth-death simulations, I sampled λ from a uniform (0, 0.1) distribution and chose a corresponding relative extinction rate (μ/λ) from a uniform (0, 0.99) distribution.

Each of the 500 simulations for each of 6 simulation scenarios was thus conducted under a potentially unique speciation-extinction parameterization. The number of taxa in each simulated tree also varied among datasets. I recorded the mean rate of speciation and extinction across each branch in each simulated tree. All simulations were conducted in C++; simulated trees are available through the Dryad data repository (doi:10.5061/dryad.hn1vn).

I analyzed each of the 3000 simulated datasets using BAMM with 3 million generations of MCMC sampling. I discarded the first half of samples from each simulation of the posterior as "burn-in" and estimated the overall "best model" as the model that was sampled most frequently by the Markov chain. I computed the mean of the posterior distribution of speciation and extinction rates on each branch for each tree. I then used OLS regression to assess the relationship between branch-specific rate estimates obtained using BAMM versus the true underlying evolutionary rates. As an additional estimate of bias, I computed the proportional error [37] in the estimated rates as a function of the true rates. This metric is computed as the weighted average of proportional rate differences across all N branches in the phylogeny, or

$$PE = \exp\left(\frac{1}{N}\sum\left[\log(r_{EST}) - \log(r_{TRUE})\right]\right) \quad (15)$$

where r_{EST} and r_{TRUE} are the estimated and true values of rates along a particular branch. A value of 2 would imply that estimated rates are, on average, equal to twice the true rate in the generating model.

Comparison with MEDUSA

I compared the performance of BAMM to that of MEDUSA [20], a maximum likelihood method for modeling among-lineage heterogeneity in speciation-extinction dynamics. Beginning with a constant rate birth-death process, MEDUSA uses a stepwise AIC algorithm to incrementally add rate shifts to phylogenetic trees until the addition of new partitions fails to improve the fit of the model to the data. Thus, MEDUSA is similar to the method described here in that it is explicitly designed to discover the number and location of distinct processes of speciation and extinction on phylogenetic trees. However, MEDUSA, as implemented and typically used, makes the assumption that rates of species diversification are constant in time within rate classes. This assumption has been rejected by studies across a range of taxonomic scales, from species-level phylogenies [16,17,18,26] to tree-of-life scale compilations of clade age and species richness [49]. However, the consequences of violating this assumption for MEDUSA analyses have not been investigated.

I analyzed each of the simulated datasets described above (500 datasets under each of 6 distinct models of diversification) using MEDUSA, using the implementation of MEDUSA available in the Geiger v1.99-3 package [50] for the R programming and statistical environment. Model selection used the default AICc criterion. I summarized the results of MEDUSA analyses in two ways. First, for each simulation scenario, I tabulated the distribution of "best fit" models, to assess the fraction of simulations for which MEDUSA was able to correctly estimate the number of processes in the generating model. Second, I used the same summary statistics described above for BAMM (e.g., proportional error) to compare branch-specific estimates of speciation rates under MEDUSA to the true rates under the generating model.

Empirical Example: Cetacean Radiation

Steeman et al. [51] provided a time-calibrated phylogenetic tree for 87 of 89 extant species of whales and dolphins (Mammalia: Cetacea). They found support for increased rates of species diversification within a major dolphin clade, the Delphinidae. They also found evidence for an increased rate of species diversification at approximately 7.5 million years before present (Ma). I used the BAMM implementation of the compound Poisson process model of diversification rate variation to investigate the tempo and mode of cetacean diversification through time. I conducted 5 million generations of MCMC sampling, with multiple independent runs to assess convergence. Finally, I assessed the sensitivity of the cetacean analyses to the choice of prior γ on the number of processes in the phylogeny.

Results

Analysis of Simulated Datasets

Figure 2 shows a representative BAMM analysis for a tree simulated under the DD3 model (see also Figure 1). BAMM results are generally robust to choice of prior on the expected number of processes (γ) in the phylogeny under the compound Poisson process model of rate variation (Figure 3). With increasing values of γ, the model with maximum *a posteriori* probability (MAP) was biased in favor of M_1, a model with two processes. However, this represents a weak tendency towards model overfitting, because the true model (M_0) was generally characterized by a posterior probability much greater than 0.05 (Figure 3E, F). This suggests that results are robust to choice of γ: even with the trend towards overfitting (Figure 3C), the method is unlikely to yield strong support for models that are more complex than the generating model (Figure 3F). The simulation results presented below are based on $\gamma = 1$.

For the five simulation scenarios with rate heterogeneity, the number of distinct processes estimated using BAMM was generally equal to the number of processes in the generating model (Figure 4). Power to infer the true number of processes decreased for the most complex models (DD3, DD4, DD5), but model overfitting was not a problem. The MAP model was no more complex than the generating model in the overwhelming majority of simulations (>95%) for all five simulation scenarios.

Estimates of speciation and extinction rates under the constant-rate model were highly correlated with rates in the generating model (Figure 5), although both rates were biased upwards for low rates. For multiprocess simulation models, branch-specific estimates of speciation rates were highly correlated with rates in the generating model (Figure 6, left). The estimated slope of the relationship between the true rates and estimated rates approached equality. However, a small percentage of simulations had estimated slopes that suggested a lack of relationship between true and estimated rates. These simulations were those where the most frequently sampled model had only a single process and thus reflect a lack of power, rather than consistent bias. In other words, branch specific estimates of rates for a multiprocess model may be poor if model underfitting has occurred. In the extreme case, a tree that is estimated to have only a single process may have very similar rate estimates on each branch; the correlation between these rates and the true rates will necessarily be low if the true model includes multiple processes and considerable rate heterogeneity across the tree.

A large fraction of the total variation in the underlying speciation rates is also explained by the estimated rates under the compound Poisson process model of diversification rate variation (Figure 6, middle). This fraction is higher for the two-process models (exp2, DD2) but remains stable across the remaining diversity-dependent scenarios (DD3, DD4, DD5). Finally, analysis of proportional error suggests that, on average, rates are not consistently over- or underestimated using BAMM; the mean proportional error between simulated and estimated rates is near 1.0 for all simulation scenarios.

I performed a similar analysis for extinction rates, with one key difference. Each simulation scenario assumed constant extinction rates within each process; hence, the number of unique extinction rates in each simulation was equal to the number of processes in the generating model. I computed the mean branch-specific extinction rate across each subclade that was governed by a distinct evolutionary process in the simulation model; I then analyzed these extinction rates across all 500 trees in a given simulation scenario together. For example, consider the phylogeny shown in Figure 1, with $k = 3$ processes. I computed the mean of the posterior distribution of all extinction estimates for branches assigned to process A, giving a single estimated rate overall for that process. I repeated this for processes B and C, such that I obtained k extinction estimates for each tree with k processes. Table 1 shows summary statistics for analyzing these sets of extinction rate estimates across all trees from a given simulation scenario. In general, relative rate differences suggest that extinction estimates are biased upwards. Nonetheless, the fraction of variance explained by the model is low in each case. Proportional error calculations were also performed as described above, although the root rate class was ignored, as it was equal to 0 in the simulation model. These values likewise indicate an upwards bias in extinction rate estimates.

Comparison with MEDUSA

I analyzed all 3000 simulated datasets using MEDUSA. The number of processes inferred for each simulation scenario are shown in Figure 7 and can be compared to the corresponding BAMM results shown in Figure 3. MEDUSA performed worse than BAMM for all scenarios with time-varying rates of species diversification. Under the DD2 model, with just two processes, MEDUSA estimated the correct number of processes in 40.2% of simulated datasets. In contrast, BAMM correctly identified the true number of processes in 90% of simulated datasets (Figure 3). For diversity-dependent scenarios with more than two processes in total, MEDUSA consistently underestimated the true number of processes in the generating model for the overwhelming majority of simulated datasets. For the DD5 scenario, MEDUSA correctly identified the generating model in fewer than 5% of simulations, versus 38.4% with BAMM (Figure 7 vs. Figure 4).

Branch-specific estimates of speciation rates under MEDUSA were, in general, extremely poor (Figure 8) when compared to the corresponding estimates under BAMM (Figure 6). The estimated slope of the relationship between the true speciation rates on each branch and the corresponding MEDUSA estimates has a modal value of zero for four of five simulation scenarios (Figure 8, left column). In contrast, BAMM estimates were far closer to the 'perfect' value of 1.0. For all simulation scenarios but exp2, the MEDUSA-estimated speciation rates explained little of the variance in the true distribution of rates (Figure 8, middle). Finally, proportional error analysis indicated that MEDUSA generally underestimates true rates of speciation when rates are time-dependent or diversity-dependent (Figure 8, right column).

Empirical Example: Cetacean Radiation

Analysis of the time-calibrated cetacean phylogeny found strong support for a two-process model (Figure 9). The posterior probability of a one-process model is $p = 0.017$, with a posterior odds ratio of 44.6 in favor of a two-process model. These results suggest a substantial increase in the rate of speciation in the ancestral lineage leading to the Delphinidae (Figure 9A), possibly excluding the killer whale *Orcinus orca*. The posterior probability of a rate shift occurring on at least one of these branches is greater than 0.975 (Figure 9B). However, we find little evidence for additional processes within the cetacean phylogeny as a whole (Figure 9 B, C).

Using output from BAMM, I computed mean rates of speciation and extinction through time during the cetacean radiation. This was done by drawing an imaginary grid of vertical lines through the time-calibrated cetacean phylogeny at equally spaced points in time. Evolutionary rates were estimated as the mean branch-specific rates for all branches that intersected the line corresponding to a specific time point. This enabled estimation of the posterior density of speciation and extinction for any point in time. These results suggest an overall decline in the background rate of whale speciation, with a large spike during the Miocene driven by the radiation of the dolphin clade (Delphinidae). Extinction rates are inferred to be relatively low overall, with a mean per-branch relative extinction rate (μ/λ) of 0.36.

Finally, I assessed the sensitivity of the cetacean dataset to choice of prior on the number of processes in the phylogeny. I used BAMM to analyze the cetacean data under four additional values of γ (0.1, 0.5, 5, and 10). In each case, the single-process model had low posterior probability and was marginally worth considering (Pr (M_1) = 0.127) only under the strongest prior ($\gamma = 0.1$). For $\gamma = 5$ and $\gamma = 10$, the posterior probability of a model with a single process was approximately 0. The MAP model had two processes under $\gamma = 0.5$ and $\gamma = 5$. For $\gamma = 10$, the MAP model had four processes, but was not substantially more probable than models with two or three processes. The posterior odds ratio for M_4 versus M_2 was merely 1.63, and for M_4 versus M_3 it was only 1.26. I estimated marginal diversification rates for each branch in the cetacean phylogeny under these prior formulations; pairwise plots for speciation rate estimates under alternative priors suggest that these rates are robust to choice of prior (Figure 10). Extinction rate estimates were sensitive to choice of prior, although estimated rates were low under all prior formulations.

Discussion

Extracting information about the tempo and mode of species diversification remains a central methodological challenge in macroevolutionary studies. I developed a Poisson process model of diversification rate variation to address several limitations of current methodological approaches for studying evolutionary dynamics on phylogenetic trees. The approach described here views phylogenetic trees as the outcome of a complex mixture of potentially dynamic evolutionary processes and enables researchers to detect rate shifts, key innovations, time-dependent speciation, and diversity-dependence within single trees. Output from the BAMM implementation of the compound Poisson process model includes (i) estimates of the number of distinct process and posterior probabilities of each possible model; (ii) estimates of locations of those processes as well as associated parameter estimates; and (iii) estimates of branch-specific rates of speciation and extinction, which can further be used to infer temporal trends in evolutionary rates (Figure 9D, E).

BAMM performed well throughout the parameter space explored here. For each of six distinct macroevolutionary scenarios, BAMM was usually able to identify the true number of processes in the generating model (Figure 3; Fig. 4). Branch-specific speciation rates estimated using BAMM are fairly accurate: relative rate differences for estimated rates are centered on 1 (Figure 6, right). Moreover, the OLS regression slope for the relationship between true and estimated branch-specific rates across individual simulation trees was generally close to 1.0; the mean of each distribution of slopes shown in Figure 6 (left column) exceeded 0.85. Surprisingly, branch-specific estimates did not decay with increasing complexity of the generating model: observed slopes (Figure 6, left) for the most complex model (DD5) were closer to 1.0 (observed mean: 0.95) and had lower variance than any other simulation scenario, including those with only two processes.

Extinction rate estimates from the model should be taken with caution. Branch specific estimates of extinction are potentially biased and, although these estimates are correlated with the true underlying rates, confidence in those estimates is low (Table 1; Figure 5). This is consistent with previous studies that have noted low power in estimating extinction rates from molecular phylogenies [23,52]. In addition, previous studies have demonstrated that extinction estimates from molecular phylogenies are exceedingly sensitive to violations of model assumptions [27,53]. Because few real-world phylogenies will conform perfectly to the assumptions of the model described here, it is likely that estimated extinction rates will be even less accurate than results in Table 1 would suggest.

By implementing an exponential change function for speciation, I was able to accurately infer diversity-dependent dynamics across a range of simulation scenarios (Figure 4; Figure 6). This is consistent with Quental and Marshall's [34] prediction that time-dependent exponential processes (equation 1) should provide good approximations to linear diversity-dependent processes. It is possible that formal diversity-dependent models [13,48,54] would provide increased power and/or precision of parameter estimates over the exponential approximation used in this study. However, fitting a full diversity-dependent model with extinction is far more computationally demanding than the exponential approximation used here. For multiprocess diversity-dependent models, computing a single likelihood currently requires numerically solving large but linear systems of ordinary differential equations. The exponential approximation implemented in BAMM results in extremely fast likelihood calculations on even the largest phylogenetic trees. No attempts have yet been made to parallelize BAMM calculations, affording additional opportunities for computational speedups.

Comparison to Existing Methods

My results suggest that MEDUSA is not robust to violations of its assumption that diversification rates are constant through time. Whereas BAMM was often able to estimate the true number of distinct processes in the generating model (Figure 4), MEDUSA consistently underestimated the number of processes (Figure 7). Furthermore, the magnitude of the underestimates became more severe with increasing model complexity. Speciation rates estimated under MEDUSA were especially poor (Figure 8) and showed little overall correspondence with true rates in the simulation model.

To be clear, the model implementation in BAMM - in contrast to MEDUSA - was explicitly designed to account for variation in evolutionary rates both through time and among lineages. However, the MEDUSA method has been applied to many empirical datasets with little attention given to the potential consequences of violating the assumption of rate-constancy through time. Using *a posteriori* simulations, Rabosky et al. [49] found that parameter estimates from MEDUSA analyses on higher taxonomic datasets largely fail to predict patterns of species richness across clades. They suggested that this failure results from MEDUSA's strong assumption of time-invariant speciation and extinction rates. It seems likely that many or most real datasets will be characterized by rate variation through time as well as among lineages. As discussed by O'Meara [55], the challenges of modeling rate heterogeneity in phylogenetic trees are likely to become more severe as we consider ever-larger phylogenetic trees [56,57,58,59]: the larger the phylogeny, the greater the likelihood that the tree is the result of a heterogeneous mixture of distinct evolutionary processes. Describing the complex mixture of dynamic processes that shape real phylogenetic trees was the primary motivation for proposing the method described in this article.

Cetacean Macroevolutionary Dynamics

The analysis of the Cetacean phylogeny provides an important window into the history of cetacean diversification through time (Figure 9) that complements results obtained by several previous studies [6,13,51]. The overall lineage accumulation curve for cetaceans is relatively flat [13], suggesting relatively little variation in speciation rates through time. However, I find strong support for a multi-process diversification model consisting of two distinct evolutionary rate regimes: a root process involving a weak slowdown in speciation through time (Figure 9A), and an explosive burst and subsequent slowdown in speciation associated with the origin of the Delphinidae (Figure 9A). Slater et al. [60] also found support for a rate shift in the crown delphinids, excluding the killer whale, using MEDUSA. It seems likely that some of the evidence in favor of the "ocean restructuring" model [51] actually reflects the independent evolutionary dynamics of delphinid and non-delphinid lineages. The increase in speciation from 13 million years ago (Ma) to 4 Ma in particular seems likely to indicate the rapid diversification of the dolphin clade. My results do not rule out the possibility that ocean restructuring contributed to this clade-specific burst and slowdown in speciation rates, but it appears equally plausible that the acceleration in rates during this interval reflects the occurrence of a key evolutionary innovation early in the history of the dolphins.

Extensions to the Model

Many extensions are possible within the framework developed here. The computational machinery for adding, moving, and deleting processes from phylogenetic trees is flexible and can easily be extended to allow alternative functional models for speciation and/or extinction rate variation through time. Another obvious future extension is to explicitly account for phylogenetic uncertainty during simulation of the posterior. As currently implemented, BAMM simulates posterior distributions of models and parameters across a fixed topology. However, phylogenetic trees are rarely (if ever) known without error. Credible intervals on parameters inferred using BAMM (Figure 9 D, E) reflect only parametric uncertainty associated with the diversification model itself and would presumably increase if we also accounted for uncertainty in tree topology and branch lengths. Finally, it would be interesting to allow joint inference on paleontological and neontological data, as there is increasing recognition that these two datatypes are frequently in conflict [61]. This objective is facilitated by theoretical advances that allow evolutionary rate estimation using both fossils and molecular phylogenies [62], although suitable datasets remain elusive.

Summary

I have described a methodological framework for inferring mixtures of processes that have influenced the structure of phylogenetic trees. By modeling phylogenies as collections of dynamic processes, the method greatly extends our ability to describe evolutionary dynamics. Most previous evolutionary studies using transdimensional MCMC on phylogenetic trees have assumed that dynamics within component processes are constant in time. By relaxing the assumption of time-homogeneous diversification, the model is better able to describe complex mixtures of both time-constant and time-varying processes. A number of recent studies have suggested that such complex dynamics might dominate speciation-extinction patterns in many empirical datasets. I suggest that the use of rjMCMC to fit time-inhomogeneous multiprocess models to phylogenetic data may have applications beyond those described here, including DNA sequence evolution, phenotypic evolution, and phylogeography.

Acknowledgments

I thank M. Alfaro. F. Bokma, R. FitzJohn, L. Harmon, J. Huelsenbeck, S. Kolokotronis, B. Moore, S. Otto, A. Pyron, J. Schraiber, J. Uyeda, and members of the Huelsenbeck lab for comments on the manuscript and/or discussion that improved the results presented here. I thank T. Nan for developing the PPSS code that facilitated parallel analysis of simulated datasets on multiprocessor computers.

Author Contributions

Conceived and designed the experiments: DLR. Performed the experiments: DLR. Analyzed the data: DLR. Contributed reagents/materials/analysis tools: DLR. Wrote the paper: DLR.

References

1. Alroy J (2010) The shifting balance of diversity among major marine animal groups. Science 321: 1191–1194.
2. Sepkoski JJ (1978) A kinetic model of Phanerozoic taxonomic diversity I. Analysis of marine orders. Paleobiology 4: 223–251.
3. Mittelbach GG, Schemske DW, Cornell HV, Allen AP, Brown JM, et al. (2007) Evolution and the latitudinal diversity gradient: speciation, extinction and biogeography. Ecology Letters 10: 315–331.
4. Ricklefs RE (1987) Community diversity - relative roles of local and regional processes. Science 235: 167–171.
5. Glor RE (2010) Phylogenetic insights on adaptive radiation. Ann Rev Ecol Evol Syst 41: 251–270.
6. Morlon H, Parsons TL, Plotkin JB (2011) Reconciling molecular phylogenies with the fossil record. Proceedings of the National Academy of Sciences of the United States of America 108: 16327–16332.
7. Paradis E (2011) Time-dependent speciation and extinction from phylogenies: a least squares approach. Evolution 65: 661–672.
8. Ricklefs RE (2007) Estimating diversification rates from phylogenetic information. Trends in Ecology & Evolution 22: 601–610.
9. Wagner CE, Harmon LJ, Seehausen O (2012) Ecological opportunity and sexual selection together predict adaptive radiation. Nature 487: 366–369.
10. Alroy J (2008) The dynamics of origination and extinction in the marine fossil record. Proc Nat Acad Sci USA 105: 11536–11542.
11. Foote M (2000) Origination and extinction components of taxonomic diversity: general problems. Paleobiology 26: 74–102.
12. Nee S, May RM, Harvey PH (1994) The reconstructed evolutionary process. Phil Trans R Soc Lond B Biol Sci 344: 305–311.
13. Etienne RS, Haegeman B, Stadler T, Aze T, Pearson PN, et al. (2012) Diversity-dependence brings molecular phylogenies closer to agreement with the fossil record. Proc R Soc Lond B Biol Sci 279: 1300–1309.
14. Rabosky DL, Lovette IJ (2008) Explosive evolutionary radiations: Decreasing speciation or increasing extinction through time? Evolution 62: 1866–1875.
15. Stadler T (2011) Mammalian phylogeny reveals recent diversification rate shifts. Proc Natl Acad Sci USA 108: 6187–6192.
16. McPeek MA (2008) Ecological dynamics of clade diversification and community assembly. Am Nat 172: E270–E284.
17. Morlon H, Potts MD, Plotkin JB (2010) Inferring the Dynamics of Diversification: A Coalescent Approach. PLoS Biology 8: e1000493.
18. Phillimore AB, Price TD (2008) Density dependent cladogenesis in birds. Plos Biology 6: e71.
19. Ruber L, Zardoya R (2005) Rapid cladogenesis in marine fishes revisited. Evolution 59: 1119–1127.
20. Alfaro ME, Santini F, Brock C, Alamillo H, Dornburg A, et al. (2009) Nine exceptional radiations plus high turnover explain species diversity in jawed vertebrates. Proc Nat Acad Sci USA 106: 13410–13414.
21. Slowinski JB, Guyer CG (1989) Testing the stochasticity of patterns of organismal diversity: an imporoved null model. American Naturalist 134: 907–921.
22. FitzJohn R (2010) Quantitative traits and diversification. Syst Biol 59: 619–633.
23. Maddison WP, Midford PE, Otto SP (2007) Estimating a binary character's effect on speciation and extinction. Syst Biol 56: 701–710.

24. Mitter C, Farrell B, Wiegmann B (1988) The phylogenetic study of adaptive zones - has phytophagy promoted insect diversification? American Naturalist 132: 107–128.
25. Paradis E (2005) Statistical analysis of diversification with species traits. Evolution 59: 1–12.
26. Rabosky DL, Glor RE (2010) Equilibrium speciation dynamics in a model adaptive radiation of island lizards. Proc Nat Acad Sci USA 107: 22178–22183.
27. Rabosky DL (2010) Extinction rates should not be estimated from molecular phylogenies. Evolution 64: 1816–1824.
28. Rabosky DL (2012) Testing the time-for-speciation effect in the assembly of regional biotas. Methods Ecol Evol 3: 224–233.
29. Jonnson KA, Fabre P-H, Fritz SA, Etienne RS, Ricklefs RE, et al. (2012) Ecological and evolutionary determinants for the adaptive radiation of the Madagascan vangas. Proc Natl Acad Sci USA 109: 6620–6625.
30. Huelsenbeck JP, Larget B, Swofford D (2000) A compound Poisson process for relaxing the molecular clock. Genetics 154: 1879–1892.
31. Green PJ (1995) Reversible jump Markov chain Monte Carlo computation and Bayesian model determination. Biometrika 82: 711–732.
32. Rabosky DL, Donnellan SC, Talaba AL, Lovette IJ (2007) Exceptional among-lineage variation in diversification rates during the radiation of Australia's most diverse vertebrate clade. Proc R Soc Lond B Biol Sci 274: 2915–2923.
33. Bartolucci F, Scaccia L, Mira A (2006) Efficient Bayes factor estimation from the reversible jump output. Biometrika 93: 41–52.
34. Quental TB, Marshall CR (2009) Extinction during evolutionary radiations: reconciling the fossil record with molecular phylogenies. Evolution 63: 3158–3167.
35. Bokma F (2008) Detection of "punctuated equilibrium" by Bayesian estimation of speciation and extinction rates, ancestral character states, and rates of anagenetic and cladogenetic evolution on a molecular phylogeny. Evolution 62: 2718–2726.
36. Silvestro D, Schnitzler J, Zizka G (2011) A Bayesian framework to estimate diversification rates and their variation through time and space. BMC Evol Biol 11: 311.
37. Eastman JM, Alfaro ME, Joyce P, Hipp AL, Harmon LJ (2011) A novel comparative method for identifying shifts in the rate of character evolution on trees. Evolution 65: 3578–3589.
38. Venditti C, Meade A, Pagel M (2011) Multiple routes to mammalian diversity. Nature 479: 393–396.
39. Hastings WK (1970) Monte Carlo sampling methods using Markov chains and their applications. Biometrika 57: 97–109.
40. Metropolis N, Rosenbluth AW, Rosenbluth MN, Teller AH, Teller E (1953) Equations of state calculations by fast computing machines. J Chem Phys 21: 1087–1091.
41. Sisson SA (2005) Transdimensional Markov Chains. J Amer Statist Assoc 100: 1077–1089.
42. Yang Z (2006) Computational Molecular Evolution: Oxford University Press.
43. FitzJohn R, Maddison WP, Otto SP (2009) Estimating trait-dependent speciation and extinction rates from incompletely resolved phylogenies. Syst Biol 58: 595–611.
44. Brock CD, Harmon LJ, Alfaro ME (2011) Testing for temporal variation in diversification rates when sampling is incomplete and nonrandom. Syst Biol 60: 410–419.
45. Cusimano N, Renner SS (2010) Slowdowns in diversification rates from real phylogenies may not be real. Systematic Biology 59: 458–464.
46. Hohna S, Stadler T, Ronquist F, Britton T (2011) Inferring speciation and extinction rates under different sampling schemes. Mol Biol Evol 28: 2577–2589.
47. Stadler T (2013) Recovering speciation and extinction dynamics based on phylogenies. J Evol Biol doi: 10.1111/jeb.12139.
48. Rabosky DL, Lovette IJ (2008) Density-dependent diversification in North American wood warblers. Proc R Soc Lond B Biol Sci 275: 2363–2371.
49. Rabosky DL, Slater GJ, Alfaro ME (2012) Clade age and species richness are decoupled across the Eukaryotic tree of life. PLoS Biol 10: e1001381.
50. Harmon LJ, Weir JT, Brock C, Glor RE, Challenger WE (2008) GEIGER: Investigating evolutionary radiations. Bioinformatics 24: 129–131.
51. Steeman ME, Hebsgaard MB, Fordyce RE, Ho SYW, Rabosky DL, et al. (2009) Radiation of Extant Cetaceans Driven by Restructuring of the Oceans. Systematic Biology 58: 573–585.
52. Nee S, Holmes EC, May RM, Harvey PH (1994) Extinction rates can be estimated from molecular phylogenies. Philos Trans R Soc Lond B 344: 77–82.
53. Rabosky DL (2010) Primary controls on species richness in higher taxa. Systematic Biology 59: 634–645.
54. Etienne RS, Haegeman B (2012) A conceptual and statistical Framework for adaptive radiations with a key role for diversity dependence. Am Nat 180: E75–E89.
55. O'Meara BC (2012) Evolutionary inferences from phylogenies: a review of methods. Ann Rev Ecol Evol Syst 43: 267–285.
56. Bininda-Emonds ORP, Cardillo M, Jones KE, MacPhee RDE, Beck RMD, et al. (2007) The delayed rise of present-day mammals. Nature 446: 507–512.
57. Jetz W, Thomas GH, Joy JB, Hartmann K, Mooers A (2012) The global diversity of birds in space and time. Nature 491: 444–448.
58. Pyron RA, Burbrink FT, Wiens JJ (2013) A phylogeny and updated classification of Squamata, including 4161 species of lizards and snakes. BMC Evol Biol 13: 93.
59. Rabosky DL, Santini F, Eastman JM, Smith SA, Sidlauskas B, et al. (2013) Rates of speciation and morphological evolution are correlated across the largest vertebrate radiation. Nature Communications 4: 10.1038/ncomms2958.
60. Slater GJ, Price SA, Santini F, Alfaro ME (2010) Diversity versus disparity and the radiation of modern cetaceans. Proc R Soc Lond B Biol Sci 277: 3097–3104.
61. Quental TB, Marshall CR (2010) Diversity dynamics: molecular phylogenies need the fossil record. Trends Ecol Evol 25: 434–441.
62. Didier G, Royer-Carenzi M, Laurin M (2012) The reconstructed evolutionary process with the fossil record. J Theor Biol 315: 26–37.
63. Pybus OG, Harvey PH (2010) Testing macro-evolutionary models using incomplete molecular phylogenies. Proc R Soc B 267: 2267–2272.

13

A Metapopulation Approach to African Lion (*Panthera leo*) Conservation

Stephanie Dolrenry[1,2]*, Jennifer Stenglein[3], Leela Hazzah[1,2], R. Scott Lutz[3], Laurence Frank[2,4]

1 Nelson Institute for Environmental Studies, University of Wisconsin, Madison, Wisconsin, United States of America, **2** Living With Lions, Mpala Research Centre, Nanyuki, Kenya, **3** Forest and Wildlife Ecology, University of Wisconsin, Madison, Wisconsin, United States of America, **4** Museum of Vertebrate Zoology, University of California, Berkeley, California, United States of America

Abstract

Due to anthropogenic pressures, African lion (*Panthera leo*) populations in Kenya and Tanzania are increasingly limited to fragmented populations. Lions living on isolated habitat patches exist in a matrix of less-preferred habitat. A framework of habitat patches within a less-suitable matrix describes a metapopulation. Metapopulation analysis can provide insight into the dynamics of each population patch in reference to the system as a whole, and these analyses often guide conservation planning. We present the first metapopulation analysis of African lions. We use a spatially-realistic model to investigate how sex-biased dispersal abilities of lions affect patch occupancy and also examine whether human densities surrounding the remaining lion populations affect the metapopulation as a whole. Our results indicate that male lion dispersal ability strongly contributes to population connectivity while the lesser dispersal ability of females could be a limiting factor. When populations go extinct, recolonization will not occur if distances between patches exceed female dispersal ability or if females are not able to survive moving across the matrix. This has profound implications for the overall metapopulation; the female models showed an intrinsic extinction rate from five-fold to a hundred-fold higher than the male models. Patch isolation is a consideration for even the largest lion populations. As lion populations continue to decline and with local extinctions occurring, female dispersal ability and the proximity to the nearest lion population are serious considerations for the recolonization of individual populations and for broader conservation efforts.

Editor: Alfred L. Roca, University of Illinois at Urbana-Champaign, United States of America

Funding: The research was funded by National Science Foundation Fellowship Grant No. 2008058924 and Panthera Kaplan Awards. The funders had no role in study design, data collection and analysis, decision to publish, or preparation of the manuscript.

Competing Interests: The authors have declared that no competing interests exist.

* E-mail: stephanie@lionguardians.org

Introduction

African lions (*Panthera leo*) once roamed the greater part of the African continent. As habitat generalists, they occupied a wide range of biomes with the exception of tropical rainforests and the interior of the Sahara desert [1]. Over the past century, it is estimated that lions' range has been reduced by approximately 75% [2]. Across the majority of their present-day range, lion populations are now primarily associated with protected areas and managed hunting areas throughout sub-Saharan Africa [3].

Several strongholds of free-roaming lions remain in the East African countries of Tanzania and Kenya, home to more than half of the remaining lion population in Africa [2]. Lions are under threat in both countries, even though both countries have a strong tourism sector and Tanzania a strong trophy hunting sector, all based largely on lions [4,5]. Lions are declining primarily due to indiscriminate killing by humans [6,7], depletion of their prey base [8,9], and overexploitation due to poor management of trophy hunting [5,10]. Habitat conversion outside of protected areas has led to increasingly fragmented lion populations that are currently under threat of further isolation [11].

Although lions subsist in a wide variety of habitats, they are most successful in areas with low to medium human densities [12,13,14]. Due to anthropogenic pressures, the once nearly continuous network of lion populations across East Africa now exhibits a metapopulation structure: distinct populations within a wider landscape with limited migration between them [15,16]. Monitoring of East African lion populations has shown that there is limited dispersal between the populations (i.e. observed dispersal between Tsavo West, Chyulu Hills and Amboseli National Parks and Namanga Forest Reserve, Kenya as well as between Serengeti National Park, Ngorongoro Conservation Area, Tanzania and the Masai Mara Reserve, Kenya [17,18,19]. Furthermore, lion populations have been observed to go extinct and recolonize (e.g. Amboseli National Park in the late 1990s [20], Sibiloi National Park (J. Harris, *pers. comm.*)) or to be 'rescued' by the immigration of new individuals after a population crash [19]. Although broader scale range maps show lions as continuous across much of East Africa, finer scale country-wide maps of known breeding or permanent lion populations reveal a distinct metapopulation pattern [2,21,22,23]. Conservation planning for lions across East Africa requires an understanding of this network of, and exchange between, lion populations (i.e., the metapopulation).

Metapopulation analysis is an effective tool to better understand broad area population dynamics and the effects of species-specific life-history traits on population connectivity [24]. Current wildlife conservation and management benefit from a metapopulation approach because numerous wildlife populations are becoming increasingly isolated with regional extinction imminent for many species [25]. A metapopulation functions at a larger scale than individual populations and can provide further insight into the dynamics of each component population in reference to the system as a whole [26,27].

Migrating individuals can re-colonize suitable patches after a local extinction has occurred, as well as enable a "rescue effect" in which immigrating individuals protect a dwindling local population from extinction [28,29]. Successful recolonization is dependent on the dispersing individuals surviving 'the matrix', an area between populations which is generally not suitable for long-term survival and reproduction [30,31,32]. Traversing the matrix poses the largest threat for dispersing migrants, especially for large carnivores that are inherently in conflict with humans [30,33,34,35].

Understanding the broader population dynamics of large carnivores can be particularly challenging given their longevity, large home ranges, and the lack of empirical data on dispersal abilities, particularly dispersal between habitat patches [30,36]. Reliable knowledge of the species' dispersal ability enables conservationists to maintain viable populations within proximity to other populations, thus ensuring sustainable exchange of individuals [37]. Although dispersal is one of the most important ecological processes, it remains one of the least understood [38]. Particularly with current rates of fragmentation and isolation, it is increasingly important to understand species' dispersal characteristics and their role in maintaining the larger metapopulation [25].

As with other polygynous mammalian species [39], male lions nearly always disperse from their natal area, often traveling two to three times farther than females [18,40,41]. Females generally stay with their natal pride or establish neighboring home ranges ([18], 'stepping stone' dispersal – [42]). Based on these generally accepted dispersal characteristics, we combine empirical data on the life-history traits and current geographic status of lion populations in a fragmented landscape with a spatially realistic modeling approach [29,43,44] to examine the lion metapopulation of Kenya and Tanzania.

We use an incidence function model (IFM) to explore the probability of occurrence of particular lion populations (i.e. patches) within a broader lion metapopulation, where both distance between patches and area of patches vary [43]. We use the IFM to consider the effect of sex-specific dispersal characteristics on metapopulation connectivity, the impact of human densities, and the threat of isolation on the remaining lion populations [45] across Kenya and Tanzania.

Materials and Methods

Modeling Approach

Reliable models are insightful for conservation management as they can help managers evaluate the current state and consider a system's future [44]. The simple and applicable incidence function model (IFM; [43,46]) has high conservation value for evaluation of populations on a broad-scale [43]. The IFM was created to use empirical data to examine the effects of patch area and isolation on patch occupancy and is one of the most commonly used metapopulation models. An IFM can be built with snapshot presence/absence data of a species at a particular site, the simplest form of data that can be collected during field studies.

Incidence function models assume that suitable habitat occurs in discrete patches surrounded by unsuitable matrix, and that occupancy of each patch is determined by local colonization and extinction events [29,45]. Extinction is negatively associated with patch area and colonization is negatively associated with patch isolation, so patch occupancy should increase with area and decrease with isolation [47]. Patch extinction and colonization are also assumed to depend on factors such as patch area (a proxy for local population size), spatial arrangement of patches, patch edge characteristics, dispersal ability of the species, and regional environmental stochasticity. In IFM models, the incidence is given as a function of connectivity and patch size [43]; in this study, factors such as lion occupancy and male and female dispersal distances, were calculated from empirical data on lion populations as well as verified metadata from various sources. Given the relatively high number of overall patches and the fraction of occupied patches during the short 'snapshot' period, we assumed the metapopulation was at equilibrium and thus executed the IFM [43].

To fit the IFM, we used a special case of generalized linear models with binomial error and logistic link function on the response of incidence (i.e. rate of occurrence) data for each habitat patch [48]. The necessary data inputs were incidence data (J_i; 1 for presence, 0 for absence), area of each patch (A_i), and coordinates of the center of each patch. We calculated all pairwise distances between patches (d_{ij}), a dispersal parameter (α), estimated from actual dispersal data equal to 1/average dispersal as well as 1/maximum dispersal, and a connectivity value (S_i). Connectivity (S_i) is a function of occupancies p_i, patch sizes A_i, patch distances d_{ij} and the species specific dispersion length parameter α. We used the negative exponential functional form of S_i as suggested by Hanski [43].

$$S_i = \sum_{i \neq j} e^{(-\alpha d_{ij}) p_i A_i}$$

Due to impacts of human density on lion populations [13,14], we included human density as a covariate in the incidence function models. Given the intense research on lion populations across Kenya and Tanzania as well as the high detectability of a breeding lion population when present (e.g. obvious tracks, roaring, livestock depredation, etc.), we believe the occurrence of false negatives (Type II error) to be negligible [15,49,50]. As discussed in Riggio et al. [2], the use of expert opinions and regional surveys for data of lion populations, particularly across East Africa, is likely to result in an inflation of lion range, not an underestimation.

Data for the Model

We used the most recent and authoritative sources to identify and map known suitable habitat patches for lion populations from both countries, Mésochina et al. [22] for Tanzania and the Conservation and Management Strategy for Lions, Kenya (Kenya Wildlife Service [21]). In the Mésochina et al. [22] report, lion populations were identified by reported frequency of observation; weekly and monthly sightings were considered permanent populations, and we omitted areas for which there was no information available. In the Kenya report, we mapped all populations labeled as 'known permanent'. Henceforth, we use the term 'patch' as synonymous with a distinct patch that was known to be suitable to support a lion population. In total, there were 25 patches with areas ranging from 86 to 127,515 km^2 (Fig. 1).

Once all patches were identified, we sent maps to regional experts (conservationists, biologists, etc.) who were familiar with current lion population status in their areas. Experts identified areas that, to the best of their knowledge, contained a breeding lion population in December 2011-January 2012; they did not include areas where only male lions or dispersers were seen

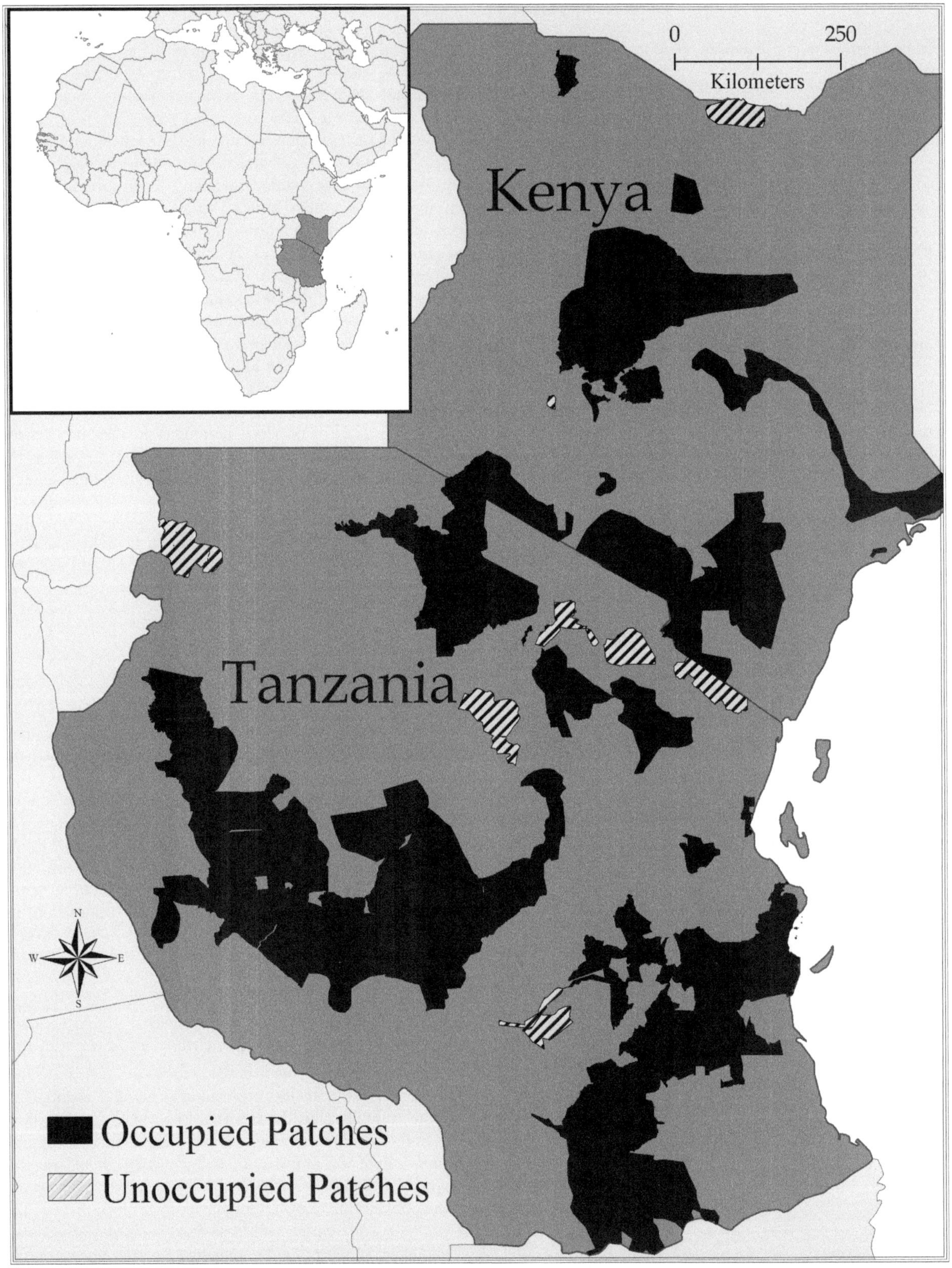

Figure 1. Map of study area in Africa. Darkened areas indicate patches of permanent lion populations (n = 25) across Kenya and Tanzania; black areas were considered occupied and striped areas were deemed unoccupied at time of survey.

infrequently. This method gave a current presence-absence scenario (i.e. the 'snapshot') for patch occupancy.

We mapped and measured (km^2) patches using ArcMap 10.1 [51], and used the centroid of each patch to calculate the pairwise

Euclidean distances between patches (d_{ij}). We subtracted the radii of the pair of patches for each pairwise distance, so the distances could be interpreted as the shortest straight-line distance between the edges of the paired patches. We gathered actual dispersal distances from the literature, unpublished reports, and the Lion Guardians program in the Amboseli ecosystem [17,40,41,Desert Lion [52,53]. We used the maximum and average reported dispersal distances of 343 and 117 km for males and 128 and 50 km for females respectively as the dispersal parameter (α). We calculated human density for each patch from the 2009 census of Kenya (Kenya Bureau of Statistics [54]) at the sub-location level and the 2002 census for Tanzania (National Bureau of Statistics Tanzania [55]) at the ward level. We averaged the human density for each patch area.

Statistical Methods

We followed the methods outlined in Hanski [43] and Oksanen [48] and performed the IFM in program R, version 2.14.0 [56]. We included the rescue effect in the model and added the mean human density for each of the patches as a covariate in the model. Therefore, our full model added one new term to Eq (9) in Oksanen [48]:

$$\log\left(\frac{J_i}{1-J_i}\right) = -\log(ey)+2\log(S_i)+x\log(A_i)+\log(H_i)$$

$$\text{logit}(J_i) = \beta_0+2\log S+\beta_1 * \log A+\beta_2 * \log H$$

where J_i is the incidence in patch i, S_i is a measure of connectivity, A_i is each patch's area, and H_i is the mean human density of each patch. When fitting the full model, the intercept, β_0 was log(ey), β_1 was an estimate of the stochasticity parameter $\hat{x}$, and β_2 was the parameter estimate for human density entering the model as a linear factor.

Additional model parameters were: A_0, the minimum suitable area, also referred to as critical patch area, and $\hat{x}$, a stochasticity parameter. The critical patch area is the minimum patch size observed for occupancy [43]; it is the smallest area found to contain a lion population. When $\hat{x}$ is large (>1), there is a range of patch sizes beyond which extinction becomes very unlikely, whereas if $\hat{x}$ is small (<1), there is no such critical patch size and even large populations in large patches have a substantial risk of extinction. Lastly, we included the parameters e, the intrinsic extinction rate, and y', the colonizing ability [43].

We used $\hat{x}$ and A_0, to find $\hat{e}$:

$$\hat{e} = A_0^x$$

Then we were able to use β_0 and $\hat{e}$ to solve for y' following Oksanen (2004):

$$-\log(ey) = \beta_0$$

$$ey = \exp(-\beta_0)$$

$$\hat{y} = \frac{\exp(-\beta_0)}{A_0^x}$$

We used Akaike's Information Criterion (AIC) to choose between nested models [57]. We ran male and female models independently, using average dispersal distance and maximum dispersal distance for each sex separately. The full models included human density additive to the patch area and the reduced models did not include human density. We used the connectivity term, $2\ln(S_i)$, as an offset in all eight models [48].

Results

There were 25 total patches with an average area of 13,898 (SD 28,639) km^2, and 60% of the patches were occupied (Fig.1). Mean human density value per patch was 162 (SD 475) per km^2. The average distance among all patches was 546 km, varying from 63 to 1,461 kilometers (Fig. 2).

The male and female lion models without the human density were only marginally stronger than the models with human density included (Table 1). We found significant differences in connectivity values between the male and female models using both maximum dispersal distance ($t = 21.16$, $p<0.001$, df = 24) and average dispersal distances ($t = 8.71$, p<0.001, df = 24) as well as between the male models (avg male to max male $t = 20.38$, p<0.001, df = 24) and female models (avg female to max female t = 9.08, p<0.001, df = 24). As was expected from the model inputs, the greater ability of males to move through the matrix (male max $a = 0.0029$, avg $a = 0.0086$ versus female max $a = 0.0078$, avg $a = 0.02$), gave the male models reliably higher connectivity (Table 1). Three patches (3, 8, and 9; Fig. S1) had the lowest connectivity values across all models (Table S1). Particularly for the two largest patches (patches 3 and 8 with areas >40,000 km^2; Fig. S1), the difference between the connectivity values of the two sexes was apparent (male max: mean (SD) = 14.78 (1.32), male avg = 0.72 (0.45); female max: mean (SD) = 1.02 (0.56), female avg = 0.01 (0.01); n = 2).

We used the estimated model parameters (Table 1) to determine probability of occupancy of the individual patches (Fig. 3). With regards to area size and its effect on patch occupancy, moderate-sized patches (2,000< area <40,000 km^2) all had a relatively high probability of incidence (patch incidence J_i: mean (SD) = 0.74 (0.26), n = 15), while the smaller patches (area ≤2,000 km^2) had lower probabilities of lion occurrence (J_i: mean (SD) = 0.40 (0.24), n = 8). Surprisingly, the largest patches (area ≥40,000 km^2) also had much lower occurrence probabilities (J_i: mean (SD) = 0.36 (0.21), n = 2). Across all scenarios, female models had a lower mean probability of patch incidence (max null models: mean (SD) = male 0.60 (0.18), female 0.57 (0.30); max human density models: male 0.55 (0.18), female 0.48 (0.30); avg null models: male 0.56 (0.31), female 0.43 (0.40); avg human density models: male 0.39 (0.31), female 0.25 (0.41)).

Adding the human density parameter did not influence either male or female models; the influence of human density in the reduced models was not significant ($p = 0.81$ and 0.32 for male models and 0.67 and 0.33 for the female models, using maximum and average dispersal distances). However, the human density parameter did decrease colonization for both sexes (colonization parameter, C; male max models $t = -2.82$, $p = 0.007$, male avg models $t = -4.49$, $p = <0.0001$; female max models $t = -2.91$, $p = 0.005$, female avg models $t = -3.32$, $p = 0.0017$, at alpha level 0.05).

Furthermore, the patches with the highest associated human density (>2,000 people/km^2, n = 2), had a decreased probability of occurrence in the models with human density added (mean = 0.60 to 0.55 in male max models; 0.57 to 0.39 in male avg models and 0.57 to 0.48 in female max models; 0.43 to 0.25 in female avg models, Fig. 3).

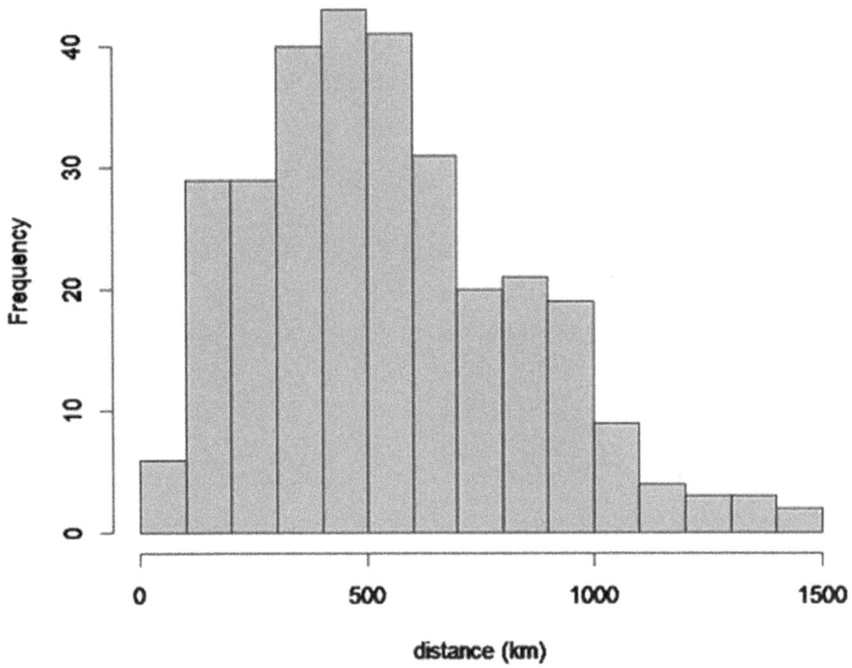

Figure 2. Pairwise distances between 25 Kenya and Tanzania lion population patches. Distance was calculated in kilometers, using data obtained during 2008–2010.

Discussion

The present-day fragmentation of the East African lion population into patches fits the metapopulation framework. Population patches are well delimited and cover only a fraction of the total area, allowing us to apply a metapopulation approach to the network of remaining lion populations across Kenya and Tanzania. In particular, we considered the impact of sex-specific dispersal characteristics on regional lion populations. With only one snapshot of population occupancy, strong differences were apparent between male and female dispersal abilities on connectivity and probabilities of occurrence.

Male lions' ability to 'rescue' declining populations over 300 km away was strong, particularly since dispersal ability is pliant [58,59]. If occupied patches, i.e. lion populations, are maintained and migration is allowed, then the males' strong rescue ability allowed for higher probabilities of patch occupancy and stronger values of connectivity [60]. Lion survival while in the matrix and a constant flow of male dispersers from stable populations are important. If male lions are not able to disperse from stable populations, as may be the case where adult male survival is low, i.e., sport hunting areas [5,61,62,63], this could result in a lower rescue effect for the broader metapopulation, causing an increased risk of extinction for local populations [28].

Fragmentation and isolation are among the key challenges to metapopulation maintenance [25,64,65]. Our results suggest that even the largest patches were negatively affected by isolation as

Table 1. Model outputs of eight African lion incidence function models.

	Mean log (Human Density) (SE)	Residual Deviance	Residual DF	AIC	$\tilde{x}$(SE)	*e*	*y′*	$\bar{S}$(SD)
Max Male (−H)	NA	35.08	23	39.08	0.44 (0.25)	7.04	3002.46	33.06 (10.52)
Max Male (+H)	−0.06 (0.24)	35.03	22	41.03	0.44 (0.24)	7.05	2381.55	33.06 (10.52)
Max Female (−H)	NA	41.61	23	45.61	0.82 (0.28)	38.40	470.13	8.53 (5.03)
Max Female (+H)	−0.11(0.25)	41.43	22	47.43	0.82 (0.28)	38.96	303.44	8.53 (5.03)
Avg Male (−H)	NA	42.95	23	46.95	0.88 (0.29)	51.14	397.48	7.92 (4.5)
Avg Male (+H)	−0.30 (0.30)	41.96	22	47.96	0.91 (0.30)	58.43	143.70	7.92 (4.5)
Avg Female (−H)	NA	69.99	23	73.99	1.83 (0.41)	3491.25	79.08	1.19 (1.08)
Avg Female (+H)	−0.40(0.41)	68.94	22	74.94	1.95 (0.49)	6008.43	23.57	1.19 (1.08)

Male and female models were calculated separately as they used different alpha values (male and female maximum (Max) and average (Avg) observed dispersal distances). Additional models were calculated with (+) and without (−) human density (H) as a covariate, added to both male and female models using both average and maximum dispersal distances. To discern model fit, human density coefficient with standard error (SE), residual deviance, residual degree of freedom and Akaike's Information Criteria (AIC) are reported for each model. Model parameters reported: $\tilde{x}$ a stochasticity parameter reported with standard error; *e*, the intrinsic extinction rate; *y′*, the colonizing ability; and a measure of connectivity $\bar{S}$, reported as an average for each model with standard deviation.

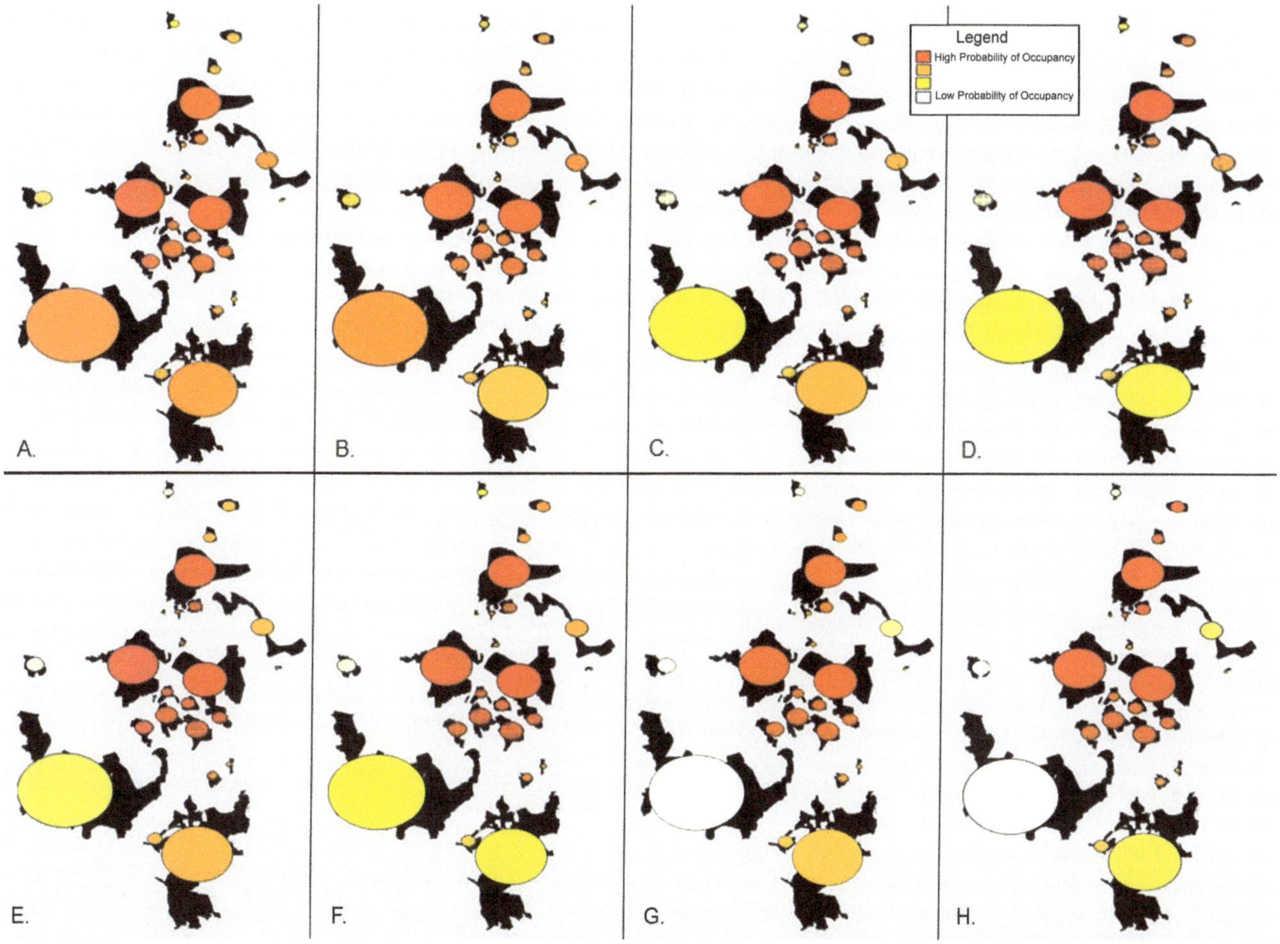

Figure 3. Probability of lion occurrence in identified lion population patches. Occurrence was calculated on 25 habitat patches in Kenya and Tanzania based on the incidence function model for A) male dispersal using maximum observed distance, without human density, B) male dispersal using maximum observed distance with human density, C) female dispersal using maximum observed distance without human density, D) female dispersal using maximum observed distance with human density model, E) male dispersal using average of observed distances, without human density, F) male dispersal using average of observed distances with human density, G) female dispersal using average of observed distances without human density, H) female dispersal using average of observed distances with human density model. In plotting the results, we used color to show the predicted incidence with warmer colors (redder) showing higher probability of incidence and yellow to white showing lower probability of incidence.

shown by the lower occurrence probabilities and the low connectivity values observed for the two largest patches (3 and 8, Fig. S1). These areas are known to hold the majority of the remaining lions found in East Africa [2,22]. However, because the IFM approach does not consider population numbers, the results are based solely on considerations of patch area and proximity to other patches.

Based on the model outputs, the higher probabilities of occurrence for the patches of northern Tanzania and southern Kenya (Fig. 3 and Table S1), concur with other studies [27] that a centrally located network of moderate-sized populations has greater probabilities of occurrence than either the few large isolated populations or the smaller well-connected patches. When examining a network of lion populations, considerations of patch isolation and fragmentation are vitally important, in conjunction with local-level considerations of individual population size and viability.

To avoid local extinction due to genetic isolation, it is necessary for a minimum of one or more dispersers in each generation to survive the matrix and recolonize other patches, allowing populations to persist in stochastic environments [27,30,31,66,67,68,69]. The models showed males to have poor 'colonizing ability' (y') in comparison to females; this 'colonizing ability' is independent of the ability to move across the matrix (S_i) [43]. Our y' estimates suggest that even though male lions can move across the matrix better than female lions (increased ability to disperse), when compared to females, they have a harder time establishing themselves in a patch. This makes sense biologically since males are more nomadic and move territories frequently whereas females disperse less often and shorter distances than males. Females choose to generally stay within their natal range or establish neighboring home ranges; once they reach a suitable patch, they tend to stay there [18,40,70].

Once patch extinction has occurred, recolonization would be dependent upon female lions, less capable dispersers, being able to reach the empty patch. Since female lion dispersal ability is limited, when populations go extinct, recolonization cannot occur if distances between patches exceed female dispersal ability or if females are not able to survive crossing the matrix. This has profound implications for the overall metapopulation of lions; the female models showed an intrinsic extinction rate from five-fold to a hundred-fold higher than the male models (Table 1).

Dispersal occurs in and is dependent upon the matrix, in this case largely non-protected, human-dominated landscapes [32]. Little is known about lion dispersal and survival in human-dominated areas. We suggest conservation efforts for lions in East Africa focus on increasing lion survival while crossing the matrix to ensure connectivity on a broad-scale. Additionally, conservation efforts could aim to improve accuracy of regional lion presence/absence data and the knowledge of occurrence and status of breeding lion populations through the establishment of a lion sightings database, as is done with other species (e.g. avian sightings databases [71,72]). Observational data on lions across East Africa would increase our understanding of the distribution of lions and provide a foundation for more effective lion conservation.

As top predators, lions are especially vulnerable and even minor human-induced changes (either negative or positive) can have large impacts [73,74,75]. Wildlife or stakeholder carrying capacity (i.e. the amount of wildife local residents are willing to 'live with' - [76,77]) is a concept that aims to increase tolerance of local communities, as a step toward increasing the number or density of lions that communities are able to tolerate or accept. Traditional-based participatory monitoring programs have been shown to increase local tolerance of lions [78], thus allowing for greater connectivity.

Improving human tolerance to allow movement of even a few individuals through non-protected or partially protected areas could play a vital role in conservation of carnivores, both to maintain populations and to allow greater exchange between them [78,79]. Even though protected areas are necessary for long-term persistence [73], alone they are not enough as most of the lands essential for supporting carnivore dispersal and thus population connectivity, are outside protected areas, in lands that are affected by humans [1,80].

In addition to suitable habitat and available prey, human tolerance for carnivores can create corridors between larger stable unfenced populations. Packer et al. [81] found fenced lion populations to occur at higher densities, but fences inhibit all exchange between populations, leading to further ecosystem fragmentation, loss of dispersal and migration routes, genetic isolation as well as reduced conservation value of wildlife in matrix areas [82]. If lions are seen as a benefit by the human communities living in the matrix, they will tolerate sharing the landscape with a challenging species allowing greater population viability, increased economic opportunities for the communities and better conservation.

Even though human density was not a significant factor for either model, we suspect that further study (i.e. subsequent surveys over longer time periods) would reveal longer-term impacts of human density on extinction and/or colonization rates. However, greater community tolerance and adaptive management policies may be more important factors than human population densities [35,79,83]. We also found the stochasticity parameter ($\hat{x}$) to be less than one for all models, implying that in this metapopulation, no lion population is immune from going extinct [84], further suggesting the need and urgency for conservation action on a broad geographic scale.

Incidence function models allowed a simple initial evaluation of lion populations and their connectivity across Kenya and Tanzania. Using only data on areas, locations and patch occupancies, we were able to analyze the effects of patch area, isolation and dispersal abilities on lion populations on a metapopulation scale. More explicit examination of extinction and colonization rates of the patches as well as further exploration of factors affecting dispersal and survival in the matrix areas would provide additional insights for making effective policy decisions about large-scale lion populations across Africa. Extirpation of lions throughout Africa is an immediate concern and the metapopulation approach and the incorporation of life-history traits are essential tools for understanding the broader networks of populations and movements between them.

Supporting Information

Figure S1 Map of study area in Africa. Darkened areas indicate patches of permanent lion populations (n = 25) across Kenya and Tanzania; black areas were considered occupied and striped areas were deemed unoccupied at time of survey. In this map, patches were numbered 1 to 25, corresponding with Patch ID of Table S1.

Table S1 Model outputs from eight incidence function models, varying the dispersal distance used by either male or female, maximum observed (Max) or average of observed (Avg) as well as the incorporation of human density (Hum) as a covariate in the models. Models without human density covariate are labeled as Null. Patches were given unique identification numbers 1 through 25. Each patch had an associated area (km^2) and human density (number people per km^2) variable. Outputs were: S_i as an estimate of connectivity associated with both male and female, maximum and average dispersal distance models; J_i as an estimate of patch occurrence, E estimates of patch extinction, and C estimate of patch colonization. All are given for the eight models and twenty-five patches.

Acknowledgments

This study would not have been possible without the knowledge and involvement of J. Cary, A. Norton, C. Packer, P. Henschel, H. Brinks, P. Schuette, A. Estes, S. Bhalla, A. Cotterill, and J. King. The manuscript was greatly improved by the valuable input of C. Ribic. The research was conducted under Kenya Research Permit No. MOEST 13/C689.

Author Contributions

Conceived and designed the experiments: SD JS. Performed the experiments: SD JS. Analyzed the data: SD JS. Contributed reagents/materials/analysis tools: SD JS LH LF. Wrote the paper: SD JS LH RSL LF.

References

1. Nowell K, Jackson P (1996) Wild cats: status survey and conservation action plan; IUCN, editor. Gland, Switzerland: IUCN/SSC Cat Specialist Group. 382 p.
2. Riggio J, Jacobson A, Dollar L, Bauer H, Becker M, et al. (2012) The size of savannah Africa: a lion's (Panthera leo) view. Biodiversity and Conservation: 1–19.
3. Bauer H, Nowell K, Packer C (2008) IUCN Red List of Threatened Species. IUCN 2012 Version 2012.1.
4. Lindsey PA, Balme GA, Booth VR, Midlane N (2012) The significance of African lions for the financial viability of trophy hunting and the maintenance of wild land. PLoS ONE 7: e29332.
5. Packer C, Kosmala M, Cooley HS, Brink H, Pintea L, et al. (2009) Sport hunting, predator control and conservation of large carnivores. PLoS ONE 4: e5941.

6. Hazzah L, Borgerhoff Mulder M, Frank L (2009) Lions and warriors: social factors underlying declining African lion populations and the effect of incentive-based management in Kenya. Biological Conservation 142: 2428–2437.
7. Ikanda D, Packer C (2008) Ritual versus retaliatory killing of African lions in the Ngorongoro Conservation Area, Tanzania. Endangered Species Research 6: 67–74.
8. Ogutu JO, Owen-Smith N, Piepho HP, Said MY (2011) Continuing wildlife population declines and range contraction in the Mara region of Kenya during 1977–2009. Journal of Zoology 285: 99–109.
9. Bauer H (2008) Synthesis of threats, distribution and status of the lion from the two lion conservation strategies. In: B. Croes RB, H. de Iongh and H Bauer, editor. Management and conservation of large carnivores in west and central Africa. Leiden: Institute of Environmental Sciences, Leiden University. 13–28.
10. Packer C, Brink H, Kissui B, Maliti H, Kushnir H, et al. (2010) Effects of trophy hunting on lion and leopard populations in Tanzania. Conservation Biology 25: 142–153.
11. Bauer H, Van Der Merwe S (2004) Inventory of free-ranging lions (Panthera leo) in Africa. Oryx 38: 26–31.
12. Riggio J (2011) The African lion (Panthera leo leo): a continent-wide species distribution study and population analysis [*Master's thesis*]: Duke University. 101 p.
13. Loveridge A, Canney S (2009) African lion distribution modeling project. Horsham, UK: WildCRU. 58 p.
14. Woodroffe R (2000) Predators and people: using human densities to interpret declines of large carnivores. Animal Conservation 3: 165–173.
15. Hanski IA, Simberloff D (1997) The metapopulation approach, its history, conceptual domain, and application to conservation. In: Hanski IA, Gilpin ME, editors. Metapopulation Biology: Ecology, Genetics and Evolution. London: Academic Press Limited. 5–26.
16. McCollough D, editor (1996) Metapopulations and wildlife conservation: Island Press.
17. Dolrenry S (2013) African lion (Panthera leo) behavior, monitoring, and survival in human-dominated landscapes [*Dissertation*]. Madison: University of Wisconsin.
18. Schaller G (1972) The Serengeti lion: a study of predator-prey relations. Chicago: The University of Chicago Press.
19. Packer C, Pusey A, Rowley H, Gilbert D, Martenson J, et al. (1991) Case study of a population bottleneck: lions of the Ngorongoro Crater. Conservation Biology 5: 219–230.
20. Chardonnet P (2002) Conservation of the African lion: contribution to a status survey. Paris, France: International Foundation for the Conservation of Wildlife.
21. Force KsNLCT (2008) Conservation and management strategy for lions and spotted hyaenas in Kenya 2009–2014. Nairobi, Kenya: Kenya's National Large Carnivore Task Force.
22. Mésochina P, Mbangwa O, Chardonnet P, Mosha R, Mtui B, et al. (2010) Conservation status of the lion in Tanzania. Paris: MNRT-WD, MLDF, TAWIRI, TAHOA & TPHA.
23. IUCN (2006) Regional conservation strategy for the lion (*Panthera leo*) in eastern and southern Africa. Gland, Switzerland: IUCN.
24. Driscoll DA (2007) How to find a metapopulation. Canadian Journal of Zoology 85: 1031–1048.
25. Wiens JA (1996) Wildlife in patchy environments: metapopulations, mosaics, and management. In: McCollough D, editor. Metapopulations and Wildlife Conservation. Washington, DC: Island Press. 53–84.
26. Hanski I, Gilpin M, editors (1997) Metapopulation biology: ecology, genetics, and evolution. San Diego, California: Academic Press. 512 p.
27. Stacey PB, Johnson VA, Taper ML (1997) Migration within metapopulations. In: Hanski IA, Gilpin ME, editors. Metapopulation biology: ecology, genetics, and evolution. San Diego: Academic Press. 215–246.
28. Brown JH, Kodric-Brown A (1977) Turnover rates in insular biogeography: effect of immigration on extinction. Ecology 58: 445–449.
29. Hanski I (1999) Metapopulation ecology. New York: Oxford University Press. 313 p.
30. Vandermeer J, Carvajal R (2001) Metapopulation dynamics and the quality of the matrix. The American Naturalist 158: 211–220.
31. Jules ES, Shahani P (2003) A broader ecological context to habitat fragmentation: why matrix habitat is more important than we thought. Journal of Vegetation Science 14: 459–464.
32. Ricketts TH (2001) The matrix matters: effective isolation in fragmented landscapes. The American Naturalist 158: 87–99.
33. Hanski I, Zhang D-Y (1993) Migration, metapopulation dynamics and fugitive co-existence. Journal of Theoretical Biology 163: 491–504.
34. Treves A, Naughton-Treves L (1999) Risk and opportunity for humans coexisting with large carnivores. Journal of Human Evolution 36: 275–282.
35. Treves A, Karanth KU (2003) Human-carnivore conflict and perspectives on carnivore management worldwide. Conservation Biology 17: 1491–1499.
36. Hellgren EC, Onorato DP, Skiles JR (2005) Dynamics of a black bear population within a desert metapopulation. Biological Conservation 122: 131–140.
37. Verner J (1992) Data needs for avian conservation biology: have we avoided critical research? Condor: 301–303.
38. Bowler DE, Benton TG (2005) Causes and consequences of animal dispersal strategies: relating individual behaviour to spatial dynamics. Biological Reviews 80: 205–225.
39. Greenwood PJ (1980) Mating systems, philopatry and dispersal in birds and mammals. Animal Behaviour 28: 1140–1162.
40. Pusey AE, Packer C (1987) The evolution of sex-biased dispersal in lions. Behaviour 101: 275–310.
41. Funston PJ (2011) Population Characteristics of Lions (Panthera leo) in the Kgalagadi Transfrontier Park. South African Journal of Wildlife Research 41: 1–10.
42. Kimura M (1953) Annual Report National Institute of Genetics; "Stepping Stone" model of population. Japan. 62–63 p.
43. Hanski I (1994) A practical model of metapopulation dynamics. Journal of Animal Ecology 63: 151–162.
44. Gilpin M (1996) Metapopulations and wildlife conservation: approaches to modeling spatial structure. In: McCollough D, editor. Metapopulations and Wildlife Conservation. Washington, DC: Island Press. 11–27.
45. Moilanen A, Hanski I (1998) Metapopulation dynamics: effects of habitat quality and landscape structure. Ecology 79: 2503–2515.
46. Hanski I, Kuussaari M, Nieminen M (1994) Metapopulation structure and migration in the butterfly (Melitaea cinxia). Ecology 75: 747–762.
47. Hanski I (1994) Patch-occupancy dynamics in fragmented landscapes. Trends in Ecology and Evolution 9: 131–135.
48. Oksanen J (2004) Incidence function model in R.
49. Mackenzie D, Nichols J, Lachman G, Droege S, Royle J, et al. (2002) Estimating site occupancy rates when detection probabilities are less than one. Ecology 83: 2248–2255.
50. Moilanen A, Nieminen M (2002) Simple connectivity measures in spatial ecology. Ecology 83: 1131–1145.
51. ESRI (2011) ArcGIS Desktop. In: Institute ESR, editor. Release 10 ed. Redlands, CA.
52. Conservation DL (2007) Dispersal and expansion. Desert Lion Conservation.
53. ALWG (2003, 2004) Lion Conservation Research. Workshops 3 & 4: From Conflict to Socioecology. In: Loveridge A, Lynam T, Macdonald DM, editors. Oxford: Widlife Conservation Research Unit. 1–95.
54. KNBS (2009) Population, Households and Density. In: Statistics KNBo, editor. Nairobi, Kenya: Kenya Open Data.
55. Statistics NBo (2002) Dar Es Salaam: National Bureau of Statistics of Tanzania.
56. Team RC (2011) R: a language and environment for statistical computing. 2.14.2 ed. Vienna, Austria: R Development Core Team.
57. Burnham KP, Anderson DR (2002) Model selection and inference: a practical information-theoretic approach. New York: Springer.
58. Björklund M (2003) The risk of inbreeding due to habitat loss in the lion (*Panthera leo*). Conservation Genetics 4: 515–523.
59. Van Dyck H, Baguette M (2005) Dispersal behaviour in fragmented landscapes: routine or special movements? Basic and Applied Ecology 6: 535–545.
60. Reed JM, Levine SH (2005) A model for behavioral regulation of metapopulation dynamics. Ecological Modelling 183: 411–423.
61. Loveridge AJ, Searle AW, Murindagomo F, Macdonald DW (2007) The impact of sport-hunting on the population dynamics of an African lion population in a protected area. Biological Conservation 134: 548–558.
62. Becker MS, Watson FGR, Droge E, Leigh K, Carlson RS, et al. (2012) Estimating past and future male loss in three Zambian lion populations. The Journal of Wildlife Management.
63. Yamazaki K (1996) Social variation of lions in a male-depopulated area in Zambia. The Journal of Wildlife Management 60: 490–497.
64. Prugh LR, Hodges KE, Sinclair ARE, Brashares JS (2008) Effect of habitat area and isolation on fragmented animal populations. Proceedings of the National Academy of Sciences 105: 20770–20775.
65. Hanski I (1997) Predictive and practical metapopulation models: the incidence function approach. Spatial ecology: the role of space in population dynamics and interspecific interactions Monographs in Population Biology 30: 21–25.
66. Mills LS, Allendorf FW (1996) The one-migrant-per-generation rule in conservation and management. Conservation Biology 10: 1509–1518.
67. Beier P, Barrett RH (1993) The cougar in the Santa Ana mountain range, California.
68. Fahrig L, Merriam G (1985) Habitat patch connectivity and population survival. Ecology: 1762–1768.
69. Hanski I, Gilpin M (1991) Metapopulation dynamics: brief history and conceptual domain. Biological journal of the Linnean Society 42: 3–16.
70. Hanby J, Bygott J (1987) Emigration of subadult lions. Animal Behaviour 35: 161–169.
71. Sullivan BL, Wood CL, Iliff MJ, Bonney RE, Fink D, et al. (2009) eBird: a citizen-based bird observation network in the biological sciences. Biological Conservation 142: 2282–2292.
72. Wood C, Sullivan B, Iliff M, Fink D, Kelling S (2011) eBird: engaging birders in science and conservation. PLoS biology 9: e1001220.
73. Woodroffe R, Ginsberg JR (1998) Edge effects and the extinction of populations inside protected areas. Science 280: 2126–2128.
74. Ray J, Redford KH, Steneck RS, Berger J, editors (2005) Large carnivores and the conservation of biodiversity. Washington, D.C.: Island Press.
75. Ray J, Hunter L, Zigouris J (2005) Setting conservation and research priorities for larger African carnivores. New York: Wildlife Conservation Society.
76. Carpenter LH, Decker DJ, Lipscomb JF (2000) Stakeholder acceptance capacity in wildlife management. Human Dimensions of Wildlife 5: 5–19.
77. Decker DJ, Purdy KG (1988) Toward a concept of wildlife acceptance capacity in wildlife management. Wildlife Society Bulletin 16: 53–57.

78. Hazzah L, Dolrenry S, Naughton L, Edwards C, Kearney F, et al. (*in press*) Participation and payments: efficacy of two lion conservation programs in Maasailand, Kenya.
79. Linnell JDC, Swenson JE, Anderson R (2001) Predators and people: conservation of large carnivores is possible at high human densities if management policy is favourable. Animal Conservation 4: 345–349.
80. Crooks KR, Burdett CL, Theobald DM, Rondinini C, Boitani L (2011) Global patterns of fragmentation and connectivity of mammalian carnivore habitat. Philosophical Transactions of the Royal Society B: Biological Sciences 366: 2642–2651.
81. Packer C, Loveridge A, Canney S, Caro T, Garnett S, et al. (2013) Conserving large carnivores: dollars and fences. Ecology letters.
82. Creel S, Becker MS, Durant SM, M'Soka J, Matandiko W, et al. (2013) Conserving large populations of lions – the argument for fences has holes. Ecology letters: n/a–n/a.
83. Allendorf TD, Allendorf K (2012) What every conservation biologist should know about human population. Conservation Biology 26: 953–955.
84. Hanski I (1992) Inferences from ecological incidence functions. American Naturalist: 657–662.

14

Crisis of Japanese Vascular Flora Shown By Quantifying Extinction Risks for 1618 Taxa

Taku Kadoya[1]*, Akio Takenaka[1], Fumiko Ishihama[1], Taku Fujita[2], Makoto Ogawa[3], Teruo Katsuyama[4], Yasuro Kadono[5], Nobumitsu Kawakubo[6], Shunsuke Serizawa[7], Hideki Takahashi[8], Masayuki Takamiya[9], Shinji Fujii[10], Hiroyuki Matsuda[11], Kazuo Muneda[11], Masatsugu Yokota[12], Koji Yonekura[13], Tetsukazu Yahara[14]

1 Center for Environmental Biology and Ecosystem Studies, National Institute for Environmental Studies, Onogawa, Tsukuba, Ibaraki, Japan, **2** Nature Conservation Society of Japan, Shinkawa, Chuo-ku, Tokyo, Japan, **3** Tokushima Prefectural Museum, Bunka-no-mori Park, Hachiman-cho, Tokushima-shi, Tokushima, Japan, **4** Kanagawa Prefectural Museum of Natural History, Odawara, Kanagawa, Japan, **5** Faculty of Science, Kobe University, Kobe, Japan, **6** Faculty of Applied Biological Sciences, Gifu University, Yanagido, Gifu, Japan, **7** Department of Biology, Aichi Kyoiku University, Igaya-cho, Kariya, Aichi, Japan, **8** Hokkaido University Museum, Sapporo, Hokkaido, Japan, **9** Department of Environmental Science, Kumamoto University, Kumamoto, Japan, **10** Division of Human Environment, University of Human Environments, Okazaki, Aichi, Japan, **11** Faculty of Natural Environment and Information Sciences, Yokohama National University, Tokiwadai, Hodogaya-ku, Yokohama city, Kanagawa, Japan, **12** Faculty of Science, University of the Ryukyus, Senbaru, Nishihara, Nakagami, Okinawa, Japan, **13** Center for Academic Resources and Archives, Botanical Gardens, Tohoku University, Aoba, Aramaki, Aoba-ku, Sendai, Miyagi, Japan, **14** Department of Biology, Kyushu University, Higashi-ku, Fukuoka, Japan

Abstract

Although many people have expressed alarm that we are witnessing a mass extinction, few projections have been quantified, owing to limited availability of time-series data on threatened organisms, especially plants. To quantify the risk of extinction, we need to monitor changes in population size over time for as many species as possible. Here, we present the world's first quantitative projection of plant species loss at a national level, with stochastic simulations based on the results of population censuses of 1618 threatened plant taxa in 3574 map cells of ca. 100 km^2. More than 500 lay botanists helped monitor those taxa in 1994–1995 and in 2003–2004. We projected that between 370 and 561 vascular plant taxa will go extinct in Japan during the next century if past trends of population decline continue. This extinction rate is approximately two to three times the global rate. Using time-series data, we show that existing national protected areas (PAs) covering ca. 7% of Japan will not adequately prevent population declines: even core PAs can protect at best <60% of local populations from decline. Thus, the Aichi Biodiversity Target to expand PAs to 17% of land (and inland water) areas, as committed to by many national governments, is not enough: only 29.2% of currently threatened species will become non-threatened under the assumption that probability of protection success by PAs is 0.5, which our assessment shows is realistic. In countries where volunteers can be organized to monitor threatened taxa, censuses using our method should be able to quantify how fast we are losing species and to assess how effective current conservation measures such as PAs are in preventing species extinction.

Editor: Eric Gordon Lamb, University of Saskatchewan, Canada

Funding: The funders had no role in study design, data collection and analysis, decision to publish, or preparation of the manuscript. This work was supported by the Ministry of the Environment of Japan and the Japanese Society of Plant Taxonomists, and in part by the Environment Research and Technology Development fund (S9-1) of the Ministry of the Environment, Japan.

Competing Interests: The authors have declared that no competing interests exist.

* E-mail: kadoya@nies.go.jp

Introduction

Listing threatened species on the basis of IUCN Red List categories plays a prominent role in conservation practices. However, the practice has received criticism owing to its reliance on qualitative data [1] and insufficient data [2,3,4]. Quantitative analysis—for example, population viability analysis (PVA)—is seldom used in determining Red List categories, since time-series data are not available for most threatened species. As a result, conservation priorities have been determined only by the distribution of threatened species, without consideration of the probability of extinction of each species [5].

Protected areas (PAs) are a cornerstone of conservation efforts [6] and now cover nearly 11% (16 263 609 km^2) of the world's land surface [7]. Various national governments have committed to expanding this coverage to 17%, as endorsed at the COP 10 of the Convention on Biological Diversity held at Nagoya in 2010 as one of the Aichi Biodiversity Targets [8]. The effectiveness of PAs should be assessed not only by their extent but also by to what degree they can reduce the risk of extinction of each species. However, assessments based on the risk of extinction of each taxon at a large scale, say the national scale, have not yet been conducted so far. The paucity of assessment of extinction risks based on data with high-enough spatial resolution hinders the effectiveness of assessment.

To overcome these difficulties, we organized more than 500 lay botanists to monitor population sizes of 1618 plant taxa throughout Japan in ca. 100-km^2 map cells (Fig. 1a). The population size and rate of change of at least one of those taxa were reported from 3574 cells (79.9% of Japan). Population size was determined on an ordinal scale of 1, 10, 100, 1000, or 10 000 individuals in 1994–1995 and 2003–2004, and changes between the census periods were used to quantify population decline. In addition, in both censuses, the rate of population decline during the prior decade was scored by the 500 lay botanists where possible. As the rate of population decline could differ among taxa and cells owing to differences in species and environmental characteristics, we need to consider not only the mean rate of population change but also its variance in order to evaluate the fate of threatened species. By taking advantage of the spatially intensive data, we quantified the uncertainty in the prediction of the viability of each taxon: we did not simply assume a constant trend for each population, but assumed that each population could be exposed to the rate of decline observed in other populations. Then, considering the uncertainty, we quantified the extinction risk for each taxon as a form of extinction probability using Monte Carlo simulations.

Using the time-series data, we also examined how effectively existing PAs can protect populations of threatened species from decline. In Japan, natural parks cover 14.3% (54 313 km^2) of the nation [7]. However, only about half of this area (7.3%, 27 747 km^2) has any regulation for biodiversity conservation and is managed by the national government. And only 0.91% (3449 km^2) is strictly conserved as core zones within natural parks. Taking the actual conservation effectiveness of the existing PAs into account, we examined how much the global PA target of 17% could reduce the number of threatened species, defined here as those having a probability of extinction within the next 100 years of ≥10%. To do so, we created software that can find a (sub)optimized solution for a spatial arrangement of PAs that would reduce the overall extinction risk (reduction of extinction risk: RER) of all species.

Methods

Database of Japanese threatened vascular plants

More than 500 lay botanists participated in censuses in 1994–1995 and 2003–2004. The surveys were conducted across Japan in each ca. 100-km^2 map cell (1° longitude by 40′ latitude). The population size (the number of reproductive individuals) at each cell was recorded as not present, 1–9, 10–99, 100–999 or ≥1000 in the first census; and as extinct, 1–9, 10–49, 50–99, 100–999, 1000–9999 or ≥10 000 in the second census. Furthermore, we asked participants to score the trend of decline (i.e., the proportion of the number in the previous period still alive 10 years later) if empirical evidence was available, as <1/100, 1/100–1/10, 1/10–1/2, 1/2–1, and >1 (i.e., increase).

Consequently, 1735 taxa were recorded in 3642 cells (81.4% of the total) that included at least one of the taxa. Of the 1735 taxa, 1618 that were recorded in 3574 map cells (79.9% of Japan) were quantifiable: Quantitative population sizes in both censuses were reported for 9115 populations covering 1089 taxa; quantitative population sizes in either census and information on decline were reported for 4755 populations covering 877 taxa; and quantitative population sizes in either census without a decline score were reported for 9600 populations covering 1184 taxa. In addition, decline scores were reported for 1413 populations of 437 taxa without population size data. In total, 24 833 populations covering the 1618 taxa were further analysed, as explained below. As some taxa consist of several populations which differ in the information types, they overlap among the information type categories. Most of the taxa had narrow ranges: 296 (18.3%) were found at only 1 cell, 374 (23.1%) at 2 or 3 cells, and 1086 (67.1%) at <10 cells.

Probabilities of extinction of each taxon

The probabilities of the extinction of each of the 1618 taxa in each decade in the next century were computed by Monte Carlo simulations with 1000 replications.

The future population size of each taxon in cell i at time $t+1$, denoted as $N_{i(t+1)}$, was estimated as

$$N_{i(t+1)} = r_{i(t)} N_{i(t)}$$

where $N_{i(t)}$ is the number of individuals and $r_{i(t)}$ is the rate of change in cell i at time step t (a multiple of 10 years).

Since the data set consists of size classes for population number and scores for rate of decline, and many records are incomplete, we considered uncertainty due to the data characteristics by including a stochastic process in choosing the initial population size, $N_{i(0)}$, and the rate of change, $r_{i(t)}$, for predictions.

For $N_{i(0)}$, we randomly drew a value for each of 1000 replications from the recorded range of population size classes in the cell in each census (with values of 1000 for the class of ≥1000 and of 10 000 for the class of ≥10 000). Therefore, $N_{i(0)}$ can be varied within the range of a size class recorded at cell i among replications.

To determine $r_{i(t)}$ for each taxon, we assumed two pools from which $r_{i(t)}$ could be drawn randomly for each time step (i.e., 10 years) for each replication: pool 1, which consists of records of population change in all cells where the taxon was reported; and pool 2, which consists of records of population change of all taxa in cell i. For each time step for each replication, a random value was drawn from pool 1 or pool 2 at a certain proportion (say 0.5 vs 0.5 from pool 1 and pool 2, respectively, such that the total probability for the two pools = 1.0) according to the scenario specified below.

Our records of the rate of change of population size are based on either (1) combinations of recorded population size classes in the first and second censuses in a cell, or (2) scores of the trend of decline (a change to <1/100, 1/100–1/10, 1/10–1/2, 1/2–1, and >1 of the value in the previous decade) reported by lay botanists. Where a type 1 record was available, we put it into the pools, treating the combination of size classes in both censuses as a single entity. Where a type 1 record was not available, we put a type 2 record into the pool as an entity instead.

The procedure to predict the probability of extinction for each taxon, based on the above settings, is as follows:

(1) Determine initial population size in cell i by randomly drawing a value from the range of a size class (say 100–999) recorded in the 2003–2004 census in cell i.

(2) Determine which pool is used to draw the rate of change according to a probability set *a priori* by a scenario specified following step 8, below.

(3) Randomly draw an entity in the selected pool. If the drawn entity is a combination of population size classes (say 100–999 in the 1994–1995 census and 1–9 in the 2003–2004 census), then randomly draw a value for each census and calculate the rate of change at time t in cell i ($r_{i(t)}$) by dividing the value in the second census by that in the first census. If the drawn entity is a score of the trend of population change (say 1/100–1/10), then determine $r_{i(t)}$ by randomly drawing a value from the range in the entity.

(4) Obtain $N_{i(t+1)}$ by multiplying the population size, $N_{i(t)}$, by $r_{i(t)}$.

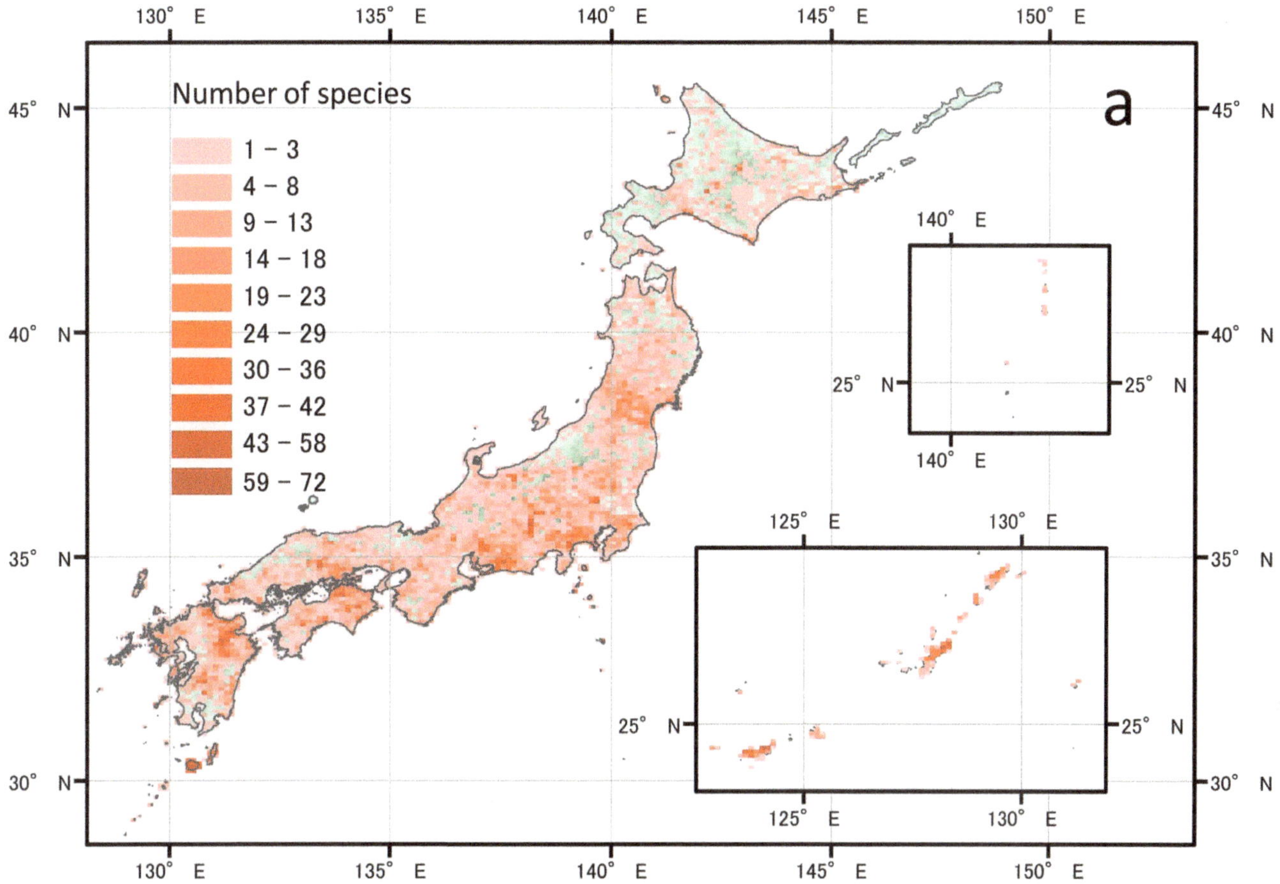
a
Number of species
1 - 3
4 - 8
9 - 13
14 - 18
19 - 23
24 - 29
30 - 36
37 - 42
43 - 58
59 - 72
130° E
135° E
140° E
145° E
150° E
45° N
40° N
35° N
30° N
25° N
125° E

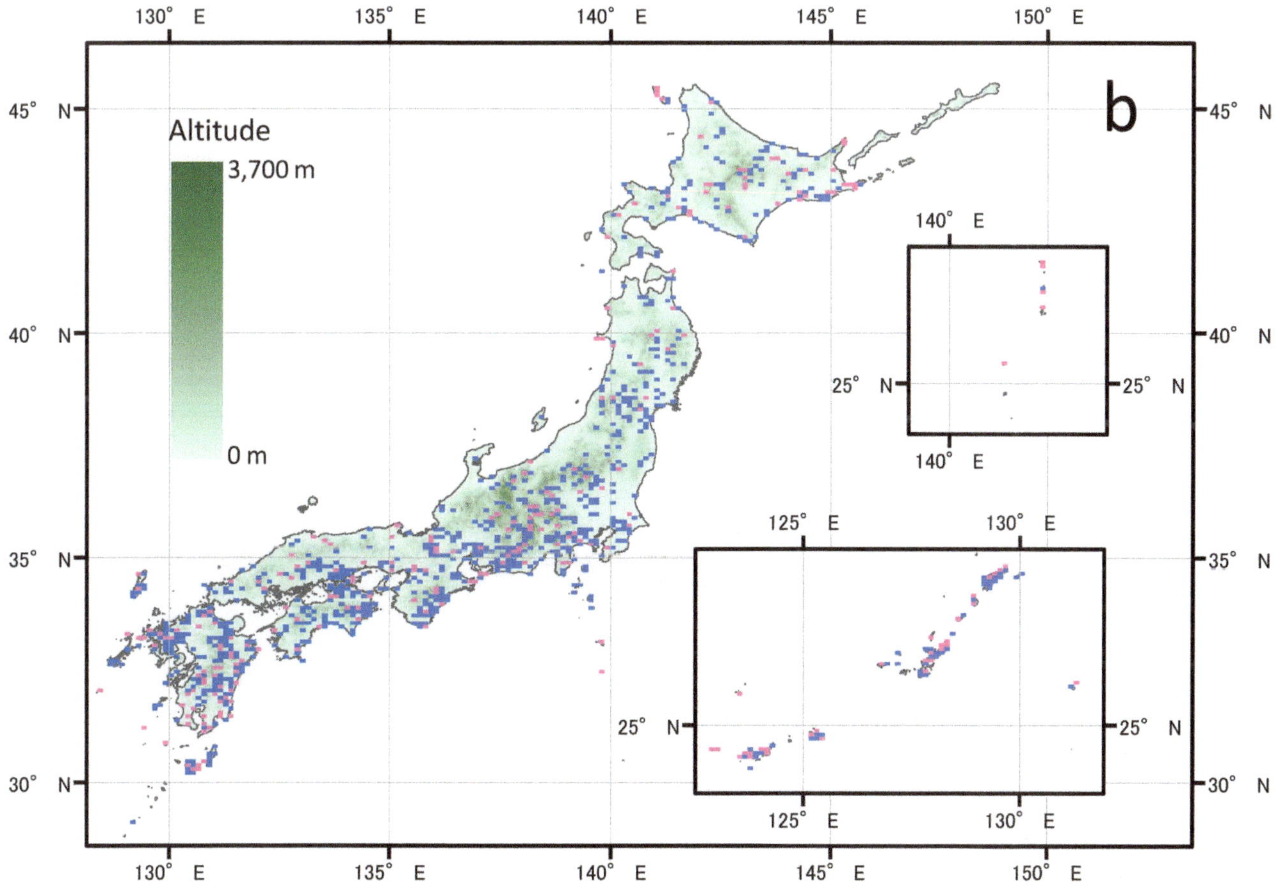
b
Altitude
3,700 m
0 m
130° E
135° E
140° E
145° E
150° E
45° N
40° N
35° N
30° N
25° N
125° E

Figure 1. Distribution of 1618 vascular plant species in 4473 terrestrial cells (ca. 100 km^2) in Japan, *a* and conservation prioritized cells, *b*. (a) One or more species were recorded in 3574 cells (79.9% of Japan). Green points represent the altitude of cells where no species were recorded with the same scale as in *b*. (b) Blue points represent 760 prioritized cells corresponding to 17% of the land area under the assumption that conservation effectiveness is 0.5 (see Fig. 2 and methods for details). Pink points represent 244 cells in which conservation effectiveness needs to be increased to conserve all threatened species; 237 of these are already included in the 760 prioritized cells. The maps were drawn using the ArcGIS 10.1 software (ESRI, Redlands, CA, USA).

(5) Perform steps 2 to 4 ten times (i.e., projection of population size over the next 100 years).
(6) Perform steps 1 to 5 for all cells where the taxon was recorded.
(7) Specify the time when the focal taxon goes extinct in all cells (i.e., the population size in every cell becomes <1).
(8) Perform steps 1 to 7 one thousand times as replications and obtain a cumulative probability of extinction at time t as the proportion of the number of extinction events by time t in the 1000 replications.

For the selection of the pool at step 3, we assumed three different scenarios: (1) $r_{i(t)}$ follows the whole-of-Japan pool of the observed rate of change for each taxon (i.e., always drawn from pool 1). (2) $r_{i(t)}$ follows the rate of change observed for all of the taxa in the focal cell (i.e., always drawn from pool 2). (3) $r_{i(t)}$ follows either the whole-of-Japan pool for each taxon or the locally observed rate of change (i.e., drawn from either pool 1 or pool 2) for all taxa at a probability of (pool 1 vs. pool 2) 0.2 vs 0.8, 0.5 vs 0.5, and 0.8 vs 0.2. The rationale of assumption 1 is that any change in a local population can happen to any other local population of that species. We call this the "what happened here may happen there" hypothesis. On the other hands, we call the assumption 2 as "what happened here will happen here again" hypothesis. Assumption 1 gives the greatest rate of extinction (Fig. 2). Although it may overestimate the risk, it makes the simplest and most precautionary assumptions about environmental and demographic stochasticity. We used this scenario in the subsequent prioritization analyses.

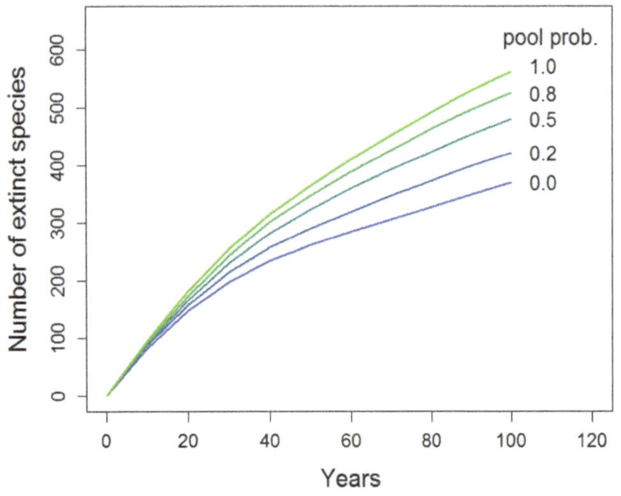

Figure 2. Number of extinct species predicted by the population viability analyses (PVAs) of 1618 vascular plant taxa in Japan. Results correspond to different scenarios in the choice of a rate of change class for the PVAs: classes were drawn (i) only from a pool of observed classes for each taxon in all of Japan (pool $P=1.0$); (ii) from both the pool and the classes observed in the same cell at a certain ratio ($P=0.2$–0.8); and (iii) only from those observed in the same cell ($P=0.0$). Results from assumption (i) always gave the greatest rate of extinction.

When we drew a value from a population size class or a score of the rate of population change, we used uniform probabilities on a logarithmic scale so as to make the expected $N_{i(t)}$ or $r_{i(t)}$ equal to the geometric mean of the upper and lower limits of the size and rate classes.

In the simulation, we assumed a few exceptions. When a population was not present in 2003–2004 in the field, the rate was defined as 1/10 000 so that any local populations go extinct during 10 years irrespective of their size. We ignored cells where population size was unknown in the second census. However, in cells where the population size was recorded in the first census, we started the simulation from the time of the 1994–1995 census. In this case, we performed step 5 eleven times to end the simulation at the same time as other populations. For 197 taxa which had populations with size information in either census but not enough information to calculate the rate of change in any population, we used the distribution of change rate classes obtained from all taxa.

Assessment of effectiveness of existing protected areas

Natural parks cover 14.3% (54 313 km^2) of Japan [7]. However, only about half of this area (7.3%, 27 747 km^2) has any regulation for biodiversity conservation and is managed by the national government. And only 0.91% (3449 km^2) is strictly conserved as core zones within natural parks, in which no human disturbance is permitted. In combination with the time-series data, we assessed how effectively existing PAs can protect populations of threatened species from decline. Because the exact locations of populations within a map cell were not recorded in the censuses, we assessed the effectiveness of the PAs as follows. First, we divided all records of the rate of change of local populations between "decreased" and "constant or increased". Then we related the category to the proportion of protected area in a cell by logistic regression. From these results, we divided the probability of avoided decline within a protected area by that of population decline outside of protected area (Fig. 3a). The result is the probability of protection success, or "conservation effectiveness". We also compared performance in terms of Akaike's information criterion (AIC) between the logistic regression which has proportion of protected area as an explanatory variable and the null model i.e., a logistic regression which has only a constant term. As an assessment of current state of populations within PAs at a whole nation scale, we were just interested in how populations within PAs have lower probability of decline than those outside of PAs. Therefore we did not include any factors other than coverage of PAs in the regression analysis.

Spatial optimization of the reserve network to reduce overall extinction risks

Development of Spatial Prioritizer based on Extinction Risks (SPERS). We developed a tool for the prioritization of spatial conservation, finding a (sub)optimized solution of PAs to reduce the overall risk of extinction (reduction of extinction risk: RER) of all species. RER is defined as:

$$RER = \sum_{1}^{S} \left(\frac{1}{T_s} - \frac{1}{T_{s-p}} \right)$$

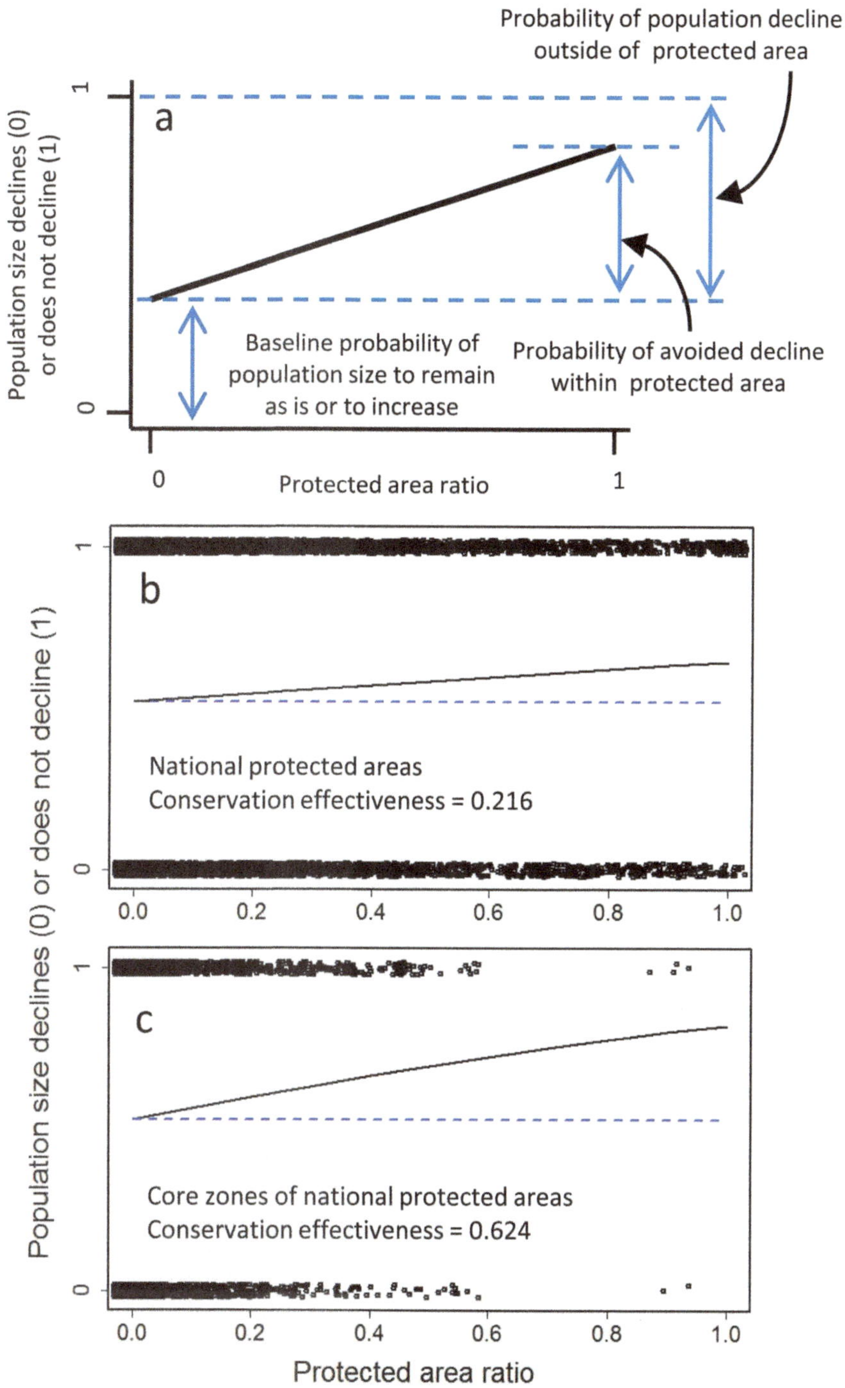

Figure 3. Relationships between the ratio of protected area in a cell and local population decline (defined as either decreased [0] or constant or increased [1]**).** (a) Conservation effectiveness is calculated as the ratio of the probability of avoided decline within a protected area to the probability of population decline outside the protected area. (b) Empirical assessment using total protected area, which corresponds to natural parks with regulations for biodiversity conservation (27 747 km^2: ca. 7.3% of Japan). (c) Empirical assessment using core zones of the national parks (3449 km^2: ca. 0.91% of Japan). Dark blue dashed line represents baseline probability of avoided decline in cells where protected area = 0%.

where S is the total number of target taxa and T_s and $T_{s\text{-p}}$ are the expected waiting times for extinction without and with protection, respectively, for a taxa s., which are quantified through the PVA simulations explained above. Our tool finds a spatial arrangement that maximizes the RER under a given number of PAs using a so-called "greedy" algorithm [9] to consider the conservation effectiveness of reserves. The maximization of the RER by a simple and honest algorithm will easily reach an infeasible computation time. This is because we must re-evaluate the risk of extinction of every species using the PVA simulation at every time when a new reserve is added. We make this feasible by using an efficient computational procedure, named SPERS (Spatial Prioritizer based on Extinction Risks).

Schematic algorithms of SPERS are shown in Figures S1–S3. The first component of SPERS prepares a database of the probabilities of local extinction (Fig. S1). SPERS calculates the probability of extinction of a local population of each taxon in each cell using the same procedure as for PVA, except that it considers conservation effectiveness. To do so, it draws on two databases of probabilities of local extinction: with no protection and with protection at an arbitrary probability of success (i.e., conservation effectiveness set *a priori*) (Fig. S2). We assumed conservation effectiveness values of 0.2, 0.5, 0.8 and 1.0 on the basis of the results of our empirical assessments explained in the previous section. When calculating the probabilities of local extinction with protection, if the rate of change of a taxon in a protected cell is $r_i(t) < 1$, SPERS alters it to 1 (no population change) with a probability that corresponds to conservation effectiveness set *a priori*.

The second component of SPERS calculates the risk of total extinction with a given set of protected sites (Fig. S2). It uses the two databases of the probabilities of local extinction in the first component. Selecting one database according to the input set of protected sites, SPERS can evaluate the local extinction of species at each site, which collectively represents the risk of total extinction of a species with the given protected sites. By omitting the re-evaluation of the risk of local extinction through the use of the two database, SPERS can efficiently evaluate the risk of total extinction in a set of protected sites.

The final component of SPERS finds an arbitrary number of sites to be protected to minimize total extinction risk (i.e., maximize RER) (Fig. S3). Relying on a greedy algorithm, SPERS incrementally adds a new site that is the most effective in terms of RER to reach the given number of sites. Although the solution obtained by the greedy algorithm is not guaranteed to be optimal, the algorithm is known to provide adequate approximation of optimal solutions [9].

Performance comparisons with existing methods. We compared the performance among spatial prioritization methods in terms of RER. As comparators, we used complementarity analysis and hotspot analysis [10]. In complementarity analysis, cells are incrementally selected using a greedy algorithm to maximize the number of represented species in the protected areas; after all species are represented in at least one cell, the algorithm proceeds so that all species are represented in more cells. In hotspot analysis, or scoring analysis, cells are incrementally selected according to species richness in them. In addition, we compared the performance of those methods with random selection, in which cells were added randomly to protected areas.

Spatial prioritization analysis using SPERS. We applied SPERS to our data, which included 757 threatened taxa, defined here as those having a probability of extinction within the next 100 years of ≥10% in the PVA simulations. First, we quantified how the RER increased as protected areas increased under an assumption of perfect conservation effectiveness (i.e., conservation effectiveness = 1.0). In this case, when we prioritized a cell, if $r_{i(t)} < 1$ at a conserved cell, it was always altered to 1 (no population change).

Then, according to the result of the assessment of conservation effectiveness, we examined how the RER increased as protected areas increased under the more realistic assumption of imperfect conservation effectiveness of 0.8, 0.5 and 0.2. In these cases, when we prioritized a cell, if $r_{i(t)} < 1$ at a conserved cell, it was altered to 1 (no population change) with a probability of 0.8, 0.5, and 0.2, according to the assumed conservation effectiveness. In each case, we examined to what degree 17% of protected areas (i.e., the Aichi target) could reduce the risk of extinction of the 757 threatened taxa.

Table 1. List of vascular plant species categorized as Extinct (EX) or Extinct in the Wild (EW) in the Red List compiled by Ministry of the Environment of Japan in 2012.

Scientific name
Extinct (EX)
Lycopodium cunninghamioides
Botrychium boreale
Ophioglossum parvifolium
Hypolepis tenuifolia
Asplenium austrochinense
Arachniodes yasu-inouei var. *angustipinnula*
Tectaria dissecta
Pyrrosia angustissima
Elatostema lineolatum var. *majus*
Ranunculus gmelinii
Rubus hatsushimae
Flemingia strobilifera
Lespedeza hisauchii
Euphrasia insignis subsp. *insignis* var. *omiensis*
Euphrasia insignis subsp. *insignis* var. *pubigera*
Euphrasia multifolia var. *kirisimana*
Cirsium toyoshimae
Aletris makiyataroi
Burmannia coelestris
Thismia tuberculata
Eriocaulon cauliferum
Eriocaulon seticuspe
Carex disperma
Cyperus diaphanus
Cyperus procerus
Fimbristylis leptoclada var. *takamineana*
Fimbristylis pauciflora
Acanthophippium striatum
Neottia kiusiana
Odontochilus poilanei
Renanthera labrosa
Zeuxine boninensis
Extinct in the wild (EX)
Thelypteris aurita
Dryopteris shibipedis
Magnolia pseudokobus
Kalanchoë spathulata
Malus hupehensis
Astragalus sikokianus
Viola stoloniflora
Lycoris sanguinea var. *koreana*
Eriocaulon heleocharioides
Liparis uchiyamae

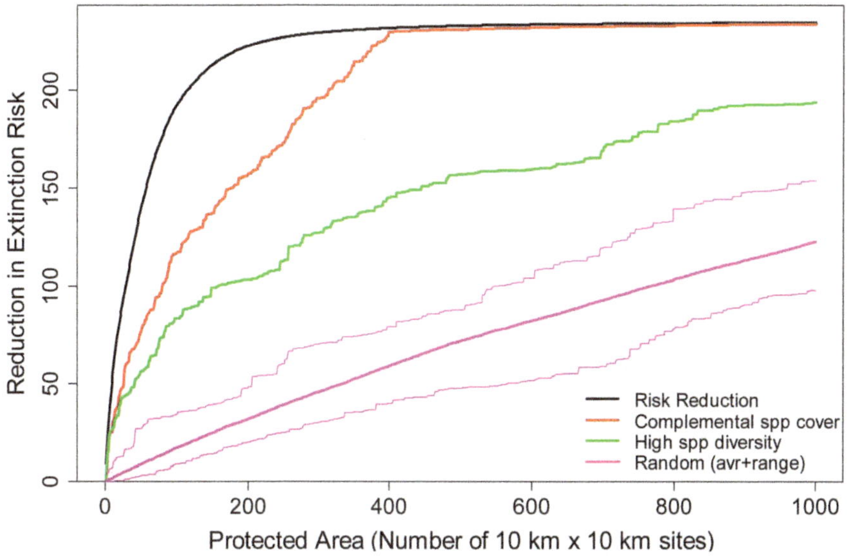

Figure 4. Comparison of performance among spatial prioritization methods in terms of reduction in extinction risk. "Risk Reduction": cell selection using SPERS. "Complemental spp. cover": complementarity analysis in which cells are incrementally selected using a greedy algorithm to maximize the number of represented species in the protected areas; after all species are represented in at least one cell, the algorithm proceeds so that all species are represented in more cells. "High spp. diversity": hotspot analysis in which cells are incrementally selected according to species richness in them. "Random": cells are added randomly to protected areas and the random selection is performed 100 times; thick pink curve = mean, thin pink curves = maximum–minimum range.

Results & Discussion

Our PVA projected that between 370 and 561 of the 1618 taxa will go extinct during the next century (Fig. 2). The worst scenario corresponded to the assumption that "what happened here may happen there" rather than "what happened here will happen here again". In contrast, only 42 taxa were lost during the past 60 years (Table 1). In Japan, 7087 vascular plant taxa have been recorded [10], making the recent extinction rate only 0.01% per year. The projected extinction rate, in contrast, is 0.05% to 0.08% per year. The number of seed plants lost in the world from 1990 to 1992 was estimated to be 163 species [11], or 0.03% per year for the

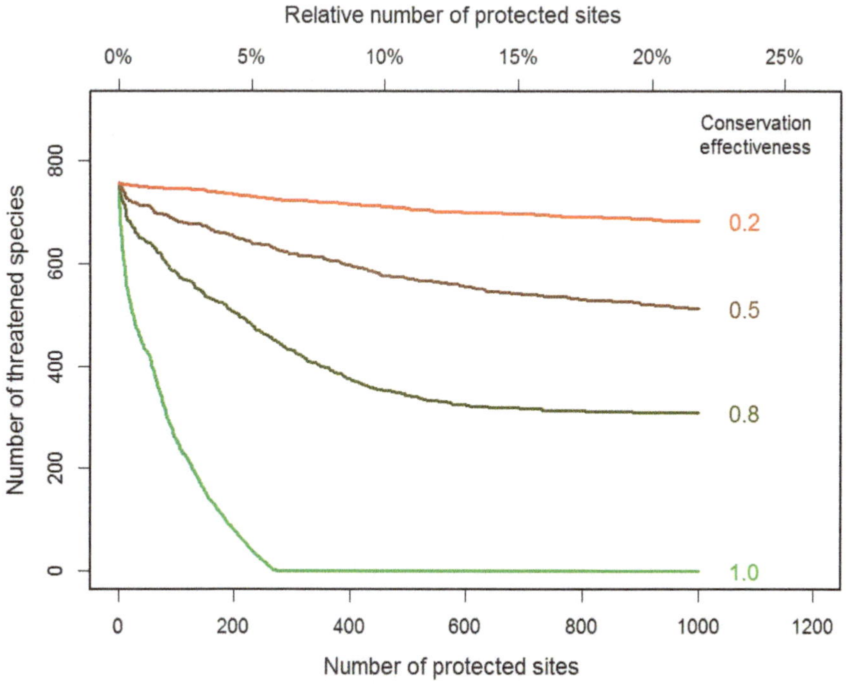

Figure 5. Relationships between number of protected cells and number of species remaining as threatened (probability of extinction in the next 100 years ≥10%). Cells are selected to maximize the reduction of extinction risk (see text for details).

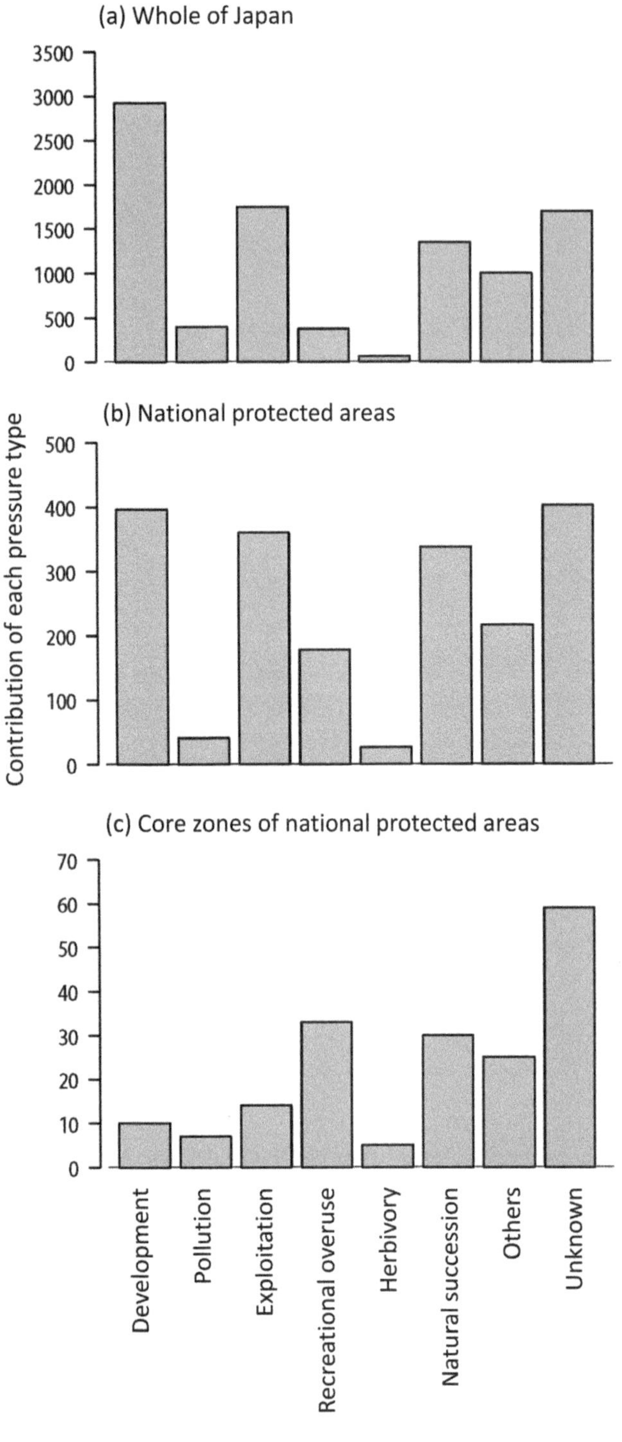

Figure 6. Contribution of pressure types causing decline of local populations in (a) whole of Japan, (b) national protected areas (cells with >20% of protected areas) and (c) core zones of national protected area (cells with >20% of core zones). For each combination of pressure type and species, the number of cells with declining population was calculated, and the numbers are summed for each pressure type.

240 000 described species [12]. The projected extinction rate in Japan is two to three times this rate.

An ideal assessment is based on spatially extensive, temporally intensive data. Although our assessment is spatially extensive, it is limited to two time points at which abundances were estimated on an ordinal scale. As this may cause uncertainty in the estimation of the risk of extinction of each taxon, we considered the uncertainty due to the limited time-series data by incorporating stochasticity in the predictions of extinction probability for each taxon: instead of directly using the observed rate of population change to predict population dynamics for a taxon at a location, we pooled the observation records across either taxa or locations and assumed that the rate of population change, r, can vary even within a cell among time steps and replications according to the frequency distribution of the records in the pools. This would be the most honest way to reflect the data uncertainty in the results of predictions and the most efficient way to take advantage of the spatially high-resolution data. As we aggregated the results of all 1618 taxa in estimating the rate of extinction of all taxa, the results are likely to capture general trends in spite of the uncertainty at each taxon level.

However, because these pooled data are themselves constrained by the temporal trend of 1994–1995 and 2003–2004, if this trend had been unique in comparison to those of the other time intervals, future projection would be biased. Therefore the predictions should be updated and improved through subsequent assessments, and we are now planning the third nationwide census for the near future. There is now an urgent need to quantify the risk of extinction of species in rapid decline by a scientifically sound method so as to facilitate awareness of the crisis and to prompt decision makers to act to prevent extinction. That is, it is important to act now, even based on imperfect data, to prevent the loss of species that would become extinct while we wait for better data to become available. This study will help meet these needs.

In the assessment of the probability of protection success, or "conservation effectiveness", compared with the null model, which assumes no effect of the PAs, models that included PAs (Fig. 3b) and core zones of PAs (Fig. 3c) were superior (i.e., AIC was smaller by 40 and 20, respectively), which suggests that existing PAs improve the status of populations to some extent. However, even in the core zones, conservation effectiveness was only 0.624 (Fig. 3c), indicating that the effectiveness of the PAs was far from perfect.

Our spatial conservation prioritization tool, SPERS, proved more efficient than the conventional complementarity analysis and hotspot analysis, especially when the number of protected cells was less than ideal so that not all taxa were represented in PAs (Fig. 4). Considering that the resources available for conservation are always limited and the extent of the protected area is often far from enough, our method is likely to outperform others in most situations. SPERS showed that the optimized PAs could reduce extinction risks effectively, in that protecting 250 cells (ca. 5.6% of Japan) rendered all of the endangered species non-threatened (Fig 5). However, when we considered more realistic conservation effectiveness values, the total RER was reduced rapidly: under the assumption of 17% PAs (i.e., protecting 760 cells), 443 (58.5%) of the 757 threatened taxa become non-threatened at an effectiveness value of 0.8, 221 (29.2%) at a value of 0.5, and only 64 (8.45%) at a value of 0.2 (Fig. 5). In these cases, even if we could achieve the 17% target with an optimized PA selection (including the relocation of existing PAs), we may be unable to achieve the target. These proportions are also below the level set by the global strategy for plant conservation endorsed at COP 10, which requires "at least 75 per cent of known threatened plant species conserved in *situ*". In addition, the relative importance of conservation effectiveness will increase as the number of protected cells is increased (beyond about 200 cells [4.5%] in this case; Fig. 5). As the spatial arrangement of existing protected sites is likely to deviate from the optimum [13], these results may still overestimate the overall effectiveness of PAs.

Our results demonstrate that mass extinction of plants is not an exaggerated scenario but a realistic possibility even in a developed country such as Japan, where conservation efforts are intensive. Under the realistic conservation effectiveness for each PA, the 17% target is likely to be inadequate. When the average conservation effectiveness is around 0.5, which our assessment shows is realistic, then to make the network of PAs adequately effective, we must improve the conservation effectiveness in about a third of the cells (244 of 760) to 1.0 (i.e., perfect conservation; Fig. 1b). This necessity indicates the need to improve the conservation effectiveness of PAs, in addition to extending the overall coverage of PAs. To improve the effectiveness of conservation areas, we need to know the types of pressures causing population decline in the areas. The data set records types of pressures which are suspected to be causing a population decline for each species in each cell. For each combination of pressure type and species, we summed the number of cells with declining population, then we summed the number of cells for each pressure type. The larger the sum, the greater the pressure on the plants. We carried out these calculations for all of Japan, national protected areas, and core zones of the protected areas. Among the pressures, land development and exploitation were less important at cells with core zones of PAs, whereas pressures of recreational overuse, herbivory and vegetation succession were increased or not improved (Fig. 6). Conservation effectiveness could be further improved by developing management schemes to focus on specific pressures. For the 17% target to be adequate, the protection success must be greatly increased in each PA or in the PAs that show the greatest risk of extinction for a given species.

As far as we know, this is the world's first reliable quantification of plant species loss and quantitative assessment of the adequacy of the 17% Aichi Biodiversity Target based on time-series population data at a national scale. Monitoring data collected by the public can play an essential role in assessing biodiversity [14,15]. Although the information we obtained covers changes only in population size classes, our approach provides a practical and quick solution to evaluating the fates of thousands of threatened species from a set of time-series observations. It should provide a basis for similar efforts to quantify species loss in other countries.

Supporting Information

Figure S1 Preparation of database of probabilities of local extinction in SPERS.

Figure S2 Estimation of extinction risks under the given set of protected sites in SPERS.

Figure S3 Selection of *N* sites to be protected to minimize the total extinction risk in SPERS.

Acknowledgments

We thank the more than 500 lay botanists who contributed to the Japanese Red Data Book project. We also thank Kazuo Somiya, of the Ministry of the Environment of Japan, and Yasuhiro Ibaragi, Masato Nagatsu, and Kaori Sato, of the Japan Wildlife Research Center, for their help and encouragement during the project.

Author Contributions

Conceived and designed the experiments: TK TY. Analyzed the data: TK FI AT. Contributed reagents/materials/analysis tools: TK FI AT TF MO TK YK NK SS HT MT SF HM KM MY KY TY. Wrote the paper: TK FI AT TY.

References

1. Rodrigues ASL, Pilgrim JD, Lamoreux JF, Hoffmann M, Brooks TM (2006) The value of the IUCN Red List for conservation. Trends Ecol Evol 21: 71–76.
2. Mace GM, Lande R (1991) Assessing extinction threats: toward a reevaluation of IUCN threatened species categories. Conserv Biol 5: 148–157.
3. Possingham HP, Wilson KA (2005) Biodiversity: Turning up the heat on hotspots. Nature 436: 919–920.
4. Possingham HP, Andelman SJ, Burgman MA, Medellin RA, Master LL, et al. (2002) Limits to the use of threatened species lists. Trends Ecol Evol 17: 503–507.
5. IUCN (2012) The IUCN Red List of Threatened Species. Version 2012.2.
6. Laurance WF, Useche DC, Rendeiro J, Kalka M, Bradshaw CJA, et al. (2012) Averting biodiversity collapse in tropical forest protected areas. Nature 489: 290–294.
7. IUCN, UNEP-WCMC (2012) The World Database on Protected Areas (WDPA): February 2012. Cambridge, UK: UNEP-WCMC.
8. CBD (2010) COP Decision X/2. Strategic plan for biodiversity 2011–2020.
9. Moore JL, Folkmann M, Balmford A, Brooks T, Burgess N, et al. (2003) Heuristic and optimal solutions for set-covering problems in conservation biology. Ecography 26: 595–601.
10. Yahara T, Kato T, Inoue K, Yokota M, Kadono Y, et al. (1998) Red list of Japanese vascular plants: summary of methods and results. Proceedings of Japanese Society for Plant Systematics 13: 89–96.
11. Moilanen A, Wilson LA, Possingham HP, editors (2009) Spatial conservation prioritization: Quantitaive methods & computational tools. Oxford: Oxford University Press.
12. Smith FDM, May RM, Pellew R, Johnson TH, Walter KS (1993) Estimating extinction rates. Nature 364: 494–496.
13. Butchart SHM, Scharlemann JPW, Evans MI, Quader S, Arico S, et al. (2012) Protecting important sites for biodiversity contributes to meeting global conservation targets. PLOS ONE 7, doi: 10.1371/journal.pone.0032529
14. Kadoya T, Ishii HS, Kikuchi R, Suda S, Washitani I (2009) Using monitoring data gathered by volunteers to predict the potential distribution of the invasive alien bumblebee Bombus terrestris. Biol Conserv 142: 1011–1017.
15. Kadoya T, Suda S, Washitani I (2009) Dragonfly crisis in Japan: A likely consequence of recent agricultural habitat degradation. Biol Conserv 142: 1899–1905.

15

A General Model of Distant Hybridization Reveals the Conditions for Extinction in Atlantic Salmon and Brown Trout

Claudio S. Quilodrán[1,2], Mathias Currat[1]*, Juan I. Montoya-Burgos[2]

1 Laboratory of anthropology, genetics and peopling history (AGP), Department of Genetics and Evolution, University of Geneva, Geneva, Switzerland, 2 Laboratory of molecular phylogeny and evolution in vertebrates, Department of Genetics and Evolution, University of Geneva, Geneva, Switzerland

Abstract

Interspecific hybridization is common in nature but can be increased in frequency or even originated by human actions, such as species introduction or habitat modification, which may threaten species persistence. When hybridization occurs between distantly related species, referred to as "distant hybridization," the resulting hybrids are generally infertile or fertile but do not undergo chromosomal recombination during gametogenesis. Here, we present a model describing this frequent but poorly studied interspecific hybridization to assess its consequences on parental species and to anticipate the conditions under which they can reach extinction. Our general model fully incorporates three important processes: density-dependent competition, dominance/recessivity inheritance of traits and assortative mating. We demonstrate its use and flexibility by assessing population extinction risk between Atlantic salmon and brown trout in Norway, whose interbreeding has recently increased due to farmed fish releases into the wild. We identified the set of conditions under which hybridization may threaten salmonid species. Thanks to the flexibility of our model, we evaluated the effect of an additional risk factor, a parasitic disease, and showed that the cumulative effects dramatically increase the extinction risk. The consequences of distant hybridization are not genetically, but demographically mediated. Our general model is useful to better comprehend the evolution of such hybrid systems and we demonstrated its importance in the field of conservation biology to set up management recommendations when this increasingly frequent type of hybridization is in action.

Editor: David L. Roberts, University of Kent, United Kingdom

Funding: This study was financed by a fellowship from CADMOS granted to JIMB and MC and partly supported by grants from the SNSF, the Canton de Genève and the G. and A. Claraz donation to JIMB. CSQ acknowledges support from CONICYT-Becas Chile and from the iGE3 student salary award. The funders had no role in study design, data collection and analysis, decision to publish, or preparation of the manuscript.

Competing Interests: The authors have declared that no competing interests exist.

* Email: mathias.currat@unige.ch

Introduction

The evolution of many plant and animal taxa has been influenced by natural interspecific hybridization [1]. However, when hybridization originates from or is intensified by anthropogenic factors, it may lead to critical consequences for species' persistence, particularly for native rare or threatened species [2]. Among other risks, interspecific hybridization can impact demography, which is of primary importance for the viability of wild populations [3].

Three types of interspecific hybridization can be defined, depending on the evolutionary closeness of parental species and the reproductive characteristics of the F_1 hybrids. The first type concerns species that hybridize but yield inviable or infertile offspring due to post-zygotic barriers, such as high difference in chromosomes homology and number. In this case, the waste of reproductive effort may threaten parental species [4]. For example, the replacement of the endangered freshwater fish *Pseudorasbora pumila* by the exotic *P. parva* in Japan is accelerated by their hybridization that produces sterile F_1 hybrids [5]. In the second type, hybrids are viable and fertile, but no recombination between homologous chromosomes occurs during their meiosis, leading to the formation of clonal or hemiclonal gametes. For example, hybrids from two European freshwater fish, the roach (*Rutilus rutilus*) and the bream (*Abramis brama*), produce non-recombinant gametes of both species [6]. Other hybrids may yield gametes containing the haploid genome of only one of the species, excluding the genome of the other parent during or before meiosis, resulting in the hemiclonal transmission of the genome of one parental species. Examples are found in many taxa, such as the *Bacillus* stick insects [7], in the teleost fish *Squalius* [8], or in frogs of the genus *Pelophylax* [9]. Finally, the third type of interspecific hybridization is characterised by F_1 hybrids undergoing recombination between homologous chromosomes during meiosis, resulting in reciprocal genetic introgression from one species into the other. This type of interspecific hybridization may lead to various outcomes, such as (i) the replacement of one or both species by a hybrid-swarm [10]; (ii) the formation of an hybrid zone more or less extended depending on the intensity of the hybrid depression [11]; or (iii) the introgression of neutral or beneficial alleles from one species to the other, impacting the evolution of the introgressed species [12,13].

The first two types are mainly the result of distant hybridization, that is, hybridization between distantly related taxa, which can belong to different species, to different genera, subfamilies or even to different orders [14,15]. In such cases, reproductive behaviour permits interspecific mating to some extent, but genetic barriers of varying intensity constraining offspring fecundity or genetic introgression between parental species exist [6]. Because types 1 and 2 have been understudied and no general model exists to predict non-trivial outcomes, our aim is to develop a simple and more general model to study those cases. We did not, however, include hybridization type 3 in the present work.

Attempts have already been made at modelling hybridization of type 1, in which hybrids are viable but infertile [16], or hybridization of type 2, in which hybrids are fertile but with gametes containing a non-recombined genome [17]. However, these models describe particular hybridization systems and are thus taxon-specific. Moreover, they do not fully address a process that is essential to investigate the demography of parental species, namely: density-dependent competition of hybrids with one or both species. Satake and Araki [18] proposed a one-gene two-alleles model that accounts for density-dependent recruitment from one to the next generation, but this model was intended to study intraspecific population interactions. These authors incorporated only panmictic mating between interacting populations, as they belong to a single species. In addition, the degree of dominance/recessivity of the alleles coding for the inherited traits in hybrids, such as resistance to diseases or to environmental disturbance, is an important parameter that can substantially modify the outcome of the system. Therefore, no current method allows to model distant hybridization systems in which assortative mating exists between the interbreeding species and which integrates the degree of dominance/recessivity inheritance and density-dependent competition.

Here we present a general model that describes the interspecific hybridization of type 1 and 2, that is, distant hybridization or the non-introgressive types. Our model considers a community composed of diploid parental species, with or without overlapping generations, and incorporates: 1) intra- and inter-specific density-dependent competition; 2) the degree of dominance/recessivity of the alleles in hybrids; and 3) assortative mating through mate choice relaxation between the interacting species. The model also considers the possibility that post-F_1 individuals can be of different polyploidy forms. Our new general model may be applied in a large range of real situations and we will illustrate its usefulness by assessing extinction risk through the study of a real case of interspecific hybridization of type 1 for which abundant literature exists.

We applied our model to assess the impact of distant hybridization on Atlantic salmon (*Salmo salar*) and brown trout (*Salmo trutta*) in Norwegian rivers, whose hybridization has been increasing due to the release of farmed fishes into the wild. Despite the high difference in chromosome number between Atlantic salmon (2n = 58) and brown trout (2n = 80), F_1 hybrids are viable and fertile [19]. However, they show differential mortality depending of the female parent (Figure 1), with high offspring survival when the female is an Atlantic salmon and the opposite when the female is a brown trout [20]. Although F_1 hybrid females produce viable offspring when they mate with an Atlantic salmon, the F_2 hybrids produce essentially inviable offspring when mating with any kind of hybrids or parental species [21,22] (Figure 1). For this reason, we consider interspecific hybridization as being of type 1, with viable but infertile hybrids.

Hybridization rates between Atlantic salmon and brown trout is increased by human accidental and deliberate releases of farmed fishes. Once in the wild, these fishes show a relaxed mate choice with frequent interspecific crosses, leading to hybrid frequency exceeding 10% [23]. Levels of up to 29% or even 60% were reported in some Norwegian rivers [24], where the hybridization rate seems to be higher in rivers hosting small and threatened populations of Atlantic salmon than in rivers with large populations [25]. This human increased hybridization rate between Atlantic salmon and brown trout may threaten local populations of parental species. Using our model, we investigated the potential consequences of this interspecific hybridization on populations of the two salmonids and identified the conditions that lead to local extinction.

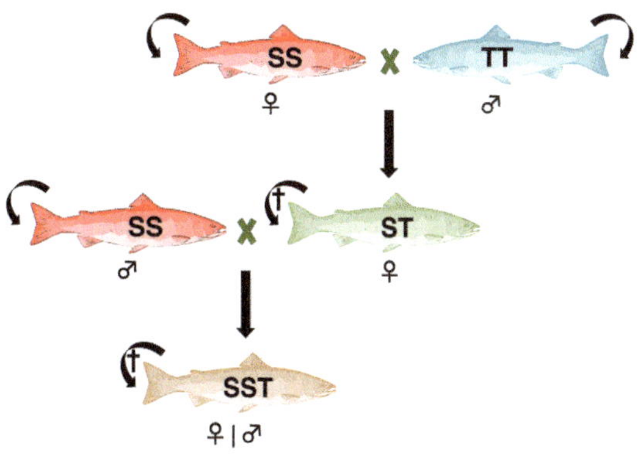

Figure 1. Fertile mating pairs of the case study. Straight and curve arrows represent heterotypic and homotypic mating, respectively. SS = Atlantic salmon; TT = brown trout; ST = first-generation hybrid; SST = second-generation hybrid (triploids). The cross symbol (†) means that mating leads to inviable offspring. Other crosses that produce high level of mortality at hatching (>95%) and malformations in the remaining offspring are not shown (see text).

Materials and Methods

Description of the model

Our model considers interspecific hybridization of diploid organisms, without chromosomal recombination in F_1 hybrids. The genotype class of parental species 0 is codified as 00 and that of parental species 1 as 11. The abundance of parental species is noted as N_0 and N_1, respectively. The number of F_1 hybrids is noted as $N_{½}$ and their genotype class is codified as 01. If crosses between F_1 hybrids and the parental species 0 and 1 generate triploid forms, these forms are codified as 001 with abundance $N_{⅓}$, and as 011 with abundance $N_{⅔}$, respectively. Additional polyploidy forms may be easily incorporated into the model following the same reasoning.

The contribution of each genotype class to the next generation is computed as the frequency of mating between individuals of a given genotype class i with individuals of genotype class j (where j can be equal or different from i), compared to all possible mating combinations. Thus, the probability M_{ij} for individuals of class i to mate with one of class j, for all $i,j \in [0,\ldots,1]$ is:

$$M_{ij(t)} = \frac{\gamma_{ij} N_{j(t)}}{\varphi_{i(t)}} \quad (1)$$

Where $\varphi_{i(t)}$ is a normalization factor such that $\Sigma_i M_{ij} = 1$. In our model, the parameter γ_{ij} is a general measure of the mating success between individuals of class i and j and is called hereafter "interbreeding success rate". The success rate can be reduced by (1) prezygotic barriers, in which case the resulting value of $1-\gamma_{ij}$ could represent a measure of assortative mating; by (2) postzygotic barriers, where γ_{ij} may be seen as a measure of hybrid viability and fertility; or by (3) a combination of both types of barriers. In any case, when $\gamma_{ij} = \gamma_{ji}$, mating success is symmetrical between both species while it is asymmetrical when $\gamma_{ij} \neq \gamma_{ji}$. When $\gamma_{ij} = \gamma_{ji} = 0$, there is no interbreeding between the two species, whereas when $\gamma_{ij} = \gamma_{ji} = 1$, the reproduction is panmictic between both species. Any other value of γ_{ij} between 0 and 1 indicates that mating is locally non-random and reproduction occurs more often between members of the same genotype class i than between individuals of genotype class i and j (see [13,26]).

To calculate the population renewal of class k, we first calculate the number of breeding pairs composed of individuals of class i and j yielding offspring of class k, weighted by the fraction of the gametes that can lead to an offspring of class k and by the relative fitness of class k, expressed as:

$$b_{ij,k(t)} = N_{i(t)} M_{ij(t)} C_{ij,k} \omega_k \quad (2)$$

where $C_{ij,k}$ is the fraction of offspring of class k resulting from a reproduction event between individuals of class i and j. Because in some cases genome exclusion before meiosis leads to the absence of particular gamete types or, alternatively, imperfect meiosis can lead to diploid gametes, the parameter $C_{ij,k}$ is used to determine the proportion of each offspring class resulting from each kind of crosses.

We introduce the parameter ω_k, which represents the fitness of a character in the offspring of class k to which parents of class i and j may contribute. For example, this can be a variable level of resistance to a disease or to environmental disturbances. For the parental species with the highest fitness has $\omega_i = 1$, while for the other parental species ω_j is a fraction of 1. In hybrids, the value of ω_k depends on the dominance degree of the character in one parental species relative to the other (ε). For hybrids of class k, it is calculated as $\omega_k = \varepsilon_{ik}\omega_i + \varepsilon_{jk}\omega_j$, with $\varepsilon_{ik} + \varepsilon_{jk} = 1$. For instance, if $\varepsilon_{ik} = 1$ and $\varepsilon_{jk} = 0$, a character with ω_i is dominant while a character with ω_j is recessive. If $\varepsilon_{ik} = \varepsilon_{jk} = 0.5$, both characters are codominant.

The final weighted number of breeding pairs yielding offspring of class k is obtained by the sum of all weighted breeding pairs generating progeny of class k:

$$n_{k(t)} = \sum_i \sum_j b_{ij,k(t)} \quad (3)$$

To calculate the population renewal of wild adult populations, we extend a version of the Ricker model [27] in which we also take into account the "lattice effects" (dynamic outcomes due to the discrete nature of the numbers of individuals in a population) by rounding off its results, with the following recursion equation [28]:

$$N_{k(t+1)} = \text{round}\left[N_{k(t)} S_k + R_k n_{k(t-\theta)} e^{\left(-\frac{\left(n_{k(t-\theta)} + \sum_{k \neq l} \alpha_{kl} n_{l(t-\theta)} \right)}{V_k} \right)} \right] \quad (4)$$

The first term on the right-hand side of equation (4) represents the fraction of adults that survive from one to the next reproductive season, in which the parameter S_k is the adult survival probability for the genotype class k. The second term of equation (4) denotes the expected amount of offspring that survives until sexual maturity after intra- and inter-specific density-dependent competition effects, where θ indicates the time to reach maturity in t+1. If S_k and θ are equal to zero, it corresponds to a non-overlapping generation model. The parameter R_k represents the population growth rate, that is, the number of progeny per breeding pair that survive until sexual maturity. The parameter α_{kl} represents the interspecific competition coefficient, with $\alpha_{kl} = 1$ indicating that individuals of class l have as much influence on individuals of class k than those of their own class k. When $\alpha_{kl} = 0$ there is no competition between individuals of class k and l, while values of α_{kl} between 0 and 1 indicate that an individual of class l exerts on an individual of class k only a fraction of the competition exerted by an individual of the same class k. Finally, V_k denotes the habitat size as introduced by Henson et al. [28], where $\sum_{k \neq l} \alpha_{kl}/V_k$ determines the interspecific density-dependent mortality before sexual maturity.

For clarity reasons, the model described above considers gonochoric organisms (the two sexes are carried by different individuals) with equal sex ratio or hermaphroditic organisms. But a simple extension of the model can account for gonochoric organisms with unequal sex ratio (see discussion).

Case study

To demonstrate the usefulness of our model we implemented it by studying a case of hybridization type 1, with viable but infertile hybrids. We assess the impact of interbreeding with asymmetrical reproductive success on populations of Atlantic salmon (*Salmo salar*) and brown trout (*Salmo trutta*) in Norwegian rivers. We considered anadromous and iteroparous populations of Atlantic salmon (noted species S with genotype SS) and brown trout (noted species T with genotype TT). According to direct estimates of parameters' values taken from populations of both species in Norwegian rivers [29], sexual maturity was set at four years ($\theta = 3$) and adult survival rate was 30% ($S = 0.3$). The parameters of growth rate (R) and habitat size (V) were estimated by a non-linear least square method (see Appendix S1).

As there is some evidence of species habitat overlap [30], we compared population dynamics with and without interspecific competition to differentiate the effects of interspecific competition from those of hybridization. However, as habitat requirement and behaviour of F_1 and F_2 hybrids have not been studied yet, we opted not to fix α_{ij} but to use a density-dependent form of competition between genotype classes i and j, calculated as:

$$\alpha_{ij(t)} = \frac{N_{j(t)}}{N_{j(t)} + N_{i(t)}} \quad (5)$$

This kind of competition depends on the number of individuals in a given habitat at a given time t [31].

We modelled the mate choice of females assuming an equal sex ratio during the mating phase. The parameter γ_{ST} is the interbreeding success rate between Atlantic salmon females (N_S) and brown trout males (N_T), whereas γ_{TS} is between brown trout females and Atlantic salmon males (see Table S1 for a list of crosses in this case study). F_1 hybrids ($N_{\frac{1}{2}}$) and F_2 allotriploids ($N_{\frac{2}{3}}$) were considered to have a panmictic reproduction ($\gamma_{\frac{1}{2}S} = \gamma_{\frac{1}{2}T} = \gamma_{\frac{2}{3}S} = \gamma_{\frac{2}{3}T} = \gamma_{\frac{1}{2}\frac{2}{3}} = \gamma_{\frac{2}{3}\frac{1}{2}} = 1$). In accordance with Galbreath and Thorgaard [32], offspring resulting from crosses between females N_S and males N_T (offspring of type $N_{\frac{1}{2}}$), and from crosses between females $N_{\frac{1}{2}}$ and males N_S (offspring of type $N_{\frac{2}{3}}$) were considered to be as fertile as offspring resulting from homotypic parental species crosses ($C_{ij,k} = 1$) All other mating combinations involving different genotype classes were considered unsuccessful ($C_{ij,k} = 0$) due to the high level of mortality at hatching (>95%) and malformations in the surviving offspring [21,22,32].

Although allotriploid individuals ($N_{2/3}$) have never been detected in the wild, we considered them here because: 1) fecundation success is high between hybrid females and Atlantic salmon males ($N_{1/2} \times N_S$) [32]; 2) allotriploid progeny was produced and grown successfully in a semi-natural stream [33]; and 3) the ploidy of hybrids and their post-F_1 status have been rarely assessed in the field [22].

Many Norwegian Atlantic salmon populations are affected by a disease caused by the monogenean ectoparasite *Gyrodactylus salaris*, which was introduced in Norway in the 1970's by Atlantic salmon transported from the Baltic sea [34]. Atlantic salmon are severely affected in most of the infected rivers, while brown trout are known to be resistant. Hybrids have an intermediate susceptibility [35]. We incorporated the effects of this disease by decreasing the relative fitness of Atlantic salmon; we tested a 20% and a 40% reduction of fitness as compared to brown trout ($\omega_S = 0.8$ and $\omega_S = 0.6$). F_1 and F_2 hybrids were considered to have an intermediate susceptibility between both species ($\varepsilon_{S1/2} = \varepsilon_{T1/2} = \varepsilon_{S2/3} = \varepsilon_{1/2\,2/3} = 0.5$).

Results

Analytical exploration of the model

We performed a theoretical description of the dynamics of the populations, first without considering the effect of interspecific hybridization.

Considering equation (4), the population N_i reaches a non-trivial equilibrium (different from zero) at:

$$\cdot N_{i(t+1)} = \frac{V_i \ln \frac{R_i \omega_i}{1-S_i} - \alpha_{ij} V_j \ln \frac{R_j \omega_j}{1-S_j}}{\omega_i \left(1 - \alpha_{ij} \alpha_{ji}\right)} \qquad (6)$$

The population size increases with higher values of growth rate (R_i) and habitat size (V_i) and decreases with the interspecific competition coefficient (α_{ij}). In cases involving fitness reduction, the density of class i increases with higher values of ω_i and decreases with ω_j, which produces an increase of competitiveness of class j. If both species do not compete, N_i is positive only if $\frac{R_i \omega_i}{1-S_i} > 0$; in this case the output is undefined when the adult survival (S_i) is equal to 1. The Ricker model produces oscillatory population sizes due to the instability of the equilibrium point. Values of growth rate $R_i > \frac{(1-S_i) e^{\frac{2}{(1-s_i)}}}{\omega_i}$ yield an unstable equilibrium and the population dynamic becomes chaotic, the output being thus strongly affected by the initial conditions of the system.

We further explored the dynamics of our model by including the effects of hybridization. Due to the additional term γ_{ij} in equation (2) and the density dependent competition effect included in equation (4), the coupled dynamics of parental and hybrid abundances are not analytically solvable. We thus analysed only a special case of interspecific hybridization $\left(M_{ij(t)} = \frac{\gamma_{ij} N_{j(t)}}{N_{i(t)} + \gamma_{ij} N_{j(t)}} \right)$, with maximum competition $\left(\frac{\alpha_{ij}}{V_i} = \frac{\alpha_{ji}}{V_j} = 1 \right)$ and with symmetric interbreeding success rate and equal demographic parameters for both parental classes ($\gamma_{ij} = \gamma_{ji}$; $R_i = R_j$; $V_i = V_j$). Here, the density-dependent effect among populations is cancelled and the dynamic depends only on the interbreeding rate and the hybrid survival probability. The proportion of parental species N_1 in a community composed by parental species N_0, F_1 hybrids ($N_{1/2}$) and F_2 hybrids ($N_{2/3}$) reaches non-zero equilibrium at:

$$\frac{\dot{N}_1}{\dot{N}_0 + \dot{N}_1 + \dot{N}_{1/2} + \dot{N}_{2/3}} = \frac{2R_1^2\left(1 - S_{2/3}\right)\left(1 - S_{2/3}\right)}{\gamma_{10} R_{1/2}(1-S_1)\left(2R_1\left(1 - S_{2/3}\right) + R_{2/3}(1-S_1)(1+\gamma_{10})\right) + 4R_1^2\left(1 - S_{2/3}\right)\left(1 - S_{1/2}\right)} \qquad (7)$$

The proportion of N_1 increases with higher values of growth rate; it decreases with increasing interbreeding rate (with N_0) and with the survival of F_1 and F_2 hybrids.

This analytical exploration of our model showed that, despite its apparent simplicity, the model is nonlinear and the outputs are not trivial, strongly depending on the input parameters. Consequently, no general conclusion can be drawn that would be valid for a wide range of situations; each case should be cautiously investigated. More complex situations, involving competition and interbreeding success rates of varying intensities are difficult to explore analytically, but may be solved numerically as illustrated by our case study.

Assessing extinction risk in salmon and trout

Using our model we analysed a case of hybridization type 1, assessing the potential effects of hybridization between Atlantic salmon and brown trout in Norwegian rivers. This interspecific cross is characterized by a sex-biased reproductive success due to high offspring mortality in crosses where the female is a brown trout. To understand the dynamics of this particular hybridization system and to identify the conditions that can lead to extinction risk, we simulated a wide range of situations by varying the values of key parameters of the model, such as interbreeding success rate, interspecific competition, habitat size and growth rate. We also evaluated the effects of a disease that reduces the fitness of salmons and hybrids.

The parameters R (growth rate) and V (habitat size) were estimated through a non-linear least square method (see Table S2). The best estimated values were $R = 3$ (SE = 0.7) and $V = 51$ (SE = 10) for both species. The same parameter values were used for F_1 and F_2 hybrids (Table 1). In the scenario where the population of Atlantic salmon is not affected by the parasitic disease ($\omega_S = 1$), we simulated the outcomes of a gradual increase of a symmetrical interbreeding success rate ($\gamma_{ST} = \gamma_{TS}$) up to a completely panmictic reproduction between both species; no changes in the proportion of salmon and trout in the community was observed. In simulations with competition we used a density-dependent form of competition between genotype classes (see methods). When interspecific competition is considered only among hybrids and parental classes ($\alpha_{ST} = \alpha_{TS} = 0$), or when competition also occurs between Atlantic salmon and brown trout ($0 < \alpha_{ST} \neq \alpha_{TS} > 0$), no extinctions were observed when the interbreeding success rate is symmetrical (Figure 2a and 2b, respectively). In simulations where the interbreeding success rate is asymmetrical ($\gamma_{ST} \neq \gamma_{TS}$), due for instance to unequal mate choice relaxation in the parental species, and when there is no interspecific competition between salmon and trout, then extinction is observed only in extreme situations with high values of interbreeding success rate (Figure 2a). Overall, these simulation results indicate that, without interspecific competition, hybridization alone is not sufficient to drive one species population to extinction. Interestingly, due to competition with hybrids (which are more abundant when interbreeding success rate is larger in salmon), the critical area of salmon extinction ($N_S = 0$) is three times larger (6%) than the area of brown trout extinction (2%, $N_S = 100$; Figure 2a). Yet, if interspecific competition is considered, these areas are equal and larger for both species (about 36%; Figure 2b). Here, a difference of interbreeding success rates larger than 12% ($\Delta(\gamma_{ST},\gamma_{TS})>0.12$) generates either salmon or trout population extinction, depending on the orientation of the deficit. This indicates that if both species are in competition for resources, the one with the highest mate choice relaxation has the lowest survival probability, due to wasted reproductive effort.

When we simulate the additional effect of the parasitic disease by reducing salmon fitness by 20% ($\omega_S = 0.8$) as compared to brown trout, and in the case of no interspecific competition, the results indicate that both species survive in the fish community at any level of symmetric interbreeding success rate ($\gamma_{ST} = \gamma_{TS}$). However, when this rate is highly asymmetric ($\gamma_{ST} \neq \gamma_{TS}$), with values of $\gamma_{ST} > 0.78$ and $\gamma_{TS} = 0$, then the salmon population become extinct. The critical area of Atlantic salmon extinction ($N_S = 0$) represents 30% of all possible combinations of asymmetric interbreeding (Figure 2c). When we consider interspecific competition (Figure 2d), salmon is completely displaced by brown trout in all simulated conditions of symmetrical interbreeding success rates ($\gamma_{ST} = \gamma_{TS}$) or when interbreeding success rates are skewed towards salmon ($\gamma_{ST} > \gamma_{TS}$). However, when interbreeding success rates are skewed towards trout ($\gamma_{ST} < \gamma_{TS}$), it allows coexistence if $\Delta(\gamma_{ST},\gamma_{TS})>0.11$, or a complete displacement of brown trout if $\Delta(\gamma_{ST},\gamma_{TS})>0.35$.

When we simulate a salmon fitness reduction of 40% ($\omega_S = 0.6$) with no interspecific competition, salmon population become extinct if $\gamma_{ST} > 0.55$ and $\gamma_{TS} = 0$. With other values of γ_{ST} and γ_{TS}, it cannot subsist at a proportion higher than 50% (Figure 2e). The critical area of extinction for the Atlantic salmon represents 51.2% of all combinations of asymmetrical interbreeding success rate. Regarding brown trout, it persists at any level of symmetric or asymmetric interbreeding success rate (Figure 2e). When we consider interspecific competition in the simulations (Figure 2f), any level of symmetric interbreeding success rates ($\gamma_{ST} = \gamma_{TS}$) or asymmetric rates skewed towards Atlantic salmon ($\gamma_{ST} > \gamma_{TS}$) leads to the displacement of salmon by brown trout, while, when skewed towards brown trout ($\gamma_{ST} < \gamma_{TS}$), it allows coexistence if $\Delta(\gamma_{ST},\gamma_{TS})>0.55$ or a complete displacement of brown trout if $\Delta(\gamma_{ST},\gamma_{TS})>0.8$. Overall, these simulations show that the parasitic disease strongly perturbs the system by threatening salmon, and this effect is enhanced by high interbreeding success rates in salmon or limited by high interbreeding success rates in trout.

The results presented above (Figure 2) remain valid when using the upper and lower limits of the 95% confidence interval of the growth rate (R) and habitat size (V) parameters (Figure S1 and Figure S2). The results with interspecific competition and symmetrical interbreeding success rates are independent of the changes in R and V. Without interspecific competition, the probability of reaching extinction is inversely proportional to both parameters R and V. We can therefore expect that without competition, the effect of hybridization, combined with the parasitic disease, would be stronger in small rivers supporting smaller and local populations, whereas the effect of hybridization would be negligible in larger rivers, with bigger populations.

We then performed a sensitivity analysis of the system regarding the population growth rate parameter (R), without considering interspecific competition (Figure 3). Under a salmon fitness reduction of 40% ($\omega_S = 0.6$), a higher value of R for all the interacting populations counteracts the negative effects that hybridization produces on the demography of salmon. With higher growth rates, higher interbreeding success rates ($\gamma_{ST} = \gamma_{TS} > 0.4$) are necessary to cause population extinction. Moreover, the dominant or recessive inheritance of resistance to pathogens in hybrids seems to have a more pronounced effect when growth rates are higher. When the trout resistance to pathogens is inherited recessively by hybrids, values of $R = 6$ allow salmon persistence even with a panmictic mate choice ($\gamma_{ST} = \gamma_{TS} = 1$). However, when resistance to pathogens is dominantly or co-dominantly inherited, then salmon extinction occurs (Figure 3a). A value of $R = 12$ generates oscillatory dynamics allowing salmon and hybrids to survive in the community even at high interbreeding success rate ($\gamma_{ST} = 1$), and even if the trout resistance to pathogens is dominantly inherited by hybrid classes (Figure 3a and 3b). With $R = 3$, an inflexion point is produced at 6 time steps (years), where the number of hybrids exceeds the number of salmons, but both classes become extinct before 23 time steps (years). A minimum of $R = 8$ is required to maintain the population of salmons, whereas values of $R > 14$ generate non-stable equilibrium in the salmonids community (Figure 4). If, in addition to the salmon fitness reduction of 40%, we add interspecific competition in our simulations, this factor drives salmon extinction even without considering interspecific hybridization (data not shown). These results indicate that hybridization alone is unlikely to cause salmon population extinction, but if it occurs in combination with competition and/or with the disease examined here, together they constitute a serious threat for salmon populations.

Discussion

Distant hybridization

We developed a general model to assess how hybridization between distant species can impact the demography of parental species. This type of hybridization occurs, on one hand, when hybrids are inviable or infertile due to post-zygotic barriers, and the risk to parental species resides in the wasted reproductive effort, as it has been reported in mammals and birds [36,37]. On the other hand, hybrids can be fertile, but their gametes may contain the non-recombined haploid genome of the two parental species (in different gametes) or a single haploid genome as the product of genome exclusion before or during meiosis. Hybrids producing clonal or hemiclonal gametes are common in plants and invertebrates [7,38]. In vertebrates, it has been frequently reported in amphibians, fish and reptiles [39–41] but not in birds nor in mammals. The model presented herein accounts for all these cases and is therefore useful to study and generate theoretical expectations in a large variety of organisms and biological issues. For instance, our model could be implemented to determine the conditions under which populations may reach a stable equilibrium in gynogenetic, parthenogenetic or hybridogenetic systems. It can also serve to understand how different polyploid forms of hybrid origin can persist over large periods of time. In the field of conservation, it is essential to determine the minimum population size and maximum hybridization rate that a species can stand before interspecific hybridization threatens its persistence. The increasing frequency of interspecific hybridization due to anthropogenic causes and global climate change is of growing concern in conservation biology, where efficient tools to project the conse-

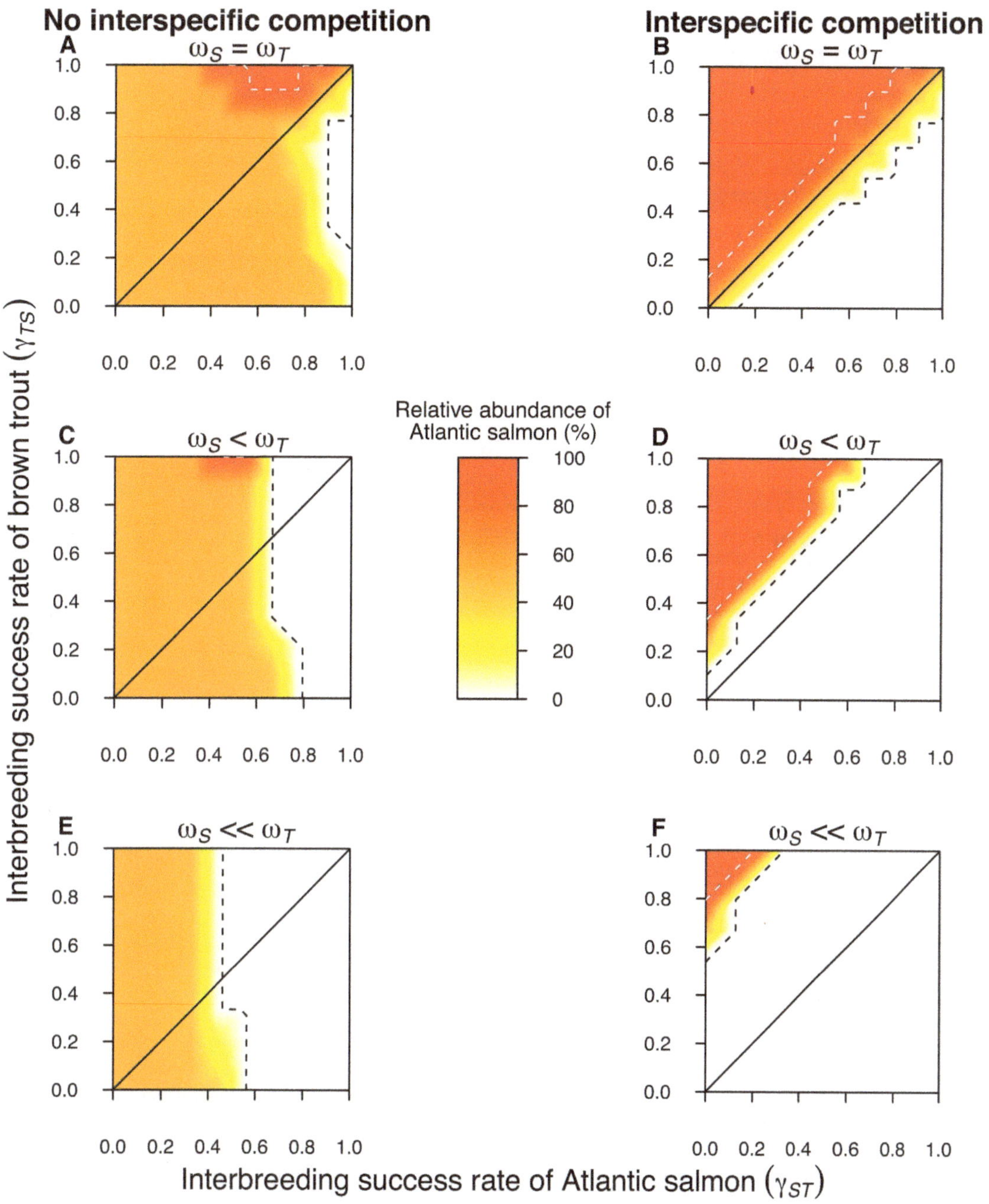

Figure 2. Relative abundance of Atlantic salmon (%) as compared to brown trout ($\frac{N_S}{N_S+N_T}$). The diagonal solid lines represent equal interbreeding success rates between Atlantic salmon (γ_{ST}) and brown trout (γ_{TS}). Black or white dotted lines delimit the extinction area of Atlantic salmon and brown trout population, respectively. $\omega_S = \omega_T$ indicates equal fitness for Atlantic salmon ($\omega_S = 1$) and brown trout. $\omega_S < \omega_T$ indicates that Atlantic salmon has a 20% fitness reduction ($\omega_S = 0.8$). $\omega_S << \omega_T$ indicates that Atlantic salmon has a 40% fitness reduction ($\omega_S = 0.6$). In a), c), and e), Atlantic salmon and brown trout do not compete. In b), d), and f), Atlantic salmon and brown trout have density-dependent competition. The data presented correspond to the situation after 100 time steps (years).

quences on the demography of parental species are particularly welcome.

The salmon and trout analysis

In natural conditions, Atlantic salmon and brown trout present low levels of interspecific hybridization, revealing efficient mechanisms of reproductive isolation between both species [20]. However, hybrids are increasingly frequent [24], in particular because escaped individuals raised in farms exhibit a relaxed species mate choice [42]. In addition, overfishing and diseases have significantly reduced salmon populations locally [24]. In such conditions, the rare species has more difficulties in finding a conspecific partner and becomes less demanding when looking for a mate, a situation known as the "*desperation hypothesis*" [43]. This situation favors hybridization, which in turn accelerates species rarity.

Table 1. List of functions and parameters with the case study values.

		Case study*
List of functions		
N_i	Number of adult individuals of genotypic class i[†]	
M_{ij}	Mating probability between genotypic class i and j	
$b_{ij,k}$	Weighted number of breeding pairs $i \times j$ resulting in offspring of class k	
n_k	Final weighted number of breeding pairs yielding offspring of class k	
Demographic parameter		
θ	Time delay from hatching to age maturity	3
S	Adult survival probability	0.3
R	Population growth rate	3
α	Interspecific competition coefficient	
V	Habitat size	51
Interbreeding parameters		
γ	Interbreeding success rate	1[a]
C	Relative offspring type produced by breeding pairs[‡]	
ω	Fitness of an inherited character	1[b]
ε	Dominance degree of parental traits on hybrids	0.5[c]

*Fixed value for: [a]$\gamma_{½S}$, $\gamma_{½T}$, $\gamma_{⅔S}$, $\gamma_{⅔T}$, $\gamma_{½⅔}$, $\gamma_{⅔½}$; [b]ω_T; [c]$\varepsilon_{S½}$, $\varepsilon_{T½}$, $\varepsilon_{S⅔}$, $\varepsilon_{½⅔}$.
[†][Initial size: $N_T = N_S = 50$; $N_{½} = N_{⅔} = 0$].
[‡][see Table S1].

Our results show that the asymmetrical interbreeding success rate between Atlantic salmon and brown trout, which might be principally due to mate choice relaxation, will yield different trends depending on the direction and intensity of the asymmetry. A higher interbreeding success rate in brown trout compared to salmon ($\gamma_{ST} < \gamma_{TS}$) produces fewer scenarios with extinction, because the offspring are inviable. In contrast, a higher interbreeding success rate in Atlantic salmon produces more potential situations with extinction, because the hybrid progeny is viable and competes with the progeny of both parental classes. Interestingly, we found that no extinction is expected if the interbreeding success rate is symmetrical between both species. Nevertheless, according to our simulations, the interbreeding success rate must be very high (>70%) and asymmetrical to drive populations of one of the two species to extinction, which means that in nature, hybridization *per se* is probably not a serious threat. However, this statement changes when hybridization is combined with an additional threat, such as the disease caused by the monogenean *G. salaries* in Atlantic salmon. In this case, an increasing interbreeding success rate of salmon increases its extinction risk.

We also show that salmon populations can reach extinction with low interbreeding success rates, but only when interspecific competition between both species is high. Although there is evidence that brown trout is a strong competitor that displaces Atlantic salmon, interspecific competition is probably not a major risk in natural sympatric populations, as they coexist in different microhabitats [30]. However, allopatric young Atlantic salmon tend to expand their space in the absence of brown trout [44], supporting the idea that brown trout outcompete young salmon in parts of its habitat. If one or both species are exotic, then interspecific competition may be enhanced. This is for instance the case in the Kerguelen Island, where both species were introduced, and brown trout is invading and displacing Atlantic salmon, with little hybridization occurring after the very initial contact [45].

The case of Atlantic salmon and brown trout is an example of human-induced environmental changes that have increased the hybridization rate between species that have historically coexisted in sympatry. However, given the low interbreeding success rates registered in the wild [45], we conclude that interspecific hybridization between Atlantic salmon and brown trout is likely not a threat *per se* for the persistence of most populations, except in very extreme situations where the interbreeding is particularly asymmetrical and high for one or both species. Such situation may be found in rivers dominated by fishes released from farms [46], as they show highly relaxed mate choice [47]. Our case study also reveales that the combined effects of interspecific hybridization with interspecific competition and/or with an additional threat, such as the parasitic disease, might seriously enhance extinction risk. In the near future, the effects of global climate change will probably call for a revision of our conclusions, as these modifications may alter habitat characteristics, migration patterns, age of maturity, reproduction time and susceptibility to diseases [48]. Our model will be the ideal tool to anticipate the impact of climate change on organisms that may undergo distant hybridization.

Model components

As compared to previous models, the one presented herein simultaneously accounts for important ecological, genetic and behavioural parameters like 1) density-dependent competition at the intra- and the inter-specific level; 2) the fact that traits can be inherited in a dominant or recessive way in hybrids; and 3) variable mate choice relaxation between interacting species leading to case specific assortative mating. This renders our model more realistic, more general and also more flexible as compared to previous attempts. Moreover, the components assembled in our model have been previously presented and validated, some being of general use in ecology and demography. The basic formula for calculating the probability M_{ij} that an individual of class i mates

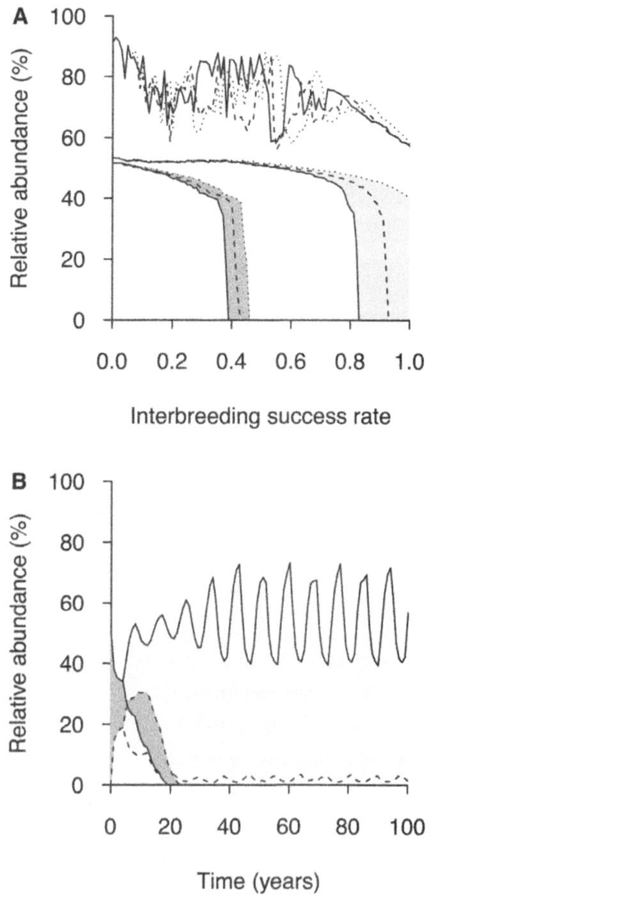

Figure 3. Relative abundance of a population of Atlantic salmon affected by a disease ($\omega_S = 0.6$). The abundance is given in percent of the total number of salmonids ($\frac{N_S}{N_S + N_T + N_{1/2} + N_{2/3}}$). Brown trout are resistant to disease ($\omega_T = 1$) and are not in competition with Atlantic salmon. (■) $R = 3$; (■) $R = 6$; (■) $R = 12$. In a) effects of varying yet symmetric interbreeding success rate ($\gamma_{ST} = \gamma_{TS}$); the trout disease resistance is inherited with the following properties: (—) dominantly in hybrids ($\varepsilon_{T\,1/2} = 1$ and $\varepsilon_{1/2\;2/3} = 1$); (...) recessively in hybrids ($\varepsilon_{T\,1/2} = 0$ and $\varepsilon_{1/2\;2/3} = 1$); and (– –) codominantly in hybrids ($\varepsilon_{T\,1/2} = 0.5$ and $\varepsilon_{1/2\;2/3} = 0.5$). In b) time series of the relative abundance of Atlantic salmon (—) and hybrid populations (...) ($(N_{1/2}+N_{2/3})/(N_0+N_1+N_{1/2}+N_{2/3})$). The data presented correspond to the situation after 100 time steps (years).

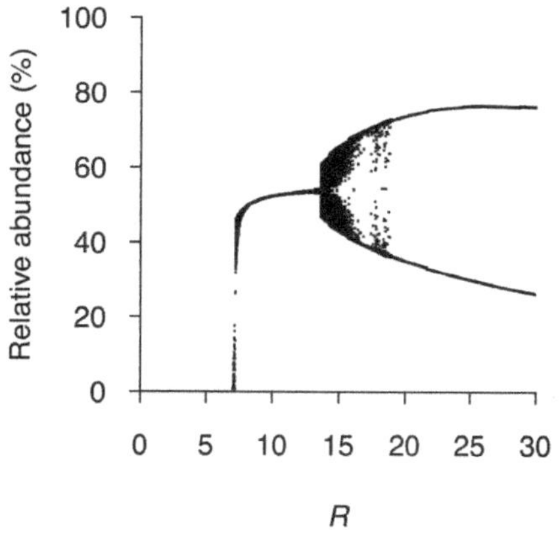

Figure 4. Bifurcation diagram of the effect of growth rate (R) on the Atlantic salmon relative abundance. When $R \geq 8$, Atlantic salmons are not threatened; when $R > 15$, then salmon density starts to be chaotic.

with one of class j in equation (1), has been proposed and used in essentially the same way as in Ferdy and Austerlitz [26] and Currat *et al.* [13]. The way we calculate the total number of offspring of a given class k, which is given by equation (3), is simply the sum of offspring of class k produced by all possible crosses that generate at least a fraction of class k in their progeny, the later being given by equation (2). To consider the temporal dynamics of wild adult populations in equation (4), we extended a version of the Ricker model [27], which is frequently used in the fields of ecology and population dynamics. Our model also takes into account the "lattice effects", in equation (4), as proposed by Henson *et al.* [28], which has been studied and validated by population ecologists. Finally, our model considers the species habitat size, the parameter V_i in our equation (4), in the same way as proposed by Henson et al. [28].

In our case study, we adapted the way of calculating one variable of the model, the interspecific competition coefficient α_{ij}. Because no prior knowledge about possible values was available, we used a density-dependent form of competition given by equation (5), as proposed by Currat and Excoffier [31]. This exemplifies the flexibility of our model, in which several parameters can be substituted by alternative ways of calculation to cope with case-specific characteristics.

Extensions of the model

While applicable to many situations in its current form, our model may be easily extended to cope with more complex or different systems. As an example of possible extension, the demographic regulation may be modified. The Riker function used here could be changed, even thought it has been designed to be used on a wide range of taxa, including fish, amphibian and insects, which are particularly subject to distant hybridization [28]. A logistic function can be an alternative way of considering demographic growth [49], which avoids the overcompensation of the exponential function that keeps adult recruitment low when spawner abundance is high [50]. It is also possible to use different times to sexual maturity for the different parental species ($\theta_i \neq \theta_j$), which may affect the dynamics of the system. The interbreeding success rate parameter incorporated in the model provides a mean to easily characterize the combination of factors that control interbreeding success, such as mate choice relaxation between different species or hybrid offspring survival. However, future applications could decompose this parameter into the interacting factors that affect the interbreeding success rate. If gonochoric organisms with unequal sex ratio are considered, the model can be simply modified by performing a separate calculation for males and female, where equation (1) and equation (2) have to be computed with class i corresponding to the subgroup of females (i^f) and class j to the subgroup of males (j^m). In a similar way, it is possible to consider female fecundity variations among parental species and hybrids, while male fecundity remains unchanged. This can be incorporated with the fitness parameter ω_k of equation (2), and by performing different calculations for males and females. An additional development could consist in incorporating genetic introgression into the model. However, further and thorough investigations would be needed to extend our model to include genetic introgression between parental species.

Limitations of the model

The use of the model is limited to the types of interspecific hybridization that do not involve genetic introgression between the parental species, that is, Type 1 and Type 2 (see introduction). The use of this model can also be limited by the amount of available knowledge about basic population and ecological parameters of the species analysed. Our model has parameters for which plausible values are required to produce accurate solutions. However, some parameters can be estimated by a linear or non-linear approach, as in our case study, if enough information about time-series of species demography is available. It is also possible to use our model to understand the role played by a specific parameter in the system by varying this parameter while keeping all other parameters unchanged.

Conclusions

The model presented herein is a tool that opens new and promising path to investigate and understand evolutionary and conservation issues, including the study of the emergence and evolution of hybrid forms and the understanding of the effects of distant hybridization on the demography of parental species. In conservation biology, our model will permit to set out management recommendations by assessing the effects of alternative strategies to reduce extinction risk, or projecting the impact of emerging threats on already affected or yet unaffected species. Moreover, our model is flexible, as it can be easily modified to accommodate additional parameters or alternative functions to better fit to taxon-specific situations. The script of our model is freely available at: http://genev.unige.ch/montoya-currat/scripts/. In the implementation presented here, we have highlighted that hybridization of type 1 between Atlantic salmon and brown trout can lead to important demographic changes in the populations, although extinction is predicted only in very peculiar and improbable situations only. The flexibility of our model enabled us to assess the influence of an additional risk factor, a parasitic disease, and showed that the combined effects of interspecific hybridization and the unequal resistance to pathogens may lead, this time, to the extinction of affected Atlantic salmon populations.

Supporting Information

Figure S1 Relative abundance of Atlantic salmon (%) as compared to brown trout ($\frac{N_S}{N_S+N_T}$). These results are obtained when using the lower limit of the 95% confident interval of growth rate ($R = 1.63$) and habitat size ($V = 31.4$) (see Figure 2, main text).

Figure S2 Relative abundance of Atlantic salmon (%) as compared to brown trout ($\frac{N_S}{N_S+N_T}$). These results are obtained when using the upper limit of the 95% confident interval of growth rate ($R = 4.37$) and habitat size ($V = 70.6$) (see Figure 2, main text).

Table S1 Mating frequencies and relative number of offspring types produced by the intercrosses among Atlantic salmon (N_S), brown trout (N_T), first-generation hybrids ($N_{1/2}$) and second-generation hybrids ($N_{2/3}$).

Table S2 Models with equal or different values of growth rate (R) and habitat size (V) for populations of Atlantic salmon (N_S) and brown trout (N_T).

Appendix S1 Estimation of the growth rate (R) and habitat size (V) parameter values by a non-linear least square method.

Acknowledgments

We thank Nicolas Ray, Claire Shea and Lara Pizurki for their constructive comments, and Christian Gillet, Eva Garcia-Vasquez and Jean-Luc Falcone for technical advices.

Author Contributions

Conceived and designed the experiments: CSQ MC JIMB. Performed the experiments: CSQ. Analyzed the data: CSQ MC JIMB. Contributed reagents/materials/analysis tools: CSQ MC JIMB. Wrote the paper: CSQ MC JIMB.

References

1. Arnold ML, Martin NH (2010) Hybrid fitness across time and habitats. Trends in Ecology & Evolution 25: 530–536.
2. Allendorf FW, Leary RF, Spruell P, Wenburg JK (2001) The problems with hybrids: setting conservation guidelines. Trends in Ecology & Evolution 16: 613–622.
3. Lande R (1988) Genetics and demography in biological conservation. Science 241: 1455–1460.
4. Rhymer JM, Simberloff D (1996) Extinction by hybridization and introgression. Annual Review of Ecology and Systematics 27: 83–109.
5. Konishi M, Takata K (2004) Impact of asymmetrical hybridization followed by sterile F-1 hybrids on species replacement in Pseudorasbora. Conservation Genetics 5: 463–474.
6. Yakovlev VN, Slyn'ko YV, Grechanov IG, Krysanov EY (2000) Distant hybridization in fish Journal of Ichthyology 40: 298–311.
7. Mantovani B, Scali V (1992) Hybridogenesis and androgenesis in the stick-insect *Bacillus rossius-Grandii benazzii* (Insecta, Phasmatodea). Evolution 46: 783–796.
8. Crespo-Lopez ME, Duarte T, Dowling T, Coelho MM (2006) Modes of reproduction of the hybridogenetic fish *Squalius alburnoides* in the Tejo and Guadiana rivers: An approach with microsatellites. Zoology 109: 277–286.
9. Kierzkowski P, Pasko L, Rybacki M, Socha M, Ogielska M (2011) Genome dosage effect and hybrid morphology - the case of the hybridogenetic water frogs of the *Pelophylax esculentus* complex. Annales Zoologici Fennici 48: 56–66.
10. McDevitt AD, Mariani S, Hebblewhite M, Decesare NJ, Morgantini L, et al. (2009) Survival in the Rockies of an endangered hybrid swarm from diverged caribou (*Rangifer tarandus*) lineages. Molecular Ecology 18: 665–679.
11. Barton NH, Hewitt GM (1985) Analysis of hybrid zones. Annual Review of Ecology and Systematics 16: 113–148.
12. Dasmahapatra KK, Walters JR, Briscoe AD, Davey JW, Whibley A, et al. (2012) Butterfly genome reveals promiscuous exchange of mimicry adaptations among species. Nature 487: 94–98.
13. Currat M, Ruedi M, Petit RJ, Excoffier L (2008) The hidden side of invasions: Massive introgression by local genes. Evolution 62: 1908–1920.
14. Liu SJ (2010) Distant hybridization leads to different ploidy fishes. Science China-Life Sciences 53: 416–425.
15. Lampert KP (2009) Clonal reproduction in freshwater fish: mechanisms, systematic overview, genetic and ecological consequences. Fundamental and Applied Limnology 174: 245–260.
16. Rodriguez DJ (1996) A model for the establishment of polyploidy in plants: Viable but infertile hybrids, iteroparity, and demographic stochasticity. Journal of Theoretical Biology 180: 189–196.
17. Som C, Reyer HU (2006) Demography and evolution of pure hybridogenetic frog (*Rana esculenta*) populations. Evolutionary Ecology Research 8: 1235–1248.
18. Satake A, Araki H (2012) Stocking of captive-bred fish can cause long-term population decline and gene pool replacement: predictions from a population dynamics model incorporating density-dependent mortality. Theoretical Ecology 5: 283–296.
19. Gray AK, Evans MA, Thorgaard GH (1993) Viability and development of diploid and triploid salmonid hybrids. Aquaculture 112: 125–142.
20. Alvarez D, Garcia-Vazquez E (2011) Maintenance of asymmetric hybridization between Atlantic salmon (*Salmo salar*) and brown trout (*Salmo trutta*) via

postzygotic barriers and paternal effects (vol 68, pg 593, 2011). Canadian Journal of Fisheries and Aquatic Sciences 68: 952–952.
21. Garcia-Vazquez E, Perez J, Ayllon F, Martinez JL, Glise S, et al. (2004) Asymmetry of post-F-1 interspecific reproductive barriers among brown trout (*Salmo trutta*) and Atlantic salmon (*Salmo salar*). Aquaculture 234: 77–84.
22. Castillo AGF, Beall E, Moran P, Martinez JL, Ayllon F, et al. (2007) Introgression in the genus Salmo via allotriploids. Molecular Ecology 16: 1741–1748.
23. Jansson H, Ost T (1997) Hybridization between Atlantic salmon (*Salmo salar*) and brown trout (*S. trutta*) in a restored section of the River Dalalven, Sweden. Canadian Journal of Fisheries and Aquatic Sciences 54: 2033–2039.
24. Urke HA, Koksvik J, Arnekleiv JV, Hindar K, Kroglund F, et al. (2010) Seawater tolerance in Atlantic salmon, *Salmo salar* L., brown trout, *Salmo trutta* L., and *S. salar*×*S. trutta* hybrids smolt. Fish Physiology and Biochemistry 36: 845–853.
25. Hindar K, Balstad T (1994) Salmonid culture and interspecific hybridization. Conservation Biology 8: 881–882.
26. Ferdy JB, Austerlitz F (2002) Extinction and introgression in a community of partially cross-fertile plant species. American Naturalist 160: 74–86.
27. Wilbur HM (1976) Density-dependent aspects of metamorphosis in *Ambystoma* and *Rana sylvatica*. Ecology 57: 1289–1296.
28. Henson SM, Costantino RF, Cushing JM, Desharnais RA, Dennis B, et al. (2001) Lattice effects observed in chaotic dynamics of experimental populations. Science 294: 602–605.
29. Jonsson B, Jonsson N (2011) Ecology of Atlantic salmon and brown trout: habitat as a template for life histories. Dordrecht: Springer-Verlag.
30. Armstrong JD, Kemp PS, Kennedy GJA, Ladle M, Milner NJ (2003) Habitat requirements of Atlantic salmon and brown trout in rivers and streams. Fisheries Research 62: 143–170.
31. Currat M, Excoffier L (2004) Modern humans did not admix with Neanderthals during their range expansion into Europe. Plos Biology 2: 2264–2274.
32. Galbreath PF, Thorgaard GH (1995) Sexual maturation and fertility of diploid and triploid Atlantic salmon×brown trout hybrids. Aquaculture 137: 299–311.
33. Garcia-Vazquez E, Ayllon F, Martinez JL, Perez J, Beall E (2003) Reproduction of interspecific hybrids of Atlantic salmon and brown trout in a stream environment. Freshwater Biology 48: 1100–1104.
34. Johnsen BO, Jensen AJ (1991) The gyrodactylus story in norway. Aquaculture 98: 289–302.
35. Bakke TA, Soleng A, Harris PD (1999) The susceptibility of Atlantic salmon (*Salmo salar* L.)×brown trout (*Salmo trutta* L.) hybrids to *Gyrodactylus salaris* Malmberg and *Gyrodactylus derjavini* Mikailov. Parasitology 119: 467–481.
36. Birkhead TR, Brillard JP (2007) Reproductive isolation in birds: postcopulatory prezygotic barriers. Trends in Ecology & Evolution 22: 266–272.
37. Rozhnov VV (1993) Extinction of the European mink: ecological catastrophe or a natural process? Lutreola 1: 10–16.
38. Gugsa L, Sarial AK, Lorz H, Kumlehn J (2006) Gynogenic plant regeneration from unpollinated flower explants of *Eragrostis tef* (Zuccagni) Trotter. Plant Cell Reports 25: 1287–1293.
39. Dawley RM, Schultz RJ, Goddard KA (1987) Clonal reproduction and polyploidy in unisexual hybrids of *Phoxinus eos* and *Phoxinus eogaeus* (pisces, cyprinidae). Copeia: 275–283.
40. MaCculloch RD, Murphy RW, Kupriyanova LA, Darevsky IS, Danielyan FD (1995) Clonal variation in the parthenogenetic rock lizard *Lacerta armeniaca*. Genome 38: 1057–1060.
41. Lode T (2001) Character convergence in advertisement call and mate choice in two genetically distinct water frog hybridogenetic lineages (*Rana kl esculenta, Rana kl grafi*). Journal of Zoological Systematics and Evolutionary Research 39: 91–96.
42. Castillo AGF, Ayllon F, Moran P, Izquierdo JI, Martinez JL, et al. (2008) Interspecific hybridization and introgression are associated with stock transfers in salmonids. Aquaculture 278: 31–36.
43. Hubbs CL (1955) Hybridization between fish species in nature. Systematic Zoology 4: 1–20.
44. Heggenes J, Saltveit SJ (2007) Summer stream habitat partitioning by sympatric Arctic charr, Atlantic salmon and brown trout in two sub-arctic rivers. Journal of Fish Biology 71: 1069–1081.
45. Horreo JL, Ayllon F, Perez J, Beall E, Garcia-Vazquez E (2011) Interspecific Hybridization, a Matter of Pioneering? Insights from Atlantic Salmon and Brown Trout. Journal of Heredity 102: 237–242.
46. Jonsson N, Jonsson B, Hansen LP (2003) The marine survival and growth of wild and hatchery reared Atlantic salmon. Journal of Applied Ecology 40: 900–911.
47. Youngson AF, Webb JH, Thompson CE, Knox D (1993) Spawning of escaped farmed Atlantic salmon (Salmo salar): hybridization of females with brown trout (Salmo trutta). Canadian Journal of Fisheries and Aquatic Sciences 50: 1986–1990.
48. Jonsson B, Jonsson N (2009) A review of the likely effects of climate change on anadromous Atlantic salmon *Salmo salar* and brown trout *Salmo trutta*, with particular reference to water temperature and flow. Journal of Fish Biology 75: 2381–2447.
49. Verhulst PF (1838) Notice sur la loi que la population poursuit dans son accroissement. Correspondance Mathematique et Physique 10: 113–121.
50. Wilbur HM (1996) Multistage life cycles. In: Rhodes OE, Jr, Chesser RK, Smith MH, editors. Population dynamics in ecological space and time. Chicago: Univ. of Chicago Press. pp. 75–108.

16

The Interaction between Selection, Demography and Selfing and How It Affects Population Viability

Diala Abu Awad[1]*, Sophie Gallina[1], Cyrille Bonamy[2], Sylvain Billiard[1]

1 UMR-CNRS 8198, Laboratoire Génétique et Évolution des Populations Végétales, Université Lille 1, Villeneuve d'Ascq, France, **2** Centre de Ressources Informatiques (CRI), Université Lille 1, Villeneuve d'Ascq, France

Abstract

Population extinction due to the accumulation of deleterious mutations has only been considered to occur at small population sizes, large sexual populations being expected to efficiently purge these mutations. However, little is known about how the mutation load generated by segregating mutations affects population size and, eventually, population extinction. We propose a simple analytical model that takes into account both the demographic and genetic evolution of populations, linking population size, density dependence, the mutation load, and self-fertilisation. Analytical predictions were found to be relatively good predictors of population size and probability of population viability when verified using an explicit individual based stochastic model. We show that initially large populations do not always reach mutation-selection balance and can go extinct due to the accumulation of segregating deleterious mutations. Population survival depends not only on the relative fitness and demographic stochasticity, but also on the interaction between the two. When deleterious mutations are recessive, self-fertilisation affects viability non-monotonically and genomic cold-spots could favour the viability of outcrossing populations.

Editor: Yury E. Khudyakov, Centers for Disease Control and Prevention, United States of America

Funding: This work benefited from the support of the Chair "Modélisation Mathémathique et Biodiversité" of Veolia Environnement - École Polytechnique - Muséum National d'Histoire Naturelle - Fondation X and the French Agence Nationale de la Recherche ANR-11-BSV7- 013-03. The funders had no role in study design, data collection and analysis, decision to publish, or preparation of the manuscript.

Competing Interests: The authors have declared that no competing interests exist.

* E-mail: diala.abu-awad@ed.univ-lille1.fr

Introduction

Population size and viability are both affected by extrinsic (e.g. environmental change and interspecific interactions) and intrinsic factors (e.g. genetic and demographic components). The genetic factors most frequently considered as contributing to population decline are the lack of adaptive potential in a changing environment, inbreeding depression and the reduction of fitness due to the accumulation of deleterious mutations (reviewed in [1]). The accumulation of deleterious mutations has often been put forth as an explanation for species extinction, especially concerning the differences in extinction rates between sexual and asexual species, or selfers and outcrossers. The relevance of the accumulation of deleterious mutations on population extinction, however, remains unclear.

Both empirical and theoretical works have insisted on the importance of deleterious mutation fixation on the decline and extinction of populations. Some models have shown that small populations can go extinct due to the acceleration of recurrent fixation of deleterious mutations with a small effect, the so called mutational meltdown [2–5]. Several empirical works have also supported this hypothesis. The fitness of experimental populations has been shown to decrease after several generations during which new deleterious mutations are fixed [6–10], and data from small highly inbred natural populations follow the same trend [11–13]. However, in these theoretical and empirical works, populations are considered to be small and isolated or, because of successive growth-dilution cycles, are subjected to recurrent and strong bottlenecks. When these conditions are not met (i.e. when populations are larger, or are not isolated or are not subject to strong recurrent bottlenecks) populations are more likely to go extinct because of other genetic and demographic factors before the fixation of deleterious mutations has an effect on population decline [1].

What about large populations, can they decline in size due to recurrent deleterious mutations? It has been suggested that the mutation load due to segregating mutations might be important in population decline [4,14]. This however remains controversial as it is generally thought that in large sexual populations deleterious mutations should be efficiently purged [15,16]. First, segregating deleterious mutations are expected to have no consequence on demography [17], especially in the presence of density-dependence, where there is compensation of the death of individuals due to selection by those that would have been lost from the population due to the lack of resources (soft selection). Empirical evidence on the other hand supports more "hard selection" [18], in which density-independent deaths due to the genetic load are not completely compensated. It is crucial to determine whether segregating mutations are important or not for determining population viability as this has direct empirical implications, especially when considering the genetic rescue of populations. Second, many aspects of population survival and extinction in presence of a high mutation load still remain unclear. When taking into account empirical estimations of genomic mutation rates (between 0.1 and more than 1 for higher Eukaryotes per genome and per generation, [19]) and the effects of deleterious mutations, population genetics theories [20] imply that we should expect extremely high mutation loads. Population genetics models consider population size as a fixed variable, and their stochastic estimations of the mutation load, even in finite populations, also agree with the existence of high mutation loads [21,22]. Can these predictions still hold if we allow selection to influence population size, or would the mutational load evolve differently? One of the earliest models, to our knowledge, that has taken into account the effect of the mutation load on mean fitness, and the latters effect on population size is a single-locus model proposed by Clarke [23]. In this model, the mutation load had a very small, almost negligible effect on population size, however Clarke [23] verbally suggested that extending the model to a whole genome could possibly lead to a more important effect.

In order to better understand the interaction between the mutation load and demography, we propose a model that combines simple deterministic population genetics and demographic models. We consider sexual reproduction, with a mean reproductive rate that depends both on population density and on the population's mutation load, and recurrent mutations segregating at an infinite number of loci. Using this model we predict the relationship between population size and the mutational load at mutation-selection equilibrium. We also predict the threshold fitness value depending on both the genetic and demographic parameters under which the population is not expected to be viable. These predictions are then verified using an individual-based simulation model in which we explicitly model the introduction of new mutations in the genome and the effect of the mean genomic recombination rate. Population size varies from one generation to the next, as it depends on individual fitness and competition. This simulation model allows for the better understanding of the mechanisms leading to population extinction, more specifically the relative importance of the polymorphism of deleterious mutations, the fixation of these mutations, and the mutational meltdown.

To illustrate the importance of such an approach (i.e. combining genetic and demographic factors) in ecology and evolution, we will address the question of the effect of self-fertilisation on population size and viability. This is indeed a long running question, as the prevalence of outcrossing species is still puzzling both in animals and plants [24]. From an evolutionary standpoint, self-fertilisation should be greatly advantaged because of Fisher's transmission advantage [25], a more efficient purge of deleterious mutations [26], and also because of their reproductive assurance (e.g. Baker's law, [27,28]). This is correlated with the empirical estimation of high transition rates from outcrossing to selfing reproductive systems, for instance in the Solanaceae [29]. Despite this transition rate, and high speciation rates in selfers compared to outcrossers, outcrossers still represent more than 40% of species in this family. Other studies in other plant families also come to this conclusion [30–32]. Goldberg *et al.* [29] show that this puzzling prevalence of outcrossers is due to higher extinction rates in selfing species than in outcrossing ones. One hypothesis to explain this difference in extinction rates is that selfers are more prone to mutational meltdown than outcrossers [4]. Empirical work on fungi, more specifically *Neurospora*, strongly supports this hypothesis, as selfing lineages accumulate more deleterious mutations and are less fit than outcrossing ones [33]. We will therefore extend our model in order to include different rates of self-fertilisation and test this hypothesis.

Model

Deterministic model and expectations

We consider a population in a constant environment, with discrete, non-overlapping generations. At generation t, the population is made up of N_t hermaphrodite individuals, where

$$N_t = N_{t-1} R_{t-1}. \tag{1}$$

R_t is the absolute multiplicative fitness of a population at a given generation t, with trait value (or relative fitness) W_t, in a population of density N_t [34] is given by

$$R_t = r_0^{1-\frac{N_t}{K}} W_t. \tag{2}$$

The density-dependent component of fitness depends on K, the carrying capacity of a population with all individuals having the optimal relative fitness, and on r_0, the intrinsic reproductive rate of such a population [34]. The second factor, W_t, corresponds to the mean relative fitness of individuals in the population, as a function of their mutation load, and so on the number of segregating or fixed deleterious mutations in the population. In this model, we consider that density dependence affects all individuals in the same way, independently of their relative fitness (or genotype).

If the mean relative fitness is at equilibrium (when the population is at mutation-selection balance), and there is no demographic stochasticity, the equilibrium of population size denoted N_{det}, can be expressed as a function of the relative fitness at equilibrium W_{eq}, r_0 and K, giving

$$N_{det} = K\left(1 + \frac{\mathrm{Ln}(W_{eq})}{\mathrm{Ln}(r_0)}\right). \tag{3}$$

By solving $N_{det} = 0$ from equation 3, we can determine the relative fitness threshold

$$W_{min} = \frac{1}{r_0}, \tag{4}$$

under which the population is not expected to be viable.

A deterministic value of W_{eq}, noted W_{det}, can be calculated at mutation-selection balance for a large population of diploid individuals with a large number of independent bi-allelic loci, where deleterious mutations with selection coefficient s and dominance h can segregate. This is done using equations for the mutation load L derived from Wright's equations for allele frequencies at equilibrium at a single locus [35,36], and gives

$$L = u \tag{5a}$$

for recessive mutations (the dominance coefficient $h = 0$), and

$$L = u\frac{4h + \alpha(1-4h)}{2h + \alpha(1-2h)} \tag{5b}$$

when $h >> 0$, where u is the deleterious mutation rate at a single locus and α is the proportion of offspring produced via self-fertilisation. If we consider that there is no epistasis and no linkage disequilibrium, then the relative population fitness when considering a genome-wide mutation load is given by $W_{det} = e^{-L}$ [20], where u is replaced by U, the haploid genomic mutation rate, when calculating L.

Finally we can calculate N_{det}, using W_{det} instead of W_{eq} in Equation 3 as an estimation of population size at equilibrium N_{eq}. Populations are not expected to be viable ($N_{det} \leq 0$) when for $h = 0$

$$r_0 \leq e^U \tag{6a}$$

and for $h >> 0$

$$r_0 \leq e^{\left(U \frac{4h+\alpha(1-4h)}{2h+\alpha(1-2h)}\right)}. \tag{6b}$$

High mutation rates and low reproductive rates are both expected to contribute to the decrease of population viability.

Simulation model

An individual-centred model with discrete non-overlapping generations was used to follow the evolution of an isolated population of variable size, made up of diploid hermaphrodite individuals.

Genomic assumptions. The genetic properties of this model, mutation and recombination, are those described in Roze's [37] model. Each individual is represented by two homologous chromosomes of length $2D$ with a potentially infinite number of loci. The map length is considered to be D from the centre of the chromosome to the edge, hence representing a chromosome with a defined centromere. The life cycle is as follows: mutation, selection, meiosis and reproduction.

The number of new deleterious mutations occurring per chromosome per generation, is sampled from a Poisson distribution with mean U, where $2U$ is the genomic mutation rate. Their position on the chromosome is sampled from a uniform distribution in $[-D,D]$. The effect of deleterious mutations on the fitness of individual i living at time t, W_t^i, is multiplicative and depends on the number of homozygous, n_{ho}, and heterozygous, n_{he}, deleterious mutations each individual carries

$$W_t^i = (1-hs)^{n_{he}}(1-s)^{n_{ho}}, \tag{7}$$

where s and h are the selective and dominance coefficients respectively and are fixed parameters. All mutations are deleterious and have the same values of s and h. The deleterious effect of these mutations is independent of population density. Recombination occurs during gamete production and is considered to be uniform along the chromosome. New individuals are a combination of two gametes, either from two different individuals for reproduction via outcrossing, or the same individual via selfing.

Demography and selection. At a given time t, population size N_t is given by

$$N_t = \sum_{i=1}^{N_{t-1}} X_{t-1}^i \tag{8}$$

where X_t^i is the number of viable offspring an individual i at time t contributes to the next generation via the female function (we consider that there is no limitation in the number of offspring an individual i can sire). X_t^i is sampled from a Poisson distribution with mean $R_t^i = W_t^i r_0^{1-\frac{N_t}{K}}$ (the individual reproductive rate).

Self-fertilisation occurs at a probability α_t^i for individual i, at time t given by

$$\alpha_t^i = \frac{\alpha_0 W_t^i}{\alpha_0 W_t^i + (1-\alpha_0)\frac{\sum_{j \neq i} W_t^j}{N_t - 1}}. \tag{9}$$

The proportion of selfed offspring depends on α_0, the proportion of an individual's male gametes that are available for self-fertilization, and on the individual's fitness W_t^i compared to the average relative fitness of the other possible fathers in the population $\left(\frac{\sum_{j \neq i} W_t^i}{N_t - 1}\right)$. The lower an individual's relative fitness as a father, the lower the proportion of offspring produced via selfing. We consider that there is no limitation in the availability of male gametes. The proportion of an individual's offspring produced by self-fertilisation is sampled from a binomial distribution with parameters α_t^i and X_t^i. When $\alpha_0 = 0$ the population is strictly outcrossing and the population is automatically considered non-viable if $N_t < 2$.

In order to facilitate the reading of the following sections, all the notations used through the text have been summarized in Table 1.

Initial conditions and simulations run. At the beginning of each simulation, we consider population size to be equal to K and that there are no deleterious mutations present in the population. The simulations are run until the population reaches equilibrium or goes extinct. We define equilibrium as the stabilisation of the mean population fitness, denoted $\overline{W}_t$; the average $\overline{W}_t$ over one thousand generations is calculated and compared to the average $\overline{W}_t$ of the previous thousand generations. If the difference between the two is lower than 1 per cent the population is considered to be at equilibrium. Population size N_{eq}, mean fitness $\overline{W}_{eq}$ and the mean number of mutations per chromosome were measured at equilibrium. Throughout the results, we will mostly be addressing the average of the mean value of the population's relative fitness across all simulations for each set of parameter values, which we note $\widehat{W}_{eq}$. If the population goes extinct, then the time to extinction is measured.

The mutational meltdown is defined as the acceleration of the decrease of population size due to the accumulation of deleterious mutations [3–5]. In order to evaluate this acceleration, once non-viable populations reach a population size of 250 individuals, the best fitting quadratic polynomial regression ($a + yt + zt^2$) is calculated to fit the decrease in $\overline{W}_t$, N_t and $\overline{R}_t$ independently of one another. When these variables are decreasing, the first order coefficient is negative. As these variables decrease with time, if they decrease in a linear fashion, then there is no mutational meltdown and the second order coefficient z is equal to 0. In the case of acceleration of the decrease of these variables with time, as expected in the case of a mutational meltdown, then the second order coefficient z, like the first order coefficient y, is negative. This second order coefficient z is calculated for each simulation run that results in population extinction. We also measure the mean population size at the fixation of the first deleterious mutation in order to detect whether population decline is associated to mutation fixation.

A wide range of values, from 0 to 1, were run for the parameter α_0, with $r_0 = 2$,4 and 10. For $r_0 = 2$, simulations were run for U between 0.1 and 0.6, mutations were mildly deleterious, moder-

Table 1. Notations.

$V, \overline{V}, \widehat{V}$	No superscript indicates deterministic values (except in the case of population size N, where it is mentioned clearly in the text), a bar indicates that the variable is the intra-population mean for one simulation run ($\overline{W}_{eq}$ is the mean relative fitness for one population) and a hat indicates that the variable is the mean across all simulations ($\widehat{W}_{eq}$ is the mean relative fitness across all simulations, conditional to survival).
$N_t, \widehat{N}_{eq}$	Population sizes at generation t and mean population size at population equilibrium across all simulations
$\sigma_N, \widehat{\sigma}_N$	Standard deviation of population size over time as a measure of the fluctuation of population size for one population and across simulations.
N_{det}	Expected population size without demographic or genetic stochasticity
r_0	The intrinsic reproductive rate.
K	The carrying capacity.
R_t	Expected number of offspring per individual at generation t.
X_t^i, R_t^i	Respectively the number of offspring produced by individual i at generation t and its expectation.
$\overline{W}_t, \overline{W}_{eq}, \widehat{W}_{eq}$	Means of the population's relative fitness at generation t and at population equilibrium for one or across simulations conditional to population survival.
W_t^i	Relative fitness of individual i at generation t.
W_{det}	Expected value of the population's component of fitness calculated using the Wright-Fisher model.
W_{min}	The threshold value of the population's mean relative fitness, under which populations are not viable.
U, D	Genomic properties: the haploid mutation rate and the recombination rate or map length.
s, h	Mutational effects: the selection coefficient and the dominance.
n_{he}, n_{ho}	The number of heterozygote and homozygote mutations on an individual's genome.
a_0	The proportion of male gametes available for selfing.
$\alpha_t^i, \overline{\alpha}_t$	Proportion of offspring produced via self-fertilisation by individual i and the population's mean proportion of offspring produced via self-fertilisation at generation t.

ately deleterious or lethal ($s=0.02$, 0.2 and 1), that were completely recessive, almost recessive, or moderately recessive ($h=0$, 0.02 and 0.2). The recombination rates taken into account reflect conditions where mutations were highly linked, moderately linked or very slightly linked ($D=0.1$, 1 and 10). Aside from the general effect of recombination rates over a whole genome, these genomic recombination rates can also reflect how mutation loads evolve within a genome in specific genomic hot and cold spots. Increasing D over 10 has very little effect on the results, which allows us to make the assumption that the mutations act as though they were independent [38]. However, it is possible, when there is selfing, that there is some linkage due to the genomic consequences of self-fertilisation. One thousand replicates were run for each group of parameter values, coming to a total of 1458000 simulations run for $r_0=2$. For $r_0=4$ and 10, one hundred replicas were run for $U=0.2$ to 1, $s=0.02$ and 0.2, $h=0$ and 0.02 and $D=10$, leading to a total of 54000 simulations run.

We compare the expected deterministic values of the mean fitness at equilibrium W_{det}, as well as the expected deterministic fitness threshold value W_{min} under which populations should not be viable, with our simulation results.

Estimating the stochastic fluctuations of population size

The stochastic fluctuation of population size from one generation to the next can be due to two mechanisms: demographic stochasticity alone or the interaction between demography and genetic selection. In order to estimate the importance of each of these two sources of stochasticity, we first estimated the fluctuations that would be observed with only demographic stochasticity and no mutations (the relative fitness $\overline{W}_t$ is a constant) and then compared these estimations with the fluctuations observed in our simulations. We use the standard deviation of population size over time σ_N as a measure of these fluctuations. We compare σ_N calculated from simulations run for 100000 generations for different constant values of $\overline{W}_t$ to the standard deviation of N_t over 100 generations when the populations in our simulations were at equilibrium. If demographic stochasticity alone can explain the fluctuation of population size, then the stochastic fluctuations calculated from the simulations with a dynamic component of relative fitness (denoted $\widehat{\sigma}_N$) should not be very different than those run with $\overline{W}_t$ as a constant.

In our simulations, population extinction is inevitable, as when time is very long, all populations go extinct due to demographic stochasticity with a probability of 1. In order to test whether populations that are expected to be viable, in other words with an expected relative fitness W_{det} greater than the threshold value W_{min} (Equation 3), go extinct because of demographic stochasticity alone, we ran stochastic simulations of Equation 8. We assume that all individuals have the same constant relative fitness W_{det} (equation 5) and the initial population size is N_{det} (Equation 3). We calculate the probability of population extinction before one order of magnitude higher than the highest time to extinction observed in our simulations with a dynamic component of relative fitness.

Results

Demographic and genetic evolution to equilibrium and extinction

Populations either evolve both demographically and genetically to a quasi-stationary equilibrium of population size and the relative fitness, noted respectively N_{eq} and $\overline{W}_{eq}$, or go extinct (illustrated in Figure 1A).

For all values of the intrinsic reproductive rate r_0, the simulation results agree with deterministic expectations of the interaction between W_{eq} and N_{eq} from Equation 3 (see Figure 1B) for viable populations. When $r_0=10$ and W_{eq} values are greater than 0.74,

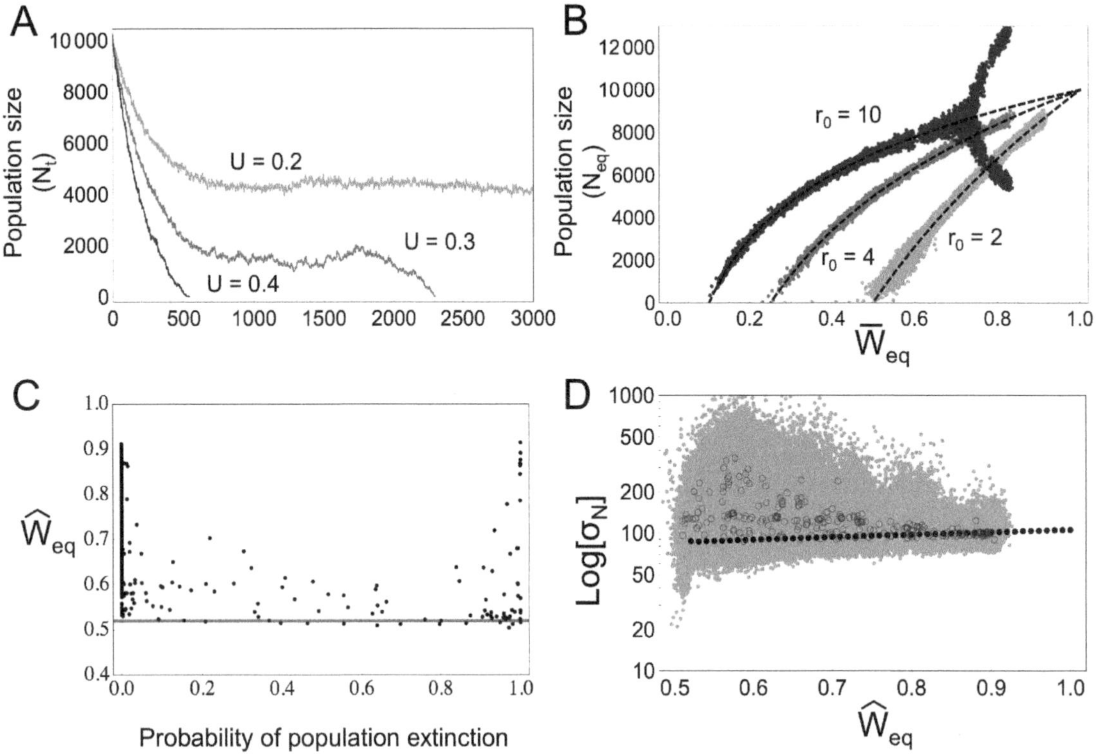

Figure 1. Demographic and genetic evolution of populations. A) Typical evolution of population size N_t with time for a viable population ($U=0.2$), one that reaches equilibrium but goes extinct due to stochasticity ($U=0.3$), and one that is not viable ($U=0.4$). $\alpha_0=0$, $s=0.02$, $h=0.2$, $D=10$, $K=10000$ and $r_0=2$. $\overline{W}_t$ follows the same pattern. **B**) Population size at equilibrium, N_{eq} as a function of mean population fitness $\overline{W}_{eq}$ for different values of r_0. The dashed line represents the expected population size N_{det} from Equation 3, and the points represent simulation results for all viable populations for all parameter sets with $D=10$ and $K=10000$. **C**) Probability of population extinction from simulations run for all sets of parameter values for $r_0=2$ and $K=10000$ as a function of the average population's mean relative fitness $\widehat{W}_{eq}$. The grey line represents the population fitness above which the probability of population extinction in less than 10^5 generations due to demographic stochasticity alone is almost null. **D**) Standard deviation of population size over time at population equilibrium σ_N as a function of the relative fitness $\overline{W}_{eq}$ from simulations run for all sets of parameter values for $r_0=2$ and $K=10000$. The light grey points each represent σ_N of a single simulation, the open circles represent the mean standard deviation across simulations per group of parameter values $\hat{\sigma}_N$ and the full circles represent results from simulations run that take into account only demographic stochasticity.

the deterministic equilibrium is oscillatory and unstable, as confirmed by a stability analysis of equation 3. This is clearly seen in the simulation results, where the populations oscillate between two states.

Very few populations are viable with a relative fitness at equilibrium $\overline{W}_{eq}$ that is lower than the estimated value of the relative fitness threshold W_{min}, and as expected from Equation 4, W_{min} decreases with an increasing intrinsic reproductive rate r_0. This strong relationship between $\overline{W}_{eq}$ and N_{eq} is not enough, however, to explain population extinction. When taking a closer look at the relationship between the probability of population extinction and mean population fitness, we do not have a positive linear relationship between the two variables, but a more bimodal distribution (Figure 1C). We observe a great range of values of $\widehat{W}_{eq}$ (i.e. the average $\overline{W}_{eq}$ of viable populations) across simulations for which all populations survive, a similar range of $\widehat{W}_{eq}$ for parameter sets for which very few populations were viable and intermediate-low values of $\widehat{W}_{eq}$ for populations with an intermediate probability of extinction.

When the deterministic value of mean fitness W_{det} is greater than the fitness threshold W_{min} but populations are not viable, extinction can be attributed to either demographic stochasticity alone, or to an interaction between both demographic and genetic stochastic processes. When we consider the mean relative fitness $\overline{W}_t$ to be a constant, so that population extinction is due only to demographic stochasticity, we find that the probability of extinction is extremely low compared to what is observed in our results. For example for $r_0=2$ and $\overline{W}_t=0.52$, the probability of extinction within a time equivalent to one order of magnitude greater than our highest time to extinction observed is almost null. In other words, these populations are expected to survive as their relative fitness is above the threshold value W_{min}, and the expected time to stochastic demographic extinction is extremely long. This however is not the case in simulations with a dynamic component of fitness, suggesting that it is the interaction between demography and genetics that leads to population extinction in such a relatively short time scale.

The importance of this interaction is observed when taking into account the standard deviation of population size over time σ_N (Figure 1D). We observe that the values of σ_N for each simulation run (grey points) fluctuate around the expected standard deviation if the change in population size from one generation to the next were due only to demographic stochasticity (full circles). Fluctuations are even higher when the relative fitness $\overline{W}_{eq}$ decreases. We note that the mean value of the standard deviation of population size over time σ_N for any group of parameter values $\widehat{\sigma}_N$ (open circles) is always greater than the expected values of standard deviation of this variable when only demographic stochasticity affects population size. This suggests that the interaction between demography, genetics and selection highly increases the stochastic fluctuations of population size. Below, we show that this increased stochasticity is important when estimating the probability of population extinction.

From this point onwards, we will mostly be addressing results with an intrinsic reproductive rate $r_0 = 2$ since that we observe the same patterns for higher values of r_0 when considering population extinction and mutational loads. For simplicity, in order to compare the effect of selfing between populations, we have chosen to compare population characteristics for different values of the proportion of male gametes available for selfing α_0 and not $\overline{\alpha}_t$, the effective mean proportion of offspring produced via self-fertilisation of the population. The difference between the two is slight (unless the population is very near extinction), with $\alpha_0 \leq \overline{\alpha}_t \leq \alpha_0(1+0.1)$, except for $\alpha_0 = 0$ and 1, where it remains fixed.

Probability of population extinction

As expected from deterministic approximations (see section "Deterministic model and expectations", Equation 6), when mutations are only slightly linked ($D = 10$) we find that for a particular intrinsic reproductive rate r_0, increasing values of the haploid mutation rate U leads to higher probabilities of extinction (Figure 2A) and lower $\widehat{W}_{eq}$ (Figure 2B). Contrary to deterministic expectations, the coefficient of selection s affects both population extinction and mean relative fitness. The effect of the coefficient of the selection and the proportion of selfed offspring both depend greatly on the dominance h of the deleterious mutations. Generally, increasing the coefficient of selection decreases the probability of extinction at higher mutation rates and increases mean relative fitness $\widehat{W}_{eq}$.

From the deterministic equations, the proportion of selfed offspring should not affect either of these variables when mutations are completely recessive and almost recessive ($h = 0$). However, self-fertilisation has a non-monotonic effect on both of these variables. Between $\alpha_0 = 0$ and 0.2 the probability of extinction increases, while $\widehat{W}_{eq}$ decreases. For α_0 between 0.2 and 0.95, the opposite tendency is generally observed, the greater α_0, the lower the probability of extinction and the higher $\widehat{W}_{eq}$. When a strictly selfing regime is imposed, no simulated populations survive when mutations are almost neutral ($s = 0.02$) and when mutations are mildly deleterious ($s = 0.2$) the probability of extinction (respectively the mean relative fitness) is the same, or slightly higher (respectively lower), as what is observed for $\alpha_0 = 0.95$. The same patterns are observed when mutations are almost recessive ($h = 0.02$). When mutations are moderately recessive ($h = 0.2$), we find that the deterministic expectations are more accurate. There is generally a monotonic effect of α_0 on both the probability of extinction viability and $\widehat{W}_{eq}$, the former decreases and the latter increases with increasing α_0. At higher mutation rates, we observe that the probability of extinction increases and $\widehat{W}_{eq}$ decreases at very high values of α_0 (>0.8). Once again, when mutations are almost neutral ($s = 0.02$), no populations survive when $\alpha_0 = 1$. For all values of h, increasing U accentuates the effect of α_0 on the decrease of $\widehat{W}_{eq}$ and increase of the probability of extinction.

At lower recombination rates ($D = 1$ and 0.1), $\widehat{W}_{eq}$ is generally not affected by s and h, but the probability of population extinction is generally greater with increasing U than what is observed for high recombination rates (for $D = 10$). This is especially true for lower proportions of selfed offspring. The effect of recombination on viability decreases with increasing selection. We observe one particular case, when mutations are very tightly linked (the recombination rate $D = 0.1$), of small effect (the coefficient of selection $s \leq 0.2$) and completely recessive (dominance $h = 0$), where increasing the haploid mutation rate U can, for low rates of α_0, decrease the probability of extinction and increase the mean relative fitness or have no effect on either (Figures 2A and 2B). This could be due to more efficient selection against deleterious mutations at higher mutation rates, as increasing the mutation rate could increase the probability that tightly linked groups of recessive mutations are found at a homozygote state and eliminated.

As mentioned above, population fitness and demographic stochasticity alone can not fully explain population viability. Figure 2C represents the mean standard deviation of population size over time at equilibrium conditional to population survival $\widehat{\sigma}_N$. We observe that, when considering parameter sets with the same mutation rate U, $\widehat{\sigma}_N$ increases with increasing probability of population extinction. It is important to note that $\widehat{\sigma}_N$ is most probably underestimated for parameter sets for which not all populations are viable, as the standard deviation of extinct populations is not taken into account during its calculation.

Accumulation and fixation of deleterious mutations

Our results show that when deleterious mutations are fixed, it is only in populations that are very small, in other words already on their way to extinction. Even though the mean number of mutations is smaller at higher proportions of selfed offspring (Figure 3A), when populations are not viable, increasing α_0 leads to the fixation of deleterious mutations at greater population sizes (Figure 3B) and with a higher probability (results not shown). The smaller the coefficient of selection and the greater the mutation rate, the greater the population size at first fixation. Therefore, if mutations are indeed fixed, they are fixed at larger population sizes when they are mildly deleterious, frequently introduced and in populations with greater proportions of selfed offspring. Mean population size at first fixation is generally relatively small compared to the carrying capacity K (see Figure 3B), except for one group of parameter values ($U = 0.1$, $s = 0.02$, $h = 0$ and $\alpha_0 = 1$, results not shown) where fixation can occur at a population size of 8000 individuals. Mean population size at first fixation is greater for high-intermediate values of α_0 when there is little recombination than when recombination rates ($D = 10$) are high, whereas outcrossing populations are not affected.

The mean number of mutations per chromosome at population equilibrium across all simulations (conditional to population survival) decreases with increasing coefficients of selection s and/ or the dominance h and for increasing proportions of male gametes available for selfing α_0 (Figure 3A shows typical results). Increasing the haploid mutation rate U increases the mean number of mutations per chromosome. The lower number of mutations per chromosome for mutations with stronger effects (either at the homo- or heterozygous state) and for higher proportions of selfed offspring can be explained by a more efficient purging [26].

Lower recombination rates generally do not affect the mean number of mutations per chromosome, except when mutations are completely recessive ($h = 0$) and there is almost no recombination ($D = 0.1$). In this case the mean number of mutations per

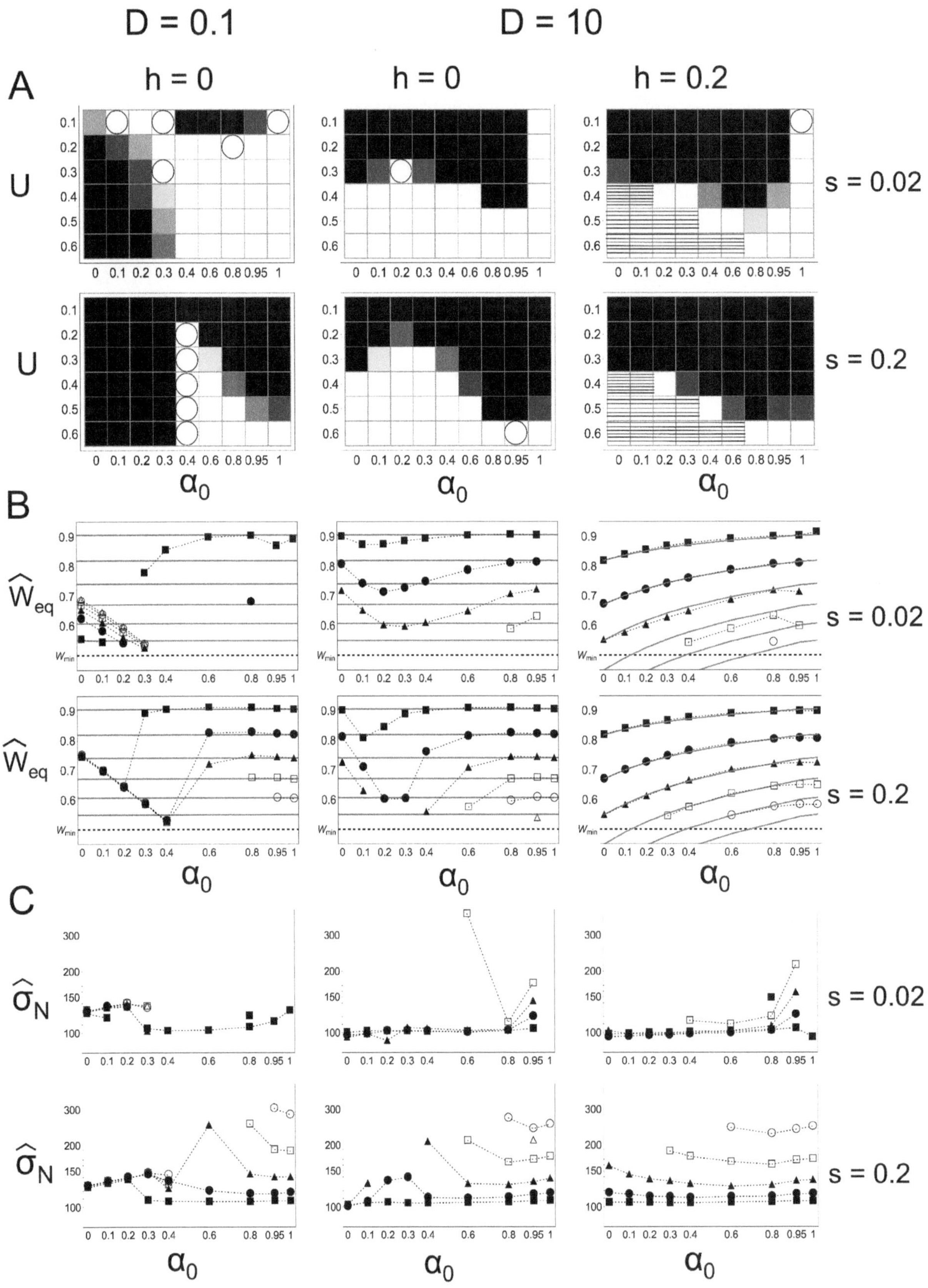

Figure 2. Population extinction and equilibrium. A) Probability of population extinction calculated from the 1000 simulations run ($r_0 = 2$ and $K = 10000$), from black (0% extinction) to white (100% extinction), as a function of α_0. The circles indicate that more than 95% of populations went extinct. The horizontal lines indicate that deterministic extinction was predicted (Equation 3). **B**) Mean values of the observed population fitness at equilibrium across simulations $\hat{W}_{eq}$ of viable populations as a function of α_0 for $r_0 = 2$ and $K = 10000$. The grey lines represent the expected mean

fitness W_{det} for (from top to bottom) $U=0.1$, 0.2, 0.3, 0.4, 0.5 and 0.6. Top row $s=0.02$ and below $s=0.2$. Missing points indicate parameter values for which none of the 1000 simulations run were viable. **C)** Standard deviation of population size over time at population equilibrium across simulations $\widehat{\sigma}_N$ with a logarithmic scale as a function of α_0, with $r_0=2$ and $K=10000$. Note that this variable is underestimated for parameter sets with less than 100% population survival, as the standard deviation of extinct populations is not taken into account. For **B** and **C:** ■ $U=0.1$, ● $U=0.2$, ▲ $U=0.3$, ○ $U=0.5$, △ $U=0.6$.

chromosome can be more than doubled for low values of α_0 independently of the coefficient of selection s, but not greatly changed for higher values of α_0 when selection is strong $s \geq 0.2$.

Mutational meltdown and time to extinction

When a population is on its way to extinction, there is a weak but significant acceleration in the decrease of the mean relative fitness $\overline{W}_t$ and in the mean reproductive rate $\overline{R}_t$, but there is a deceleration in the decrease of population size N_t for high recombination rates ($D=10$, see Figure 4A). Higher mutation rates U, lower dominance of mutations h and lower proportions of selfed offspring α_0 contribute to the acceleration (respectively the deceleration) of the decrease of $\overline{W}_t$ and $\overline{R}_t$ (respectively N_t). Even though both the mean relative fitness and the reproductive rate show an overall tendency to decrease at an accelerating rate, the low population density allows for the deceleration of the decrease of population size as the smaller the population size, the more resources available per individual for reproduction, as $\overline{R}_t$ is density dependent (Equation 2).At low recombination rates, there is neither an acceleration or a deceleration in the decrease of $\overline{W}_t$, $\overline{R}_t$ and N_t.

The mean time to extinction for non-viable populations has a complex relationship with the proportion of male gametes available for selfing α_0 (Figure 4B). The general patterns that are observed, however, are that the effect of α_0 on the time to extinction is attenuated with increasing values of U. Increasing selection (s and h) reduces the mean time to extinction for populations with low proportions of offspring produced via self-fertilisation, but increases it for higher α_0 (Figure 4B). The contrary is observed when mutations are very mildly deleterious ($s=0.02$) and recessive ($h=0$ and 0.02), with longer times to extinction for outcrossing populations than for selfing populations. The pattern of the effect of α_0 and selection on time to extinction remains unchanged when recombination rates are low.

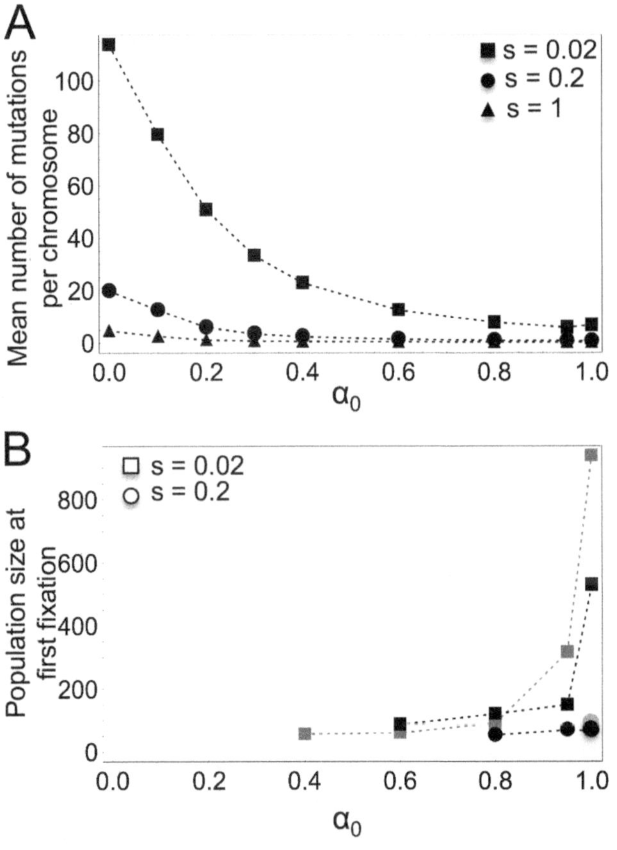

Figure 3. Accumulation and fixation of deleterious mutations. **A)** Mean number of mutations per chromosome equilibrium across simulations as a function of α_0. $U=0.1$, $h=0.02$, $D=10$ and $r_0=2$. Missing points present parameter values for which no populations were viable. **B)** Mean population size at first fixation of deleterious mutations for populations greater than 50 individuals as a function of α_0. $U=0.5$ in grey, $U=0.6$ in black. $h=0.2$, $D=10$ and $r_0=2$.

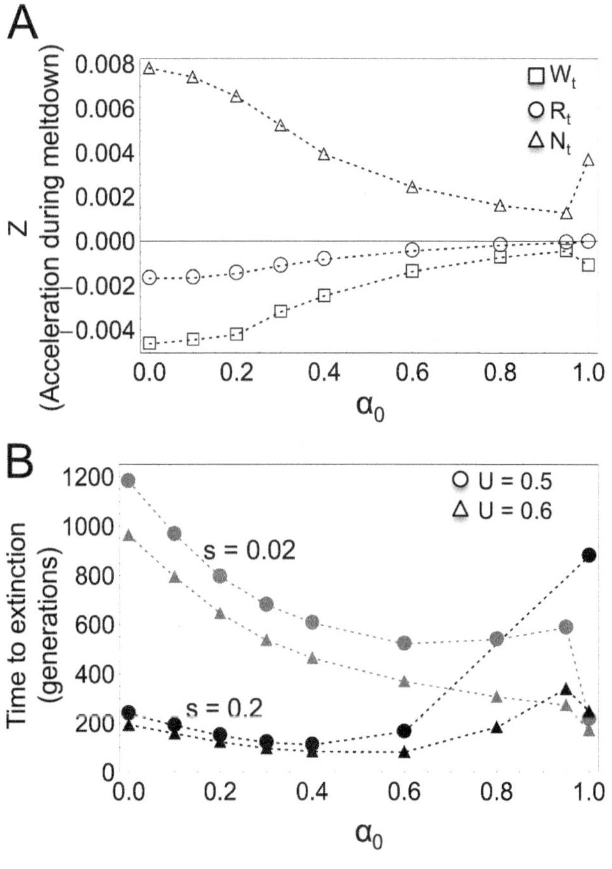

Figure 4. Mutational meltdown and time to extinction. A) Median value of acceleration in the rate of decrease of $\overline{W}_t$, $\overline{R}_t$ and N_t, noted z, for non-viable populations from $N_t=250$ until extinction as a function of α_0. $U=0.6$, $s=0.02$, $h=0.02$, $D=10$, $r_0=2$ and $K=10000$. **B)** Mean time to extinction as a function of α_0. $D=10$, $r_0=2$ and $K=10000$.

Discussion

It is generally accepted that selection is less effective in small populations, which could lead to their extinction due to mutational meltdown [2–5], whereas large populations are able to purge recurrent deleterious mutations and remain at mutation-selection balance [15,39]. Our results suggest that there are values of genetic parameters for which even large populations cannot purge deleterious mutations fast enough to reach mutation-selection balance and go to extinction due to the increasing number of segregating mutations, which in turn increase the mutational load. This implies two things: 1) Mutation selection balance is not attainable for all genetic parameters as hypothesized by population genetics models and 2) Populations can go to rapid extinction due to segregating deleterious mutations. Self-fertilisation, while expected to allow for more efficient purging of deleterious mutations [26], does not always allow for lower probability of extinction. Population fitness as well as the amplitude of the fluctuation of population size both contribute to the eventual fate of a population, with lower population fitness and greater fluctuations leading to higher probabilities of extinction. More specifically, our results show that there is a synergistic interaction between genetics and demography, which increases the stochastic fluctuations of population size.

Empirical estimations of the genetic parameters used in our model are now becoming available. The idea that there is a strong correlation between h and s has often been shown in empirical works ([40–42]; also see [43]). As the dominance coefficient has been estimated as being around 0.25 [44] and most of the deleterious mutations that make up an individual's mutational load are of small effect, this implies that the most realistic parameters we have run are for $h=0.2$ and $s=0.02$. However, a new approach using a phenotypic landscape model has shed doubt on this interpretation; the dominance and the coefficient of selection of mutations might well be independent of one another [44]. This justifies our choice to study several values and combinations of these values of s and h values. We have chosen to consider mutations with constant and only deleterious effects, which is one of the limits of this model. Mutations found in natural populations have variable coefficients of selection s and dominance h and the distributions of these variables are still under debate as they can vary not only between species [45], but also between populations [46]. How the variability of mutation affects the evolution of populations is still unclear and yet to be fully taken into account in theoretical models.

The genomic mutation rates (equivalent to $2U$ in our model), estimated empirically in various organisms range from 0.01 to 1 [19] and even greater [17] in eukaryotes. We therefore explore realistic mutation rates, even though in our case we consider that all new mutations are deleterious.

Population size, viability and the mutational meltdown

Generally, if we are able to predict demographic factors (such as the intrinsic reproductive rate, the carrying capacity and density dependence) and the mean relative fitness of populations (or the mutational load), we are able to predict population size and a threshold value of mean relative fitness, below which a population is not viable (Figure 1A). However, predicting a relatively large (or non-null) population size is not enough to guarantee population survival within a relatively long time scale. In our model we find that population extinction is not due to demographic stochasticity alone, but to increased levels of stochasticity that result from the interaction between demography and genetics.

The importance of this interaction on population extinction has already been suggested in literature [47]. Our results find that it is indeed non-negligible as shown by the fluctuation of population size over time in our model, which is a result of demographic stochasticity and a dynamic mutation load or relative fitness, which are not independent of one another (Figure 1D). The mean fluctuation of the population size for a given set of parameter values (open circles) are too great to be due to demographic stochasticity alone (full circles). Even though stochastic processes affect the reproductive rate, contemporary stochasticity alone does not account for this variance. Past stochastic events, or the mutational history of the population (where the deleterious mutations are in the genome, and at what frequencies), can also influence this variable, as observed in Figure 1D, where the standard deviation of population size σ_N (grey points) varies greatly around the mean standard deviation observed $\widehat{\sigma}_N$ (open circles). For parameter values where the probability of extinction is different than 0 and 1, the importance of past stochastic events, is even more evident as the fate of a population is not sealed. The importance of past mutational events has also been observed in experimental mutation accumulation lines, where replicate populations with the same genetic origin do not all go to extinction (e.g. for yeast populations [8]). We propose, that in order to predict the probability of extinction of populations, it is not only important to predict the mean fitness, but also the fluctuation of population size. Further exploration of this model is required to estimate how the genetic and demographic parameters affect the amplitude of the fluctuations of population size. No theoretical work, to our knowledge, has taken on this question from a demo-genetic point of view.

Clarke's [23] work highlights the importance of the effect of selection on demography, and his prediction that taking into account the accumulation of deleterious mutations throughout the genome would allow a significant decrease of population size is confirmed by our model, where in some cases populations go to extinction. It has been suggested that the timing of selection is crucial in order to assess the effect of the mutation load on population size [17,23]. In our model we have chosen relative fitness to affect only the reproductive rate. However, it is possible that selection that affects the consumption of resources (K) could lead to different results [23]. It is often considered that selection has no effect on demography (e.g [17]), but if the mutation load has a direct effect on an individual's reproductive capacity, as is the case in our model, then the effect of selection on population size cannot be ignored. This has important implications on how data on population size from natural populations should be interpreted (see below).

The importance of the reproductive rate concerning population extinction has already been suggested by other models, where populations with higher intrinsic reproductive rates have longer times to extinction [4,48] or lower probabilities of extinction [14]. These predictions are an inherent property of our model, as populations with high intrinsic reproductive rates r_0 are expected to be viable at higher mutational loads (Equation 6), and are supported by our stochastic simulations (Figure 1B). In our model, we consider a stable environment, which is an unrealistic hypothesis. It is therefore probable that we overestimate population viability, as shown by Higgins and Lynch's model [49], which, upon taking environmental stochasticity into account, suggests that it increases the rate of accumulation of deleterious mutations.

To what extent are we capable of estimating the mean relative fitness (or the mutational load)? In spite of explicitly considering demography, we find that the simplified deterministic predictions of the mutation load are reliable when mutations have a strong effect ($s=0.2$ and $h=0.2$). However, when mutations are almost neutral, mean fitness is overestimated, especially when the mutations are recessive ($h=0$) and the genomic recombination rate is low as the number of mutations per chromosome is increased [50], affecting the purging process. The deterministic expectations of mean fitness when comparing them to simulated results have already been shown to be reliable by population genetics models [21,22], however none of these models explicitly include the effect of demography. This interaction, between genetics and demography, could in fact be important, as shown by the unexpected non-monotonic relationship between population fitness and the proportion of selfed offspring in our model (Figure 2B), which we discuss below.

It is important to note that a high mutation rate and a large number of segregating deleterious mutations do not necessarily lead to a higher mutation load. In the case of very little recombination ($D=0.1$), increasing the mutation rate increases mean fitness when mutations are recessive and of small effect, and hence decreases the mutation load. Therefore, the mutation rate in itself is not sufficient, and depending on the effects of deleterious mutations and the recombination rates, increasing the genomic mutation rate does not always lead to an increase in the mutational load. The extreme case of very little recombination could also be translated as the existence of genomic regions with low recombination rates known as cold-spots (reviewed in [51]), allowing for the accumulation of deleterious mutations ([52], p. 555). The existence of such genomic regions could in fact have an important influence on the survival of populations. Low rates of recombination are expected to reduce population fitness and increase the rate of fixation of deleterious mutations [53]. Though our results confirm this for mutations that are moderately recessive ($h=0.2$), it is not the case when considering recessive mutations, where the contrary is observed. At such low recombination rates, the high levels of linkage-disequilibrium lead to the formation of tightly linked groups of mutations. These groups of mutations act as a single "super locus". When the mutations are recessive, this load remains silent at the heterozygous state, but once at the homozygous state, the multiplicative effects of these small mutations are expressed and lead to a very deleterious effect. The relative fitness of individuals that become homozygous for only one of these different super loci is extremely low. In this case, the advantage of outcrossing is much higher, as outcrossed offspring have a higher probability of being heterozygous at these loci than selfed offspring.

From our simulated results, we conclude that when populations are on their way to extinction, whether we observe a mutational meltdown depends not only on the values of the genetic parameters, but also on the variable measured. Due to the nature of the density dependence in this model, the decrease in population size decelerates when population density is low: the smaller the population size, the more resources available to the few remaining individuals. Even though both the reproductive rate and the mean relative fitness do show an acceleration in their decrease when populations are on their way to extinction, we find that the existence of this mutational meltdown depends greatly on the effect of the deleterious mutations, the mutation rate, and the proportion of offspring produced via self-fertilisation.

The importance of segregating mutations has already been suggested by Lynch *et al.* [4], but this work concluded that fixation had a greater effect on the meltdown. Our model suggests that the fixation of deleterious mutations is a consequence rather than a cause of decline towards population extinction, this however could be due to the fact that in Lynch *et al.*'s [4] model the genetic load affected offspring survival, whereas in our model there is a direct link between the reproductive rate and the mutational load. The effect of the accumulation of segregating deleterious mutations has been considered to be negligible when considering large populations [15,16], even more so when considering their eventual extinction over a short time scale because of this process (but see [14]). This does not seem to be the case when considering the interaction between demography and genetics. A feed-back loop between these two properties seems to lead to a "cost of purging": Unfit individuals do not reproduce, decreasing population size at the next generation, which in turn increases the effect of drift, leading to a lower efficiency of purging and more unfit individuals. This continual increase of the number of segregating deleterious mutations eventually leads to a demo-genetic extinction. Mutation-selection balance is therefore not the automatic fate of initially large populations, and the cost of purging can lead to a very rapid extinction [14].

How does selfing affect population size and viability?

Our results indicate that selfing has an effect both on population size and viability. We often observe that selfing populations have lower probability of extinction than outcrossing populations at higher mutation rates (see Figure 2A), especially when there is strong selection, in which case selfers are always expected to have larger population sizes. However, when selection is weak, we find that strict-selfing and low levels of selfing (but not strictly outcrossing) hinder both size and viability. As this has not been noted in other models, even when genetic drift is taken into account [21,36], it is possibly a consequence of the interaction between genetics and demography. A possible hypothesis to explain this observation is that the viability of populations concerning the accumulation of deleterious mutations depends on two opposing properties 1) The purge of these mutations and 2) The avoidance of expressing them. Outcrossers accumulate deleterious mutations [26], but avoid the cost of inbreeding depression [54], with most of their mutations being at a heterozygous state. Selfers purge these mutations [26] and even though they do not accumulate as many as outcrossers do, many are at a homozygous state [55]. We therefore propose that populations with low proportions of offspring produced via self-fertilisation suffer from both the inconveniences of outcrossers and selfers, not only do they accumulate deleterious mutations, as purging is not as efficient as for high proportions of selfed offspring, they also express them, and suffer from the demographic cost of purging.

The effect of self-fertilisation on the extinction of populations due to the accumulation of deleterious mutations has long been debated [4,28,56]. Our results suggest that the accumulation, hence fixation, of deleterious mutations is probably an insufficient explanation for higher extinction rates. In spite of this model's limitations, we find that even though self-fertilisation does affect population extinction due to genetic deterioration, the effects of the genetic parameters are complex and do not result in a simple pattern of the effect of selfing on the time to extinction. When selection is weak, strict outcrossers are less likely to go extinct than strict selfers, agreeing with Lynch *et al.*'s [4] results. However, allowing for a small amount of outcrossing (e.g. a proportion of male gametes available for selfing $\alpha_0=0.95$) is enough to greatly decrease the probability of extinction, even allowing for a higher probability of population survival at higher mutation rates than for strict outcrossing (see Figure 2A). Strong selection reverses this observation, with strict outcrossers being more prone to a mutational meltdown than selfers, in accordance with Bernardes' [14] results. What rate of selfing is more likely to cause extinction or lead to population vulnerability is not clear and greatly depends on both the genetic (mutation rates, genomic recombination rates, deleterious effects of mutations) and demographic (intrinsic reproductive rate) parameters. It has already been suggested that selfing has a greater effect on population extinction when considering the fate of beneficial mutations and their effects after environmental change. For instance Glémin and Ronfort [57] showed that if adaptation is due to standing variation, then outcrossers are less prone to extinction than selfers. Their model, however, is not demographically explicit.

When considering genomic cold-spots with low recombination rates, outcrossers are greatly advantaged when mutations are recessive, as they do not express these accumulated mutations. The lower viability of selfing populations in our results for such low recombination rates supports the observation that selfing species could be more likely to evolve higher recombination rates, in order to avoid the hitchhiking of deleterious mutations [58]. The difference in extinction rates between outcrossers and selfers observed empirically [29] could perhaps be due to such genomic regions where mutations of small effect segregate.

The non-monotonic effect of selfing on the probability of extinction in our results, could offer a possible explanation to the differences in extinction rates between selfers and outcrossers within the same family. For a transition to be successful, the transition in the reproductive mode has to be of large effect, going from complete outcrossing to high proportions of selfed offspring, for in some cases the mutational load that an outcrossing population could put up with could prove lethal for a reproductive mode with low proportions of offspring produced via self-fertilisation. The observed high extinction rates could therefore be related to the transition process and not the reproductive mode in itself.

Empirical implications

How can the correlation between population size and population fitness be interpreted? In most empirical works, a positive correlation between the two is translated as the negative effect of a small population size on population fitness due to inbreeding, the fixation of deleterious mutations or a lack of reproductive assurance (for example [59,60]). Another possible interpretation which is not often considered is simply that population size is a consequence and not the initial cause of a high mutation load, just as in some cases small population size does not seem to lead to a decline in fitness [61].

Generally, small populations are considered to be most at risk of extinction within a relatively short time frame due to inbreeding depression, mutational meltdown and demographic stochasticity [1,15,47]. Empirical experiments have therefore concentrated on the extinction of small populations, through the accumulation of deleterious mutations [6–10]. Even though the fixation of deleterious mutations can lead to the mutational meltdown of small populations [3–5], our results suggest that the interaction between demography and genetics can lead to the extinction in large populations due to segregating mutations alone and at relatively fast rates. In initially large populations, once the "mutational meltdown" is underway, fixation is rare and is a consequence rather than a cause of population decline. The importance of segregating mutations in population decline could have implications in conservation biology, as in most empirical studies, it is automatically assumed that the load leading to population decline is fixed or almost fixed [62,63]. If population decline is mostly due to segregating mutations, then there exists a real potential of purging the deleterious load through conservation efforts. Small populations are also expected to be more prone to demographic stochasticity, which should act more rapidly on population extinction than genetic factors [1,47], as has been shown through empirical experiments [64]. The higher levels of stochasticity in the variation of population size observed due to this interaction compared to the expected effect of demographic stochasticity alone in our results (Figure 1D) indicate that stochastic events (that are not due to external factors such as environmental stochasticity) are not only detrimental in small populations, but can also be so in large populations.

Our results suggest that measuring the decline in population size could be misleading when attempting to asses whether a population is going into a mutational meltdown or not. Depending on the density dependence, a population on the way to extinction could seem to be demographically stable, as the decrease of population size could potentially decelerate with time, becoming barely detectable. As the mutational load is not accessible, measuring the acceleration of the decline of the mean relative fitness is not possible in natural populations. However, measuring the acceleration in the decline of the mean reproductive rate over several generations could be a more informative measure concerning population extinction, or the mutational meltdown, and is empirically more accessible. The lack or rarity of a mutational vortex when our simulated populations are on the decline could indicate that, if the segregating deleterious mutations can be purged at any time, then, as there is very little or no increase in the rate of reduction of population size, conservation efforts could be applied successfully even when populations reach relatively small sizes.

In conservation biology, Population Viability Analyses (PVA) are the most frequently used tool for estimating the probability of population extinction. PVAs take mostly demographic data and parameters into consideration and do not take into account genetics explicitly. They have proved useful and accurate when considering external pressures (i.e. over-fishing, fragmentation, etc.) that affect population demography. However, the effectiveness of PVA's remains ambiguous, as even though they can be relatively accurate predictors of the evolution of population demography[65], in other cases the population growth rates can be over-estimated [66]. Could the overestimation of growth rates of growth rates be due to the omission of the genetic effects on population demography?

The direct relationship between higher intrinsic reproductive rates and greater population sizes for the same relative fitness in our model (Equation 3, Figure 1B) leads to a relationship between the reproductive and mutational rates in viable populations (Equation 6). This could mean that for a population with a high genomic mutation rate to be viable, it must have a large enough intrinsic reproductive rate. This relationship has not been studied either theoretically or empirically, and it could be interesting to perform a comparative analysis between species to test if there is such a relationship. We suggest that there could be a correlation between the genomic mutation rates of a species and the number of gametes produced, which could represent the intrinsic reproductive capacity.

Acknowledgments

Numerical results presented in this paper were carried out using the regional computational cluster supported by Université Lille 1, CPER Nord-Pas-de-Calais/FEDER, France Grille, CNRS and on the Biomed virtual organization of the European Grid Infrastructure (http://www.egi.eu) via DIRAC (http://diracgrid.org). We thank the technical staff of the

CRI-Lille 1 centre, the European Grid Infrastructure and the supporting National Grid Initiatives for providing the technical support and infrastructure. We would also like to thank Denis Roze, Sylvain Glémin, Xavier Vekemans and Luis-Miguel Chevin for their input and helpful comments.

Author Contributions

Conceived and designed the experiments: DAA SB. Performed the experiments: DAA. Analyzed the data: DAA SB. Contributed reagents/materials/analysis tools: DAA SG CB SB. Wrote the paper: DAA SB.

References

1. Frankham R (2005) Genetics and extinction. Biological Conservation 126: 131–140.
2. Lynch M, Gabriel W (1990) Mutation load and the survival of small populations. Evolution 44: 1725–1737.
3. Lande R (1994) Risk of population extinction from fixation of new deleterious mutations. Evolution 48: 1460–1469.
4. Lynch M, Conery J, Burger R (1995) Mutational meltdowns in sexual populations. Evolution 49: 1067–1080.
5. Coron C, Porcher E, Méléard S, Robert A (2013) Quantifying the mutational meltdown in diploid populations. American Naturalist in press (accepted).
6. Newman D, Pilson D (1997) Increased probability of extinction due to decreased genetic effective population size: Experimental populations of Clarkia pulchella. Evolution 51: 354–362.
7. Frankham R, Gilligan D, Morris D, Briscoe D (2001) Inbreeding and extinction: Effects of purging. Conservation Genetics 2: 279–285.
8. Zeyl C, Mizesko M, de Visser J (2001) Mutational meltdown in laboratory yeast populations. Evolution 55: 909–917.
9. Vassilieva L, Lynch M (1999) The rate of spontaneous mutation for life-history traits in Caenorhabditis elegans. Genetics 151: 119–129.
10. Baer C, Shaw F, Steding C, Baumgartner M, Hawkins A, et al. (2005) Comparative evolutionary genetics of spontaneous mutations affecting fitness in rhabditid nematodes. Proceedings of the National Academy of Sciences of the United States of America 102: 5785–5790.
11. Packer C, Pusey A, Rowley H, Gilbert D, Martenson J, et al. (1991) Case-study of a population bottleneck - lions of the ngorongoro crater. Conservation Biology 5: 219–230.
12. Westemeier R, Brawn J, Simpson S, Esker T, Jansen R, et al. (1998) Tracking the long-term decline and recovery of an isolated population. Science 282: 1695–1698.
13. Gelatt T, Davis C, Stirling I, Siniff D, Strobeck C, et al. (2010) History and fate of a small isolated population of Weddell seals at White Island, Antarctica. Conservation Genetics 11: 721–735.
14. Bernardes A (1995) Mutational meltdown in large sexual populations. Journal de Physique I France 5: 1501–1515.
15. Hedrick P (2004) Recent developments in conservation genetics. Forest Ecology and Management 197: 3–19.
16. Whitlock MC BR (2004) Fixation of new mutations in small populations. In: Ferrière R CD Dieckmann U, editor, Evolutionary conservation biology, Cambridge University Press, chapter 9. pp. 155–177.
17. Agrawal AF, Whitlock MC (2012) Mutation load: The fitness of individuals in populations where deleterious alleles are abundant. Annual Review of Ecology, Evolution, and Systematics 43: 115–135.
18. Saccheri I, Hanski I (2006) Natural selection and population dynamics. Trends in Ecology and Evolution 21: 341–347.
19. Sniegowski P, Gerrish P, Johnson T, Shaver A (2000) The evolution of mutation rates: separating causes from consequences. Bioessays 22: 1057–1066.
20. Haldane J (1937) The effect of variation on fitness. American Naturalist 71: 337–349.
21. Bataillon T, Kirkpatrick M (2000) Inbreeding depression due to mildly deleterious mutations in finite populations: size does matter. Genetical Research 75: 75–81.
22. Haag C, Roze D (2007) Genetic load in sexual and asexual diploids: Segregation, dominance and genetic drift. Genetics 176: 1663–1678.
23. Clarke B (1973) Mutation and population size. Heredity 31: 367–379.
24. Jarne P, Auld J (2006) Animals mix it up too: The distribution of self-fertilization among hermaphroditic animals. Evolution 60: 1816–1824.
25. Fisher R (1941) Average excess and average effect of a gene substitution. Annals of EugenicsA 11: 53–63.
26. Glémin S (2003) How are deleterious mutations purged? Drift versus nonrandom mating. Evolution 57: 2678–2687.
27. Baker H (1955) Self compatibilty and establishment after long distance dispersal. Evolution 9: 347–349.
28. Stebbins G (1957) Self fertilization and population variability in the higher plants. American Naturalist: 337–354.
29. Goldberg E, Kohn J, Lande R, Robertson K, Smith S, et al. (2010) Species selection maintains self-incompatibility. Science 330: 493–495.
30. Schoen D, Johnston M, LHeureux A, Marsolais J (1997) Evolutionary history of the mating system in Amsinckia (Boraginaceae). Evolution 51: 1090–1099.
31. Perez-Barrales R, Vargas P, Arroyo J (2006) New evidence for the Darwinian hypothesis of het-erostyly: breeding systems and pollinators in Narcissus sect. Apodanthi. New Phytologist 171: 553–567.
32. Sakai S, Wright SJ (2008) Reproductive ecology of 21 coexisting Psychotria species (Rubiaceae): when is heterostyly lost? Biological Journal of the Linnean Society 93: 125–134.
33. Gioti A, Stajich J, Johannesson H (2013) Neurospora and the dead-end hypothesis: Genomic consequences of selfing in the model genus. Evolution Accepted.
34. Chevin L, Lande R (2010) When do adaptive plasticity and genetic evolution prevent extinction of a density-regulated population? Evolution 64: 1143–1150.
35. Caballero A, Hill W (1992) Effects of partial inbreeding on fixation rates and variation of mutant- genes. Genetics 131: 493–507.
36. Roze D, Rousset F (2004) Joint effects of self-fertilization and population structure on mutation load, inbreeding depression and heterosis. Genetics 167: 1001–1015.
37. Roze D (2009) Diploidy, Population Structure, and the Evolution of Recombination. American Naturalist 174: S79–S94.
38. Roze D (2012) Spatial heterogeneity in the strength of selection against deleterious alleles and the mutation load. Heredity 109: 137–145.
39. Wright S (1931) Evolution in Mendelian populations. Genetics 16: 97–159.
40. Simmons M, Crow J (1977) Mutations affecting fitness in drosophilia populations. Annual Review of Genetics 11: 49–78.
41. Steinmetz L, Scharfe C, Deutschbauer A, Mokranjac D, Herman Z, et al. (2002) Systematic screen for human disease genes in yeast. Nature Genetics 31: 400–404.
42. Agrawal A, Whitlock M (2011) Inferences about the distribution of dominance drawn from yeast gene knockout data. Genetics 187: 553–U274.
43. Phadnis N, Fry J (2005) Widespread correlations between dominance and homozygous effects of mutations: Implications for theories of dominance. Genetics 171: 385–392.
44. Manna F, Martin G, Lenormand T (2011) Fitness landscapes: An alternative theory for the dom- inance of mutation. Genetics 189: 923–U303.
45. Eyre-Walker A, Keightley PD (2007) The distribution of fitness effects of new mutations. Nature Reviews Genetics 8: 610–618.
46. Keightley PD, Halligan DL (2009) Analysis and implications of mutational variation. Genetica 136: 359–369.
47. Lande R (1988) Genetics and demography in biological conservation. Science 241: 1455–1460.
48. Robert A, Couvet D, Sarrazin F (2002) Fitness heterogeneity and viability of restored populations. Animal Conservation 5: 153–161.
49. Higgins K, Lynch M (2001) Metapopulation extinction caused by mutation accumulation. Proceedings of the National Academy of Sciences of the United States of America 98: 2928–2933.
50. Kondrashov A (1982) Selection against harmful mutations in large sexual and asexual populations. Genetical Research 40: 325–332.
51. Petes T (2001) Meiotic recombination hot spots and cold spots. Nature Reviews Genetics 2: 360–369.
52. Charlesworth B, Charlesworth D (2010) Elements of Evolutionary Genetics. Roberts and Company.
53. Charlesworth D, Morgan M, Charlesworth B (1993) Mutation accumulation in finite outbreeding and inbreeding populations. Genetical Research 61: 39–56.
54. Charlesworth D, Charlesworth B (1987) Inbreeding depression and its evolutionary consequences. Annual Review of Ecology and Systematics 18: 237–268.
55. Charlesworth D (2003) Effects of inbreeding on the genetic diversity of populations. Philosophical transactions of the Royal Society of London Series B-Biological Sciences 358: 1051–1070.
56. Takebayashi N, Morrell PL (2001) Is self-fertilization an evolutionary dead end? revisiting an old hypothesis with genetic theories and a macroevolutionary approach. American Journal of Botany 88: 1143–1150.
57. Glémin S, Ronfort J (2013) Adaptation and maladaptation in selfing and outcrossing species: new mutations versus standing variation. Evolution 67: 225–240.
58. Roze D, Lenormand T (2005) Self-fertilization and the evolution of recombination. Genetics 170: 841–857.
59. Fischer M, Matthies D (1998) Experimental demography of the rare Gentianella germanica: seed bank formation and microsite effects on seedling establishment. Ecography 21: 269–278.
60. Dostalek T, Munzbergova Z, Plackova I (2010) Genetic diversity and its effect on fitness in an endangered plant species, Dracocephalum austriacum L. Conservation Genetics 11: 773–783.
61. Costin B, Morgan J, Young A (2001) Reproductive success does not decline in fragmented populations of Leucochrysum albicans subsp albicans var. tricolor (Asteraceae). Biological Conservation 98: 273–284.
62. Fredrickson R, Siminski P, Woolf M, Hedrick P (2007) Genetic rescue and inbreeding depression in Mexican wolves. Proceedings of the Royal Society B-Biological Sciences 274: 2365–2371.
63. Heber S, Varsani A, Kuhn S, Girg A, Kempenaers B, et al. (2013) The genetic rescue of two bottlenecked South Island robin populations using translocations of inbred donors. Proceedings of the Royal Society B-Biological Sciences 280.

64. Wootton J, Pfister C (2013) Experimental separation of genetic and demographic factors on extinction risk in wild populations. Ecology 94: 2117–2123.
65. Schoedelbauerova I, Tremblay R, Kindlmann P (2010) Prediction vs. reality: Can a PVA model predict population persistence 13 years later? Biodiversity and Conservation 19: 637–650.
66. Bell T, Powell K, Bowles M (2013) Viability Model Choice Affects Projection Accuracy and Reintroduction Decisions. Journal of Wildlife Management 77: 1104–1113.

17

Rebuilding Biodiversity of Patagonian Marine Molluscs after the End-Cretaceous Mass Extinction

Martin Aberhan[1]*, Wolfgang Kiessling[1,2]

1 Museum für Naturkunde, Leibniz Institute for Evolution and Biodiversity Science, Berlin, Germany, 2 GeoZentrum Nordbayern, Paläoumwelt, Universität Erlangen–Nürnberg, Erlangen, Germany

Abstract

We analysed field-collected quantitative data of benthic marine molluscs across the Cretaceous–Palaeogene boundary in Patagonia to identify patterns and processes of biodiversity reconstruction after the end-Cretaceous mass extinction. We contrast diversity dynamics from nearshore environments with those from offshore environments. In both settings, Early Palaeogene (Danian) assemblages are strongly dominated by surviving lineages, many of which changed their relative abundance from being rare before the extinction event to becoming the new dominant forms. Only a few of the species in the Danian assemblages were newly evolved. In offshore environments, however, two newly evolved Danian bivalve species attained ecological dominance by replacing two ecologically equivalent species that disappeared at the end of the Cretaceous. In both settings, the total number of Danian genera at a locality remained below the total number of late Cretaceous (Maastrichtian) genera at that locality. We suggest that biotic interactions, in particular incumbency effects, suppressed post-extinction diversity and prevented the compensation of diversity loss by originating and invading taxa. Contrary to the total number of genera at localities, diversity at the level of individual fossiliferous horizons before and after the boundary is indistinguishable in offshore environments. This indicates an evolutionary rapid rebound to pre-extinction values within less than *ca* 0.5 million years. In nearshore environments, by contrast, diversity of fossiliferous horizons was reduced in the Danian, and this lowered diversity lasted for the entire studied post-extinction interval. In this heterogeneous environment, low connectivity among populations may have retarded the recolonisation of nearshore habitats by survivors.

Editor: Donald J. Colgan, Australian Museum, Australia

Funding: The study was supported by the Deutsche Forschungsgemeinschaft (grant number: Ab 109/8-1; www.dfg.de). The funders had no role in study design, data collection and analysis, decision to publish, or preparation of the manuscript.

Competing Interests: The authors have declared that no competing interests exist.

* Email: martin.aberhan@mfn-berlin.de

Introduction

The study of mass extinction events can reveal patterns of selective survival and the dynamics of ecosystem recovery over evolutionary time scales [1–3]. Apart from developing models of the processes involved in recovery dynamics [2], [4], such information can be gleaned from empirical studies of the geological past.

The mass extinction at the Cretaceous-Palaeogene (K-Pg) boundary is among one of the best studied Phanerozoic bioevents [5]. Geographically, the scope of studies encompasses local, regional, and global scales [6–8]. Yet, our understanding is still limited about how global signals of diversity dynamics arose as a result of the accumulation of events at local and regional scales. A general challenge is the reciprocal relationship between the geographical extent of a study and its temporal resolution. The 11-myr binning resolution of global occurrence compilations of marine macrofauna [8] is too coarse to uncover the dynamics of post-extinction diversity rebounds, whereas local-scale analyses may lack general validity.

Here we construct high-resolution diversity trajectories from benthic marine molluscs at two fossiliferous K-Pg boundary sites in Patagonia and incorporate information on the evolutionary and biogeographical characteristics of the fauna and on species abundances. The two studied sites represent marginal to shallow marine (nearshore) and middle shelf environments (offshore). We focus on evolutionary patterns and longer-term diversity trends across the K-Pg interval. In particular, we test two null hypotheses: Within the first *ca* 4 million years (myr) after the extinction event (1) the diversity of mollusc assemblages is indistinguishable from that of pre-extinction faunas; (2) the transient loss of diversity through global extinctions and local extirpations was compensated for by gains via immigrating and newly originating taxa. We demonstrate that hypothesis (1) is only partly supported, and hypothesis (2) can be rejected by empirical evidence.

Materials and Methods

No permits were required for the described study, which complied with all relevant regulations.

Sampling localities

We analysed two stratigraphic sections in the Argentine part of Patagonia. The first one is a composite section from the adjacent localities Bajada del Jagüel (BJG, 38°6′S, 68°23′W) and Opaso (38°8′S, 68°24′W), located in the Neuquén Basin north of the city of Neuquén. The second section, close to Estancia San Ramón in the middle Chubut River valley (42°41′S, 69°51′W), is located in the Cañadón Asfalto Basin. During latest Cretaceous–earliest

Palaeocene times the sea in both areas formed narrow embayments, which opened to the Atlantic Ocean to the east [9].

The sections at BJG and Opaso consist of monotonous, middle shelf siliciclastic mudstones of the Jagüel Formation, representing open marine sedimentation below, or close to, the storm wave base [10]. In the following they are referred to as 'offshore'. Because the section at Opaso virtually lacks a Maastrichtian record to which the Danian samples could be compared, the faunal data from BJG and Opaso were arranged chronologically according to their distance from the K-Pg boundary and integrated into a single succession (i.e. our offshore section). We thereby assumed equal sedimentation rates at these close-by localities. Faunal samples were derived from 11.4 metres below the K-Pg boundary to 27.5 metres above the boundary.

In San Ramón, faunal sampling extends over 188 m of varied sedimentary rocks of the Lefipán Formation. The Maastrichtian to early Palaeocene succession comprises siliciclastic mudstones, siltstones, sandstones, and conglomerates. Litho- and biofacies indicate deposition in a marginal to shallow marine, tide-dominated deltaic setting with tidal bars, prodelta, and delta plain environments, which are represented in both the Maastrichtian and the Danian [9]. In this paper, the aforementioned deltaic setting is referred to as 'nearshore'. Salinity was an important factor affecting the composition and ecological structure of the fauna in these environments. The uppermost Maastrichtian is characterized by virtually monospecific assemblages of the bivalve genus *Corbicula* owing to oligohaline to mesohaline salinity, whereas more fully marine conditions existed below and above this interval [9]. These very low-diversity brackish-water assemblages were excluded from our analyses because they represent a different environment.

Biostratigraphy and interval lengths

In BJG, the K-Pg boundary is at the erosive base of a coarse-grained, tuffaceous sandstone bed [10], [11]. Calcareous nannofossils prove the presence of the first Cenozoic nannofossil zone (NP1 or NNTp1) in the upper part of the sandstone layer, and younger subzones in the mudstones overlying the sandstone [10]. The mudstones below the sandstone are late Maastrichtian in age [10], [11]. According to Keller et al. [11], the 11.4 metres of section below the K-Pg boundary, which yielded the late Maastrichtian quantitative samples in our study, are characteristic of the nannofossil zone CC26 and the *A. mayaroensis* and *P. hariaensis* foraminiferal zones (zonal names according to [12]). In Opaso, the K-Pg boundary sandstone is exposed about half a metre above the base of the sampled section. Micro- and nannofossils indicate the presence of early (NP1-NP3?/P1a-b) to late Danian (NP4/P1c) biozones [13]. Applying the ages given in Gradstein et al. [12], the samples of BJG and Opaso extend from ~2 myr before the K-Pg boundary to ~4 myr after the boundary. Because the earliest Danian planktonic foraminiferal zones P0 and Pα, the latest Maastrichtian *P. hantkeninoides* foraminiferal zone, and impact tracers are missing, the sections at BJG and Opaso are stratigraphically incomplete [10], with possibly up to 500 kyr of the earliest Danian and 300 kyr of the latest Maastrichtian missing [10], [11]. In summary, the amount of time represented by the Danian strata is about twice that of the Maastrichtian strata.

In the nearshore San Ramón section, a distinct boundary layer is not preserved. However, the presence of latest Maastrichtian and earliest Danian dinoflagellate markers and age-diagnostic palynomorphs indicate a very minor hiatus [14]. A more detailed biostratigraphical subdivision is not available in this very shallow-water setting. Because depositional environments are the same below and above the boundary, and because the sampled Maastrichtian succession (100 metres) is similar in thickness to that of the Danian (88 metres), we tentatively assume roughly equal time spans for the studied pre- and post-extinction intervals.

Analytical methods

We quantitatively sampled 157 single beds of uniform lithology for macroinvertebrates, comprising a total of 5,547 individuals. Bivalves and gastropods dominate the macrofauna. Taxa were identified as precisely as possible, mostly at species level. Our analyses focus on genus diversity of bivalves and gastropods. We elevated subgenera to genus rank and omitted all taxa not resolvable at least to the genus level. Because most genera are represented by a single species at the studied localities, the number of taxa is almost identical at genus and species level. Following standard procedures [15] the number of individuals of bivalves was determined as the sum of articulated specimens plus the larger number of right or left valves of disarticulated specimens plus one-half the number of fragments which could not be assigned to either right or left valves but make up more than 50% of a valve. In one special case – *Entolium membranaceum*, an equivalved and equilateral pectinoid with symmetrical auricles – distinction of the right and left valves was impossible, thus numbers of specimens were halved to yield numbers of individuals. Gastropods were counted when more than 50% of the shell was preserved.

To avoid large variations in sample size, we pooled samples of successive fossiliferous beds until at least 100 individuals were reached or until the lithology changed markedly. In the offshore section, this resulted in 24 pooled samples with a mean size of 129.1±7.1 (standard error) individuals. The 28 pooled nearshore samples had a mean of 63.1±5.5 individuals.

We focus on two measures of diversity. The first one determines the standardised number of genera for each of the pooled samples. We refer to this measure as sample-level diversity [16]. We standardised sample-level diversity using the 'shareholder quorum' subsampling method developed by Alroy [8], [17]. This method involves random sampling of individuals to equal frequency coverage, the shareholder quorum. In our analyses the quorum was set to 0.8 and our values are averages of 100 iterations. The shareholder quorum is superior to traditional subsampling methods (e.g., rarefaction) in that variations in evenness are explicitly taken into account. Therefore, it is less biased by different abundance distributions and represents a nearly unbiased estimate of richness.

Our second measure of diversity aims at comparing the total number of Danian genera in a section with that of Maastrichtian genera in the same section. This measure is termed total (Maastrichtian or Danian) diversity. To this end, the genera of the pooled and standardised samples were summed up over consecutively younger samples for the Maastrichtian and for the Danian part of the section, respectively. Genera are added to the count of total diversity at the sampling level at which they first appeared in the succession and are considered to persist up to the K-Pg boundary (for Maastrichtian total diversity) or to the youngest Danian sample (for Danian total diversity). This procedure constructs sample-based accumulation curves [18], whereby the sequence of samples is ordered in time via stratigraphy. This approach assumes that diversity accumulates during both the Maastrichtian and the Danian. While confidence intervals on stratigraphic ranges of genera support this approach, we emphasize that this procedure is thought to provide theoretical 'genus accumulation curves', which allow for a comparison of Maastrichtian and Danian diversity at various sampling levels.

With reference to the K-Pg boundary, we assigned each genus to one of five categories: (1) going extinct globally; (2) disappearing in Patagonia but continuing elsewhere; (3) surviving; (4) immigrating; and (5) originating. The reference area for categories 2–5 is Patagonia because geographical restriction to localities may produce strong sampling biases. We categorised genera as immigrants when they occurred in the Danian sections of our study area, had no Cretaceous record from Patagonia, and had been reported outside of Patagonia either before the Danian or within a part of the Danian which is stratigraphically older than our own field records. We categorised genera as originating in Patagonia when they are endemic to this region and first appeared in the Danian, or when a Danian occurrence in our sections is stratigraphically older than records from anywhere else in the world. For each pooled and standardised sample we calculated proportional diversity of the various categories. For the purpose of assigning genera to the five evolutionary/biogeographical categories, geographical and stratigraphical range data were retrieved from the Paleobiology Database (http://paleobiodb.org), Sepkoski's compendium [19], and numerous monographs on Cretaceous and Palaeogene molluscs. The main sources of information and specific references that were used for the categorisation of taxa are provided in Tables 1 and 2. A full documentation of primary data is presented in Tables S1 and S2.

Extinction intensities are expressed as percent extinctions of genera using the data presented in Tables 1 and 2. For comparison with our Patagonian sites, percent global extinctions of molluscan genera were also calculated for three other well-studied K-Pg boundary localities: Brazos River, Texas [20]; Braggs, Alabama [21]; and Stevns Klint, Denmark [22]. These calculations used the primary data provided in the respective publications.

Finally, we compared the relative abundances of the most common Maastrichtian species with their abundances in the Danian and vice versa at both sections. We summed up the total abundances of each species in the Danian and Maastrichtian parts of our respective sections and calculated their proportional abundance in each pooled sample. The proportional abundance of the most common species (those in the upper 90th percentile of all proportional abundances) was then ordered by decreasing proportional abundance and compared with the proportions in the stratigraphically older (for Danian abundances) or younger (for Maastrichtian abundances) parts of the sections. The bivalve *Corbicula*, which is very abundant in the low-diversity brackish-water assemblages, was excluded from this analysis.

All analyses were performed in the R programming environment (www.r-project.org). The field-collected Patagonian specimens are housed at the Departamento de Ciencias Geológicas, Universidad de Buenos Aires, Argentina.

Results

Genus-level extinction rates

Intensities of end-Cretaceous extinctions may vary markedly across local, regional, and global scales. In our nearshore section the percentage of global extinctions of benthic molluscan genera is 21.7%. In the offshore section 32.1% of genera became extinct. These values are comparable to those of well-studied Northern Hemisphere boundary sections such as Braggs, Alabama (30.5% extinction of bivalves and gastropods [21]) or Stevns Klint, Denmark (22.5% extinction of bivalves [22]), but markedly higher than at Brazos River, Texas (10.7% extinction of bivalves and gastropods [20]). Local extirpations (60.9% in our nearshore section, 53.6% in the offshore section) are much greater than global extinctions. These higher values are the cumulative result of local extirpations, global extinctions, and sampling effects.

Maastrichtian and Danian diversity compared

Sample-level diversity in offshore environments fluctuated without a consistent temporal trend (Figure 1). The diversity of even the first quantitative Danian samples is mostly within the range of Maastrichtian assemblages. In contrast, Danian sample-level diversity in nearshore habitats is, on average, distinctly lower than that of pre-extinction assemblages (Figure 1). This pattern is evident in both fine-grained and coarse-grained lithologies and persists until the end of the studied time interval, suggesting contrasting diversity patterns between nearshore and offshore environments. Despite comparable extinction intensities, recovery of sample-level diversity was only achieved offshore.

Total diversity in the Danian is significantly and consistently lower than total Maastrichtian diversity at the same sampling level (Figure 2). This pattern holds for both nearshore and offshore environments, and points to a distinct difference in diversity structure between Maastrichtian and Danian faunas.

The contributions of surviving, immigrating, and originating genera

In both environments, the most diverse components of Maastrichtian assemblages are genera that survived the K-Pg boundary event (Figure 3). Immigrating and newly evolving genera play only subordinate roles in the composition of Danian samples. Their occurrence is often sporadic and without obvious trends (Figure 3). The reduced total diversity in the Danian (Figure 2) indicates that immigrants and newly evolved taxa could not compensate for the loss of Maastrichtian diversity at the K-Pg boundary.

Abundance shifts of dominant species across the K-Pg boundary

In both environmental settings, the dominant Maastrichtian species became extinct at the K-Pg boundary (Figure 4). In the nearshore environment, this affected the five most abundant Maastrichtian species. Offshore, three of the four most abundant species did not survive the boundary, only the oyster *Pycnodonte vesicularis*, the second most common Maastrichtian species in the offshore environment, survived. However, with only a single specimen found in the Danian, it can be considered as ecologically extinct. It is best classified as a 'Dead Clade Walking', a term coined to characterise lineages that survived mass extinctions only to remain marginal or decline in the aftermath [23].

A second intriguing abundance pattern is seen in species that dominate the Danian assemblages. These were usually rare in the Maastrichtian part of the respective stratigraphic section (Figure 5). The three most abundant nearshore species, collectively making up almost 60% of Danian individuals, are survivors with low abundances during the Maastrichtian. In the offshore environment, three of the five most abundant Danian species are survivors that also changed their relative abundances from rare to common. The two dominant Danian offshore species – the oyster *Pycnodonte burckhardti* and the pectinid *Delectopecten neuquenensis* – are Patagonian endemics that originated in the early Danian.

Discussion

Incumbency effects and recovery with minor origination

The persistent dominance of survivors (Figure 3) points to the effectiveness of survival mechanisms [24] and contrasts with recovery models in which a short-term survival phase is followed by diversification owing to origination of new taxa [25]. At any

Table 1. Categorisation of molluscs at San Ramòn.

Class	Genus	Range in section	Category	Additional notes and references on records in the	
				Cretaceous	Palaeogene
Bivalvia	*Acesta*	Maa	2		
Gastropoda	*Acteon*	Maa	3		[50]
Bivalvia	*Anomia*	Maa	3		See Table 2
Bivalvia	*Aphrodina*	Danian	3	Valanginian of Argentina [51], [52]	
Bivalvia	*Arcomytilus*	Danian	4		
Bivalvia	*Astarte*	Danian	3	Hauterivian of Argentina [53]	Palaeocene of Argentina [54]
Bivalvia	*Australoneilo*	Maa – Danian	3		
Gastropoda	*Austroaporrhais*	Maa	3		Danian of Patagonia [54], [55]
Bivalvia	*Austrotrigonia*	Maa	1		
Bivalvia	*Bicorbula*	Maa – Danian	3		
Bivalvia	*Brachidontes*	Maa	3		Eocene of Patagonia [56], [57]
Bivalvia	*Camptochlamys*	Maa	1	Casadío et al. [58]	
Bivalvia	*Corbicula*	Maa – Danian	3		
Gastropoda	*Cryptorhytis*	Maa	3		See Table 2
Bivalvia	*Cucullaea*	Maa	3		Danian of Patagonia (unpublished data of authors)
Bivalvia	*Cucullastis*	Danian	4	Late Cretaceous of New Zealand [59]	
Bivalvia	*Cuspidaria*	Danian	4		
Bivalvia	*Eriphyla*	Maa – Danian	3		
Bivalvia	*Etea*	Maa	2		
Gastropoda	*Euspira*	Maa – Danian	3		
Bivalvia	*Glycymeris*	Danian	3	Maastrichtian of Argentina (unpublished data of authors from Sierra Huantraico)	
Gastropoda	*Heteroterma*	Maa – Danian	3		
Bivalvia	*Inoperna*	Maa	1		
Bivalvia	*Lahillia*	Maa	3		Eocene of Patagonia [60]
Bivalvia	*Ledina*	Maa – Danian	3		
Bivalvia	*Leionucula*	Maa – Danian	3		
Bivalvia	*Linearia*	Maa	2		Eocene of USA [61]
Bivalvia	*Lyriochlamys*	Maa	1		
Bivalvia	*Meretrix*	Maa – Danian	3		
Bivalvia	*Mesocallista*	Maa	1		
Bivalvia	*Modiolus*	Maa – Danian	3		
Gastropoda	*Nacella*	Danian	3	See Table 2	
Bivalvia	*Neilo*	Maa	3		Danian [62], Eocene [60] and Oligocene [50] of Patagonia
Bivalvia	*Nucula*	Maa	3		Danian of Patagonia [26]
Bivalvia	*Ostrea*	Maa – Danian	3		
Bivalvia	*Pachymya*	Maa	1		
Bivalvia	*Pacificor*	Maa – Danian	3		
Bivalvia	*Pacitrigonia*	Maa	1		
Bivalvia	*Panopea*	Maa – Danian	3		
Gastropoda	*Parasyrinx*	Maa	2		
Bivalvia	*Pholadomya*	Maa	2		
Bivalvia	*Phygraea*	Maa	3		See Table 2
Bivalvia	*Pinna*	Maa – Danian	3		
Gastropoda	*Priscaphander*	Maa	3		See Table 2

Table 1. Cont.

Class	Genus	Range in section	Category	Additional notes and references on records in the	
				Cretaceous	Palaeogene
Bivalvia	*Protagelus*	Maa	1		
Gastropoda	*Pseudamaura*	Danian	4		[54]
Bivalvia	*Pseudolimea*	Maa	3		See Table 2
Gastropoda	*Pseudotylostoma*	Danian	5		Genus endemic to Patagonia
Bivalvia	*Pteromyrtea*	Maa - Danian	3		
Gastropoda	*Pyropsis*	Maa	2		
Bivalvia	*Rinetrigonia*	Maa	1		
Bivalvia	*Saccella*	Danian	3	Maastrichtian of Argentina (unpublished data of authors from Sierra Huantraico)	
Bivalvia	*Solyma*	Maa	1		
Bivalvia	*Spineilo*	Danian	5		Danian records from Patagonia are from the lower part of the Danian (this study; foraminiferal zone P1b of [63]), whereas oldest records from New Zealand [64] are of late Danian age [26]. Palaeocene records from Russia appear to be late Danian or younger [65]; see also [66]
Gastropoda	*Struthioptera*	Maa - Danian	3		
Bivalvia	*Tikia*	Maa - Danian	3		
Gastropoda	*Turritella*	Maa - Danian	3		

Alphabetical list of genera (including subgenera) of bivalves and gastropods from the late Maastrichtian and Danian at San Ramón, western Argentina, and their assignment to evolutionary/biogeographical categories.
Evolutionary/biogeographical categories: 1 = going extinct globally in the late Maastrichtian; 2 = disappearing regionally from Patagonia in the late Maastrichtian; 3 = surviving the K-Pg boundary in Patagonia; 4 = immigrating to Patagonia in the Danian; 5 = originating in Patagonia during the Danian. Maa = Maastrichtian.

given sampling level, originating and invading genera could not compensate for the loss of genera caused by extinctions (Figure 2). A similar observation was reported for Antarctic molluscs at the K-Pg boundary, where few new genera were observed, and only later in the Danian did the number increase [26].

Delayed recovery is frequently explained by unfavourable environments, either as protracted impact-related effects, or as specific regional effects [27–29]. However, local environmental stress fails to explain the Danian *versus* Maastrichtian difference in total diversity in the offshore section where there is no evidence for a shift to more unfavourable environments. Similarly, in the nearshore section the range of sedimentary environments and overall environmental heterogeneity in the Danian and in the Maastrichtian match well [9]. Furthermore, previous work from our working group that used the same raw data from the same nearshore section showed that even Danian assemblages from stable, subtidal environments, characterised by diverse trophic and life habit groups, did not reach Maastrichtian diversity levels [9]. This is exemplified by comparing the Danian *Struthioptera-Venericardia* faunal association with the Maastrichtian *Struthioptera-Panopea* association of Scasso et al. [9]. Both associations are dominated by infaunal guilds but also have relatively high proportions of epifaunal molluscs. In both associations, trophic resources have been subdivided among suspension-feeders, deposit-feeders, and carnivores, suggesting biologically controlled communities [9]. Yet, the diversity of samples forming the Danian association remains distinctly below the diversity of the Maastrichtian samples. Thus, unfavourable habitats cannot be held responsible for the differences among pre- and post-extinction diversity at either of the two studied sections.

An alternative explanation for the delayed recovery would be low immigration rates caused either by geographic isolation of the basins or by a general shortage in the supply of taxa from outside Patagonia. However, there is no evidence of isolation because the Patagonian marine basins opened to the South Atlantic Ocean to the east, which was a large open ocean with pelagic sedimentation during K-Pg boundary times [9], [10]. Based on general physical oceanographic principles, wind driven surface currents, tidally induced amphidromic circulation, and thermohaline circulation are assumed to have been well developed in the studied areas, all of which would have contributed to the exchange of waters with surrounding regions. The presence in the Neuquén Basin of diverse planktonic foraminifera [11], a plankton group only found in high diversity in marine waters of normal salinity, provides direct evidence for such exchange with open marine waters. Palaeogeographic reconstructions of the marine basins in Patagonia for the Maastrichtian–Palaeocene interval show direct shallow water connections between the basins [9]. Also, with the closely located South American shelf to the north, the Antarctic shelf to the south, and refugia along the margin of the southern-Pacific [26], several potential areas were available as a source for new immigrants. Therefore, geographic isolation and a shortage in the supply of larvae are unlikely scenarios for the delayed recovery in diversity.

We suggest that the numbers of new taxa and of taxa invading from outside Patagonia were low due to the presence of survivors

Table 2. Categorisation of molluscs at Bajada del Jagüel and Opaso.

Class	Genus	Range in sections	Category	Additional notes and references on records in the	
				Cretaceous	**Palaeogene**
Gastropoda	*Acteon*	Maa	3		[50]
Bivalvia	*Ambigostrea*	Maa	1		
Bivalvia	*Amphidonte*	Maa	1		
Bivalvia	*Anomia*	Maa - Danian	3		
Bivalvia	*Aphrodina*	Danian	3	Valanginian of Argentina [51]	
Bivalvia	*Arcomytilus*	Danian	4		
Bivalvia	*Australoneilo*	Maa - Danian	3		
Bivalvia	*Bicorbula*	Danian	3	See Table 1	
Bivalvia	*Brachidontes*	Maa	3		Eocene of Patagonia [56], [57]
Bivalvia	*Camptochlamys*	Maa	1	[58]	
Bivalvia	*Cosmetodon*	Maa - Danian	3		
Gastropoda	*Cryptorhytis*	Maa - Danian	3		
Bivalvia	*Cucullastis*	Danian	4	Late Cretaceous of New Zealand [59]	
Bivalvia	*Delectopecten*	Danian	4	Early Cretaceous of Greenland [67]	Danian of Patagonia [62], [68]
Bivalvia	*Disparilia*	Maa - Danian	3		
Bivalvia	*Entolium*	Maa	1		
Bivalvia	*Glycymeris*	Danian	3	Maastrichtian of Argentina (unpublished data of authors from Sierra Huantraico)	
Bivalvia	*Gryphaeostrea*	Danian	3	Aguirre-Urreta [69]	del Río et al. [62], [66]
Gastropoda	*Heteroterma*	Danian	3	See Table 1	
Bivalvia	*Jagolucina*	Danian	5		Thanetian of Australia [70]; Eocene of France [71]
Bivalvia	*Leionucula*	Danian	3	See Table 1	[66]
Bivalvia	*Lima*	Maa	2		
Bivalvia	*Linearia*	Maa	2		Eocene of USA [61]
Bivalvia	*Lyriochlamys*	Maa	1		
Bivalvia	*Mesocallista*	Maa	1		
Bivalvia	*Modiolus*	Maa - Danian	3		
Gastropoda	*Nacella*	Maa - Danian	3		
Bivalvia	*Neilo*	Maa	3		Eocene [60] and Oligocene [50] of Patagonia
Bivalvia	*Nicaniella*	Danian	4		
Bivalvia	*Nuculana*	Maa - Danian	3		
Bivalvia	*Ostrea*	Maa - Danian	3		
Bivalvia	*Pacificor*	Danian	3	See Table 1	
Bivalvia	*Panopea*	Danian	3	See Table 1	
Gastropoda	*Perissodonta*	Danian	5		Danian records from Patagonia are from the lower part of the Danian (this study; foraminiferal zone P1b of [63]), whereas oldest Palaeocene records from New Zealand [64] are late Danian [26]
Bivalvia	*Phelopteria*	Maa	1	[72]	
Bivalvia	*Phygraea*	Maa - Danian	3		
Bivalvia	*Pinna*	Danian	3	See Table 1	
Gastropoda	*Priscaphander*	Danian	3	See Table 1	
Bivalvia	*Protagelus*	Maa	1		
Bivalvia	*Pseudolimea*	Maa - Danian	3		

Table 2. Cont.

Class	Genus	Range in sections	Category	Additional notes and references on records in the	
				Cretaceous	Palaeogene
Bivalvia	*Pteromyrtea*	Maa - Danian	3		
Gastropoda	*'Pyropsis'*	Danian	4		Species belonging to a probably new genus referred to as *'Pyropsis'* have been discussed in [26], [73], [74]; oldest records are from Antarctica
Bivalvia	*Roudairia*	Maa	1		
Bivalvia	*Saccella*	Danian	3	Maastrichtian of Argentina (unpublished data of authors from Sierra Huantraico)	
Bivalvia	*Spineilo*	Danian	5		See Table 1
Bivalvia	*Spondylus*	Maa	2		
Gastropoda	*Struthioptera*	Danian	3	See Table 1	
Gastropoda	*Taioma*	Danian	4	Late Cretaceous of Antarctica [74]	[74], [75]
Bivalvia	*Tikia*	Maa - Danian	3		
Gastropoda	*Tornatellaea*	Danian	4		Palaeocene of Argentina [54], [75]

Alphabetical list of genera (including subgenera) of bivalves and gastropods from the late Maastrichtian and Danian at Bajada del Jagüel and Opaso, western Argentina, and their assignment to evolutionary/biogeographical categories.
Evolutionary/biogeographical categories: 1 = going extinct globally in the late Maastrichtian; 2 = disappearing regionally from Patagonia in the late Maastrichtian; 3 = surviving the K-Pg boundary in Patagonia; 4 = immigrating to Patagonia in the Danian; 5 = originating in Patagonia during the Danian. Maa = Maastrichtian.

within Patagonia impeding their progress. This phenomenon is known as incumbency or priority effects, in which existing organisms prevent the successful establishment of newly evolving or immigrating species [30], [31]. Resistance to invasions or to the establishment of newly evolved taxa occurs because these 'would-be usurpers' are initially less well adapted than established species, and the latter keep incipient populations of new species from taking hold [32], [33]. A closely related concept is that of pre-emptive competition [34], emphasizing the importance of being 'first in the field'. Incumbency can be expected to play a role when resources are limited and communities are saturated with species [15]. Mass extinctions are generally thought to reduce or remove the incumbency effect [35]. Our analyses, however, suggest that despite significant extinctions, the removal of incumbents did not entail an increase in immigration or origination rates to counterbalance diversity reduction. Rather, the number of

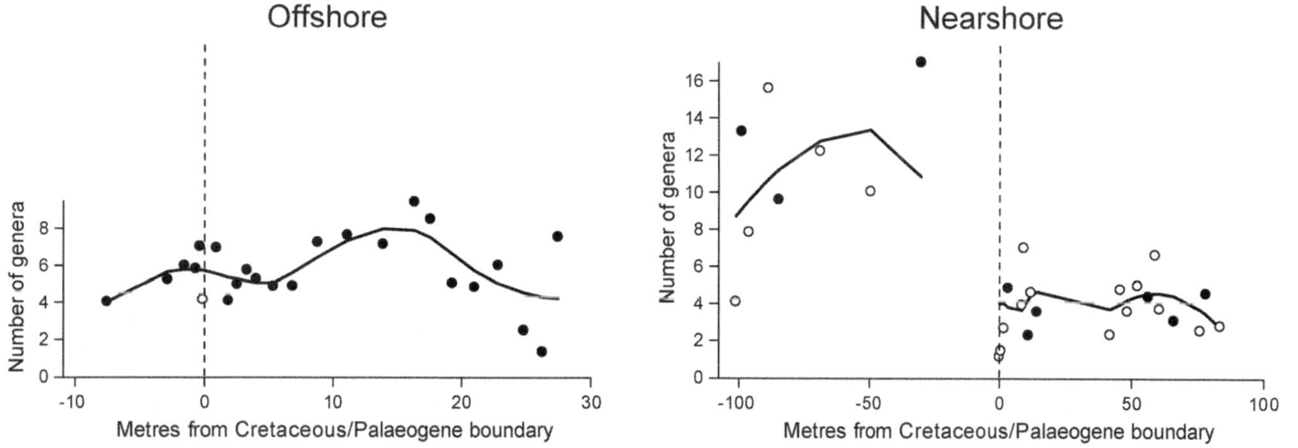

Figure 1. Sample-level diversity of benthic marine molluscs in two Cretaceous-Palaeogene boundary sections in Patagonia. Standardised genus diversity is based on the shareholder quorum subsampling method [17]. Dashed vertical lines indicate position of the Cretaceous-Palaeogene boundary; values on x-axis refer to metres before (Cretaceous, negative values) or after the boundary. Curved solid line is LOESS smoothing with a smoothing parameter of 0.5. Black dots indicate samples from fine-grained sedimentary rocks (siliciclastic mudstones and siltstones), open circles indicate samples from coarse-grained sedimentary rocks (sandstones and conglomerates). Nearshore, the uppermost *ca* 26 metres of the Cretaceous are characterised by low-diversity brackish-water assemblages, which are not included in the figure.

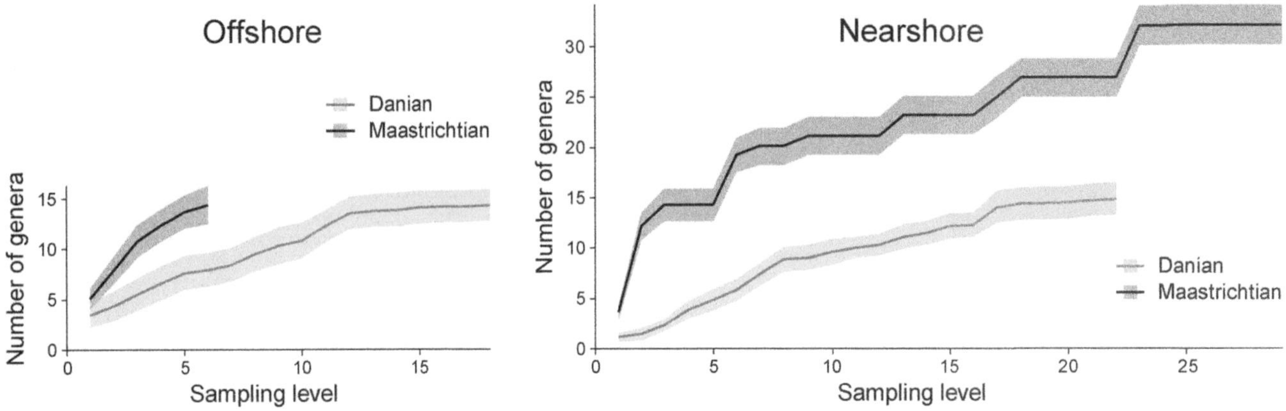

Figure 2. Comparing Cretaceous (Maastrichtian) and Palaeogene (Danian) accumulation curves for the estimated number of molluscan genera. Sampling level refers to consecutively numbered fossiliferous samples, arranged from stratigraphically oldest to youngest, and shows the cumulative number of genera at the respective level (see text for details on pooling and standardising of samples). Grey areas indicate the standard deviations around the means. In both offshore and nearshore habitats total diversity in the Danian is consistently lower than total Maastrichtian diversity at any given sampling level.

survivors seems to have remained sufficiently high to impose constraints on the establishment of immigrants and originating taxa, consistent with incumbency. We consider incumbency as an

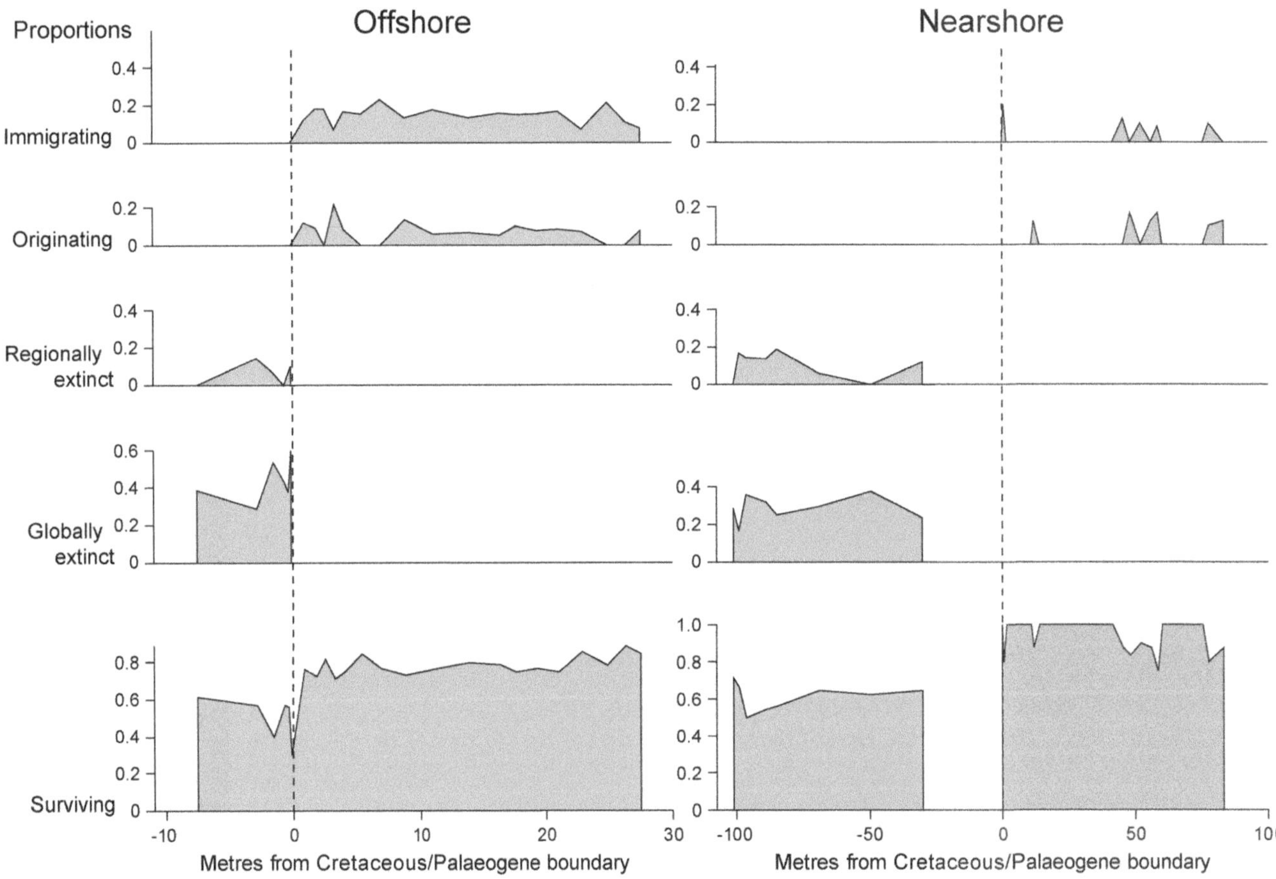

Figure 3. Proportional diversity of molluscan genera in five evolutionary/biogeographical categories across the Cretaceous-Palaeogene boundary. Proportions are shown for an offshore (panels on the left) and a nearshore boundary section (panels on the right) in Patagonia. Dashed vertical lines indicate position of the Cretaceous-Palaeogene boundary; values on x-axis refer to metres before (Cretaceous, negative values) or after the boundary. Nearshore, the uppermost *ca* 26 metres of the Cretaceous are characterised by low-diversity brackish-water assemblages, which are not included in the figure. Post-extinction faunas are dominated by survivors. Immigrating and originating genera play subordinate roles in the composition of Danian samples.

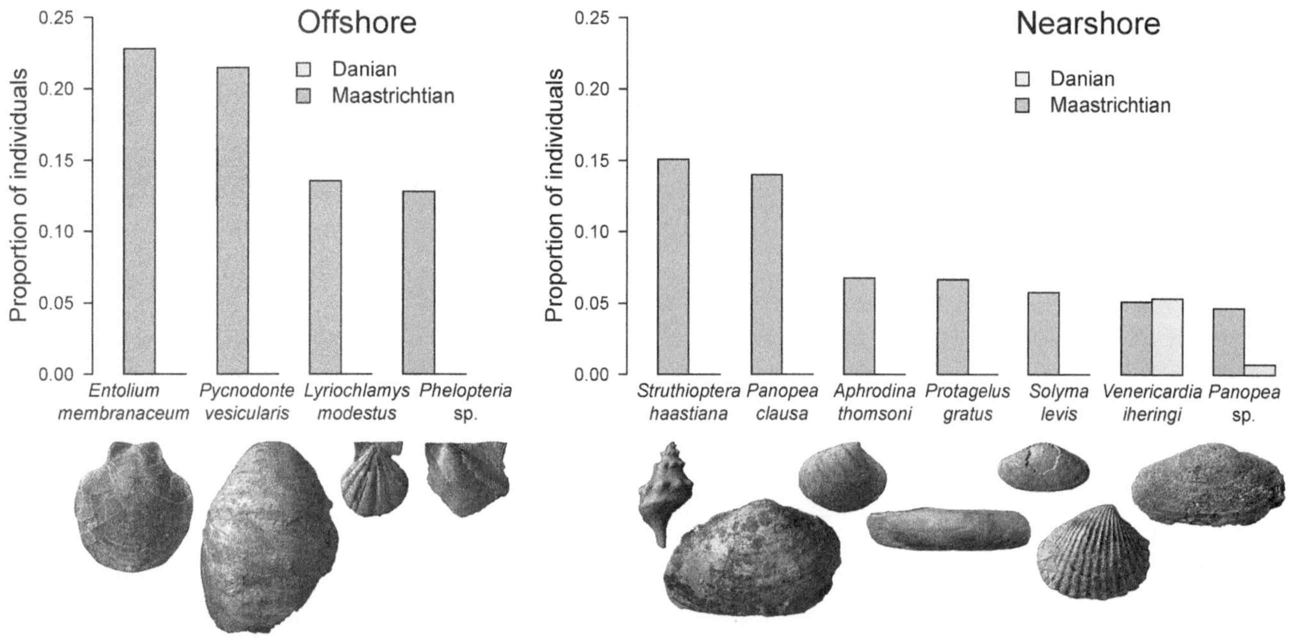

Figure 4. Proportional abundance of most common Maastrichtian molluscs contrasted with their abundances in the Danian. The displayed Maastrichtian species are those in the upper 90th percentile of the pooled samples of that time. Their Danian proportions are plotted alongside. The majority of the dominant Maastrichtian species did not survive into the Danian.

important mechanism for the delayed recovery in both onshore and offshore environments of Patagonia.

The extinction of dominants and the rise of rare species

In both studied environments the most common Maastrichtian species became extinct at the K-Pg boundary (Figure 4). This observation appears to be in contrast with ecological studies which suggest that high abundance promotes survival [36]. It can, however, be reconciled with the finding that post-Palaeozoic marine bivalves with high abundances exhibit extinction rates elevated over those of moderately common genera [37]. Also, results of a regional study of North American bivalves showed that

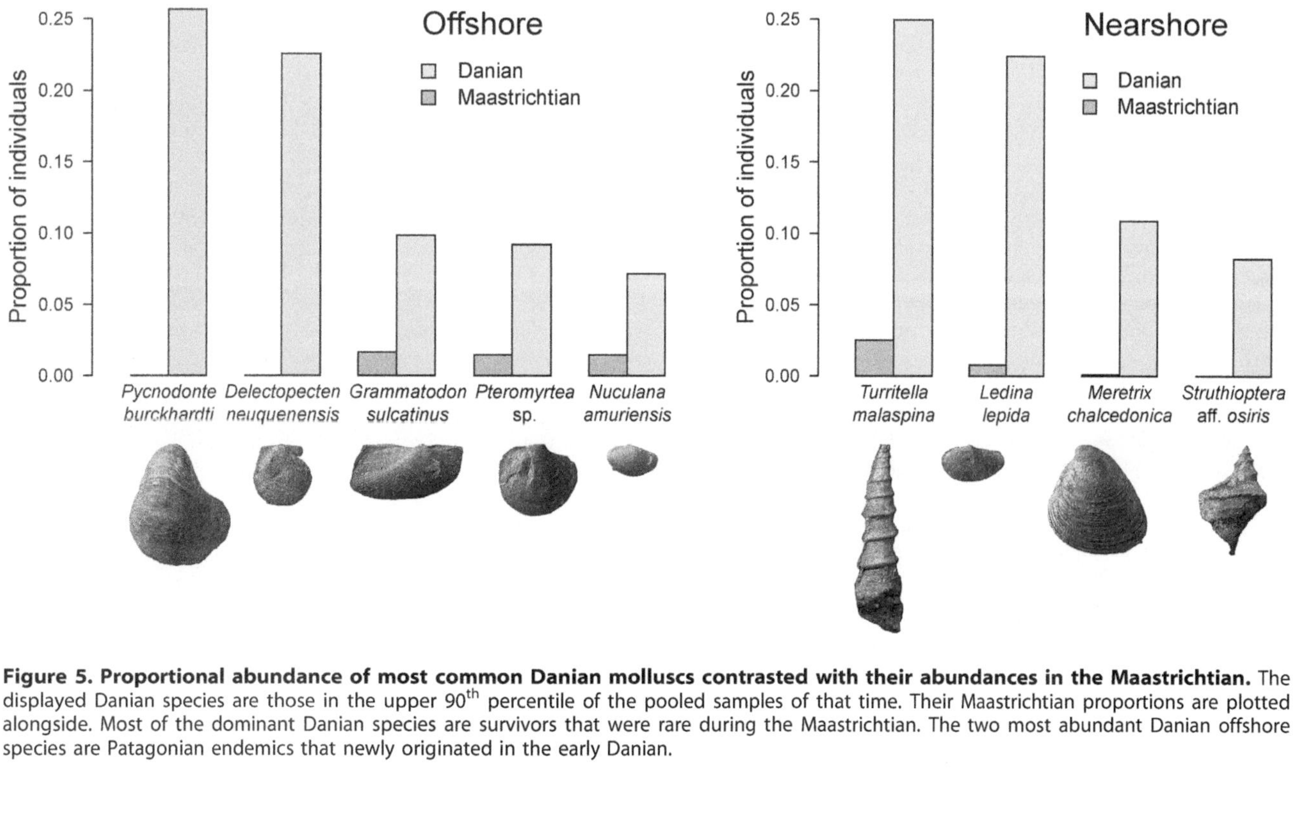

Figure 5. Proportional abundance of most common Danian molluscs contrasted with their abundances in the Maastrichtian. The displayed Danian species are those in the upper 90th percentile of the pooled samples of that time. Their Maastrichtian proportions are plotted alongside. Most of the dominant Danian species are survivors that were rare during the Maastrichtian. The two most abundant Danian offshore species are Patagonian endemics that newly originated in the early Danian.

abundant taxa were no more likely to survive the K-Pg mass extinction than rare ones [38].

Interestingly, survivors that dominate the Danian assemblages were rare during the Maastrichtian (Figure 5). We suggest that although being resistant to environmental stress, the new dominants were poor competitors before the boundary event. Once they were released from competition by the removal of dominants, the relative competitive ability of survivors increased. This suggests a particular type of incumbency, where numerically subordinate taxa arose to ecological dominance in the aftermath of a major environmental perturbation. Modelling of competitive interactions provides a potential mechanism for such reorganisations of ecosystems [39]: The loss of a competitively superior species releases otherwise unavailable resources, allowing the competitively inferior species to expand its realised niche. Thus, less abundant species may not only provide invasion resistance by their aggregate effects on resource utilization [40], but can be important by gaining dominance after environmental turmoil. Because distinct inversions in abundance across mass extinctions are known from other ecosystems [41] this may indicate a more general macroecological pattern.

The diversity-based conclusions stress the importance of survivors in post-extinction assemblages (Figure 3). This finding is generally confirmed by our abundance-based analyses with one important modification. In the offshore environment, the two top-ranking Danian species are newly evolved bivalves that originated in the early Danian of Patagonia (Figure 5). Their dominance indicates that newly originating post-extinction taxa can have high numbers of individuals even though overall numbers of originating taxa are low. There is a striking taxonomical, morphological, and ecological analogy between these two Danian species and the two most abundant Maastrichtian species in the offshore environment (Figures 4, 5). All four species are epifaunal suspension feeders and it seems that, after its extinction, the Maastrichtian pectinoid *Entolium membranaceum* was replaced by the Danian pectinoid *Delectopecten neuquenensis*, whereas the oyster *Pycnodonte vesicularis* was replaced by the congeneric *Pycnodonte burckhardti*. Although end-Cretaceous extinctions did not remove entire ecological guilds it is plausible that they created enough amount of open ecospace into which taxa with equivalent ecological traits could evolve.

The pace of recovery in sample-level diversity

The mean sample-level diversity remained unchanged across the K-Pg boundary in the studied Patagonian offshore environments at BJG and Opaso (Figure 1). The early Danian hiatus of *ca* 500 kyr at BJG [11] provides an upper limit for the duration necessary to restore sample-level diversity. The stratigraphically more complete section at Brazos River, east Texas, represents an environment similar to our offshore setting. Sample-level diversity at Brazos started to rebound *ca* 1 myr after the boundary [6] and full local diversity recovery to Cretaceous levels was attained after about 2 myr [42]. Thus, complete recovery of sample-level diversity to pre-extinction values took longer there than in Patagonia. Similarly, another study concluded that the local richness of offshore mollusc assemblages of the U.S. Gulf Coastal Plain recovered to Cretaceous values *ca* 2.7 myr after the event [43]. In conclusion, the time necessary to recover to pre-extinction diversity in mollusc-dominated offshore ecosystems varies and may take less than *ca* 0.5 myr according to our results from Patagonia and up to *ca* 2.7 myr according to data from the U.S. Gulf Coastal Plain. In contrast to offshore settings, Danian sample-level diversity did not recover to Maastrichtian levels in the Patagonian nearshore environments. We have shown from the data that there is a prolonged period of incomplete recovery in nearshore environments, although a lack of more detailed chronostratigraphic information prevents us from quantifying its duration more precisely.

Divergence of Maastrichtian and Danian diversity structure

Accumulation of total Danian diversity slowed down over time (Figure 2), seemingly compatible with a model of diversity-dependent niche-filling and a gradual approach to carrying capacity. These curves, however, should not be taken as accurate reflections of the rate at which post-extinction diversity builds up because they build on the assumption that (1) all genera found in any sample in the Danian are also present in all younger Danian samples, and (2) any genera not found in older samples represent true absences.

Geological time is critical for our interpretation that Figure 2 indicates systematic differences between Maastrichtian and Danian biodiversity. For example, the underlying (re-)population dynamics may be the same for each interval, but an apparent difference could be generated if Maastrichtian strata represented a much longer time span than Danian strata. Because the opposite is true in the offshore section (Danian samples encompass about twice as much time as Maastrichtian samples) any potential bias would operate only in the reverse direction. Interval lengths are less clear in the nearshore section but we assume roughly equal amounts of time (see above).

The reduced total diversity in the Danian and the longer-term stability of sample-level diversity across the boundary in offshore habitats are no contradiction. Jointly these patterns imply that, on average, Danian offshore assemblages retained, or rapidly regained, their Maastrichtian sample-level diversity despite being recruited from an overall smaller pool of Danian taxa. This indicates a shift in the partitioning of sample-level diversity and total diversity.

Erwin [44] pointed out that positive feedbacks may increase carrying capacity during biotic recovery. An example of a positive feedback in mollusc-dominated ecosystems may be the Cenozoic radiation of predatory gastropods and the resulting escalatory trends ([43] and references therein). If such effects were present in our area, they must have occurred subsequent to the studied time interval.

Nearshore–offshore differentials in recovery

In contrast to the offshore environment, the reduction in nearshore Danian sample-level diversity was pronounced and lasted over the entire studied interval. Differential evolutionary dynamics along onshore–offshore gradients – well-known in post-Palaeozoic benthic marine invertebrates, with preferential origination occurring onshore [45], [46] – is inadequate to account for the observed difference in our study. Not only would this mechanism favour the recovery of onshore diversity – which is just the opposite of the pattern we observe – but originations in general played only a minor role during recovery in our study areas. The nearshore–offshore difference might have arisen from unpredictable, stochastic processes, consistent with the contingent recovery model [47] which emphasises the importance of chance in the recovery of marine plankton from the end-Cretaceous extinction. A more mechanistic explanation focuses on the role of survivors as the most diverse group. We suggest that the recolonisation by survivors from nearby localities may have been facilitated in the spatially and temporally more homogeneous and continuous offshore environment. Under such conditions we expect that populations are well connected with each other. In ecology, high levels of connectivity within metapopulations are often considered to be beneficial for recovery after local depletion or extirpation [48]. In contrast, species inhabiting the more variable nearshore environments tend to have more patchy distributions. Their supposedly lower connectivity among populations may have hampered the recolonisation by survivors in nearshore settings.

Conclusions

We analysed so far underrepresented K-Pg boundary sections from the Southern Hemisphere, which contribute to a more complete understanding of recovery processes. Overall, our results corroborate and extend earlier work finding interregional and environmental variation in recovery dynamics in benthic [3], [7] and planktonic marine systems [28], [29]. Diversity of the post-extinction interval was controlled by taxa that survived the extinction event in Patagonia. We emphasise the importance of biotic interactions and conclude that survivors had negative effects on the potential origination of new taxa and the invasion of exotic genera through incumbency. Furthermore, survivors constituted the pool of taxa from which the majority of new dominants were recruited after the most abundant Maastrichtian species went extinct at the boundary. Such distinct shifts in the abundances of survivors are a feature these benthic marine ecosystems share with terrestrial plant communities across mass extinction boundaries [49]. Restoration of sample-level diversity of mollusc assemblages can be rapid (< *ca* 500 kyr in offshore habitats). In contrast, total Danian diversity remained below that of the Maastrichtian, despite a marked Palaeogene increase in the global diversity of gastropods and bivalves [8].

Supporting Information

Table S1 Primary data of San Ramón.

Table S2 Primary data of Bajada del Jagüel and Opaso.

Acknowledgments

We thank Dave Jablonski, Dave Lazarus, Jocelyn Sessa, Sonny Walton, and an anonymous reviewer for valuable discussions and constructive comments on the manuscript. We thank Francisco Medina, Roberto Scasso, Henning Scholz, and Sven Weidemeyer for assistance in the field, and Elke Siebert for support in figure preparation. This is Paleobiology Database publication number 200.

Author Contributions

Analyzed the data: WK MA. Wrote the paper: MA. Designed the study: MA. Interpreted results: MA WK. Prepared figures: WK MA. Improved final manuscript: WK.

References

1. McKinney ML (1997) Extinction vulnerability and selectivity: Combining ecological and paleontological views. Annu Rev Ecol Syst 28: 495–516.
2. Erwin DH (2001) Lessons from the past: Biotic recoveries from mass extinctions. Proc Natl Acad Sci USA 98: 5399–5403. (doi:10.1073/pnas.091092698).
3. Aberhan M, Weidemeyer S, Kiessling W, Scasso RA, Medina FA (2007) Faunal evidence for reduced productivity and uncoordinated recovery in Southern Hemisphere Cretaceous/Paleogene boundary sections. Geology 35: 227–230. (doi:10.1130/G23197A.1).
4. Solé RV, Saldaña J, Montoya JM, Erwin DH (2010) Simple model of recovery dynamics after mass extinction. J Theor Biol 267: 193–200. (doi:10.1016/j.jtbi.2010.08.015).
5. Schulte P, Alegret L, Arenillas I, Arz JA, Barton PJ, et al. (2010) The Chicxulub asteroid impact and mass extinction at the Cretaceous-Paleogene boundary. Science 327: 1214–1218. (doi:10.1126/science.1177265).
6. Hansen TA, Farrell BR, Upshaw B (1993) The first 2 million years after the Cretaceous-Tertiary boundary in east Texas: Rate and paleoecology of the molluscan recovery. Paleobiology 19: 251–265.
7. Jablonski D (1998) Geographic variation in the molluscan recovery from the end-Cretaceous extinction. Science 279: 1327–1330.
8. Alroy J (2010) The shifting balance of diversity among major marine animal groups. Science 329: 1191–1194. (doi:10.1126/science.1189910).
9. Scasso RA, Aberhan M, Ruiz L, Weidemeyer S, Medina FA, et al. (2012) Integrated bio- and lithofacies analysis of coarse-grained, tide-dominated deltaic environments across the Cretaceous/Paleogene boundary in Patagonia, Argentina. Cretaceous Research 36: 37–57. (doi:10.1016/j.cretres.2012.02.002).
10. Scasso RA, Concheyro A, Kiessling W, Aberhan M, Hecht L, et al. (2005) A tsunami deposit at the Cretaceous-Paleogene boundary in the Neuquén Basin of Argentina. Cretaceous Research 26: 283–297. (doi:10.1016/j.cretres.2004.12.003).
11. Keller G, Adatte T, Tantawy AA, Berner Z, Stinnesbeck W, et al. (2007) High stress late Maastrichtian – early Danian palaeoenvironment in the Neuquén Basin, Argentina. Cretaceous Research 28: 939–960. (doi:10.1016/j.cretres.2007.01.006).
12. Gradstein FM, Ogg JG, Schmitz MD, Ogg GM (2012) The geologic time scale 2012. Oxford: Elsevier. 1144 p.
13. Papú OH, Prámparo MB, Náñez C, Concheyro A (1999) Palinología y micropaleontología de la formación Jagüel (Maastrichtiano-Daniano), perfil Opaso, cuenca Neuquina, Argentina. Servicio Geológico Minero Argentino, Anales 33: 17–31.
14. Barreda VD, Cúneo NR, Wilf P, Currano ED, Scasso RA, et al. (2012) Cretaceous/Paleogene floral turnover in Patagonia: drop in diversity, low extinction, and a *Classopollis* spike. PLoS One 7: 1–8. (doi:10.1371/journal.pone.0052455).
15. Patzkowsky ME, Holland SM (2007) Diversity partitioning of a Late Ordovician marine biotic invasion: controls on diversity in regional ecosystems. Paleobiology 33: 295–309.
16. Kowalewski M, Kiessling W, Aberhan M, Fürsich FT, Scarponi D, et al. (2006) Ecological, taxonomic, and taphonomic components of the post-Paleozoic increase in sample-level species diversity of marine benthos. Paleobiology 32: 533–561.
17. Alroy J (2010) Fair sampling of taxonomic richness and unbiased estimation of origination and extinction rates. In: Alroy J, Hunt G, editors. Quantitative Methods in Paleobiology. The Paleontological Society Papers 16. 55–80.
18. Gotelli NJ, Colwell RK (2001) Quantifying biodiversity: procedures and pitfalls in the measurement and comparison of species richness. Ecol Lett 4: 379–391.
19. Sepkoski JJ (2002) A compendium of fossil marine animal genera. Bulletins of American Paleontology 363: 1–563.
20. Hansen T, Farrand RB, Montgomery HA, Billman HG, Blechschmidt G (1987) Sedimentology and extinction patterns across the Cretaceous-Tertiary boundary interval in east Texas. Cretaceous Research 8: 229–252. (doi:10.1016/0195-6671(87)90023-1).
21. Bryan JR, Jones DS (1989) Fabric of the Cretaceous–Tertiary marine macrofaunal transition at Braggs, Alabama. Palaeogeogr Palaeoclimatol Palaeoecol 69: 279–301. (doi:10.1016/0031-0182(89)90170-3).
22. Heinberg C (1999) Lower Danian bivalves, Stevns Klint, Denmark: Continuity across the K/T boundary. Palaeogeogr Palaeoclimatol Palaeoecol 154: 87–106. (doi:10.1016/S0031-0182(99)00088-7).
23. Jablonski D (2002) Survival without recovery after mass extinctions. Proc Natl Acad Sci USA 99: 8139–8144. (doi:10.1073/pnas.102163299).
24. Harries PJ, Kauffman EG, Hansen TA (1996) Models for biotic survival following mass extinction. In: Hart MB, editor. Biotic recovery from mass extinction events. London: The Geological Society. 41–60.
25. Erwin DH (1998) The end and the beginning: recoveries from mass extinctions. Trends Ecol Evol 13: 344–349.
26. Stilwell JD (2003) Patterns of biodiversity and faunal rebound following the K–T boundary extinction event in Austral Palaeocene molluscan faunas. Palaeogeogr Palaeoclimatol Palaeoecol 195: 319–356. (doi:10.1016/S0031-0182(03)00364-X).
27. Coxall HK, D'Hondt S, Zachos JC (2006) Pelagic evolution and environmental recovery after the Cretaceous-Paleogene mass extinction. Geology 34: 297–300. (doi:10.1130/G21702.1).
28. Jiang S, Bralower TJ, Patzkowsky ME, Kump LR, Schueth JD (2010) Geographic controls on nannoplankton extinction across the Cretaceous/Palaeogene boundary. Nat Geosci 3: 280–285. (doi:10.1038/ngeo775).
29. Hull PM, Norris RD (2011) Diverse patterns of ocean export productivity change across the Cretaceous-Paleogene boundary: New insights from biogenic barium. Paleoceanography 26: 1–10. (doi:10.1029/2010PA002082).
30. Valentine JW, Jablonski D, Krug AZ, Roy K (2008) Incumbency, diversity, and latitudinal gradients. Paleobiology 35: 169–178.
31. Algar AC, Mahler DL, Glor RE, Losos JB (2013) Niche incumbency, dispersal limitation and climate shape geographical distributions in a species-rich island adaptive radiation. Glob Ecol Biogeogr 22: 391–402. (doi:10.1111/geb.12003).

32. Sheehan PM (2008) Did incumbency play a role in maintaining boundaries between Late Ordovician brachiopod realms? Lethaia 41: 147–153.
33. Vermeij GJ, Roopnarine PD (2013) Reining in the Red Queen: the dynamics of adaptation and extinction reexamined. Paleobiology 39: 560–575. (doi:10.1666/13009).
34. Hallam A (1990) Biotic and abiotic factors in the evolution of early Mesozoic marine molluscs. In: Ross RM, Allmon WD, editors. Causes of evolution – a paleontological perspective. Chicago: The University of Chicago Press. 249–269.
35. Bambach RK, Knoll AH, Sepkoski JJ (2002) Anatomical and ecological constraints on Phanerozoic animal diversity in the marine realm. Proc Natl Acad Sci USA 99: 6854–6859. (doi:10.1073/pnas.092150999).
36. Gaston KJ, Blackburn TM, Greenwood JJD, Gregory RD, Quinn RM, et al. (2000) Abundance–occupancy relationships. J Appl Ecol 37 (Suppl. 1): 39–59.
37. Simpson C, Harnik PG (2009) Assessing the role of abundance in marine bivalve extinction over the post-Paleozoic. Paleobiology 35: 631–647.
38. Lockwood R (2003) Abundance not linked to survival across the end-Cretaceous mass extinction: Patterns in North American bivalves. Proc Natl Acad Sci USA 100: 2478–2482. (doi:10.1073/pnas.0535132100).
39. Roopnarine PD (2006) Extinction cascades and catastrophe in ancient food webs. Paleobiology 32: 1–19.
40. Lyons KG, Brigham CA, Traut BH, Schwartz MW (2005) Rare species and ecosystem functioning. Conserv Biol 19: 1019–1024. (doi:10.1111/j.1523-1739.2005.00106.x).
41. McElwain JC, Popa ME, Hesselbo SP, Haworth M, Surlyk F (2007) Macroecological responses of terrestrial vegetation to climatic and atmospheric change across the Triassic/Jurassic boundary in East Greenland. Paleobiology 33: 547–573.
42. Hansen TA, Kelley PH, Haasl DM (2004) Paleoecological patterns in molluscan extinctions and recoveries: comparison of the Cretaceous–Paleogene and Eocene–Oligocene extinctions in North America. Palaeogeogr Palaeoclimatol Palaeoecol 214: 233–242. (doi:10.1016/j.palaeo.2004.06.017).
43. Sessa JA, Bralower TJ, Patzkowsky ME, Handley JC, Ivany LC (2012) Environmental and biological controls on the diversity and ecology of Late Cretaceous through early Paleogene marine ecosystems in the U.S. Gulf Coastal Plain. Paleobiology 38: 218–239.
44. Erwin DH (2007) Increasing returns, ecological feedback and the Early Triassic recovery. Palaeoworld 16: 9–15. (doi:10.1016/j.palwor.2007.05.013).
45. Jablonski D (2005) Evolutionary innovations in the fossil record: the intersection of ecology, development, and macroevolution. J Exp Zool 304B: 504–519. (doi:10.1002/jez.b.21075).
46. Kiessling W, Aberhan M (2007) Environmental determinants of marine benthic biodiversity dynamics through Triassic-Jurassic times. Paleobiology 33: 414–434.
47. Hull PM, Norris RD, Bralower TJ, Schueth JD (2011) A role for chance in marine recovery from the end-Cretaceous extinction. Nat Geosci 4: 856–860. (doi:10.1038/ngeo1302).
48. Hughes TP, Carpenter S, Rockström J, Scheffer M, Walker B (2013) Multiscale regime shifts and planetary boundaries. Trends Ecol Evol 28: 389–395. (doi:10.1016/j.tree.2013.05.019).
49. McElwain JC, Punyasena SW (2007) Extinction events and the plant fossil record. Trends Ecol Evol 22: 548–557. (doi:10.1016/j.tree.2007.09.003).
50. del Río CJ (2004) Tertiary marine molluscan assemblages of eastern Patagonia (Argentina): a biostratigraphic analysis. J Paleontol 78: 1097–1122.
51. Lazo DG (2003) Taxonomy, facies relationships and palaeobiology of bakevelliid bivalves from the Lower Cretaceous of west-central Argentina. Cretaceous Research 24: 765–788. (doi:10.1016/j.cretres.2003.09.001).
52. Lazo DG (2007) Early Cretaceous bivalves of the Neuquén Basin, west-central Argentina: notes on taxonomy, palaeobiogeography and palaeoecology. Geol J 42: 127–142. (doi:10.1002/gj.1080).
53. Kauffman EG, Leanza HA (2004) A remarkable new genus of Mytilidae (Bivalvia) from the Lower Cretaceous of Southwestern Gondwanaland. J Paleontol 78: 1187–1191. (doi:10.1666/0022-3360(2004)078).
54. del Río CJ (2002) Molluscos del Terciario marino. Geología y Recursos Naturales de Santa Cruz 2: 1–22.
55. Zinsmeister WJ, Griffin M (1995) Late Cretaceous and Tertiary aporrhaid gastropods from the southern rim of the Pacific Ocean. J Paleontol 69: 692–702.
56. Chiesa JO, Camacho HH (2001) Invertebrados marinos Eocenos de la parte inferior de la Formacíon Río Foyel, Provincia de Río Negro, Argentina. Rev Esp Paleontol 16: 299–316.
57. Chiesa JO, Parma SG, Camacho HH (1995) Invertebrados marinos de la Formación El Chacay (Eoceno), provincia de Santa Cruz, Argentina. Sistemática y bioestratigrafia. Academia Nacional de Ciencias Exactas, Físicas y Naturales, Monografias 11: 17–68.
58. Casadío S, Griffin M, Parras A (2005) *Camptonectes* and *Plicatula* (Bivalvia, Pteriomorphia) from the Upper Maastrichtian of northern Patagonia: palaeobiogeographic implications. Cretaceous Research 26: 507–524. (doi:10.1016/j.cretres.2005.01.009).
59. Woods H (1917) The Cretaceous faunas of the north-eastern part of the South Island of New Zealand. New Zealand Geological Survey, Palaeontological Bulletin 4: 1–41.
60. Casadío S, Griffin M, Marenssi S, Net L, Parras A, et al. (2009) Paleontology and sedimentology of Middle Eocene rocks in Lago Argentino area, Santa Cruz Province, Argentina. Ameghiniana 46: 27–47.
61. Squires RL, Goedert JL (1994) New species of Eocene small to minute mollusks from the Crescent Formation, Black Hills, southwestern Washington. Veliger 37: 253–266.
62. del Río CJ, Concheyro A, Martínez SA (2011) The Maastrichtian – Danian at General Roca (Patagonia, Argentina): a reappraisal of the chronostratigraphy and biostratigraphy of a type locality. Neues Jahrb Geol Paläontol Abh 259: 129–156. (doi:10.1127/0077-7749/2011/0103).
63. Berggren WA, Kent DV, Swisher CC, Aubry M-P (1995) A revised Cenozoic geochronology and chronostratigraphy. In: Berggren WA, et al., editors. Geochronology, time scales and global stratigraphic correlation. SEPM Special Publications 54: 129–212.
64. Harrington HJ (1958) Geology of Kaitangata Coalfield. New Zealand Geological Survey Bulletin 59: 1–131.
65. Kafanov AI, Ogasawara K (2004) Neogene and Paleogene molluscan (Bivalvia) cenozones of the Russian Northeast. Science Reports of the Institute of Geoscience, University of Tsukuba, Section B 25: 9–45.
66. del Río CJ, Martínez SA, Stilwell JD, Concheyro A (2007) Paleontology of the Cerros Bayos section, Roca Formation (Danian), La Pampa Province, Argentina. Alcheringa 31: 241–269. (doi:10.1080/03115510701484713).
67. Waller TR, Marincovich L (1992) New species of *Camptochlamys* and *Chlamys* (Mollusca: Bivalvia: Pectinidae) from near the Cretaceous/Tertiary boundary at Ocean Point, North Slope, Alaska. J Paleontol 66: 215–227.
68. del Río CJ, Beu AG, Martínez SA (2008) The pectinoidean genera *Delectopecten* Stewart, 1930 and *Parvamussium* Sacco, 1897 in the Danian of Northern Patagonia, Argentina. Neues Jahrb Geol Paläontol Abh 249: 281–295.
69. Aguirre-Urreta MB (1989) The Cretaceous decapod Crustacea of Argentina and the Antarctic Pensinsula. Palaeontology 32: 499–552.
70. Darragh TA (1997) Gastropoda, Scaphopoda, Cephalopoda, and new Bivalvia of the Paleocene Pebble Point Formation, Victoria, Australia. Proceedings of the Royal Society of Victoria 109: 57–108.
71. Chavan A (1946) Nouvelles notes sur les *Jagonia* (Lamellibranches). Bulletin du Muséum National d'Histoire Naturelle Paris, sér. 2, 18: 345–347.
72. Lazo DG, Damborenea SE (2011) Barremian bivalves from the Huitrín Formation, west-central Argentina: taxonomy and paleoecology of a restricted marine association. J Paleontol 85: 721–749.
73. Stilwell JD (2003) Macropalaeontology of the *Trochocyathus-Trematotrochus* band (Paleocene/Eocene boundary), Dilwyn Formation, Otway Basin, Victoria. Alcheringa 27: 245–275. (doi:10.1080/03115510308619107).
74. Stilwell JD, Zinsmeister WJ, Oleinik AE (2004) Early Paleocene mollusks of Antarctica: systematics, paleoecology and paleobiogeographic significance. Bulletins of American Paleontology 367: 1–89.
75. Griffin M, Hünicken MA (1994) Late Cretaceous-Early Tertiary gastropods from southwestern Patagonia, Argentina. J Paleontol 68: 257–274.

18

Constraints upon the Response of Fish and Crayfish to Environmental Flow Releases in a Regulated Headwater Stream Network

Edwin T. Chester[1], Ty G. Matthews[2], Travis J. Howson[2], Kerrylyn Johnston[3], Jonathon K. Mackie[2], Scott R. Strachan[1], Belinda J. Robson[1]*

1 Environmental and Conservation Sciences, Murdoch University, Perth, Western Australia, Australia, **2** School of Life and Environmental Sciences, Deakin University, Warrnambool, Victoria, Australia, **3** Marine and Freshwater Research Laboratory, Murdoch University, Perth, Western Australia, Australia

Abstract

In dry climate zones, headwater streams are often regulated for water extraction causing intermittency in perennial streams and prolonged drying in intermittent streams. Regulation thereby reduces aquatic habitat downstream of weirs that also form barriers to migration by stream fauna. Environmental flow releases may restore streamflow in rivers, but are rarely applied to headwaters. We sampled fish and crayfish in four regulated headwater streams before and after the release of summer-autumn environmental flows, and in four nearby unregulated streams, to determine whether their abundances increased in response to flow releases. Historical data of fish and crayfish occurrence spanning a 30 year period was compared with contemporary data (electrofishing surveys, Victoria Range, Australia; summer 2008 to summer 2010) to assess the longer-term effects of regulation and drought. Although fish were recorded in regulated streams before 1996, they were not recorded in the present study upstream or downstream of weirs despite recent flow releases. Crayfish (*Geocharax* sp. nov. 1) remained in the regulated streams throughout the study, but did not become more abundant in response to flow releases. In contrast, native fish (*Gadopsis marmoratus, Galaxias oliros, Galaxias maculatus*) and crayfish remained present in unregulated streams, despite prolonged drought conditions during 2006–2010, and the assemblages of each of these streams remained essentially unchanged over the 30 year period. Flow release volumes may have been too small or have operated for an insufficient time to allow fish to recolonise regulated streams. Barriers to dispersal may also be preventing recolonisation. Indefinite continuation of annual flow releases, that prevent the unnatural cessation of flow caused by weirs, may eventually facilitate upstream movement of fish and crayfish in regulated channels; but other human-made dispersal barriers downstream need to be identified and ameliorated, to allow native fish to fulfil their life cycles in these headwater streams.

Editor: Christopher J Fulton, The Australian National University, Australia

Funding: This research was funded by the Glenelg-Hopkins Catchment Management Authority (www.ghcma.vic.gov.au) and Wannon Water (www.wannonwater.com.au). The funders had no role in study design, data collection and analysis, decision to publish, or preparation of the manuscript.

Competing Interests: The authors have the following interests: This study was funded in part by Wannon Water (?www.wannonwater.com.au) which is a publicly owned statutory corporation. There are no patents, products in development or marketed products to declare. Receipt of this funding does not alter the authors' adherence to all the PLOS ONE policies on sharing data and materials.

* E-mail: b.robson@murdoch.edu.au

Introduction

Much has been published on the consequences of river regulation for populations of fish and other large consumers, however much less work has focussed on the impacts of regulating headwater streams [1]. In semi-arid and mediterranean climate zones, first to third order streams are often regulated for water extraction [2], and this may lead to perennial streams becoming intermittent or the prolongation of dry periods in intermittent streams [3,4]. Regulation of headwater streams for water extraction often involves the construction of small weirs, which act as barriers to the movement of freshwater animals [4]. Therefore, such regulation may have a negative impact on abundances by two mechanisms: reducing habitat area through streambed drying downstream of the weir, and by creating barriers to movement.

Biodiversity in seasonally flowing streams largely depends on the viability and persistence of refuges (such as pools) within the channel, and upon a cycle of recolonization from perennially flowing parts of the stream or the greater catchment [5]. Maintaining drought refuges is one key action that can sustain biodiversity in seasonally flowing streams [6]. Flow releases from weirs could be targeted to sustain refuge pools, by increasing the total duration of flow downstream during the year [4], and through selective shorter-term releases. Connectivity through water flow down the stream channel allows colonization and movement downstream from perennial reaches (especially from above weirs) and upstream from within the greater catchment. Fish species found in headwater streams can recolonise streams quickly following disturbances like drought [7]. Their (generally) small bodies, short life spans, and early-aged reproduction (e.g. Galaxiidae species mature in their first year of life and have the highest gonadosomatic index of all Australian fish) permit rapid population growth [8], facilitating that process, but even small barriers may inhibit dispersal [9]. The benthic native crayfish may disperse more slowly than fish, but channel obstructions are less likely to be barriers because they may travel overland for short distances.

Flow releases are a commonly used tool for managing stream and river health; however there are relatively few evaluations of their success in non–perennial streams (but see [10,11]). Such evaluations are required for adaptive management, to improve the targeting of flows for maximum benefit. Here, we studied summer flow releases from weirs in headwater streams and their effect on fish and crayfish assemblages. There were three reasons why we expected a positive response to summer flow releases: (1) increased flows may increase the connectedness of streams and facilitate movement; (2) summer releases may sustain refuge pools by improving water depth, water quality and survivorship; (3) summer releases may keep riffles located downstream of weirs wetted for longer, providing a greater duration of spawning habitat for some species (e.g. *G. oliros* [8]). Therefore, the aim of this project was to determine whether native fish and crayfish responded to these releases in headwater streams of the Victoria Range, Australia, where stream regulation for potable use is common practise. To do this, species assemblages were sampled in four regulated streams and four unregulated streams, before and after the initiation of summer flow releases.

Methods

Ethics statement

All research in this study was conducted under permits from the Department of Sustainability and Environment and the Department of Fisheries, Victoria (DSE permit number 10003269, DPI permit number RP896). Electrofishing was carried out in accordance with the Australian Code of Electrofishing Practice (1997) and Australian Code of Practice for the Care and Use of Animals for Scientific Purposes (7th edition 2004) under approval by Deakin University Animal Welfare Committee (A15-2007). All streams and species sampled are protected within the Grampians National Park, Victoria, Australia. All captured animals were returned to the streams alive and no mortality was recorded. Data may be obtained by contacting the corresponding author.

Study sites and flow regimes

The Victoria Range (maximum altitude 979 m) is the westernmost of a series of mountain ranges running north–south in the Grampians National Park in western Victoria, Australia. With a mediterranean to semi–arid climate (400–600 mm average annual rainfall), streams vary from perennially flowing to seasonally dry with considerable interannual variation in flow regime [5]. Riparian vegetation is dry sclerophyll woodland and forest, and all streams have sandstone boulders and cobbles with areas of bedrock, descending to a sandplain where channels may become indistinct (described in [6]) (Figure 1). From 2000 to 2010, stream flows were affected by one of the longest and most intense droughts historically experienced [12]. Compared to historical norms, flow periods were generally shorter, with a prolonged dry summer–autumn, and reduced spring rains.

Four of the eight study streams are regulated for town water supply (Figure 1), with weirs that extract water and deliver it to storage via underground pipes (described in [4]). Four of the streams have perennial flow: Honeysuckle Creek (unregulated), Gap Creek (regulated in 1914 [13]) and Camp Creek (regulated in 1960 [13]), which both have intermittent flow downstream of their weirs, and Brown Creek (regulated in 1939 [13]), which has two branches that are regulated. The other four streams flow seasonally: Cultivation Creek, Deep Creek and Hut Creek (unregulated with perennial pools over summer–autumn) and Number 1 Creek (regulated and can dry completely in summer–autumn). Further descriptions of the streams, stream locations and flow regimes may be found in [3] and [5].

An environmental flow release of up to 0.4 ML/day was released into Camp and Gap Creeks during April 2009 and in both branches of Brown Creek: increased flows were passed through the weirs after modifications for that purpose. These comprised valves in ~8 cm diameter pipes, installed above the base flow height downstream of the weirs, and not specifically designed to allow animal movement. The passing flows were continued until May 2009 in all creeks except for Camp Creek, which continued until early June; however, there was no interruption in flow in Camp Creek due to the contemporaneous onset of winter rain. In contrast, cessation of the passing flows in May led to a short period of interrupted flow (and dry stream beds) directly below the weirs in Gap Creek and both branches of Brown Creek. Gap Creek did not resume flow until early June 2009, from diffuse sources after winter rain; Number 1 Creek also started flowing at this time (it did not receive an environmental flow release during April 2009, because there was no water upstream of the weir). Temporally patchy flow also resumed in parts of Brown Creek during June, but again not in reaches immediately downstream of the weirs, which only held water following spates later in winter. Passing flow releases began again at the end of September 2009 in all four regulated streams and continued into winter 2010 (but again, there was no water upstream in Number 1 creek from early summer).

Study species

Historically, river blackfish (*Gadopsis marmoratus* (Richardson)), obscure galaxias (*Galaxias oliros* Raadik) [14], and common jollytails (*Galaxias maculatus* (Jenyns)) were the native fish species recorded most often in Victoria Range streams [15]. *G. marmoratus* in these streams is probably the as yet undescribed northern species [16]. The southern pygmy perch (*Nannoperca australis* Günther) and the eastern dwarf galaxias (*Galaxiella pusilla* (Mack)) have been recorded in vegetated downstream reaches and in riverine wetlands [15]. These *Galaxiella* are probably the western, undescribed species [17]. Exotic fish are rare in these streams, but the redfin perch (*Perca fluviatilis* Linnaeus) has been recorded in low numbers in Cultivation Creek [15]. An undescribed species of freshwater crayfish (*Geocharax* sp. nov.1) also occurs commonly in these streams [5,18]. This assemblage is typical of western Victorian upland streams and is less rich in fish species (including of exotic species) than lowland streams [19] and rivers [20].

Sampling design

The hypotheses tested in this study were: (1) native fish are less abundant in regulated streams than unregulated streams; (2) fish occur less frequently in regulated streams receiving summer passing flows than in nearby unregulated streams; (3) species composition (including crayfish) differs between regulated streams receiving passing flows during summer and nearby unregulated streams. Fish and crayfish were sampled over two years (from summer 2008 to summer 2010) to compare four regulated streams (Gap Creek, No. 1 Creek, Camp Creek, Brown Creek) with four unregulated streams (Cultivation Creek, Hut Creek, Deep Creek, Honeysuckle Creek) using single pass electrofishing surveys.

Electrofishing surveys were done in summer (March) and spring (October) 2008 to provide "before treatment" data; then in spring (October) 2009, and summer (March) 2010, after passing flows

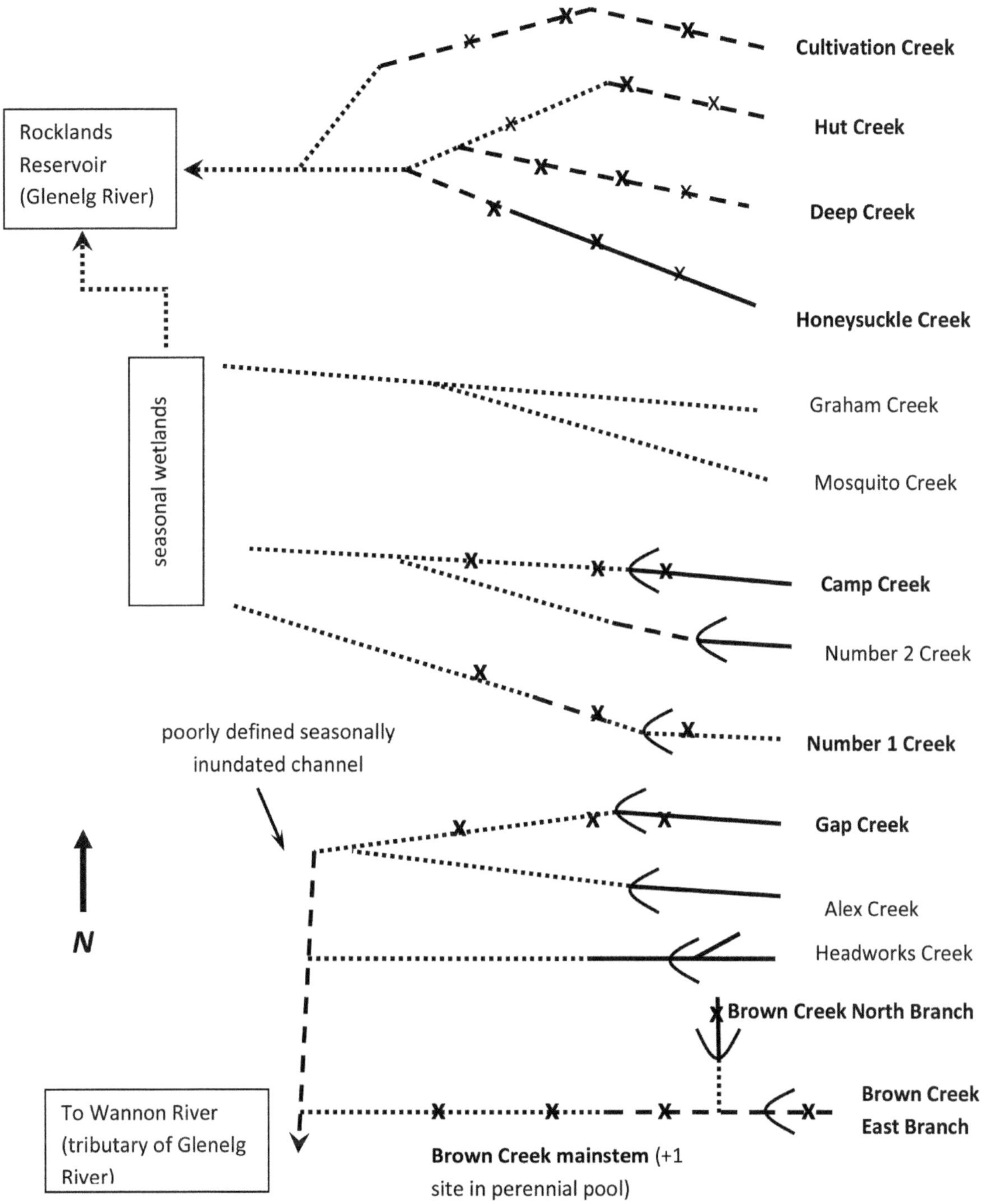

Figure 1. Diagram of the study streams (named in bold) in the Victoria Range, showing flow regimes and representing conditions at peak dryness, without summer passing flows (solid lines = perennial flow, dashed lines = seasonal flow with perennial pools and dotted lines = seasonal, completely dry) and weirs ('U' shapes). Survey sites denoted by '**x**': primary sites were upstream and downstream of weirs, and at similar elevations on unregulated streams; all potential fish/crayfish refuges were sampled. Sites with faint 'x' were in surveys prior to this study, and were visited in summer 2008; they were subsequently dropped as they were not needed to characterize the unregulated streams for comparison with the regulated ones. General flow direction is from right to left. Note that this is a schematic diagram and is not to scale.

had been in operation (Table 1). These sampling times should have detected both fish migrating in spring (most likely spawning galaxias) and fish and crayfish that had remained in refuge pools further downstream over the summer. Individuals were counted and, in the summer sampling, measured (total length and weight (g) for fish, occipital carapace length (OCL) for crayfish); crayfish were not always netted, so estimates include counts for animals seen and not captured.

Electrofishing equipment was customized, based around a SAMUS 725M 12 volt unit fitted into a ruggedized, wearable vest with additional safety features and output meters for the operator. An 11" (280 mm) circular stainless steel anode was used, along with a 3 m cathode of 6 mm woven stainless steel cable. Waveform was varied systematically through the survey, to target animals (and species) of different size and sensitivity, while minimizing error from alarmed individuals escaping detection as well as the chance of mortality. Settings (frequency 99 Hz, pulse width 0.05–0.25 ms) were based on previous trials with this equipment, and took into account the very different reactions of fish and crayfish. All sites had relatively narrow (1–5 m) and shallow channels with substantial woody debris, and riffle/run–pool sequences which could be efficiently searched by the team of three operators as they moved upstream. However, since streams varied in width, depth, presence of pools and woody debris, and intermittency, total effort in each stream also varied to maintain a consistent sampling efficiency. At all reaches, care was taken to maximize the probability of detection of any animals present, whether in a flowing run or in isolated refuge pools.

Hundred metre reaches — a substantial part of these short streams, and in some cases most that was accessible — were sampled upstream and downstream of weirs in regulated streams, and at similar positions in the catchment in unregulated streams (Figure 1). Additionally, and because no fish were initially found in the regulated streams, reaches were also sampled well downstream of the weirs to exclude the possibility that they were present but had not made their way upstream. There are weirs on two branches of Brown Creek, so both were sampled, as well as a near perennial spring–fed pool between the confluence and the downstream site. Two sites were sampled in each unregulated stream after 2009, apart from the smaller Hut Creek where only one reach, in the most reliably flowing section, was sampled (so, total 25 sites in 2008, 20 in 2009, Figure 1). Confidence in the ability to characterize these streams was bolstered by the consistency of results compared with the 2007 pilot study, which had shown a high degree of distinctiveness among them. For these reasons, total linear metres sampled (as measured by GPS, and surveyed along the bank) differed somewhat among streams; therefore, total catch was standardized to 100 m for comparison.

Historical data on fish and crayfish distributions (presence/absences) from [13,15] and also Tunbridge (1989; as reported in 13) were compared with our data to determine longer term temporal changes in populations (Table 1). Jackson and Davies [15] used a range of gear including electrofishing (number of passes not specified), dip–netting in backwaters and gill nets in deep pools and surveyed sites between 30 and 60 m in length in spring and early summer, 1979. Raadik [13] used single pass electrofishing and occasional dip–netting in 1995 and 1996 and sampled stream lengths up to 100 m. We had collected data on fish and crayfish presences prior to 2008 using baited traps, while sampling invertebrates, and from initial (winter 2007) electrofishing in unregulated streams. While there are some methodological variations among these studies, they provide definite presence records of fish and crayfish species in Victoria Range streams over three decades. This provided a context in which to assess the effects of environmental flow releases.

Data analysis

Hypothesis 1 and 2 could not be formally tested because no fish were captured in the four regulated streams. That is, obtaining zeroes for half of the cells in the design meant that formal statistics

Table 1. Presence of fish and crayfish in the eight streams at each sampling time, compared with historical presences.

Stream	Type	Historical (1979, 1989, 1995-96) (2006-2007)[#]	Summer 2008	Spring 2008	Spring 2009	Summer 2010
Brown Ck	**Regulated**	*Geocharax, G. marmoratus, G. oliros*	*Geocharax*	*Geocharax*	*Geocharax*	*Geocharax*
Gap Creek	**Regulated**	*Geocharax, G. oliros*	*Geocharax*	*Geocharax*	*Geocharax*	*Geocharax*
No. 1 Creek	**Regulated**	*Geocharax, G. oliros*	<u>*Geocharax*</u>			
Camp Creek	**Regulated**	*Geocharax, G. marmoratus, G. oliros*	<u>*Geocharax*</u>			*Geocharax*
Hut Creek	**Unregulated**	*Geocharax#, G. marmoratus, G. oliros, G. maculatus*	*Geocharax, G. oliros,* <u>*G. maculatus,*</u> <u>*N. australis*</u>	*Geocharax*	*Geocharax*	*Geocharax, G. oliros, G. maculatus*
Cultivation Ck	**Unregulated**	*Geocharax#, G. marmoratus, G. oliros, G. maculatus, P. fluviatilis*	*Geocharax,* <u>*G. marmoratus*</u>*, G. oliros, G. maculatus,* <u>*N. australis*</u>*,* <u>*G. pusilla*</u>	*Geocharax,* <u>*G. marmoratus,*</u> *G. oliros, G. maculatus*	*Geocharax,* <u>*G. marmoratus,*</u> *G. oliros, G. maculatus*	*Geocharax,* <u>*G. marmoratus,*</u> *G. oliros, G. maculatus,* <u>*P. fluviatilis*</u>
Deep Creek	**Unregulated**	*G. marmoratus#, G. oliros*	<u>*Geocharax,*</u> *G. marmoratus, G. oliros, G. maculatus, N. australis*	*G. marmoratus*	*G. marmoratus*	*G. marmoratus, G. oliros, G. maculatus*
Honeysuckle Ck	**Unregulated**	*Geocharax#, G. marmoratus#, G. oliros, G. maculatus, N. australis*	<u>*Geocharax,*</u> *G. marmoratus, G. oliros, G. maculatus,* <u>*N. australis*</u>	*G. marmoratus, G. oliros*	*G. marmoratus,* <u>*G. oliros*</u>	*G. marmoratus, G. oliros*

Compiled from Raadik (1996) and Jackson and Davies (1983) as well as author's unpublished data from 2006–2007. Bold indicates species found upstream of weirs; underlined indicates rare or encountered infrequently (i.e. very patchily distributed, in one place in channel: not given for historical occurrences); [#] denotes almost all animals caught electrofishing the unregulated streams in 2007 were these species.

could not be used. This result does provide clear confirmation of both hypotheses because fish were collected in large numbers in all four unregulated streams (see below), demonstrating the efficacy of the fish capture method and therefore the veracity of the zero values. Hypothesis 3, that species composition would differ between the regulated and unregulated streams was therefore also confirmed without further need for statistical testing, although the crayfish *Geocharax* sp. nov. 1 was present in the regulated streams.

Differences in the fish/crayfish assemblage composition among unregulated streams were analysed using fish counts per 100 m from each stream, untransformed, in a two factor ANOSIM without replication: stream (n = 4) by time (n = 4: summer 2008, spring 2008, spring 2009, summer 2010) in PRIMER version 6 [21]. Chi-square tests were used to determine whether crayfish size frequencies (OCL, n = 5: <10 mm, 11–15 mm, 16–20 mm, 21–25 mm, >26 mm) differed among streams (counts per 100 m, n = 5, Hut Creek, Brown Creek east, Brown Creek north, Brown Creek mainstem, Gap Creek). Chi-square tests for homogenous counts of crayfish were used to determine whether crayfish responded to flow releases in Gap Creek and Brown Creek, separately, by comparing capture frequency per 100 m across the four sampling times.

Results

Species occurrences: flow releases and historical data

There are four clear results from this dataset. Firstly, although fish were recorded in the regulated streams prior to 1996, no fish had been recorded upstream of weirs in any streams and only a single fish (*G. marmoratus*) was observed downstream of the weir in Camp Creek in winter 2006 (Table 1). Secondly, fish were consistently present in unregulated streams (except Hut Creek in spring 2008 and summer 2009) despite prolonged drought conditions during 2006–2010. Thirdly, flow releases in 2010 were not associated with the presence of fish species in regulated streams, because no fish were detected in any of the regulated streams (Table 1). Fourthly, the crayfish *Geocharax* sp. nov. 1 remained in the regulated streams throughout the study, and occupied reaches downstream of weirs that were inundated by the flow releases. However, its occurrence in streams where it was already rarely detected before summer flow releases had not changed by 2010.

Several species were rare and only occasionally encountered in the unregulated streams (Table 1): the pygmy perch (*Nannoperca australis*), the dwarf galaxias (*Galaxiella pusilla*) and the non-native redfin perch (*Perca fluviatilis*). The latter was also captured in the same stream (Cultivation Creek) in 1979 [15]. These occurrences may be partly explained by the greater effort (i.e. more sites) in summer 2008, although the *P. fluviatilis* was caught in 2010.

Patterns of abundance in fish and crayfish

The four unregulated streams differed markedly in species composition and abundance (R = 0.5, *P* = 0.008; Figure 2) but assemblage composition did not differ consistently among sampling times, because the stream to stream variation led to there being greater levels of variation within times than between them (R = −0.18, *P* = 0.87). *G. marmoratus* were most abundant in Honeysuckle and Deep Creeks but also occurred in Cultivation Creek and Hut Creek; those in Cultivation Creek were restricted to a single refuge pool sampled upstream. *G. oliros* was most abundant in Honeysuckle Creek but also occurred in the other three streams. *G. maculatus* was rare in all creeks except Cultivation Creek where it occurred in large numbers (Figure 2d). While the two *Galaxias* species were more abundant in summer 2008 than summer 2010 in Deep and Honeysuckle Creeks (i.e. those with the most perennial water), Cultivation and Hut Creeks (which dried to only a few pools over this period) held more in 2010, although there were many *G. maculatus* in Cultivation Creek in spring 2008 when it was flowing. Very few juvenile *G. maculatus* were recorded (Table 2). Adult *G. oliros* were widespread in the unregulated streams but again, juveniles were rare (Table 2). However, juvenile *G. marmoratus* were relatively common in Honeysuckle and Deep Creeks (Table 2).

A single crayfish *Geocharax* sp. nov. 1 was recorded from Deep Creek. Only a few were found in Honeysuckle Creek, in a pool well downstream (Figure 2a), and none were ever found upstream, although they had been recorded there in 1989 (Table 1). *Geocharax* occurred alone and in low abundances in regulated Number 1 and Camp Creeks (Figure 3). Pools upstream of weirs in Brown and Gap Creeks were refuges for crayfish, although they were not found upstream in Brown Creek in spring 2009.

Fish and crayfish co-occurred in unregulated streams, but rarely in significant numbers other than in Cultivation Creek. Crayfish were relatively abundant in Hut Creek, where fish were uncommon, and in spring 2008 and 2009 only crayfish were recorded there (Figure 2c). In contrast, crayfish were uncommon in Honeysuckle and Deep Creeks where fish abundances were highest, especially *G. marmoratus*. Where large crayfish populations were found, size frequency differed among streams ($\chi^2_{16} = 47$, $P < 0.01$). Juveniles were rare in Hut Creek but frequent in Gap Creek, and Brown Creek (Table 2). *Geocharax* OCL ranged between 7–33 mm, and an increased proportion of crayfish of all sizes was captured following flow releases in Gap ($\chi^2_3 = 32.8$, $P<0.01$) and Brown Creeks ($\chi^2_3 = 16.75$, $P<0.05$; Figure 3).

Discussion

Responses to environmental flow releases by native fish and crayfish

These results clearly showed that a single year of environmental flow releases in headwater streams in the Victoria Range was insufficient to produce a response in fishes. Crayfish were detected in formerly dry streambeds downstream of weirs in Gap Creek and Brown Creek, but they had apparently moved only limited distances within each stream to occupy wetted channel close to source areas. The apparent disappearance of crayfish upstream of the weir in Brown Creek in spring 2009 was anomalous, and may have been due to animals dispersing into expanded habitat, or even because of increased predation from birds in a weir pool made shallow by flow release.

Macroinvertebrates responded quickly to these flow releases [4], but it is probable that a single year is too short a time or that the flows were of insufficient magnitude to allow native fish to recolonise the regulated streams. It is also possible that dispersal barriers downstream are also preventing recolonisation by fish. Such barriers may or may not be important for crayfish dispersal; but movement from refuges downstream may be slow, and terrestrial 'shortcuts' between streams would seem to be the quickest route to colonize streams like Number 1 and Camp Creeks, if survival on land is possible for *Geocharax*. However, there was no evidence of this happening up to 2010.

Galaxias maculatus is known to be a weak swimmer [9] but *G. oliros* appears to be more competent [22]. We have observed *G. oliros* in lower reaches of the Brown Creek system, but they were not observed downstream of the weirs during the present study. This suggests that some road crossings, fords and culverts, on the regulated streams may be impassable to small fish, even during flow releases. Recent research shows that it is possible to modify

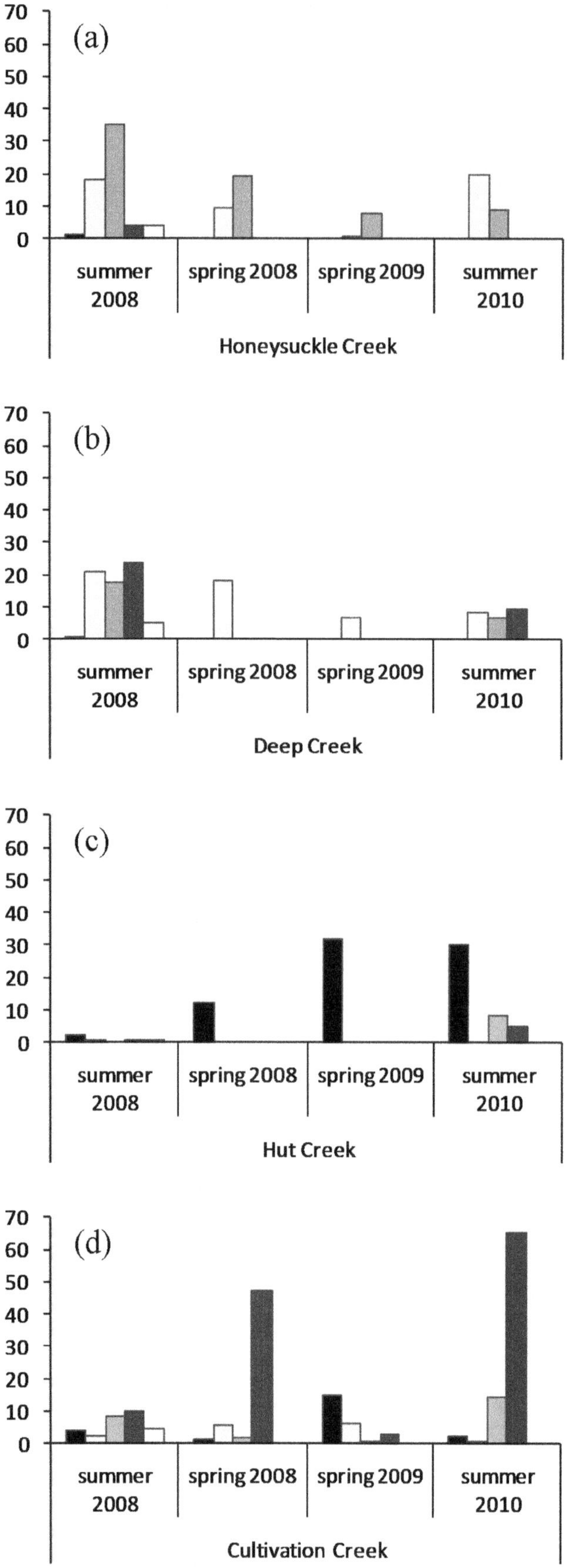

Figure 2. Total abundances per 100 m sampling effort in the four unregulated streams from 2008 to 2010, (a) Honeysuckle Creek; (b) Deep Creek; (c) Hut Creek; (d) Cultivation Creek. Black = *Geocharax* sp., pale grey = *G. marmoratus*, mid-grey = *G. oliros*, dark grey = *G. maculatus*, white = *N. australis.*

most culvert and road crossing designs to facilitate migration by small species of native fish [9,23] and such modifications should be considered as an adjunct to environmental flow releases to improve ecosystem connectivity and ecological outcomes [24].

The weirs themselves are certainly substantial barriers, so fish absences upstream are not surprising. The only opportunity for fish to access channels upstream of weirs would have been during large rainfall events, across the weir wall. This is probably beyond the capabilities of the small–bodied native fish species found in headwater streams [25]. Also, flow releases approximately equivalent to baseflow (<0.4 ML/day/channel) at the weirs did not necessarily result in surface flows lower in the catchment, so there were probably still reaches of dry streambed at times that would limit upstream movement by fish [22]. Lastly, weirs tend to create three discrete levels of flow downstream (baseflow arising from the flow release, no flow, or weir–topping spates) in contrast to a more gradual transition between flow states in unregulated streams.

Native fish populations in regulated streams declined, whereas crayfish remained

Clear patterns were evident in native fish occurrences and fish were either absent from the regulated streams or were present in such low numbers that they were not detected. Fish were neither observed nor collected upstream of weirs from 2006 to 2009, yet with the same sampling methods, large numbers of fish were collected in the unregulated streams from 2007 onwards, suggesting that if fish were present in the regulated streams, they would have been caught. Native fish appear likely to have become locally extinct in the regulated streams, not only upstream, but also downstream of weirs, in the decade between 1996 and 2006 and fish abundances had probably declined steadily in these streams prior to that time. Therefore, the hypothesis that native fish were less abundant in regulated streams than unregulated streams in the Victoria Range was confirmed. In contrast, populations of the crayfish *Geocharax* sp. nov. 1 did not decline in the regulated streams.

Hut Creek is unregulated but fish were not observed there between spring 2008 and spring 2009, showing that these headwater streams may be temporarily naturally fishless. *Galaxias oliros* were observed in Hut Creek both before (winter 2007) and after (summer 2010) this period, showing that fish populations in some unregulated streams were affected by the "millennium drought" [12], which peaked in 2006 in the Grampians [5], followed by recovery. Indeed, assemblages in the other three unregulated streams were consistent across sampling times and also consistent with historical records back to 1979.

The reasons for these patterns probably lie in the autecology of each species of fish and crayfish. Some crayfish species are able to walk overland in search of new suitable habitat when water bodies dry up [26,27]. If *Geocharax* sp. nov. 1 has this ability, then constructed weirs may not present a barrier to their persistence in regulated streams (particularly when dispersing *downstream* from refuges in weir pools). Furthermore, *Geocharax* sp. nov. 1 is able to aestivate during dry periods in a chamber that it constructs in the stream bed [5] and so it is better able to resist drying than native fish. *Geocharax* reproduction does not involve migration and juveniles co-occur spatially with adults so life cycles can, as far as we know, be completed within a small spatial extent. This

Table 2. Numbers of adult and juvenile fish and crayfish recorded for each stream in summer, per 100 m stream length.

			Geocharax sp. nov. 1		*G. marmoratus*		*G. oliros*		*G. maculatus*		*N. australis*	
Regulation	**Streams**	**Sampling period**	**Adults > 15 mm**	**Juveniles < 15 mm**	**Adults > 150 mm**	**Juveniles < 123 mm**	**Adults > 42 mm**	**Juveniles < 42 mm**	**Adults > 38 mm**	**Juveniles < 38 mm**	**Adults > 30 mm**	**Juveniles < 30 mm**
Unregulated	Hut Creek	summer 2008	0.4	2.0	0	0.8	0	0	0.4	0	0.4	0
		summer 2010	26.0	4.0	0	0	8	0	5.0	0	0	0.0
	Cultivation Creek	summer 2008	2.7	1.3	1.7	0.7	7.3	1	9.7	0	1.0	3.7
		summer 2010	2.0	0	0.5	0	14	0.5	64.5	0.5	0	0
	Deep Creek	summer 2008	0.3	0	15.7	5	17.7	0	23.3	0	5.0	0
		summer 2010	0	0	5.3	2.7	6.7	0	9.3	0	0	0
	Honeysuckle Creek	summer 2008	P	P	5.7	12.7	34	1	4.0	0	3.7	0
		summer 2010	0	0	3	17	8.5	0.5	0	0	0	0
Regulated	Brown Creek	summer 2008	5.6	5.6								
		summer 2010	7.8	7.8								
	Gap Creek	summer 2008	P	P								
		summer 2010	13.5	13.5								
	Number 1 Creek	summer 2008	P	P								
		summer 2010	0	0								
	Camp Creek	summer 2008	P	P								
		summer 2010	1.3	0.3								

Adult and juvenile fish sizes (*G. oliros* based on *G. olidus*) from [20,30,51]. Adult and juvenile crayfish sizes (OCL) for *Geocharax falcata* were used from [52]. P = present, not measured (too few or seen and not caught).

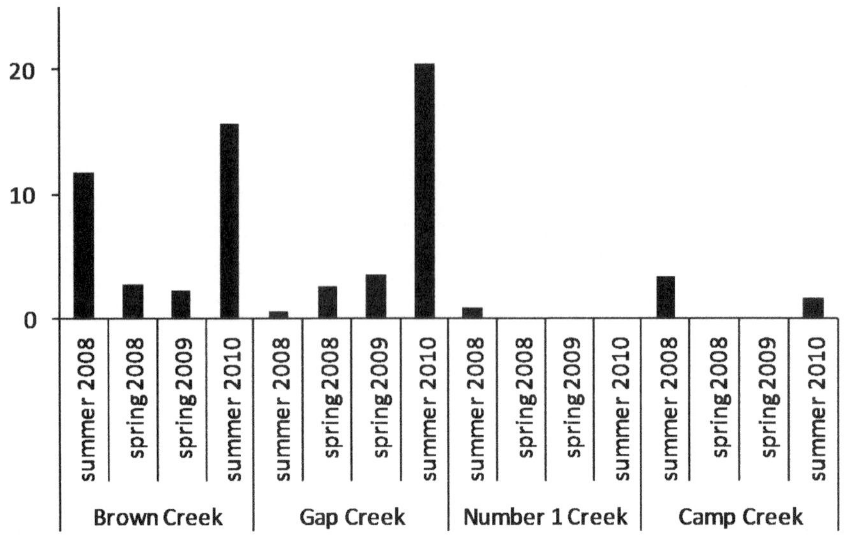

Figure 3. ***Geocharax*** **sp. nov. 1 abundance for each stream per 100 m sampling effort.**

species may also be a flexible omnivore like its congener *G. falcata* [28], meaning that when it emerges from aestivation, it can feed on leaf litter that is immediately available in the stream following the summer leaf fall. The native fish species with which it shares these streams cannot survive in the absence of surface water and are predators, so they may suffer from starvation if they enter newly inundated streams, potentially further increasing fish mortality.

Although *G. oliros* is tolerant of harsh physicochemical conditions in stream pools [29,30] it cannot withstand loss of surface water and so is likely to be lost from stream sections that dry frequently [22]. However, it is possible that *G. oliros* populations could persist in the perennially flowing sections of the regulated streams upstream of weirs in Camp, Gap and Brown Creek because it did so between 1979 and 1996, although population sizes declined during this time [13]. It may be unable to maintain small self–sustaining populations in headwater streams and hydrological connectivity may be required to allow recruitment to resupply populations [22], in particular because most *Galaxias* species are relatively short–lived. The size frequency data showed that most individuals were small, supporting the presence of younger fish in the unregulated streams.

Galaxias maculatus also tolerates high temperatures, low pH and low dissolved oxygen levels in stream pools [31,32], and may even tolerate emersion for short periods [33]. It lives for $\approx$ 2 years [30], and is probably potamodromous in Rocklands Reservoir and its unregulated tributaries, migrating upstream to spawn in streams during spring flows, as has been observed for lake populations of this species in Western Australia [31], Victoria [32] and South America [34]: the presence of mostly larger fish in Cultivation Creek, where its abundances were highest, supports this notion. However, in the Wannon River it may be diadromous, but has never been recorded in the regulated tributaries upstream. Barriers against upstream dispersal are likely to be important in determining the mode of movement for reproduction in these native fish. While Camp and Number 1 Creeks flow into heavily vegetated wetlands, potentially isolating the headwaters from Rocklands Reservoir, Brown and Gap Creek flow into the Wannon River (which joins with the Glenelg many kilometres downstream of Rocklands). The historical and present day absence of this species might then be explained by obstructions in the lower and middle reaches, which are likely to prevent the migration of diadromous fish [35]. If this is the case, *G. maculatus* may not respond to further years of summer–autumn flow releases in these streams. Both galaxid fish clearly have a considerable capacity to survive drying in perennial stream pools, sufficient to withstand the millennium drought in the unregulated streams. Downstream of weirs, increased loss of stream pools due to extreme drying has probably caused local extinctions, while upstream of weirs, small populations in headwaters were probably not self–sustaining across decades in the absence of re-supply by migrating individuals.

Gadopsis marmoratus is a much longer–lived native fish and is territorial [36] and was sustained in the unregulated stream network by perennial pools. However, increasing temperatures as climate change progresses are likely to limit adult, juvenile and gamete survival of *G. marmoratus*, despite its tolerance of long periods without stream flow [37,38]. Results suggest that it was recruiting in the unregulated streams. Abundances of this species in the unregulated streams were more consistent across times than the other species and its size frequency distributions were less skewed. *G. marmoratus* was also found both above and below the weirs in Brown Creek prior to 2006, but no individuals were collected during the present study. If *G. marmoratus* in Brown Creek were spring–summer breeders [36,37], then, in the regulated streams prior to environmental flow releases, juveniles would have hatched when the stream channel was dry downstream of the weirs on both branches. As juveniles are probably the dispersing stage in this species [38] they may have been unable to leave their natal reaches during summer and may have been consumed by the territorial adults once they left the guarded nests, limiting both local and downstream recruitment. The known low fecundity in this species [30] may have assisted their decline prior to environmental flow releases. To recolonise the regulated streams, juveniles would need to be available downstream and able to move upstream into formerly dry reaches and through or over weirs, and this may be unlikely during spring–summer; although it is possible, as proven by the presence of a single juvenile collected downstream in Camp Creek in 2006.

Persistence of native fish and crayfish in unregulated and regulated streams during supraseasonal drought

The fish assemblages in the unregulated streams have been remarkably consistent since they were first surveyed in 1979; although the fish fauna was not diverse, even for headwater streams, which typically have low alpha and higher beta diversity [39]. The consistent patterns observed during the present study differ from another long–term study of headwater stream fish, in Virginia, USA [7]. This study used similar methods to sample a more species-rich fish fauna in fewer streams, but over a longer time period (≈69 y), finding a high species turnover through time, especially in two streams which had been impacted by agriculture. The unregulated Victoria Range streams have not been impacted by development since 1979 and this may partially explain the temporal consistency of their assemblages. The presence of Rocklands Reservoir downstream might also act to prevent species turnover through limiting immigration. However, so few fish species occurred that it is difficult to assess turnover at decadal scales, with the exception of probable extinctions in the regulated streams. Nevertheless, the consistency of the differences in assemblages between these small streams, which was most marked at the height of the drought, is interesting, especially since the study reaches on different channels were often separated, through confluences, by only a few kilometres. At that time, there were conspicuous exclusions of species from different parts of the system: no galaxiids were found in Deep and Cultivation creeks; only a few *G. marmoratus* were collected from one pool in Cultivation Creek; a single *G. oliros* and few *G. marmoratus* were in Hut Creek; a single *G. oliros* was in Honeysuckle Creek (Table 1).

Regardless, the relative stability of the fish fauna in the unregulated streams suggests that native fish species coped fairly well with seasonal flows, and survived the millennium drought. Models have predicted increased abundances of *G. olidus* when cease–to–flow periods were reinstated in artificially perennial streams [19], and our results support the idea that the ecologically similar *G. oliros* tolerates seasonal drying in unregulated streams. In their review, Matthews and Marsh-Matthews [40] found that droughts often left few detectable effects on the fish fauna of the south-west USA. They suggested that climate change impacts may cause local extinctions through increased water temperature. Species distribution models for *G. olidus* showed varying results for different climate change scenarios [42], but some field observations suggest an increased frequency of local extinction in south–eastern Australia [41]. Species distribution models for *G. maculatus* showed minimal changes in distribution [42]. All three species also persisted in regulated streams with prolonged drying for several decades (regulation occurred between 1914 and 1960) prior to the onset of drought [13,15]. It appears likely then, that the combination of the millennium drought and regulation led to the complete decline of native fish in the regulated streams.

Relationship between fish and crayfish distributions

In the unregulated streams, high fish and crayfish counts did not overlap spatially. This pattern may be due to predation upon *Geocharax* by fish, especially by the larger *G. marmoratus*, since freshwater crayfish may form an important component of fish diets [43–45]. Therefore, in streams where fish abundances are lower due to prolonged drying, pools or dispersal barriers may comprise predation refuges for crayfish. Other studies have shown negative relationships between fish and crayfish and/or higher crayfish abundances in intermittent streams and wetlands, particularly following drought [46–49]. These patterns may have broader significance for other invertebrate species and detritus processing, because omnivorous crayfish have been shown to be ecosystem engineers in headwater streams, as well as direct predators on smaller invertebrates [50].

Conclusions

A single year of summer–autumn environmental flow releases did not produce a clear, detectable response in fish or crayfish in headwater streams in the Victoria Range, Victoria, Australia. Strong evidence suggests that fish became extinct both upstream and downstream of weirs in four of these streams, between 1996 and 2008. Therefore, a detectable response to environmental flow releases would require recolonization of the streams, and this was not observed. Possible explanations for this include: 1) a single year of flow releases probably did not provide sufficient time for recolonisation by fish, in contrast to invertebrates which had not become extinct upstream of weirs; 2) the volume of flow releases did not provide sufficient flow for fish to migrate from downstream reaches; 3) dispersal barriers, including breaks in surface flow, may exist downstream of the weirs that prevent recolonization. For crayfish, the rate of dispersal is probably too slow to have allowed significant increases in density across all regulated streams during this study. An indefinite continuation of flow releases annually, and reducing the unnatural breaks in flow imposed by weirs, may eventually facilitate dispersal up regulated channels. But, other potential barriers also need to be identified and ameliorated, to provide the best conditions for fish and crayfish to fulfil their life cycles in these streams. Environmental flow releases in headwater streams should be accompanied by the identification and management of anthropogenic barriers to fish dispersal. Furthermore, the probable extinction of fish in the regulated streams but not in the unregulated streams, in response to decadal-scale drought, shows that human alterations to stream flow regimes may pose a greater threat than long-term drought or climate change impacts on flow regime.

Acknowledgments

The authors wish to thank Parks Victoria and staff from the Glenelg-Hopkins Catchment Management Authority and Wannon Water for their support during the research project and Deakin University and Murdoch University for their support of the preparation of this work for publication.

Author Contributions

Conceived and designed the experiments: ETC TGM TJH KJ JKM SRS BJR. Performed the experiments: ETC TGM TJH KJ JKM SRS BJR. Analyzed the data: ETC TGM BJR. Contributed reagents/materials/analysis tools: ETC TGM TJH KJ JKM SRS BJR. Wrote the paper: ETC TGM TJH KJ JKM SRS BJR.

References

1. Murchie KJ, Hair KPE, Pullen CE, Redpath TD, Stephen SHR, et al. (2008) Fish response to modified flow regimes in regulated rivers: research methods, effects and opportunities. River Res Appl 24: 197–217.
2. Gasith A, Resh VH (1999) Streams in mediterranean climate regions: abiotic influences and biotic responses to predictable seasonal events. Ann Rev Ecol Syst 30: 51–81.
3. Robson BJ, Matthews TG, Lind PR, Thomas NA (2008) Pathways for algal recolonization in seasonally–flowing streams. Freshw Biol 52: 2385–2401.
4. Mackie JK, Chester ET, Matthews TG, Robson BJ (2012) Macroinvertebrate response to environmental flows in headwater streams in western Victoria, Australia. Ecol Eng 53: 100–105.

5. Chester ET, Robson BJ (2011) Drought refuges, spatial scale and the recolonization of invertebrates in non-perennial streams. Freshw Biol 56: 2094–2104.
6. Robson BJ, Chester ET, Mitchell BD, Matthews TG (2013) Disturbance and the role of refuges in mediterranean climate streams. Hydrobiol 719: 77–91.
7. Hitt NP, Roberts JH (2012) Hierarchical spatial structure of stream fish colonization and extinction. Oikos 121: 127–137.
8. King AJ, Humphries P, McCasker NG (2013) Reproduction and Early Life History. In: Humphries P and Walker K, editors. Ecology of Australian Freshwater Fishes. Melbourne: CSIRO Publishing. pp. 159–193.
9. Doehring K, Young RG, McIntosh AR (2011) Factors affecting juvenile galaxiid fish passage at culverts. Mar Freshw Res 62: 38–45.
10. Lind PR, Robson BJ, Mitchell BD (2007) Multiple lines of evidence for the beneficial effects of sustaining environmental flows in two lowland rivers in Victoria, Australia. River Res Appl 23: 933–946.
11. Robson BJ, Mitchell BD (2010) Metastability in a river subject to multiple disturbances may constrain restoration options. Mar Freshw Res 61: 778–785.
12. Bond NR, Lake PS, Arthington AH (2008) The impacts of drought on freshwater ecosystems: an Australian perspective. Hydrobiol 600: 3–16.
13. Raadik TA (1996) Aquatic fauna survey (fish and decapod crustaceans) of the Hamilton water supply system, Grampians National Park. Melbourne: Arthur Rylah Institute. 38 p.
14. Raadik TA (2011) Systematic revision of the mountain galaxias, *Galaxias olidus* Gunther, 1866 species complex (Teleostei: Galaxiidae) in eastern Australia. Canberra: University of Canberra. 505 p.
15. Jackson PD, Davies JN (1983) Survey of the fish fauna in the Grampians region, south-western Victoria. Proc R Soc Vic 95: 39–51.
16. Miller AD, Waggy G, Ryan SG, Austin CM (2004) Mitochondrial 12S rRNA sequences support the existence of a third species of freshwater blackfish (Percicthyidae: *Gadopsis*) from south-eastern Australia. Mem Mus Vic 61: 121–127.
17. Coleman RA, Pettigrove V, Raadik TA, Hoffman AA, Miller AD, et al. (2010) Microsatellite markers and mtDNA data indicate two distinct groups in dwarf galaxias, *Galaxiella pusilla* (Mack) (Pisces: Galaxiidae), a threatened freshwater fish from south-eastern Australia. Cons Gen 11: 1911–1928.
18. Schultz MB, Smith AA, Richardson AMM, Horwitz P, Crandall KA, et al. (2007) Cryptic diversity in *Engaeus* Erichson, *Geocharax* Clark, and *Gramastacus* Riek (Decapoda: Parastacidae) revealed by mitochondrial 16S rDNA sequences. Invert Syst 21: 569–587.
19. Bond NR, McMaster D, Reich P, Thomson JR, Lake PS (2010) Modelling the impacts of flow regulation on fish distributions in naturally intermittent lowland streams: an approach for predicting restoration responses. Freshw Biol 55: 1997–2010.
20. Howson TJ, Robson BJ, Matthews TG, Mitchell BD (2012) Size and quantity of woody debris affects fish assemblage structure in a sediment-disturbed lowland river. Ecol Eng 40: 144–152.
21. Clarke KR, Gorley RN (2006) PRIMER v6: User Manual/Tutorial. Plymouth: PRIMER-E. 190 p.
22. Dexter T, Bond N, Hale R, Reich P (2013) Dispersal and recruitment of fish in an intermittent stream network. Austral Ecology: in press.
23. Feurich R, Boubee J, Olsen NRB (2012) Improvement of fish passage in culverts using CRD. Ecol Eng 47: 1–8.
24. Robson BJ, Mitchell BD, Chester ET (2011) An outcome-based model for predicting recovery pathways after flow restoration in rivers: the Recovery Cascade Model. Ecol Eng 37: 1379–1386.
25. Beatty S, Allen M, Lymbery A, Storer T, White G et al. (2013) Novel methods for managing freshwater refuges against climate change in southern Australia: evaluating small barrier removal to improve refuge connectivity — a global review of barrier decommissioning and a process for southern Australia in a drying climate. Gold Coast: National Climate Change Adaptation Research Facility, Australia. 74 p.
26. Gherardi F, Barbaresi S (2000) Invasive crayfish: activity patterns of *Procambarus clarkii* in the rice fields of the Lower Guadalquivir (Spain). Arch Hydrobiol 150: 153–168.
27. Kerby JL, Riley SPD, Kats LB, Wilson P (2005) Barriers and flow as limiting factors in the spread of an invasive crayfish (*Procambarus clarkii*) in southern California streams. Biol Cons 126: 402–409.
28. Johnston K, Robson BJ, Fairweather PG (2011) Trophic positions of omnivores are not always flexible: evidence from four species of freshwater crayfish. Austral Ecol 36: 269–279.
29. McMaster D, Bond N (2008) A field and experimental study on the tolerances of fish to *Eucalytpus camaldulensis* leachate and low dissolved oxygen concentrations. Mar Freshw Res 59: 177–185.
30. Harris JH, Bond NR, Closs GP, Gehrke PC, Nicol SJ, et al. (2013) Dynamics of Populations. In: Humphries P, Walker K, editors. Ecology of Australian Freshwater Fishes. Melbourne: CSIRO Publishing. pp. 223–244.
31. Chapman A, Morgan DL, Beatty SJ, Gill HS (2006) Variation in life history of land-locked lacustrine and riverine populations of *Galaxias maculatus* (Jenyns 1842) in Western Australia. Env Biol Fish 77: 21–37.
32. Laurenson LJB, French RP, Jones P, Ierodiaconou D, Gray S, et al. (2012) Aspects of the biology of *Galaxias maculatus*. J Fish Biol 81: 1085–1100.
33. Urbina MA, Forster ME, Glover CN (2012) Leap of faith: voluntary emersion behaviour and physiological adaptations to aerial exposure in a non-aestivating freshwater fish in response to aquatic hypoxia. Physiol Behav 103: 240–247.
34. Barriga JP, Battini MA, Cussac VC (2007) Annual dynamics variation of a landlocked *Galaxias maculatus* (Jenyns 1842) population in a Northern Patagonian river: occurrence of juvenile upstream migration. J Appl Ichthyology 23: 128–135.
35. Lassalle G, Crouzet P, Rochard E (2009) Modelling the current distribution of European diadromous fishes: an approach integrating regional anthropogenic pressures. Freshw Biol 54: 587–606.
36. Allen GR, Midgley SH, Allen M (2002) Field guide to the freshwater fishes of Australia. Perth: Western Australian Museum. 394 p.
37. Jackson PD (1978) Spawning and early development of the river blackfish *Gadopsis marmoratus* Richardson (Gadopsiformes: Gadopsidae) in the McKenzie River, Victoria. Australian J Mar Freshw Res 29: 193–198.
38. Koster WM, Crook DA (2007) Diurnal and nocturnal movements of river blackfish (*Gadopsis marmoratus*) in a south-eastern Australian upland stream. Ecol Freshw Fish 17: 146–154.
39. Finn DS, Bonada N, Murria C, Hughes JM (2011) Small but mighty: headwaters are vital to stream network biodiversity at two levels of organization. J North Am Benthol Soc 30: 963–980.
40. Matthews WJ, Marsh-Matthews E (2003) Effects of drought on fish across axes of space, time and ecological complexity. Freshw Biol 48: 1232–1253.
41. Morrongiello JR, Beatty SJ, Bennett JC, Crook DA, Ikedife DNEN, et al. (2011) Climate change and its implications for Australia's freshwater fish. Mar Freshw Res 62: 1082–1098.
42. Bond NR, Thomson J, Reich P, Stein J (2011) Using species distribution models to infer potential climate change-induced range shifts of freshwater fish in south-eastern Australia. Mar Freshw Res 62: 1043–1061.
43. Stein RA (1977) Selective predation, optimal foraging and the predator-prey interaction between fish and crayfish. Ecol 58: 1237–1253.
44. Garvey JE, Stein RA, Thomas HM (1994) Assessing how fish predation and interspecific prey competition influence a crayfish assemblage. Ecol 75: 532–547.
45. Jackson PD (1978) Benthic invertebrate fauna and feeding relationships of brown trout, *Salmo trutta* Linnaeus and river blackfish *Gadopsis marmoratus* Richardson, in the Aberfeldy River, Victoria. Australian J Mar Freshw Res 29: 725–742.
46. Englund G (1999) Effects of fish on the local abundance of crayfish in stream pools. Oikos 87: 48–56.
47. Flinders CA, Magoulick DD (2003) Effects of stream permanence on crayfish community structure. Am Midl Nat 149: 134–147.
48. Dorn N (2008) Colonization and reproduction of large macroinvertebrates are enhanced by drought-related fish reductions. Hydrobiol 605: 209–218.
49. Chessman BC, Jones HA, Searle NK, Growns IO, Pearson MP (2010) Assessing effects of flow alteration on macroinvertebrate assemblages in Australian dryland rivers. Freshw Biol 55: 1780–1800.
50. Creed RP, Reed JM (2004) Ecosystem engineering by crayfish in a headwater stream community. J North Am Benthol Soc 23: 224–236.
51. Barbee NC, Hale R, Morrongiello J, Hicks A, Semmens D, et al. (2011) Large-scale variation in life history traits of the widespread diadromous fish, *Galaxias maculatus*, reflects geographic differences in local environmental conditions. Mar Freshw Res 62: 790–800.
52. Johnston K, Robson BJ, Austin CM (2010) Population structure and life history characteristics of *Gramastacus insolitus* and *Geocharax falcata* (Parastacidae) in the Grampians National Park, Australia. Freshw Crayfish 17: 245–253.

19

Likelihood Inference of Non-Constant Diversification Rates with Incomplete Taxon Sampling

Sebastian Höhna*

Department of Mathematics, Stockholm University, Stockholm, Sweden

Abstract

Large-scale phylogenies provide a valuable source to study background diversification rates and investigate if the rates have changed over time. Unfortunately most large-scale, dated phylogenies are sparsely sampled (fewer than 5% of the described species) and taxon sampling is not uniform. Instead, taxa are frequently sampled to obtain at least one representative per subgroup (e.g. family) and thus to maximize diversity (diversified sampling). So far, such complications have been ignored, potentially biasing the conclusions that have been reached. In this study I derive the likelihood of a birth-death process with non-constant (time-dependent) diversification rates and diversified taxon sampling. Using simulations I test if the true parameters and the sampling method can be recovered when the trees are small or medium sized (fewer than 200 taxa). The results show that the diversification rates can be inferred and the estimates are unbiased for large trees but are biased for small trees (fewer than 50 taxa). Furthermore, model selection by means of Akaike's Information Criterion favors the true model if the true rates differ sufficiently from alternative models (e.g. the birth-death model is recovered if the extinction rate is large and compared to a pure-birth model). Finally, I applied six different diversification rate models – ranging from a constant-rate pure birth process to a decreasing speciation rate birth-death process but excluding any rate shift models – on three large-scale empirical phylogenies (ants, mammals and snakes with respectively 149, 164 and 41 sampled species). All three phylogenies were constructed by diversified taxon sampling, as stated by the authors. However only the snake phylogeny supported diversified taxon sampling. Moreover, a parametric bootstrap test revealed that none of the tested models provided a good fit to the observed data. The model assumptions, such as homogeneous rates across species or no rate shifts, appear to be violated.

Editor: Art F. Y. Poon, British Columbia Centre for Excellence in HIV/AIDS, Canada

Funding: The author has no support or funding to report.

Competing Interests: The author has declared that no competing interests exist.

* E-mail: Sebastian.Hoehna@gmail.com

Introduction

Patterns of biodiversity reflected in phylogenetic estimates indicate that (1) rates of diversification are not constant over time or across the tree and (2) taxonomic sampling is both incomplete and non-random. Furthermore, knowing the sampling strategy is crucial for unbiased estimation of diversification rates [1–3]. Researchers sample species in a manner that is not uniform. Instead, taxa are often selected so that the diversity is maximized, e.g. sampling at least one species per family [4,5]. This strategy is called *diversified* sampling [2].

Several extensions have been proposed to model non-constant diversification rates [6–9] but the combination of incompletely sampled phylogenies and non-constant rates has attracted less attention. It is well known how to accommodate uniform taxon sampling (also called random sampling), where every taxon has the same probability to be included in the dataset, in inference based on the birth-death process [6,10]. The birth-death process with uniform taxon sampling has been extended to time-dependent rates [8] and diversity-dependent rates [11]. Diversified taxon sampling has only been considered in the context of constant rates [2] and, to my knowledge, the corresponding likelihood functions for non-constant rates have not been available previously.

In the present paper I derive the likelihood function for the birth-death process with diversified taxon sampling and time-dependent diversification rates. Thus, the diversification rates may be defined as any function that only depends on the variable time e.g. an exponentially decaying speciation rate. The focus of the paper lies on frequentist inference of the parameters of the birth-death process by means of Maximum Likelihood Estimation (MLE) and model selection by means of the Akaike Information Criterion (AIC) and Bayesian Information Criterion (BIC).

It has been claimed that MLEs of speciation and extinction rates are unbiased if the correct model is used for the inference [8,12] although the MLE is not always an unbiased estimator [13]. I will investigate by means of simulations if the MLE is biased.

Selection of different birth-death models using the AIC score has been widely applied and used to reject constant-rate models [9,14–16] although Rabosky [7] showed that the AIC score may be misleading. Therefore I investigate the power of the AIC and BIC scores to recover the true model.

Besides the use in Maximum Likelihood inference as in this study, the derived likelihood functions are also important for Bayesian phylogenetic inference. The birth-death process assigns a probability distribution on phylogenetic trees and divergence times. Thus, the birth-death process is frequently used in Bayesian phylogenetic inference as a prior probability distribution [10,17–20]. This prior probability distribution influences divergence times estimates and more realistic birth-death models improve the estimation of divergence times and rates of molecular evolution [21–23].

Eventually, I applied six different diversification rate models on three large-scale phylogenies: an ant phylogeny [24], a mammal phylogeny [4] and a snake phylogeny [25]. These three phylogenies have in common that (1) they are relatively large (149, 164 and 41 sampled taxa), (2) are sparsely sampled (fewer than 5% of the known species included) and (3) taxa were included as representatives of different clades (diversified sampling). I conclude by investigating the adequacy of the used models (model fit) and discuss model shortcomings and potential model improvements.

Materials and Methods

The birth-death process with non-constant diversification rates and diversified taxon sampling

Following the notation of Nee et al. [6], I define the birth-death process with non-constant rates and diversified sampling for rooted, strictly bifurcating trees. At any given point in time t, the number of living species is denoted by $N(t)$. The number of species $N(t)$ is modeled by a Markovian birth and death process with individual birth and death rates given by $\lambda(t)$ and $\mu(t)$.

The process starts with two species at time t_1 – which is the time of the most recent common ancestor and therefore $N(t_1)=2$. For each species, a speciation or an extinction event occurs randomly with rates $\lambda(t)$ and $\mu(t)$ respectively. At a speciation event, the ancestral species splits into two descendant species and at an extinction event the species dies. The events happen instantaneously and speciation and extinction rates are equal for all species. At the present time T the process is stopped. An example of this process is illustrated in Figure 1.a. Then, the extinct lineages are removed because only the extant taxa are observed (see Figure 1.b). Let $m=N(T)$ denote the number of extant species at time T in the complete tree (see Figure 1.b) out of which n species are sampled. Under uniform taxon sampling, every extant species at the current time T has the same probability ρ to be sampled and included in the reconstructed tree (see Figure 1.c). Under diversified sampling, all species that arose from the last $m-n$ speciation events are removed (see [2] for a more detailed description). The process is illustrated in Figure 1 and Figure 2.

Starting with one species at time t ($N(t)=1$) there are three key probability functions: the probability of survival ($N(T)>0$), exactly k species ($N(T)=k$) and exactly one species at time T ($N(T)=1$) given by [6,26]

$$P(N(T)>0|N(t)=1)=\left(1+\int_t^T(\mu(s)\exp(r(t,s)))ds\right)^{-1} \tag{1}$$

$$P(N(T)=k|N(t)=1)=(1-P(N(T)>0|N(t)=1)\exp(r(t,T)))^{k-1}$$

$$\times P(N(T)>0|N(t)=1)^2\exp(r(t,T)) \tag{2}$$

$$P(N(T)=1|N(t)=1)=P(N(T)>0|N(t)=1)^2\exp(r(t,T)) \tag{3}$$

where $r(t,s)=\int_t^s\mu(x)-\lambda(x)dx$. Equation (1) is derived from Equation (24) in [6] and Equation (2) and Equation (3) are derived from Equation (3) in [6] (see also Höhna (2013) Equation (2–4) [27]). These three probability functions are used to define the probability density of the observed reconstructed tree.

The probability density of all speciation events $\{t\}=\{t_1,\ldots,t_{n-1}\}$ in the reconstructed tree under complete sampling (f_C) is

$$f_C(\{t\}|N(t_1)=2)=P(N(T)=1|N(t_1)=1)^2$$

$$\times\prod_{i=2}^{n-1}(i\times\lambda(t_i)\times P(N(T)=1|N(t_i)=1)) \tag{4}$$

which was derived by Thompson for constant rates (Equation (3.4.6) though with a different combinatorial factor which represents the probability of the tree topology)[28]. Additionally, I modified the likelihood to use the time-dependent rate functions of [6].

The probability density under diversified taxon sampling can be derived by recognizing that all speciation events are independent and identically distributed [2]. The definition of diversified sampling dictates that all unobserved speciation events leading to extant species occurred after the last observed speciation event. Hence, the probability of each unobserved speciation event is $1-F(t_{n-1}|N(t_1)=1,t_1\leq t_{n-1}\leq T)$ where $F(t_{n-1}|N(t_1)=1,t_1\leq t_{n-1}\leq T)$ is the probability that a speciation event has occurred prior to t_{n-1}, see Figure 2 for an illustration.

The probability density function of the divergence times for non-constant rates is (adopted from [27] Equation (8))

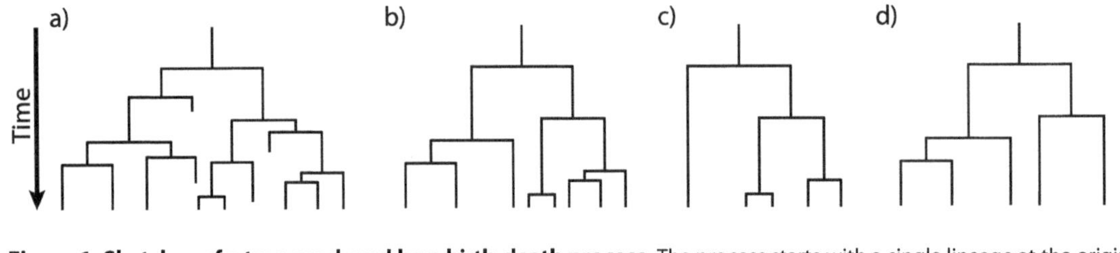

Figure 1. Sketches of a tree produced by a birth-death process. The process starts with a single lineage at the origin. At each speciation event the ancestral lineage is replaced by two descendant lineages. At an extinction event the lineage simply terminates. a) A complete tree including the extinct lineages. b) The reconstructed tree of tree a) without the extinct lineages. c) A randomly sampled tree of the same reconstructed tree. Every taxa had the same probability ρ to be sampled. d) A five-taxon tree where taxa are selected to maximize diversity (diversified sampling).

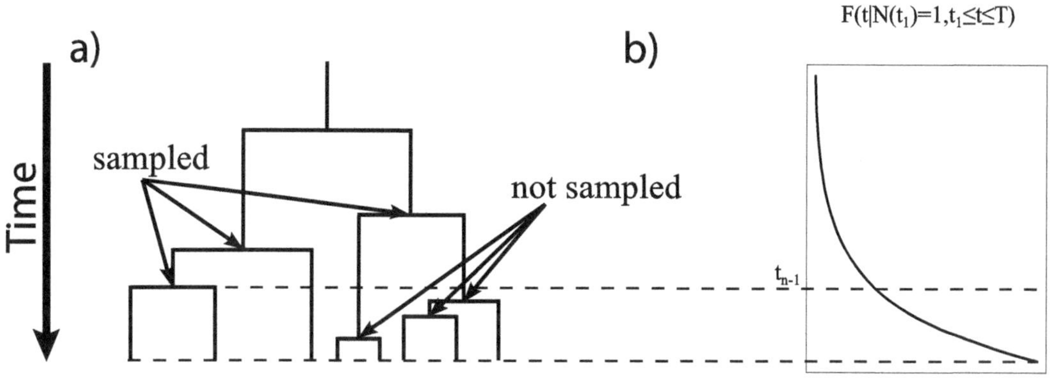

Figure 2. Illustration of diversified sampling and the probability density. a) A reconstructed tree where three speciation events are sampled and last three speciation events are not. b) The cumulative probability density of the time of a speciation event. All non-sampled speciation events occurred some time after the dashed line (t_{n-1}) and thus have the probability $1-F(t_{n-1})$.

$$f(t|N(t_1)=1,t_1\leq t\leq T)=\frac{\lambda(t)P(N(T)=1|N(t)=1)}{1-P(N(T)>0|N(t_1)=1)\exp(r(t_1,T))} \quad (5)$$

and thus the distribution function being

$$F(t|N(t_1)=1,t_1\leq t\leq T)=1-\frac{1-P(N(T)>0|N(t)=1)\exp(r(t,T))}{1-P(N(T)>0|N(t_1)=1)\exp(r(t_1,T))}\ . \quad (6)$$

A short proof of Equation (6) is given in the Appendix.

From [2] Equation (A.1) in a slightly different notation we have the probability density function of the speciation events in the reconstructed tree given the time of the most recent common ancestor (t_1) and the total number of species m (sampled and missing species) at the present time

$$f_D(\{t\}|N(t_1)=2,N(T)=m)=\frac{1}{n-1}\binom{m-2}{n-2}(1-F(t_{n-1}|t_1\leq t_{n-1}\leq T))^{m-n}$$

$$\times\prod_{i=2}^{n-1}f(t_i|t_1\leq t_i\leq T)$$

$$=\frac{1}{n-1}\binom{m-2}{n-2}(1-F(t_{n-1}|t_1\leq t_{n-1}\leq T))^{m-n}$$

$$\times\prod_{i=2}^{n-1}\frac{\lambda(t_i)P(N(T)=1)|N(t_i)=1)}{1-P(N(T)>0)|N(t_1)=1)\exp(r(t_1,T))} \quad (7)$$

The probability density conditioned only on the time since the most recent common ancestor is obtained using Equation (12) in [27]

$$f_D(\{t\}|N(t_1)=2)=f_D(\{t\}|N(t_1)=2,N(T)=m)\times P(N(T)=m|N(t_1)=2)$$

$$=\binom{m-2}{n-2}(1-P(N(T)>0|N(t)=1)\exp(r(t,T)))^{m-n}$$

$$\times(P(N(T)>0)|N(t_1)=1)\exp(r(t_1,T)))^2$$

$$\times\prod_{i=2}^{n-1}\lambda(t_i)P(N(T)=1)|N(t_i)=1)\ . \quad (8)$$

In the remainder of the paper I will use Equation (8) for the computation of the likelihood under diversified taxon sampling. Equation (4) with the rate function $r(t,s)=\int_t^s\mu(x)-\lambda(x)dx$ if $s<T$ and $r(t,s)=\int_t^s\mu(x)-\lambda(x)dx-\log(\rho)$ if $t\leq T\leq s$ will be used for the likelihood function under uniform taxon sampling (see [6] Equation (31)). For the analysis I condition on starting with two species at the most recent common ancestor and on survival of the process.

Parameters of the Birth-Death Process

I evaluated six different birth-death models with time-varying rates, proceeding from the simplest model to the most general and

Table 1. The six different birth-death models with the corresponding parameters.

Model	$\lambda(t)$	$\mu(t)$
Model 1	λ_0	0
Model 2	$\lambda_1*\exp(-\alpha*t)$	0
Model 3	λ_0	μ
Model 4	$\lambda_0+\lambda_1*\exp(-\alpha*t)$	0
Model 5	$\lambda_1*\exp(-\alpha*t)$	μ
Model 6	$\lambda_0+\lambda_1*\exp(-\alpha*t)$	μ

biologically realistic (Table 1). Other models with perhaps more complex patterns of rate variation over time are possible and could be of interest in future research. The first three models have been used previously but the last three models are new contributions.

- Model 1: A constant-rate pure birth process, i.e. the Yule process [29]. The number of species is monotonically and exponentially increasing.
- Model 2: A decreasing pure birth process with the speciation rate declining towards zero. This process is equivalent to the decreasing rate pure birth process used in Rabosky and Lovette [14]. The number of species is monotonically increasing.
- Model 3: A constant-rate birth-death process, as used in Thompson [28]. The expected number of species increases exponentially with time.
- Model 4: A pure birth process with a decaying rate of speciation but a constant, non-zero speciation rate the longer the process continues ($\lambda(t) = \lambda_0 + \lambda_1 * \exp(-\alpha * t)$). Thus, the process does not stop producing new species after the initial burst, as it is the case in Model 2. As for the other two pure birth processes, the number of species increases monotonically.
- Model 5: A birth-death process with an initial phase of expansion (higher speciation rate than extinction rate) and afterwards converging to a critical branching process, i.e. the speciation rate and the extinction rate are equal ($\lambda(t) = \mu + \lambda_1 * \exp(-\alpha * t)$ and $\mu(t) = \mu$). Although one might assume the expected number of species to remain constant for a critical branching process, this does not hold if the process is conditioned on survival (see Figure 3).
- Model 6: A birth-death process with a constant extinction rate, a constant part of the speciation rate and a decreasing part of the speciation rate. It represents the situation when diversity is continuously increasing starting with rapid radiation, followed by a steady expansion and some constant species turnover over time ($\lambda(t) = \lambda_0 + \lambda_1 * \exp(-\alpha * t)$ and $\mu(t) = \mu$).

The parametrization of the models can be found in Table 1. Figure 3 shows the expected number of species ($E[N(T)]$) alive at time T. $E[N(T)]$ is obtained analytically using the fact that $N(T)$ is geometrically distributed, see Equation (5) in [27]. Note that the process is conditioned on survival and thus $E[N(T)]$ increases even if $\lambda(t) = \mu(t)$.

Accuracy of Parameter Estimation

Using simulation, I investigated the accuracy of parameter estimation under birth-death models – including pure birth models – and the impact of tree size on the accuracy. Additionally, I tested the effectiveness of the AIC [30] and BIC [31] for choosing between models. Finally, I applied the models on three empirical large-scale phylogenies.

The MLE is an unbiased estimator of the true parameters in the limit (if the number of observation goes to infinity) which has been shown using the Central Limit Theorem ([13], Chapter 5, page 118). Hence, for very large trees the MLE of the diversification rates is unbiased, and may or may not be biased for small samples. Condamine et al. [12] claimed that extinction rate estimates are unbiased if the inference model is the same as the simulation model and refer to [16] and [8]. I numerically investigated if the MLE is indeed unbiased if the model assumptions hold true. I simulated 1000 trees under complete taxon sampling for the time of the process $T \in \{0.25, 0.5, \ldots, 5\}$ and conditioning on survival of

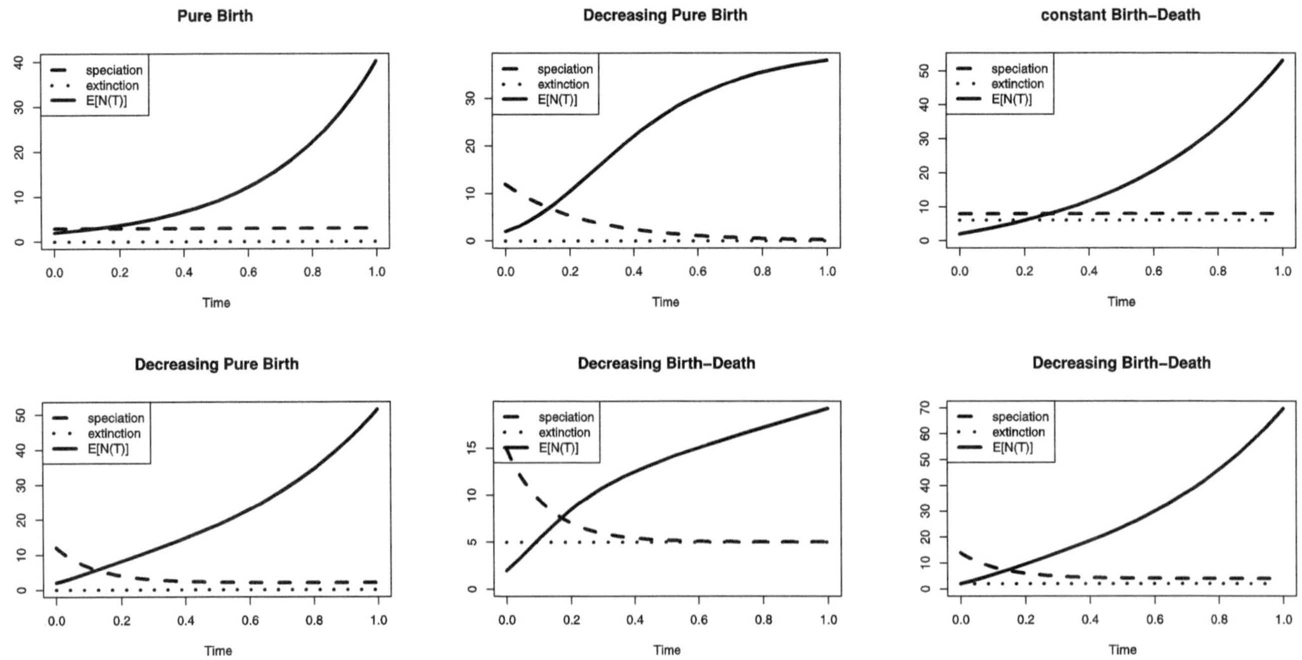

Figure 3. The six birth-death models used in this study. Each plot shows the speciation rate (dashed) and the extinction rate (dotted) over time, as well as the expected number of species ($E[N(t)]$) if the process endured for a time t. Model 1: A constant-rate pure birth process. Model 2: A decreasing pure birth process with the speciation rate declining towards zero as time continues. Model 3: A constant-rate birth-death process. Model 4: A pure birth process with a constant part of the speciation rate and a decreasing part of the speciation rate. Model 5: A birth-death process with an initial phase of expansion (higher speciation rate than extinction rate) and afterwards converging to a critical branching process. Model 6: A birth-death process with a constant extinction rate, a constant part of the speciation rate and a decreasing part of the speciation rate.

the process under (1) a constant-rate pure birth process ($\lambda = 1.0$) (2) a constant-rate birth-death process ($\lambda = 1.6$, $\mu = 0.8$) and (3) a birth-death process with a decreasing speciation rate ($\lambda(t) = 1 + 4 * \exp(-1 * t)$, $\mu = 1$). Then, I estimated the model parameters λ, μ and α for each tree choosing the true model. These parameter settings were chosen so that the expected diversity after time $t = 5$ was approximately 200. Here, I varied the time of the process and thus also the size of the simulated tree to study the impact of the tree size on the bias of the estimated parameters.

I assumed a speciation rate larger than the extinction rate ($\lambda(t) > \mu(t)$). Hence the expected number of species in the phylogenetic tree ($E[N(T)]$) increases the longer the process takes. Therefore, I was able to observe if the bias decreases or increases with more taxa by running the birth-death process for different times. The simulations were performed in R using the package *TESS* [27].

Model selection using the AIC and BIC

In frequentist inference the best model can be selected out of a set of (non-nested) candidate models via the AIC [30] and BIC [31]. The model with the lowest AIC (or BIC) score is selected as the best model. Here I used the AIC corrected (AICc) for small sample sizes,

$$\text{AICc} = -2\log(L) + 2k + \frac{2k(k-1)}{n' - k - 1} \quad (9)$$

where k denotes the number of parameters in the model, n' the number of speciation events ($n' = n - 2$) and L the likelihood score of the MLE [32,33].

The AICc needs the number of observations (sample size) for the computation. One may ask what the sample size of one phylogenetic tree is: either just one single observation (i.e. one tree) or $n - 2$ observations (the number of speciation events except the root). The choice will strongly influence the AICc score. I argue that a single phylogenetic tree represents a sample of size $n - 2$. The number of species alive at time T is geometrically distributed [6] and all speciation events in the reconstructed tree are independent and identically distributed (see [28] and [10]). Thus we have one observation from a geometric distribution and $n - 2$ observation from some other distribution (Equation (6)) but the $n - 2$ speciation events are dependent on the age of the tree and the number of sampled taxa. Therefore a single tree from a birth-death process resembles a sample of size $n - 2$ (see also [34]).

Rabosky [7] showed that the model selection by means of the AIC can be misleading. In Rabosky's study up to 50% of the simulations under a constant-rate birth-death process favored a more complex, non-constant rate model. Nevertheless, many analyses of birth-death models have been based on the ability of the AIC score to distinguish between models [14,15]. Rabosky[17] used the AIC but Burnham and Anderson[32] showed that the AIC will favor more complex models for small sample sizes and the AICc should be used instead because it gives a higher penalty to more complex models. Similarly, the BIC is more conservative than the AIC. Therefore, I investigated the performance of the AICc and BIC to select the true model under which the data were simulated using the following setup.

I simulated 100 trees with $n = 100$ taxa under (1) a constant-rate pure birth process, (2) a decreasing-rate pure birth process and (3) a constant-rate birth-death process with $\rho \in \{0.05, 0.15, \ldots, 0.95\}$ once under uniform taxon sampling and once under diversified taxon sampling applied to all three models. For the constant-rate pure birth process I choose the rate $\lambda = 1.0$; for the decreasing rate pure birth process I choose the rate function $\lambda(t) = 4.0 * \exp(-0.5 * t)$ and for the constant-rate birth-death process I choose the rates $\lambda = 1.0$ and $\mu = 0.75$. The tree size remained fixed for this study to resemble the size of empirical datasets. For each tree the best model out of the six mentioned models in Table 1 was selected.

Analysis on Empirical Phylogenies

Sampling to maximize taxonomic representation is an acknowledged common practice in phylogenetic systematics (e.g. systematic studies that use a single species per family or genus [35–37]). However, the prevalence of diversified sampling and its importance in studies of lineage diversification are not well understood. I applied the birth-death models on three large-scale phylogenies of: ants, mammals and snakes. All three phylogenies are examples of diversified sampling.

Moreau et al. [24] built a phylogeny of extant ants sampled from nearly half of the described genera (139 of 288). The phylogeny contains 149 samples of the approximately 11,800 described species which gives a sampling probability of just above 1%. The estimated most recent common ancestor (MRCA) was 168 million years ago (Ma). Moreau et al. [24] did not estimate diversification rates but observed that most new lineages arose 100 to 70 Ma. I removed duplicate species per genus and the outgroup species to obtain a phylogeny with sampling as close as possible to diversified sampling.

Pyron and Burbrink [25] reconstructed a phylogeny of all known families and subfamilies of snakes. The phylogeny contains 41 samples of the nearly 3,500 snake species which again resulted into a sampling fraction of just above 1%. Pyron and Burbrink [25] studied clade-dependent diversification rates and found four diversification rate shifts. Furthermore, clade ages and diversification rates seemed to be negatively correlated.

Meredith et al. [4] constructed a family-level mammal phylogeny with 1–3 samples per family. The phylogeny contains 164 samples of the approximately 5,400 described species, a sampling fraction of approximately 3%. Their analysis revealed two rate increases, at ~ 100 Ma and/or ~ 83 Ma, and a decrease at ~ 78 Ma. The original analysis only estimated rates until 40 Ma because of the sampling bias. I reduced the sample size so that exactly one species per family remained.

For each dataset the best model was selected by means of the AICc score and the BIC score. Each of the six models was applied under the assumption of uniform sampling and diversified sampling, giving a total of 12 different models. The empirical estimates of the sampling probability ρ were used. The MLEs were obtained using the optimization routine *optim* in R repeated 10 times with different initial values to check for convergence. The R script and the phylogenies are available in the Dryad data repository at http://doi.org/10.5061/dryad.rd2s3.

Model Adequacy Testing. The adequacy of a model, often called model fit, is the probability of the observed data under the model [38]. Analytical computation of the probability of the data is infeasible in most situation and a simulation method, such as the parametric bootstrap, is necessary [39]. The parametric bootstrap computes the probability of the observed data (model adequacy) by simulating datasets, computing summary statistics for the simulated datasets and testing whether the summary statistic of the observed data falls within the 95% interval of the simulated summary statistics.

For each model and dataset I simulated 10000 trees under the MLE parameters conditioning on the time of the process equals the observed age of the tree and another 10000 trees under the MLE parameters conditioning on the number of taxa being equal to the number of observed (sampled) taxa. For the first 10000 trees I computed the empirical distribution of the γ−statistic [40] and the number of taxa sampled and for the second 10000 trees I computed the empirical distribution of the age of the trees. Finally I computed the probability of the summary statistics of the observed tree.

Results and Discussion

Accuracy of Parameter Estimation

Despite the claim of Condamine et al. [12] I observed in the simulation study that the MLEs were biased (see Figure 4) even under the true model. As expected, the bias decreased for larger trees and one may get unbiased estimates if the tree were very large. Even under a constant-rate pure birth process the MLE was biased for trees with fewer than 50 taxa compared with the results of Morlon et al. who found no bias, see Figure S4 in [8]. Specifically, the speciation rate was underestimated (see Figure S1 in File S1). However, under a constant-rate birth-death process the speciation rate was overestimated and the extinction rate underestimated (see Figure S2 in File S1).

Under the decreasing speciation rate birth-death model, the extinction rate μ and the rate α at which the speciation rate declines were also biased (see Figure 4). In these cases, μ and α were overestimated. For small trees, the common observation of rapid radiations ($\alpha>0$, see[41]) could be due to the bias in the maximum likelihood estimates. Rabosky [7] observed this preference towards more complex models. Nevertheless, the parameters defining the time-dependent speciation and extinction rates of a birth-death process can be estimated with trees that are large enough.

Morlon et al. investigated if the estimated net-diversification rate ($\lambda-\mu$) is biased, which is equivalent to my analysis under the constant-rate pure birth process. However, the bias is small for the simpler models, such as the constant-rate pure birth model. It is therefore conceivable that Morlon et al. attributed the bias to the rather large uncertainty in the parameter estimate (see the 95% confidence interval in Figure 4). Nevertheless, using extensive simulations the bias can be observed and reproduced and is increasingly important for more complex models.

Unfortunately the bias changes with different models and parameter values (see Figures S1 and S2 in File S1). A correction for the bias was not performed in this study because the empirical phylogenies are relatively large but could be important for smaller phylogenies or in future research. For a single study it is possible to simulate under the inferred parameter values and re-estimate the parameters in order to quantify the bias.

Model Selection by means of AIC

My results indicate that the correct model can be inferred if the AICc score is used (see Figure 5). For the two constant-rate models the error rate did not depend on the sampling probability ρ. However, with increasing sampling probability the error rate increased for the decreasing rate pure birth model (Model 2).

These results are more optimistic than the results of [7]. The improvements are due to the advantage of the AICc over the AIC. Nonetheless the AICc is not flawless and the accuracy depends on the sample size and model complexity.

Furthermore, I tested, using the same simulation scheme, if the correct sampling method could be inferred. The ability to identify the correct sampling strategy based on a likelihood method has not been investigated previously. In [2] we only investigated the effect of the sampling strategy on the inferred parameters. The results show that the sampling method can clearly be recovered (Figure 6). The error rate starts to increase sharply only for almost complete sampling. This is to be expected because uniform taxon sampling and diversified taxon sampling are identical for $\rho=1$ [2].

The same exercise was repeated using the BIC score for model selection. The results are very similar with the difference that the error rate of the constant-rate pure birth model was lower and the error rate of the decreasing rate pure model was higher (see Figure S3 and S4 in File S1). This result is not surprising because the BIC is more conservative and thus penalizes the more complex model (the decreasing rate pure birth model) more intensively.

Empirical Analyses

Diversified sampling was identified as the best-fitting model for the snake phylogeny. The best model was a constant-rate birth-death process with rates $\lambda=2.3488$ and $\mu=2.3488$, which resembles a critical branching process (see Figure 7). Note here that I conditioned on $\lambda\geq\mu$ and with unconstrained parameters the extinction rate is estimated to be larger than the speciation rate. The difference in the AICc score between the best model under uniform sampling versus diversified sampling was $\Delta_{AICc}=12.812$, showing strong support for diversified sampling.

The inferred extinction rate being as high as the speciation rate ($\lambda\approx\mu$) is surprising. Under a critical branching process, the net-diversification rate $\lambda-\mu=0$ and the expected diversity remains constant. However, the snake clade accumulated as many as 3400 species in 169.7261 million years. The estimated extinction rate might thus be an artifact of the condition on survival of the process. Indeed, if the same inference procedure was repeated without conditioning on survival of the process, the extinction was estimated to be lower than the speciation rate (data not shown).

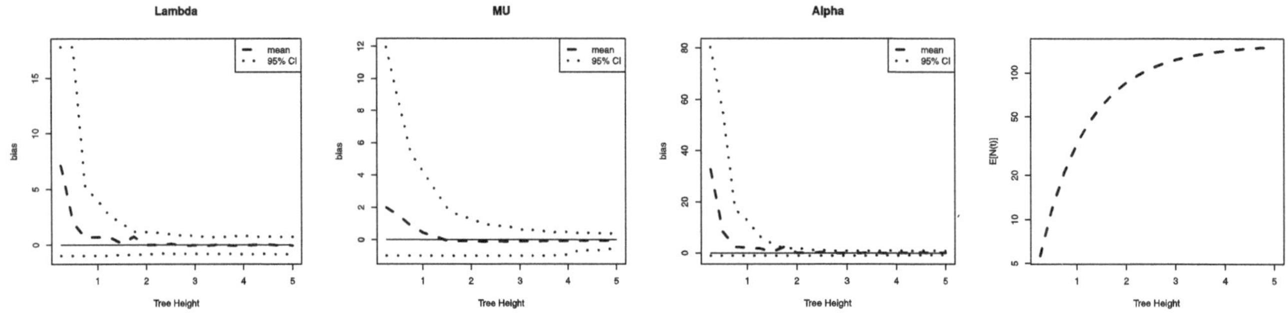

Figure 4. The bias in the maximum likelihood estimates of the speciation and extinction rate. The true parameters were $\lambda=4$, $\mu=1$ and $\alpha=1$. The figure shows that the bias decreases with larger trees (by simulating trees with a larger time T). The expected number of species ($E[N(T)]$) is presented to illustrate the increase in diversity over time.

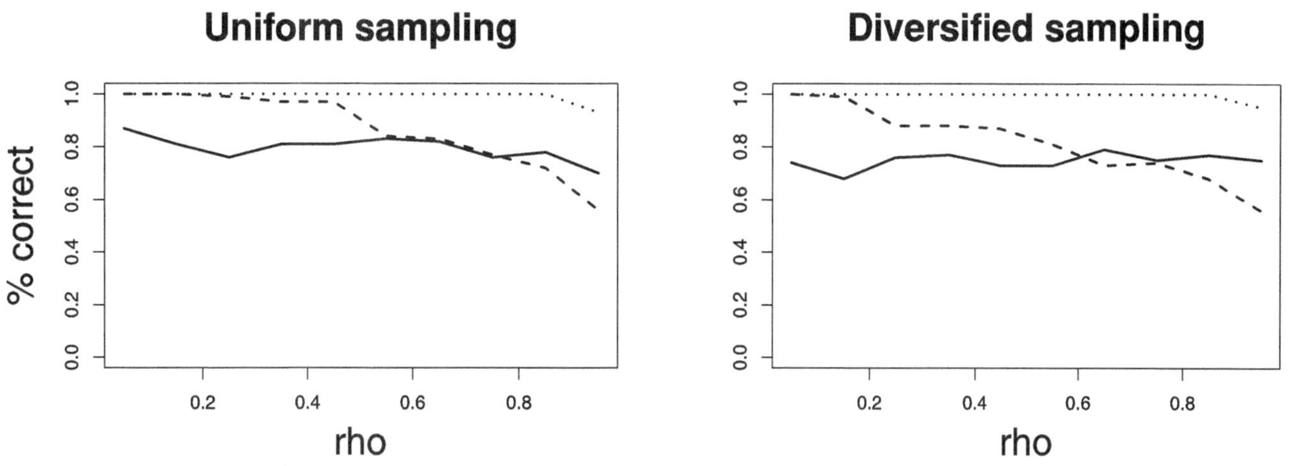

Figure 5. Sensitivity analysis of the success of the sample-size corrected Akaike Information Criterion to select the correct model. Trees were simulated under three different models: constant-rate pure birth (solid line), decreasing rate pure birth (dashed line) and constant-rate birth-death (dotted line). The x-axis shows simulations for different sampling probabilities ρ.

The mammal and the ant phylogenies favored a model with uniform sampling instead. The best model for the mammal phylogeny was the constant-rate birth-death model with rates $\lambda=0.1247$ and $\mu=0.0637$. The best model for the ant phylogeny was a constant-rate pure birth process with $\lambda=0.0567$, although a decreasing rate pure birth process gave a higher likelihood but not significant improvement (difference in likelihood scores: $\Delta_L=1.6837$).

Allowing non-constant diversification rates improved the model fit for the ant phylogeny but not for the mammal and snake phylogenies. The chosen time-dependent rate functions used in this study are likely too simplistic, as Meredith et al. [4] observed first an increase and then a decrease in speciation rates in the mammal phylogeny, a scenario that was not considered here. The derived likelihood equations can be used readily for other time-dependent rate functions. For example, it is possible to use the derived likelihood equations for a birth-death shift process with diversified taxon sampling [7,9].

The parametric bootstrap analyses revealed that most models are not adequate for the observed data and it is very improbable (p-value smaller than 0.01) that the data were generated under these models (see Table S1–S3 in File S1). This results emphasizes that the models are too simplistic and some model assumptions are violated, e.g. a constant rate function. Only the constant-rate pure birth model could not be rejected according to its model fit for the ant phylogeny. Note that the constant-rate birth-death model was also not rejected but the estimated extinction rate was zero. However, the constant-rate pure birth model was not the best model according the AICc and BIC. Thus, the AICc and BIC may favor a model with a worse fit and a model adequacy test, as performed here, should always accompany a model selection procedure.

Possible Model Violations

The mathematical description I used here of diversified taxon sampling is rather crude. It requires that all speciation events before a given time are included and all speciation events after this time are excluded (see Figure 2). A relaxation of this assumption may yield much better estimates. For instance, very old speciation events that define splits of families or other higher-order taxa have a higher probability to be included. Still, the probability of a recent speciation event to be included is unlikely to be zero and some old speciation events may occasionally be missed.

For example, Meredith et al. [4] reconstructed the mammal phylogenetic tree based on 1–3 species per family. Including at least one species per family is clearly an example of diversified taxon sampling, although not as strict as in the mathematical definition used here. The present strict definition of diversified sampling would require the missing species (more than 5200) to have speciated after the last sampled speciation event, which was

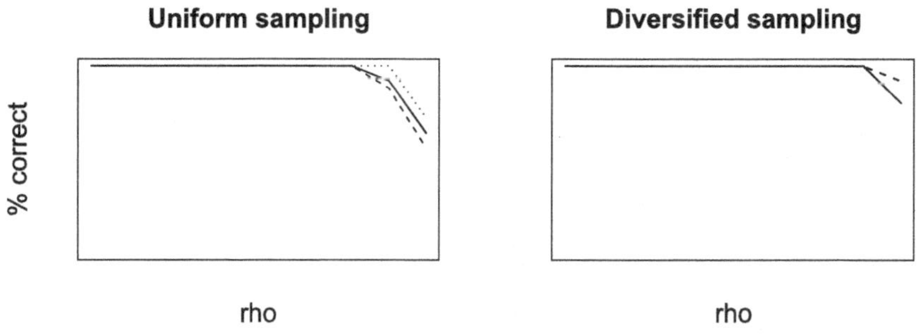

Figure 6. The sensitivity analysis testing whether the sampling strategy can be inferred. Trees were simulated under three different models: constant-rate pure birth (solid line), decreasing rate pure birth (dashed line) and constant-rate birth-death (dotted line). The x-axis shows simulations for different sampling probabilities ρ.

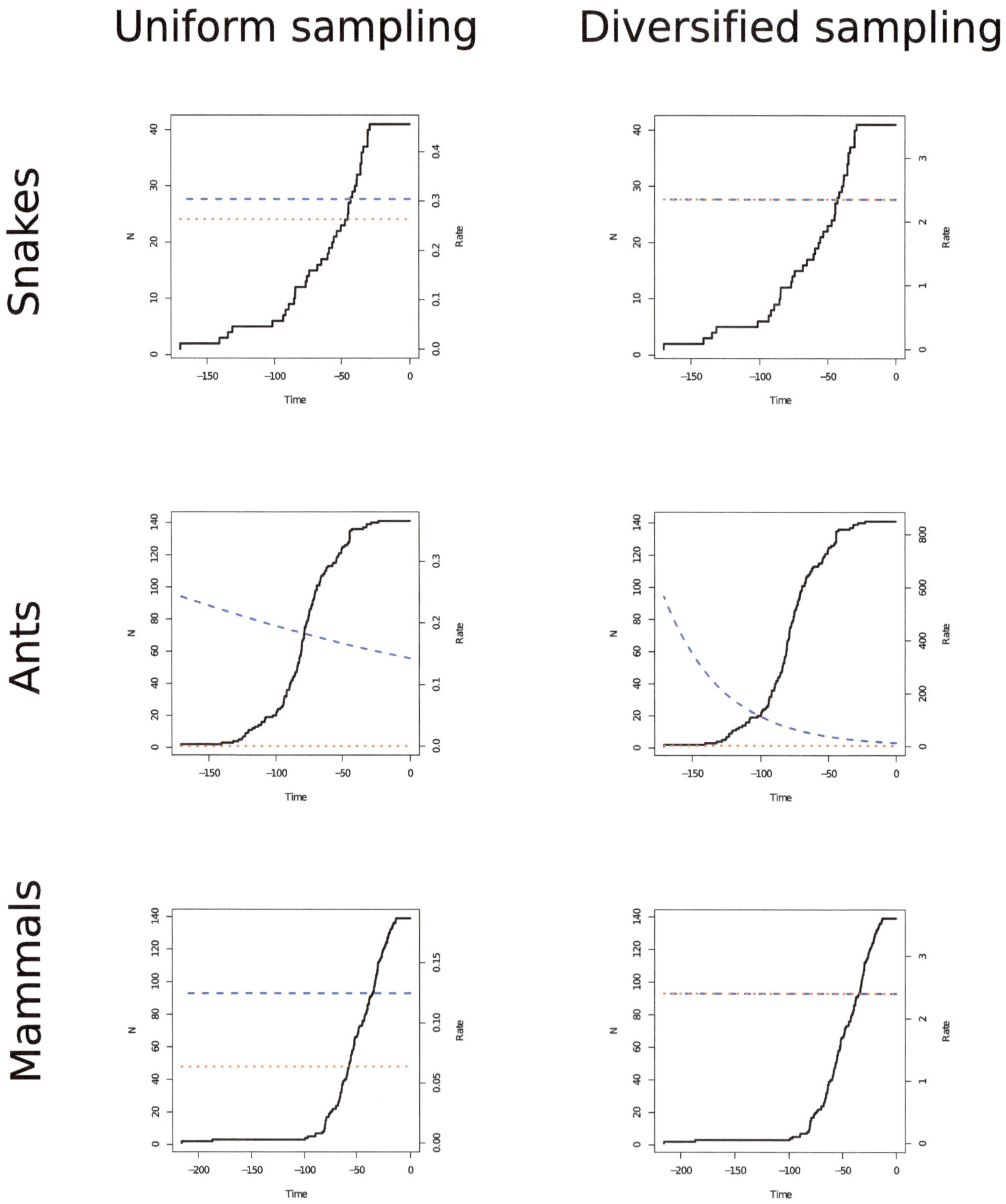

Figure 7. The estimated diversification rates for the snake, ant and mammal dataset. The plots show the change over time in total number of surviving lineages (solid black line), the speciation rate (dashed blue line) and the extinction rate (dotted red line) is given.

6.73 million years ago. Instead, the diversified sampling could be used as a process for higher-order phylogenies, as described by [42]. A suitable cut-off is needed after which all speciation events are discarded and then the diversified sampling can be applied. The advantage of the method in the present paper is that it includes non-constant diversification rates in contrast to the method in [42].

A simple test showed that assuming a broader time interval when the missing speciation events could have occurred improved the likelihood score enormously. I removed the last 20% of the speciation events because these are likely to violate the diversified sampling scheme. Diversified sampling was then preferred over uniform sampling on these pruned phylogenies (data not shown). Hence, diversified sampling is very sensitive to small violation of the sampling procedure. However, this data manipulation is treacherous and may mislead conclusions.

It seems promising to extend diversified sampling by softening the constraint that all missing speciation events must have occurred after the last observed speciation event. A possible extension is to allow speciation events to be sampled with probability ρ_1 before some given time t_ρ and with probability ρ_2 after this time. For example, we could have sample 80% of the speciation event before the last observed speciation event and 0% after the last speciation event. The diversified sampling method here is a special case of this extension when $\rho_1 = 1$, $\rho_2 = 0$ and $t_\rho = t_{n-1}$.

Conclusions

In the present paper I derived the probability density function for a birth-death process with non-constant rates and diversified taxon sampling. The birth-death process with diversified sampling allows biologists to analyze diversification rates of large-scale phylogenies that are sparsely sampled. I analyzed three such phylogenies including ants, mammals and snakes respectively.

The snake phylogeny supported the diversified taxon sampling model whereas the mammal and the ant phylogeny supported uniform taxon sampling. Nevertheless, it was obvious from the description of both the mammal and the ant study that diversified taxon sampling had been used but likely in a weaker sense than in the strict mathematical definition. The models presented in the current paper show improvements over the currently available models by allowing for non-constant rates and diversified sampling. Relaxing the mathematical definition of diversified taxon sampling used here and finding an intermediate between diversified taxon sampling and uniform taxon sampling appears to be the next step in future research.

In a simulation study I showed that the MLE of the diversification rates is biased, but if the trees are large enough the effect can be neglected. The AICc and the BIC can select the correct model and have a low false-negative rate. However, these results only apply under perfect condition when the true model is known and the model assumptions are not violated. Thus, it proved essential to investigate the fit of the models by means of parametric bootstraps. Here none of the models could provide a satisfactory fit when simulated trees were compared with the observed trees. Hence I recommend to simulate trees under the inferred parameters and compare the simulates tree with the observed trees instead of relying simply on the AICc and BIC to choose the best model.

The likelihood functions described in this paper are implemented in the R package *TESS* [27] and are freely available. The simulations of reconstructed trees were also performed with *TESS* and the R scripts for the model selection are available in the Dryad data repository at http://doi.org/10.5061/dryad.rd2s3.

Acknowledgments

I would like to thank Tracy A. Heath, Bastien Boussau, Michael Landis and Fredrik Ronquist for comments on the manuscript and Tom Britton for help with the proof of Equation (6).

Author Contributions

Conceived and designed the experiments: SH. Performed the experiments: SH. Analyzed the data: SH. Contributed reagents/materials/analysis tools: SH. Wrote the paper: SH.

References

1. Cusimano N, Renner S (2010) Slowdowns in diversification rates from real phylogenies may not be real. Systematic biology 59: 458.
2. Höhna S, Stadler T, Ronquist F, Britton T (2011) Inferring speciation and extinction rates under different species sampling schemes. Molecular Biology and Evolution 28: 2577–2589.
3. Cusimano N, Stadler T, Renner SS (2012) A new method for handling missing species in diversification analysis applicable to randomly or nonrandomly sampled phylogenies. Systematic biology 61: 785–792.
4. Meredith R, Janečka J, Gatesy J, Ryder O, Fisher C, et al. (2011) Impacts of the Cretaceous Terrestrial Revolution and KPg extinction on mammal diversification. Science 334: 521–524.
5. Fiz-Palacios O, Schneider H, Heinrichs J, Savolainen V (2011) Diversification of land plants: insights from a family-level phylogenetic analysis. BMC evolutionary biology 11: 341.
6. Nee S, May RM, Harvey PH (1994) The Reconstructed Evolutionary Process. Philosophical Transactions: Biological Sciences 344: 305–311.
7. Rabosky D (2006) Likelihood methods for detecting temporal shifts in diversification rates. Evolution 60: 1152–1164.
8. Morlon H, Parsons T, Plotkin J (2011) Reconciling molecular phylogenies with the fossil record. Proceedings of the National Academy of Sciences 108: 16327–16332.
9. Stadler T (2011) Mammalian phylogeny reveals recent diversification rate shifts. Proceedings of the National Academy of Sciences 108: 6187.
10. Yang Z, Rannala B (1997) Bayesian phylogenetic inference using DNA sequences: a Markov Chain Monte Carlo Method. Molecular Biology and Evolution 14: 717–724.
11. Etienne R, Haegeman B, Stadler T, Aze T, Pearson P, et al. (2012) Diversity-dependence brings molecular phylogenies closer to agreement with the fossil record. Proceedings of the Royal Society B: Biological Sciences 279: 1300–1309.
12. Condamine FL, Rolland J, Morlon H (2013) Macroevolutionary perspectives to environmental change. Ecology letters.
13. Pawitan Y (2001) In all likelihood: statistical modelling and inference using likelihood. Oxford University Press, USA.
14. Rabosky D, Lovette I (2008) Explosive evolutionary radiations: decreasing speciation or increasing extinction through time? Evolution 62: 1866–1875.
15. Alfaro M, Santini F, Brock C, Alamillo H, Dornburg A, et al. (2009) Nine exceptional radiations plus high turnover explain species diversity in jawed vertebrates. Proceedings of the National Academy of Sciences 106: 13410–13414.
16. Morlon H, Potts M, Plotkin J (2010) Inferring the dynamics of diversification: a coalescent approach. PLoS Biology 8: e1000493.
17. Rannala B, Yang Z (1996) Probability distribution of molecular evolutionary trees: A new method of phylogenetic inference. Journal of Molecular Evolution 43: 304–311.
18. Lartillot N, Lepage T, Blanquart S (2009) Phylobayes 3: a bayesian software package for phylogenetic reconstruction and molecular dating. Bioinformatics 25: 2286.
19. Drummond A, Suchard M, Xie D, Rambaut A (2012) Bayesian phylogenetics with beauti and the beast 1.7. Molecular Biology and Evolution 29: 1969–1973.
20. Ronquist F, Teslenko M, van der Mark P, Ayres DL, Darling A, et al. (2012) Mrbayes 3.2: efficient bayesian phylogenetic inference and model choice across a large model space. Systematic Biology 61: 539–542.
21. Lepage T, Bryant D, Philippe H, Lartillot N (2007) A general comparison of relaxed molecular clock models. Molecular biology and evolution 24: 2669.
22. Heath TA (2012) A hierarchical bayesian model for calibrating estimates of species divergence times. Systematic biology 61: 793–809.
23. Heath T, Holder M, Huelsenbeck J (2012) A dirichlet process prior for estimating lineage-specific substitution rates. Molecular Biology and Evolution 29: 939–955.
24. Moreau CS, Bell CD, Vila R, Archibald SB, Pierce NE (2006) Phylogeny of the ants: diversification in the age of angiosperms. Science 312: 101–104.

25. Pyron RA, Burbrink FT (2012) Extinction, ecological opportunity, and the origins of global snake diversity. Evolution 66: 163–178.
26. Kendall DG (1948) On the generalized "birth-and-death" process. The Annals of Mathematical Statistics 19: 1–15.
27. Höhna S (2013) Fast simulation of reconstructed phylogenies under global time-dependent birth–death processes. Bioinformatics 29: 1367–1374.
28. Thompson E (1975) Human evolutionary trees. Cambridge University Press Cambridge.
29. Yule G (1925) A mathematical theory of evolution, based on the conclusions of dr. jc willis, frs. Philosophical Transactions of the Royal Society of London Series B, Containing Papers of a Biological Character 213: 21–87.
30. Akaike H (1974) A new look at the statistical model identification. Automatic Control, IEEE Transactions on 19: 716–723.
31. Schwarz G (1978) Estimating the dimension of a model. The annals of statistics 6: 461–464.
32. Burnham KP, Anderson DR (2004) Multimodel inference understanding aic and bic in model selection. Sociological methods & research 33: 261–304.
33. Sullivan J, Joyce P (2005) Model selection in phylogenetics. Annual Review of Ecology, Evolution, and Systematics 36: 445–466.
34. Lambert A (2010) The contour of splitting trees is a lévy process. The Annals of Probability 38: 348–395.
35. Brusatte SL, Benton MJ, Desojo JB, Langer MC (2010) The higher-level phylogeny of archosauria (tetrapoda: Diapsida). Journal of Systematic Palaeontology 8: 3–47.
36. Davis R, Baldauf S, Mayhew P (2010) The origins of species richness in the hymenoptera: insights from a family-level supertree. BMC Evolutionary Biology 10: 109.
37. Ronquist F, Klopfstein S, Vilhelmsen L, Schulmeister S, Murray DL, et al. (2012) A total-evidence approach to dating with fossils, applied to the early radiation of the hymenoptera. Systematic Biology 61: 973–999.
38. Bollback J (2002) Bayesian model adequacy and choice in phylogenetics. Molecular Biology and Evolution 19: 1171–1180.
39. Huelsenbeck JP, Hillis DM, Jones R (1996) Parametric bootstrapping in molecular phylogenetics: applications and performance. Molecular zoology: advances, strategies, and protocols 3: 19–45.
40. Pybus O, Harvey P (2000) Testing macro–evolutionary models using incomplete molecular phylogenies. Proceedings of the Royal Society of London Series B: Biological Sciences 267: 2267–2272.
41. Pennell M, Sarver B, Harmon L (2012) Trees of Unusual Size: Biased Inference of Early Bursts from Large Molecular Phylogenies. PloS one 7: e43348.
42. Stadler T, Bokma F (2013) Estimating speciation and extinction rates for phylogenies of higher taxa. Systematic Biology 62: 220–230.

20

Assessing Fishing and Marine Biodiversity Changes Using Fishers' Perceptions: The Spanish Mediterranean and Gulf of Cadiz Case Study

Marta Coll[1,2]*, Marta Carreras[3], Cristina Ciércoles[4], Maria-José Cornax[3], Giulia Gorelli[1], Elvira Morote[5], Raquel Sáez[1]

1 Institut de Ciències del Mar (ICM-CSIC). Barcelona, Spain, 2 UMR EME 212 Centre de Recherche Halieutique Méditerranéenne et Tropicale. IRD - IFREMER & Université Montpellier II. Sète Cedex, France, 3 OCEANA, Madrid, Spain, 4 Investigación, Planificación y Desarrollo S.A., Madrid, Spain, 5 Universidad de Almería (UAL). Almeria, Spain

Abstract

Background: The expansion of fishing activities has intensively transformed marine ecosystems worldwide. However, available time series do not frequently cover historical periods.

Methodology: Fishers' perceptions were used to complement data and characterise changes in fishing activity and exploited ecosystems in the Spanish Mediterranean Sea and Gulf of Cadiz. Fishers' interviews were conducted in 27 fishing harbours of the area, and included 64 fishers from ages between 20 to >70 years old to capture the experiences and memories of various generations. Results are discussed in comparison with available independent information using stock assessments and international convention lists.

Principal Findings: According to fishers, fishing activity substantially evolved in the area with time, expanding towards deeper grounds and towards areas more distant from the coast. The maximum amount of catch ever caught and the weight of the largest species ever captured inversely declined with time. Fishers (70%) cited specific fishing grounds where depletion occurred. They documented ecological changes of marine biodiversity during the last half of the century: 94% reported the decline of commercially important fish and invertebrates and 61% listed species that could have been extirpated, with frequent mentions to cartilaginous fish. Declines and extirpations were in line with available quantitative evaluations from stock assessments and international conventions, and were likely linked to fishing impacts. Conversely, half of interviewed fishers claimed that several species had proliferated, such as cephalopods, jellyfish, and small-sized fish. These changes were likely related to trophic cascades due to fishing and due to climate change effects. The species composition of depletions, local extinctions and proliferations showed differences by region suggesting that regional dynamics are important when analysing biodiversity changes.

Conclusions/Significance: Using fishers' perceptions, fishing and ecological changes in the study area were documented. The recovery of local ecological knowledge provides valuable information complementing quantitative monitoring and evaluation surveys.

Editor: Konstantinos I. Stergiou, Aristotle University of Thessaloniki, Greece

Funding: MC was supported by the Marie Curie EC CIG grant to BIOWEB and the Spanish Research Program Ramon y Cajal. GG was funded through the EC program Science in Society EC. FP7 (ref # 266544-GAP2). The funders had no role in study design, data collection and analysis, decision to publish, or preparation of the manuscript.

Competing Interests: The authors have declared that no competing interests exist.

* E-mail: mcoll@icm.csic.es

Introduction

From coastal areas to the open seas, human activities impact marine ecosystems [1,2]. In the past, human impacts on marine ecosystems were mainly due to exploitation of marine resources and fishing. These activities diversified with time and currently include other impacts such as habitat degradation, pollution and eutrophication, climate change, and the invasion of non-indigenous species [3–6]. These impacts have an effect on the biodiversity and abundance of marine resources [7–9], and on marine ecosystem structure and functioning [10,11].

The Mediterranean Sea is one of the most impacted marine ecosystems worldwide [5,12–14]. Exploitation and habitat loss are the main human activities shaping its biodiversity changes [12]. Fishing impacts are notable and several commercial species are highly exploited or overexploited [15,16]. Non-commercial species have also shown important declines [16,17].

Additionally, the impact of widely distributed factors such as climate change or the invasion of non-indigenous species is steadily increasing [18–20]. Human impacts are distributed in a heterogenic way along the Mediterranean Sea and many areas are subjected to multiple impacts simultaneously [13]. Some of the most impacted areas are located in the western Mediterranean Sea region [13].

The information on past ecosystems and changes in marine biodiversity is quickly lost because of perception change (this is known as "shifting baselines" syndrome) [21–23]. This is even more evident in ecosystems where changes have occurred centuries ago, such as the Mediterranean Sea [24]. Therefore, several studies have aimed at recovering knowledge from past Mediterranean ecosystems to reconstruct ecosystem baselines [14,25]. The loss of information from the past is also due to the availability of long time series of data is frequently scarce. Data only started to be regularly available from the early 1990s [e.g. 26], after important changes in marine ecosystems had already occurred [14,17,25].

To overcome some of the limitations in data availability and to recover historical information, scientists have documented changes in marine ecosystems using fishers' perceptions (also called Local Ecological Knowledge, LEK, or Traditional Ecological Knowledge, TEK) [27–31]. In the Mediterranean Sea, applications to recover this knowledge have yield interesting results, such as information on depletion of large predators [32] and on changes in marine fish diversity due to meridionalization [33].

In this study we used fishers' perceptions to complement available information on (i) fishing changes with time, and (ii) ecological changes of marine ecosystems in the most western Mediterranean area: the Spanish Mediterranean Sea. We extended our study area to include the Gulf of Cadiz (Atlantic adjacent waters to the Mediterranean Sea) due to the fact that both fisheries and ecosystems share similar features [34]. In addition, and to contrast the information obtained with traditional knowledge, we discussed our results in comparison with available independent quantitative information using stock assessments and international convention lists. This comparison allowed us to evaluate the utility of this traditional knowledge with the aim to reconcile it with data formally collected by fisheries scientists.

Materials and Methods

Ethics statement

The study followed all the guidelines of the ethics committee of CSIC and available legislation (*Comité de Ética CSIC*, http://www.csic.es/web/guest/normativa). The interviews included a first page with the objectives of the research and contact information, which was read out loud to the participants before starting the interview. All interviews took place within the country and participants provided verbal consent to participate in the study by agreeing to participate in the interview. Participants could decide either to have their names written down in the first page of the interview or to leave it anonymous.

Study area

Our study area includes the Spanish Mediterranean (Figure 1), and it is extended to the Gulf of Cadiz, in the immediate Atlantic waters outside the Mediterranean Sea. This region is located in the Balearic FAO area 37.1.1 and the South-East Atlantic area 27.IXa. The area is limited by the Gulf of Lions in the north, Moroccan waters in the south and the Sardinia FAO area 37.1.3, including the Tyrrhenian Sea and adjacent waters to the east.

The continental shelf of the Spanish Mediterranean and Gulf of Cadiz regions is generally narrow and the 50 m depth is usually found approximately three nautical miles from shore (Figure 1). Fishing activities are mainly coastal or littoral, and they last approximately 12 hours a day [34]. The main fishing fleets operating in the area are bottom trawling (~24% of the fleet), purse seine (~9%), bottom and surface long liners (~3%), and a diverse artisanal fleet, including gillnets (~64%) [34]. The area contains 95 fishing harbours, more than 90 first-source fishing markets (where catches that are unloaded in the harbour are commercialized), and several fishers brotherhoods [fishers labour organizations unique to Spain that date back to early Medieval times, 35].

In 2008–2009, the total number of fishing vessels in the study region was 4 140, with total declared power of 453 593 Kw and a mean total length of 12 meters [36]. In late 2000s, more than 13 000 people directly worked as fishers in the area [36]. The area is inhabited by approximately 24 million people.

Major species in the official landings of the area include sardine (*Sardina pilchardus*), anchovy (*Engraulis encrasicolus*), and other small and medium-sized pelagic fish such as round sardinelle (*Sardinella aurita*), horse mackerel (*Trachurus* spp.) and mackerel (*Scomber* spp.), and demersal species, such as common hake (*Merluccius merluccius*), red mullets (*Mullus* spp.), anglerfish (*Lophius* spp.) and blue whiting (*Micromesistius poutassou*). Invertebrate catches are also economically important such as, for example, that of red shrimp (*Aristeus antennatus*), European spiny lobster (*Palinurus elephas*), rose shrimp (*Parapenaeus longirostris*), Norway lobster (*Nephrops norvegicus*) and various cephalopod species [37,38]. The Mediterranean Sea also yields important catches of tunas and associated species, especially of Atlantic Bluefin tuna (*Thunnus thynnus*) and swordfish (*Xiphias gladius*) [34].

Fishers' interviews

This study was conducted within a larger project aiming at collecting available information to better estimate total fisheries removals from the area. Individual interviews with selected fishers were developed and executed in order to get information on different aspects of fishing activities and mechanisms to commercialize landed catches, and including recreational and artisanal fishing, discarding and illegal fishing, and black market practices [34,39]. In addition, we retrieved information on fishers' perceptions regarding (i) fishing changes with time, and (ii) ecological changes of marine ecosystems.

Fishers were interviewed from 27 harbors distributed in four different regions (Figure 1): the Gulf of Cadiz (in the Atlantic waters adjacent to the Mediterranean Sea) (with n = 14 fishers interviewed), the Mediterranean Andalusia region (n = 12), Tramontana region (in the northern Spanish Mediterranean area) (n = 20) and the Balearic Islands (n = 18). Fishers were all professional skippers and crew of the fishing fleet. Interviewees were chosen due to they were deemed reliable to provide objective information since previous relationships had been directly or indirectly established with them. This was deemed important because the aim was to minimize uninformative and low quality answers.

The interviews included a series of open or fixed questions, some of them using multiple-choice and with quantitative or qualitative (yes-no) answers. First a series of questions targeted to retrieve specific information from the fisher (such as age, year of starting fishing, fishing gear used, species targeted) were posed. Secondly, fishers were asked a series of questions regarding fishing activity and the following questions were posed: (1–2) What was the depth/distance from the coast you used to fish when you started fishing, and what is the depth/distance you fish at present; (3) What was your highest catch in a day of fishing and when was it (year)? (4) What was the weight of the largest species that you ever caught, what was it (species), and when was it (year)? (5) Do you know of any locations where you fish that are depleted? and (6–7) What was the weight of the largest common hake ever caught and

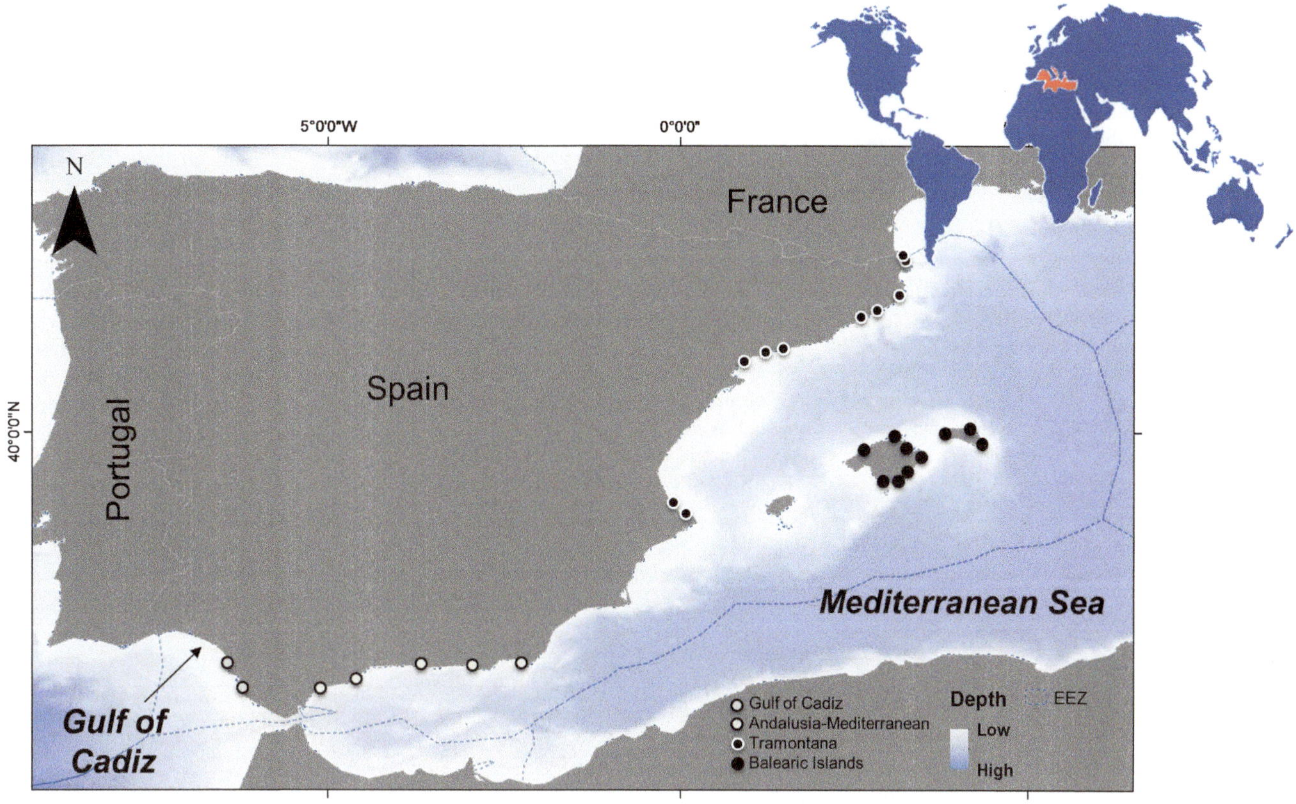

Figure 1. Study area in the Spanish Mediterranean Sea and Gulf of Cadiz. Harbours where interviews were conducted are indicated.

when was it (year)? Finally, and regarding ecosystem changes, fishers were asked: (8–9) Do you know of any species in the area that had declined/disappeared during the time you have been fishing? And (10) Do you know of any species in the area that had increased during the time you have been fishing? Faunistic guides or species landed in the harbor at the same time of the interviews were used to confirm the identification of species mentioned by fishers during the interviews.

To contrast the information obtained with traditional knowledge, we compared results with available independent quantitative information using stock assessments and international convention lists from: (1) The International Union for Conservation of Nature and Natural Resources (IUCN) [16], (2) The Bern Convention (or Convention on the Conservation of European Wildlife and Natural Habitats, 1979) [40], (3) The Barcelona Convention (or Convention for the Protection of the Marine Environment and the Coastal Region of the Mediterranean, 1976, amended in 1995, and the Protocol Concerning specially Protected Areas and Biological Diversity in the Mediterranean, SAP-Bio, 1995) [41], (4) the Bonn Convention (or Convention on the Conservation of Migratory Species of Wild Animals 1983) [42], (5) The CITES list (or Convention on International Trade in Endangered Species of Wild Fauna and Flora, 1975) [43], and (6) Assessments from the General Fisheries Commission for the Mediterranean Sea - Scientific Advisory Committee [44].

Statistical analysis

We used three non-parametric tests to analyze our results with the aim to avoid problems with the asymmetrical distribution of the data, to maximize the robustness of our results and to minimize the probability to commit type I errors. The non-parametric Mann–Whitney U test was used to compare the data obtained regarding fishing at different depths and distances. This analysis was developed with all data from all fishers regardless their age, and by separating the age in two groups: younger than 40 years or older. The non-parametric Spearman rank's correlation was used to detect correlation between quantitative data, such as weight of catch or caught species and year. Information about the year was used as year reporting the specific quantity (reported year) and year that fishers started fishing. This was due to the fact that in some cases fishers did not remember the reported year, but they did remember when they started fishing.

To investigate differences in the species composition list of depletions, local extinctions and proliferations between regions, the non-parametric multivariate permutational analysis of variance (PERMANOVA, in PRIMER with PERMANOVA+ v. 6, PRIMER-E Ltd., Plymouth, UK) on the Euclidean distance matrix was employed. PERMANOVA calculates a pseudo-F statistic that is directly analogous to the traditional F-statistic for multifactorial univariate ANOVA models, but uses permutation procedures (here 9999 permutations) to obtain p-values for each term in the model [45]. We performed a 1-way analysis with the region factor and the list of depletions, local extinctions and proliferations, separately. We considered that each interview from the same location was an independent sample and the location factor was divided in four levels: the Andalusia-GoC (Gulf of Cadiz in the Atlantic region), the Andalusia-Mediterranean (southern Mediterranean region), the Tramontana area (northern Spanish coast) and the Balearic Islands (Figure 1). The Similarity Percentages Procedure (SIMPER) [46] was used when results from PERMANOVA were significant to identify the species that were most important in each region regarding depletions, local extinctions and proliferations.

Results

Interviewed fishers

During 2011 and 2012, a total of 64 fishers were interviewed. Fishers' ages were from 20 to >70 years old (Figure 2a). Fishers started fishing from 1939 to 2005, thus covering more than 70 years of observations as a result of the experience of the interviewees (Figure 2a).

Most of interviewed fishers had been involved in different fishing segments during their lives and most of them were fishing in the trawling fleet, but had also participated in other fishing activities (Figure 2b). Fishers were targeting a variety of highly commercial species such as red mullets (52%), common hake (45%), anglerfish (42%), red shrimp (42%), European anchovy (31%) and European sardine (30%). Overall, 37 fish and invertebrate species were listed as targeted (Table S1). This highlights the multispecificity of the fishing activity in the area (Figure 2b).

Fishing activity

a) **Fishing depth and distance**: Results showed a clear and significant difference between the depth and distance that fishers used to fish when they started fishing, compared to where they fish at present (Figures 3a and b). Thus, both the present depth and distance from the coast were significantly larger now than when fishers started fishing (p-value $=0.005$ and p-value $=0.023$, respectively for depth and distance). The range of values at present was wider than in the past for both depth and distance.

The spatial representation of these results illustrates how the fishing activity has potentially expanded from coastal and shelf areas to deeper areas that are more distant from the coast (Figure 4). While in the past the mean value of fishing was 100 m depth, at present fishers are working at mean depths of 200 m. The range values also expanded substantially, from 1–600 m in the past to 2–1000 m at present (minimum and maximum values reported, respectively). Similarly, the mean distance reported by fishers in the past was 19 km, while this value increased to 36 km

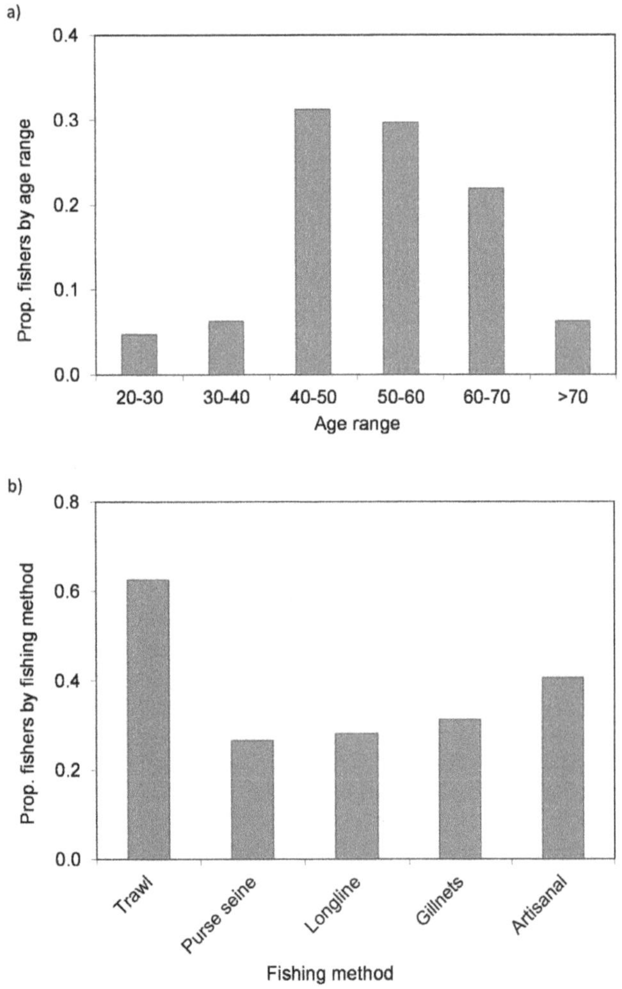

Figure 2. Proportion of a) fishers interviewed by age range (n = 64), and b) fishing methods used by fishers. In Figure 2b fishers can be represented more than once due to results integrate their knowledge through their entire working life, thus the total proportions do not sum to 1.

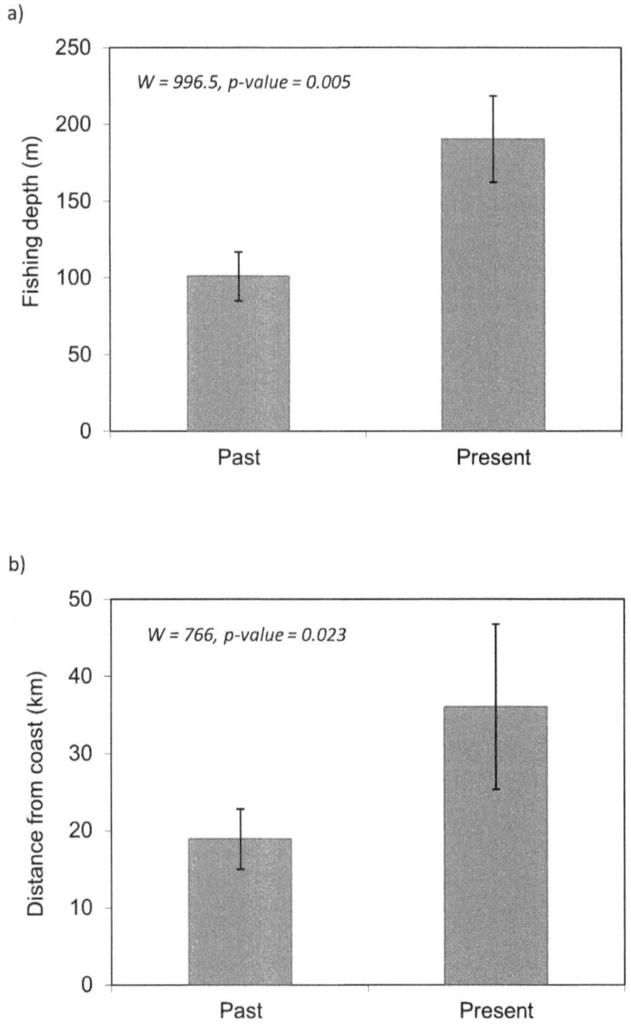

Figure 3. Mean and standard deviation of a) depth (m) and b) distance from the coast (km) between past and present fishing activities.

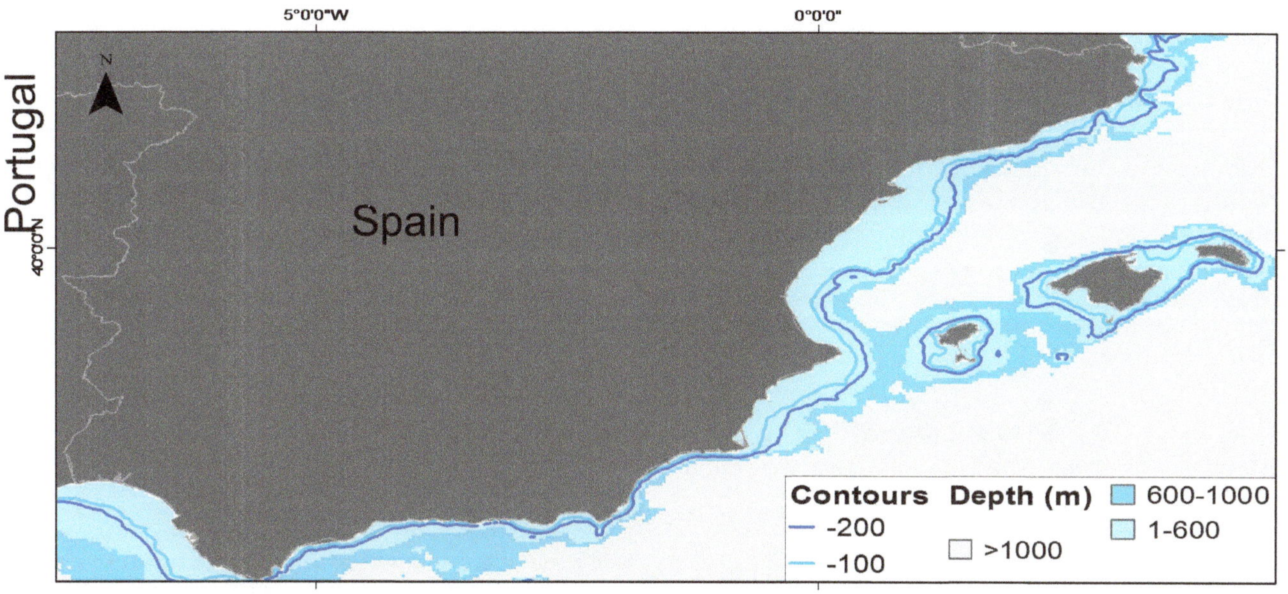

Figure 4. Spatial representation of mean depth and range values between past (mean = 100, range = 1–600 m) and present (mean = 200, range = 1–1000 m) fishing activities.

in the present. The range of distances also shows an expansion of the fishery, from 1–111 km to 1–370 km in the present.

Interestingly, when separating results by the age of fishers in two groups (younger or older than 40 years), significant difference between the depth and distance that fishers used to fish in the past compared to where they fish at present was significant only for fishers with older ages (Figures 5a and b).

b) **Highest catch in a day of fishing and largest species ever caught:** The fishers were asked to estimate the highest catch in metric tonnage (t) in a day of fishing and when was it (year). Results showed a significant decline (Spearman $r = -0.37$, p-value $= 0.01$) with time on the largest amount caught in a day of fishing (best catch) analysed against the year that fishers started fishing (Figure 6a). Since fishers frequently mentioned the catch of small pelagic fish and squids as the total amount of boxes being landed, these boxes had to be converted into total amount of tonnes with approximations given by fishers. When those values reported in boxes of small pelagic fish and squids were removed from the analysis the declining pattern was even clearer (Spearman $r = -0.43$, p-value <0.00) (Figure 6b). This may be due to the fact that small pelagic fish and squids depend more strongly than demersal organisms on the effects of environmental fluctuations. Results were similar (Spearman $r = -0.51$, p-value < 0.00) when these data was compared against the year when the highest catch occurred (Figure 6c). Both results showed that while from the 1950s to 1980s the best catch could be as high as 10 tonnes per day, in the 1990s and 2000s it was always below 5 tonnes.

The fishers were also asked to estimate the weight (t) of the largest species ever caught, what was it (species), and when was it (year). Results when looking at the largest species ever caught against the year of start fishing showed no clear results (Figure 7a). However, when removing individuals of whale shark *Rhincodon typus* and basking shark *Cetorhinus maximus* (both large filter feeders) from the analysis, a negative trend emerged (Spearman $r = -0.28$,

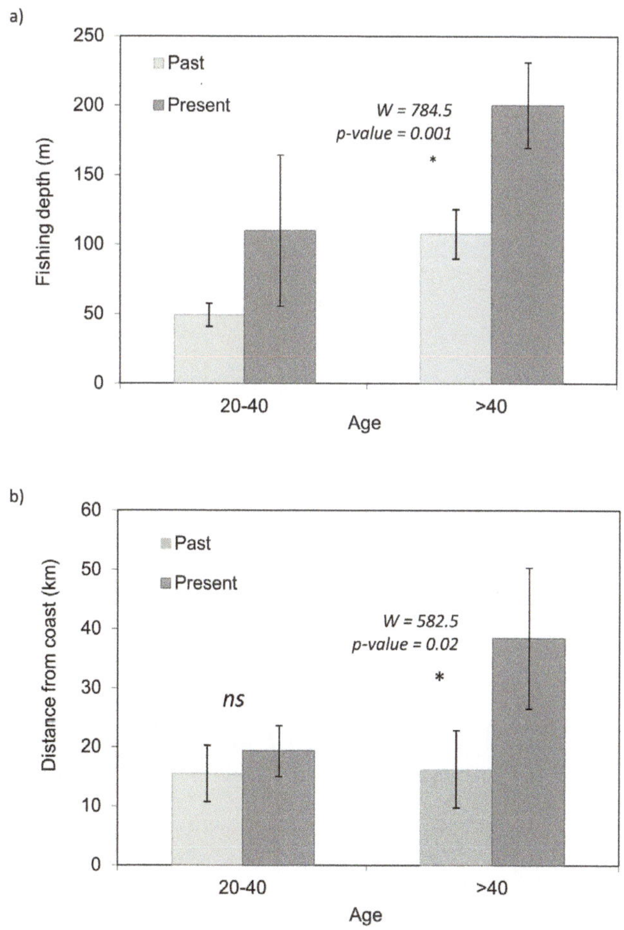

Figure 5. Mean and standard deviation of a) depth (m) and d) distance from the coast (km) between past and present fishing activities by age group (fishers younger or older than 40 years old).

p-value = 0.05) (Figure 7b). Filter feeders may have benefited in the past from larger amounts of prey in the ecosystem due to trophic cascades or environmental fluctuations and thus this second analysis wanted to remove this potential effect. When analyzing these results by decade to aggregate the data, results illustrate that while fishermen starting fishing from the 1940s to the 1960s had caught species of half a ton, the largest species ever caught by young fishers in the area is of less than 200 kg. This result illustrates that the biggest species ever caught seemed to be smaller in recent years.

The composition of the species that were reported to be the largest ever caught changed with time (Figure 8a). Shark species were the most cited until the 1980s. Fish species increased from the 1940s until the 1980s, but declined afterwards. Marine mammals were only mentioned in the 1960s and invertebrate species in the 1980s. In addition, the mean size composition of the largest fish and sharks ever caught [considering the mean sizes give in 47] mentioned by fishers also changed to smaller sizes with time, from larger sharks (such as the common thresher shark *Alopias vulpinus* and hammerhead sharks *Sphyrna* sp.) and fish (bluefin tuna) to smaller sharks (such as the bluntnose sixgill shark *Hexanchus griseus*

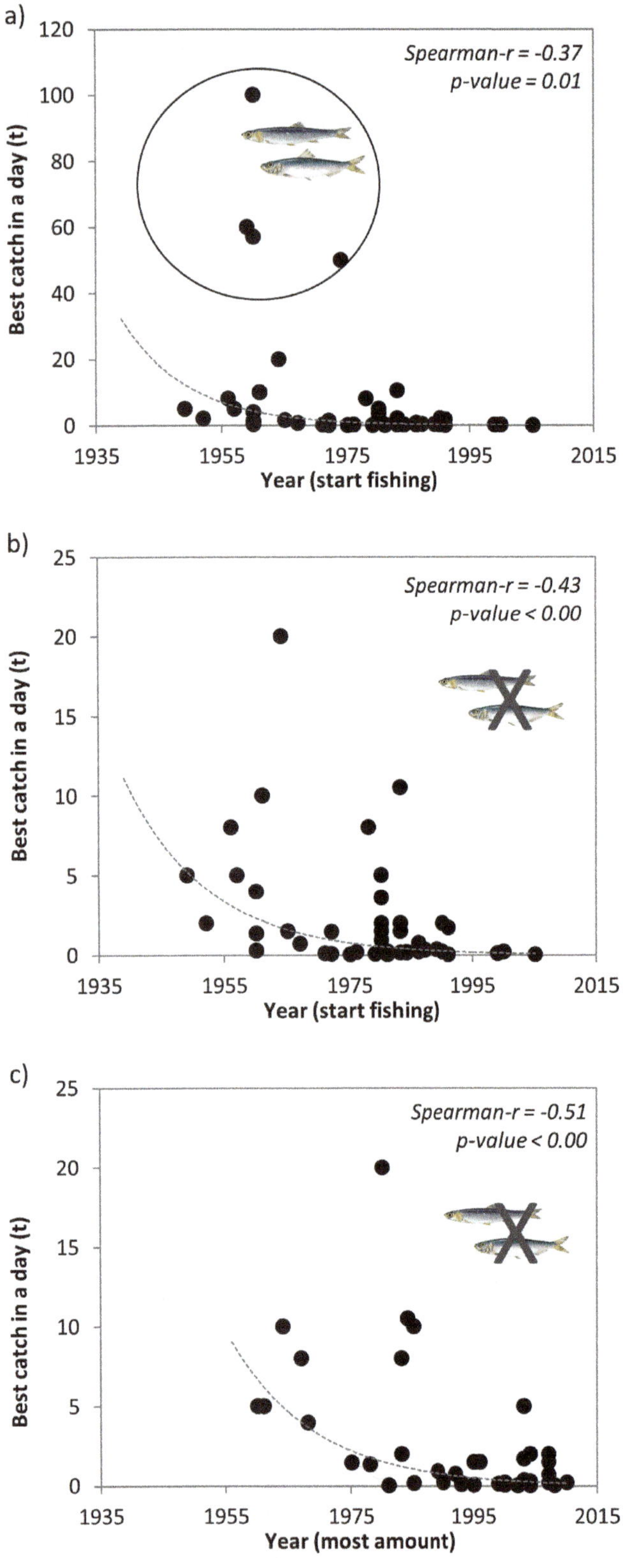

Figure 6. Assessment of the a) best catch in a day of fishing (t) and year of starting fishing (n = 47), b) best catch in a day of fishing (t) and year of starting fishing, excluding small pelagic fish and squids (n = 43), and c) best catch in a day of fishing (t) (n = 43) and year of the best catch. (Species drawings: www.gencat.cat).

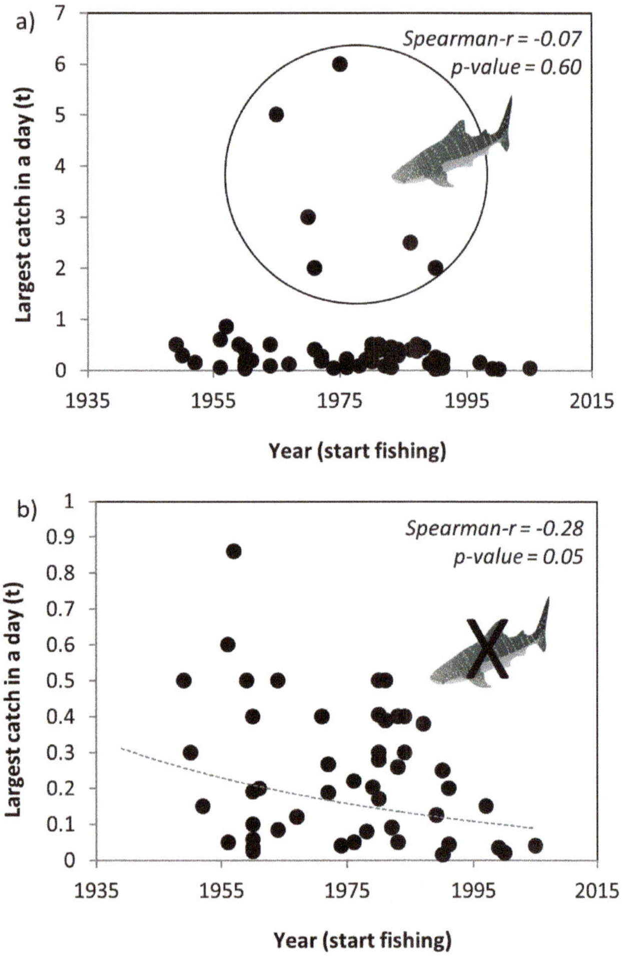

Figure 7. Assessment of the a) largest caught species in a day of fishing (t) and year of starting fishing (n = 58), and b) largest caught species in a day of fishing (t) and year of starting fishing, excluding whale and basking sharks (n = 56). (Species drawings: http://ian.umces.edu/symbols).

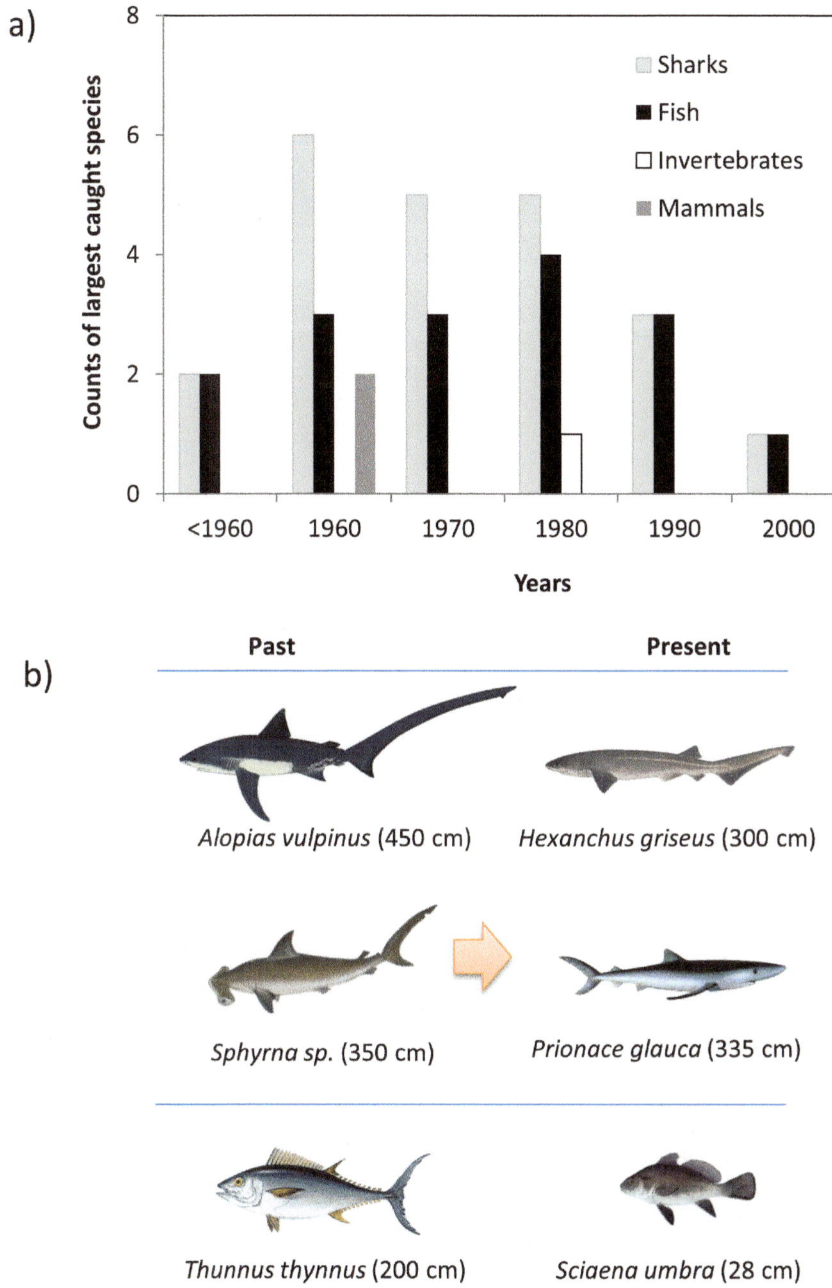

Figure 8. Composition of the largest caught species listed by fishers a) by decade and main species group, and b) main cartilaginous and bony fish species reported during the first and last decades and their mean length (species drawings: http://ian.umces.edu/symbols & www.gencat.cat).

and the blue shark *Prionace glauca*), and smaller fish (such as brown meagre *Sciaena umbra*) (Figure 8b).

Common hake is highly targeted in the study areas by trawlers but also by long liners and artisanal fisheries. Results on the weight of the largest common hake ever caught (kg) plotted against the year of starting fishing showed no clear trend (Figure 9a). However, a negative relationship emerged (Spearman-r = −0.62, p-value <0.00) when plotting the data with the year of the largest hake ever caught, despite this dataset was shorter because several fishermen did not remember the year they caught their largest hake.

c) **Expansion, depletion and shifting baselines:** Overall, the majority of fishers knew several sites that had been depleted (Table 1). Of those knowing depleted sites, 34% of them highlighted that "all" sites where they fished were depleted and 5% mentioned that "many" sites had been depleted. Another 31% listed specific locations in their fishing grounds that had been depleted.

When comparing the year of the largest catch (best catch) with the year of the largest caught species (largest catch) reported by fishers, a positive and significant relationship (Spearman-r = 0.64, p-value <0.00) highlighted that years when fishers captured the

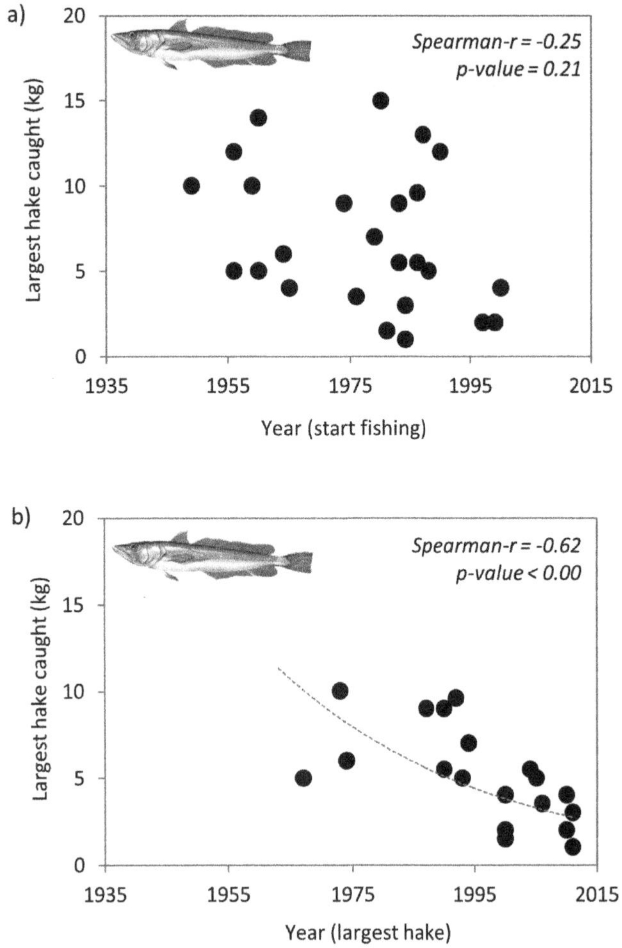

Figure 9. Largest common hake individuals caught (kg) by a) years of starting fishing (n = 30) and b) year of the largest caught individual (n = 25). (Species drawings: www.gencat.cat).

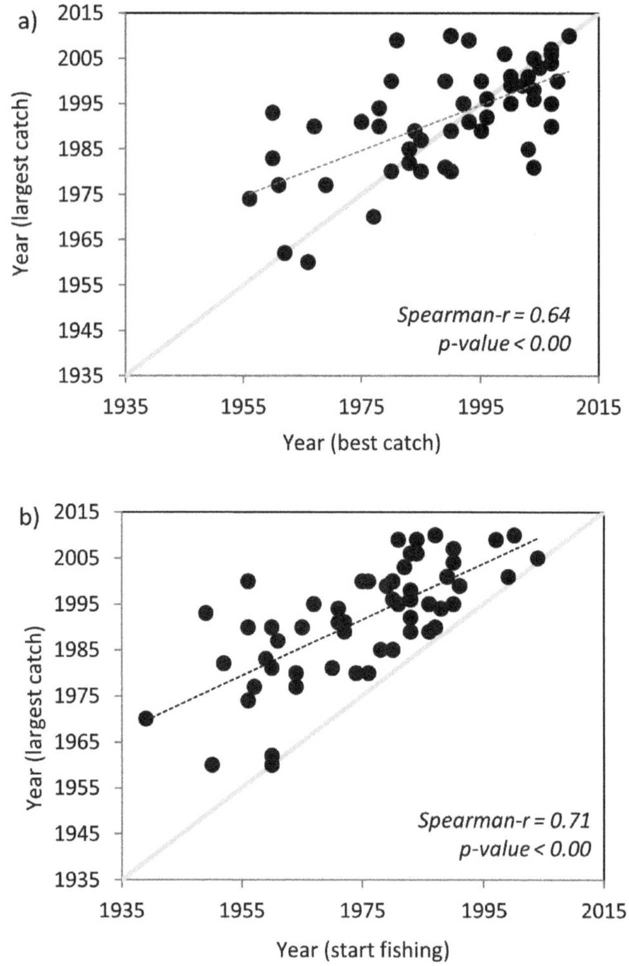

Figure 10. Assessment of the a) year of the best catch and of the largest caught species, and b) year at starting fishing and the largest caught species. Diagonal grey line represents the x = y function.

largest catch tended to coincide with years of the best catch (Figure 10a). However, when the best catch was made before the 1990s, the largest catch was later in fishers' career, while after the 1990s fishers first recorded a largest catch and then a best catch.

When comparing the years of start fishing with the year of the largest catch a positive and also significant relationship (Spearman-r = 0.71, p-value <0.00) was observed (Figure 10b). This relationship approached the relationship that one would expect under the hypothesis that there is a shifting in the baseline: the year that fishers start fishing is the best year, and afterwards the largest catch declines (a similar and also significant relationship was observed between years of start fishing and best catch, not shown here). These patterns may highlight an expansion of the fishery, followed by depletion and a shift in the ecological baseline of fishers.

Table 1. Summary of answers from fishers' interviews.

Sites	**Yes**	**No**	**NK-NA**
Depletions	70%*	2%	28%
Species	**Yes**	**No**	**NK-NA**
Depletions	94%	2%	4%
Extinctions	61%	27%	12%
Proliferations	53%	36%	11%

*"All": 34%, "Many": 5%, Specific sites: 31%. NK-NA: not known, no answer.

Ecosystem changes

a) **Depleted species:** Ninety-four percent of the fishers listed one or more species or groups of species as depleted (Table 1). In total, 76 taxa were listed (Table S2). Of those most frequently listed, the most common were commercial species such as common hake, sardine, anchovy and red mullets (Table 2). Other commercial species followed, such as mackerel, anglerfish, common seabream *Pagrus pagrus* and flatfish *Solea solea*. Fishers also listed invertebrate species, such as the spinous spider crab *Maja squinado* and the spiny lobster, and various species of sharks such as the nursehound *Scyliorhinus stellaris*, the smooth-hound *Mustelus mustelus* and the longnose spurdog *Squalus blainville* (Table 2).

PERMANOVA analysis highlighted significant differences by region of species identified as depleted (Table 3). SIMPER results showed that while the category "all", "shrimps", sardine, common seabream, and blue whiting were the main species identified as depleted in the Gulf of Cadiz, common hake and anchovy were important in the Mediterranean Andalusian side and in the Tramontana region (Table 4). The Balearic Islands showed different species such as two species of sharks (smooth-hound and longnose spurdog), garfish *Belone belone*, and spiny lobster as most cited by fishers to be depleted.

b) **Local extinctions**: Sixty-one percent of the fishers listed one or more species or groups of species that had locally disappeared (Table 1). In total, 33 taxa were listed (Table S2). Of those most frequently listed, one was an invertebrate, the spinous spider crab, five were sharks (the angelsharks *Squatina squatina* and *S. oculata*, the common guitarfish *Rhinobatos rhinobatos*, the nursehound and the smooth-hound *Mustelus asterias*), and three were fish species (seahorses *Hippocampus* sp., the blackspot seabream *Pagellus bogaraveo*, and the brown meager *Sciaena umbra*). Marine mammals and turtles were listed as generic groups (Table 2).

PERMANOVA analysis highlighted significant differences by region of species identified as locally extinct (Table 3). While in the Gulf of Cadiz and the Balearic Islands main species listed were cartilaginous fish (common guitarfish, angelsharks, smooth-hound and nursehound), in the Tramontana region main species listed were bony fish such as brown meagre, common dolphinfish *Coryphaena hippurus* and European sprat *Sprattus sprattus* (Table 4). Fishers from the Balearic Islands mainly listed the spinous spider crab as locally extinct.

c) **Proliferations:** Fifty-three percent of fishers listed species or groups of species that had proliferated (Table 1). In total, 37 taxa were listed (Table S2). Of those species most frequently listed, three highly commercial fish were listed the most: mackerels, red mullets and sardines (Table 2). In addition, round sardinella and "octopuses" were listed several times. Other fish species were listed, such as horse mackerels and John Dory *Zeus faber*, five groups of invertebrates, such as "jellyfish", "shrimps" and "squids", and three small-size shark species, such as the small-spotted catshark *Scyliorhinus canicula* and the common stingray *Dasyatis pastinaca*.

PERMANOVA analysis highlighted significant differences by region of species identified as proliferating by fishermen (Table 3). In the Gulf of Cadiz and the Andalusian Mediterranean site, mackerel species were the ones that contributed the most to the differences between regions (Table 4). The rest of species with a high contribution were different in each region. In the Tramontana region two groups of invertebrates were especially relevant ("octopus" and "cuttlefish").

Discussion

Our study evidences that fishers perceptions from the Spanish Mediterranean Sea and Gulf of Cadiz contain very useful information about fishing activity and ecosystem changes in the area, in agreement with previous studies conducted in different locations of the Mediterranean Sea [32,33].

Regarding fishing activities, our results show that fishing in the study area has gone deeper and farther from shore, in line with a general expansion of fishing activities documented worldwide [2,48,49] and in the study area [50–52]. This was more evident when considering the answers from the elder fishers, suggesting that the most important changes in the ecosystem occurred a few decades ago. Fishers were also able to identify areas where depletion had happened in the study areas, which coincides with other studies documenting changes in marine ecosystems of the Mediterranean Sea [12] and is consistent with the exponential fishing effort growth observed in the study area [50,51,52, and see Figure 11a-b and Annex S1 and S2]. Both results suggesting depletion of resources and the expansion of fishing activities toward deeper and farther waters highlight a progressive depletion of coastal areas in the region.

Fishers reported that the species abundance, measured as the best catch ever and the largest caught species, and the species and size composition of some species in the catch, have changed significantly with time (Figures 6, 7, 8, 9 and Tables 1 and 2). This is in agreement with other studies that use quantitative or qualitative fisheries information in specific areas from the western, central and eastern Mediterranean Sea [e.g., 14,25,53,54]. In addition, and despite our results regarding common hake are not so clear since the weight of the largest common hake ever caught (kg) plotted against the year of starting fishing showed no clear trend, there are evidences that hake has been increasingly subjected to overfishing since the 1980s and nowadays the population is mainly supported by young age classes [55,56]. In fact, in 2011 the International Union for Conservation of Nature (IUCN) listed the common hake as a vulnerable species in the Mediterranean fish assessment [16] and, hake is assessed as being overexploited in several areas of the Mediterranean [15,44,57,58].

Our results also provide valuable information regarding changes at the species composition level of the exploited ecosystem. Fishers listed several commercial and non-commercial species as depleted. In several cases, these results are in line with various quantitative information available (Table S2) and other regional lists and bibliographic references [59,60]. Of the 43 species of fish listed as depleted, ten species were listed as data deficient (DD), near threatened (NT), vulnerable (VU) or endangered (EN) by IUCN [16] (Table S2). This represents the 24% of the species listed by fishers. Three of those species were also listed in the Bern and Barcelona conventions [40,41], and one was assessed as overexploited by the General Fisheries Commission of the Mediterranean Sea (GFCM) [44] (Table S2). Of the shark species listed as depleted, nine of the ten species listed had been listed as DD, NT, VU, EN or critically endangered (CR) by IUCN [16] (Table S2). This represents the 90% of the species listed by fishers and is consistent with the fact that sharks are highly vulnerable to fishing activity both globally [61,62] and in the Mediterranean Sea [63,64], and that fishing effort in the study area has grown exponentially (Figure 11a–b, and Annex S1 and S2). Of the invertebrate species listed, three species had been included in the Bern and Barcelona convention [40,41] as species with exploita-

Table 2. Summary of taxa and groups listed by fishers as being depleted (D), locally extinct (E), and as having proliferated (P).

Species	Depletion	Extinction	Proliferation
Aristeus antennatus	1		1
Auxis rochei	1		1
Belone belone	2		
Coryphaena hippurus	1	2	1
Dasyatis pastinaca			2
Dentex dentex			2
Dicentrarchus labrax	2		
Engraulis encrasicolus	10		1
Epinephelus spp.	2		
Helicolenus dactylopterus	1	1	
Hippocampus sp.		3	
Homarus gammarus	1	1	
Lithognathus mormyrus	2		
Lophius spp.	5		1
Maja squinado	4	10	
Merluccius merluccius	12		1
Micromesistius poutassou	3		
Mullus spp.	8		6
Mustelus asterias		3	
Mustelus mustelus	3	2	1
Nephrops norvegicus	2		1
Oxynotus centrina	2	1	
Pagellus bogaraveo	1	3	
Pagellus erythrinus	2	1	
Pagrus auriga	2		
Pagrus pagrus	5		
Palinurus elephas	4		
Plectorhinchus mediterraneus	2	1	
Pomadasys incisus	2		
Prionace glauca	1	1	
Psetta maxima	2		
Rajiformes	1		2
Rhinobatos rhinobatos		6	
Sarda sarda	2		
Sardina pilchardus	11	1	3
Sardinella aurita			3
Sciaena umbra	2	3	1
Scomber spp.	6		7
Scorpaena scrofa	2		
Scyliorhinus canicula			2
Scyliorhinus stellaris	5	8	
Scyllarides latus	1		1
Seriola dumerili	3		
Solea solea	5		
Solea sp.	2		
Sparisoma cretense	2		
Sparus aurata	2		1
Spicara spp.	2		
Sprattus sprattus		2	

Table 2. Cont.

Species	Depletion	Extinction	Proliferation
Squalus acanthias	2	1	
Squalus blainvillei	3		
Squatina oculata	2	4	
Squatina squatina		9	
Torpedo spp.		2	
Trachurus spp.	1		2
Trichiurus lepturus	1	1	
Trisopterus minutus	2		
Umbrina cirrosa		2	
Zeus faber	2		2
Groups	**Depletion**	**Extinction**	**Proliferation**
ALL	11		
Cephalopods	1		
Commercial invertebrates	1	1	
Cuttlefish	1		2
Jellyfish			2
Octopus	1		5
Marine turtles	2	1	
Commercial shrimps	3		2
Squids	1	1	2
Whales	1	1	

This table lists species that were mentioned at least twice in a group (D, E or P), or that were listed in two or more groups (Table S2 contains the full list of species).

tion being regulated, and three additional ones were assessed as overexploited by the GFCM [44] (Table S2). Most of these species listed as being depleted by fishers are known to have been highly impacted by fishing [15,16]. In some cases though, some species may have declined due to additional effects of climate change, as it is the case of sardine and European sprat [65] and the increase sea surface temperature trend in the area (Figure 11c). Changes in the environmental conditions in the study area with time include an increase of sea surface temperature and declines in river runoffs (as exemplified in Figure 11c for the Ebro River) and, in general, but warmer and drier conditions (Figure 11d), with clear effects on marine ecosystems in the region [66].

Regarding locally extinct or extirpated species listed by fishers, the spinous spider crab and the European lobster *Homarus gammarus*, both highly commercial species, were included in the Bern and Barcelona conventions [40,41]. Of the finfish species listed, three were also listed by IUCN as NT and VU: this is the case of the brown meager, the shi drum *Umbrina cirrosa*, and seahorses [16]. Of shark species, nine of the ten species of sharks listed by fishers were in the IUCN list and three of those in other international agreements [16]. When looking at all fish species listed, 12 of the 21 fish species were IUCN listed [16], and of those six were also listed in other international agreements (so 50% of fish) and one more (sardine) was included in the GFCM assessment as fully exploited or overexploited [44] (Table S2). Locally extinctions or extirpations can mainly be related with an increase of fishing intensity in the area to target commercial species (Figure 11a–b and Annex S1 and S2) and the high vulnerability of some target and non-target species to fishing [15,16].

Table 3. PERMANOVA analysis by region. D: depletion, E: local extinction, P: proliferation.

Analysis	df	SS	MS	Pseudo-F	P (perm)
D - region	3	19375	6458.2	1.49	**0.02**
E - region	2	37345	18672	6.20	**0.00**
P - region	3	26987	8995.5	2.36	**0.00**

It is interesting to note that fishers listed a larger group of species being depleted or locally extinct than official conventions and assessments (Table S2). This may be due to fishers are reporting local depletions and extirpations and thus these results are relevant at a small regional scale. It could also be related to the fact that fishers are the immediate observers of the sea, thus they may list depletions that have not been yet observed by the scientists or are not reported by official documents yet. On the other hand, stock assessment datasets in the Mediterranean Sea are still limited to few highly commercial species (such as common hake, anchovy, sardine, red mullets, Norway lobster and red shrimp) [15,44] and thus assessments are still missing for many exploited species. It is also recognised that commercial fishes, and also invertebrate species, are more difficult to list in official conventions due to unavailability of data but also economic interests to keep exploiting the species [67].

Fishers also listed species that became more abundant with time, in agreement with independent information from the area

Table 4. SIMPER analysis by region and species or groups (Table 2) by depletion, local extinction, and proliferation lists.

Regions	Species or groups	Av. contribution (%)	Cum. contribution (%)
A. Depletion			
Gulf of Cadiz	"ALL"	34.7	34.7
	Sardina pilchardus	15.2	49.8
	Pagrus pagrus	9.8	59.6
	"Shrimps"	9.6	69.2
	Micromesistius poutassou	7.7	76.9
Andalusia-Med	*Merluccius merluccius*	54.8	54.8
	Engraulis encrasicolus	23.7	78.5
Tramontana	*Sardina pilchardus*	22.1	22.1
	Merluccius merluccius	19.7	41.8
	"ALL"	19.3	61.0
	Engraulis encrasicolus	11.8	72.8
	Mullus spp.	9.9	82.7
Balearic Islands	*Mustelus mustelus*	25.0	25.0
	Squalus blainvillei	18.0	42.9
	"ALL"	12.8	55.7
	Belone belone	7.7	63.4
	Mullus spp.	7.7	71.1
	Palinurus elephas	7.7	78.7
B. Extinctions			
Gulf of Cadiz	*Rhinobatos rhinobatos*	61.3	61.3
	Squatina squatina	17.3	78.6
	Mustelus asterias	7.1	85.7
Tramontana	*Sciaena umbra*	67.4	67.4
	Coryphaena hippurus	15.0	82.4
	Sprattus sprattus	11.2	93.6
Balearic Islands	*Maja squinado*	49.7	49.7
	Scyliorhinus stellaris	32.1	81.8
	Squatina squatina	17.2	99.0
C. Proliferations			
Gulf of Cadiz	*Scomber* spp.	50.0	50.0
	Mullus spp.	37.5	87.5
Andalusia-Med	*Scomber* spp.	84.0	84.0
	Sardinella aurita	16.0	100.0
Tramontana	"Octopus"	87.5	87.5
	"Cuttlefish"	6.8	94.3
Balearic Islands	*Zeus faber*	55.6	55.6
	Rajiformes	44.4	100.0

[52]. This may be due to these species had benefited from a reduction of predators in some areas (due to predation release or competition decrease), as can be the case of cephalopods or small-sized fish [53,68]. Other species may have also benefited from changes in physical and oceanographic conditions related to climate change, as it is the case of the round sardinella [69], cephalopods [68] and jellyfish [70] (Figure 11c–d). However, the way that multiple impacts (such as fishing and climate change) interact and combine to impact species and productivity patterns of marine ecosystems is still hardly known [71] and difficult the interpretation of some biodiversity changes. In addition, some benthopelagic and pelagic species listed as proliferating, such as John Dory, could be perceived as having increased due to modifications of the fishing technique (for example, by increasing the height of the bottom trawling net). This issue needs further investigation.

Our results showed some specificity by region as well. This can be due to our study area presents a north to south gradient in the Mediterranean Sea (Figure 1), which is characterized by some differential oceanographic and productivity conditions that can affect the presence and production of species [72]. In addition, fishing effort differences between the Spanish peninsula and the

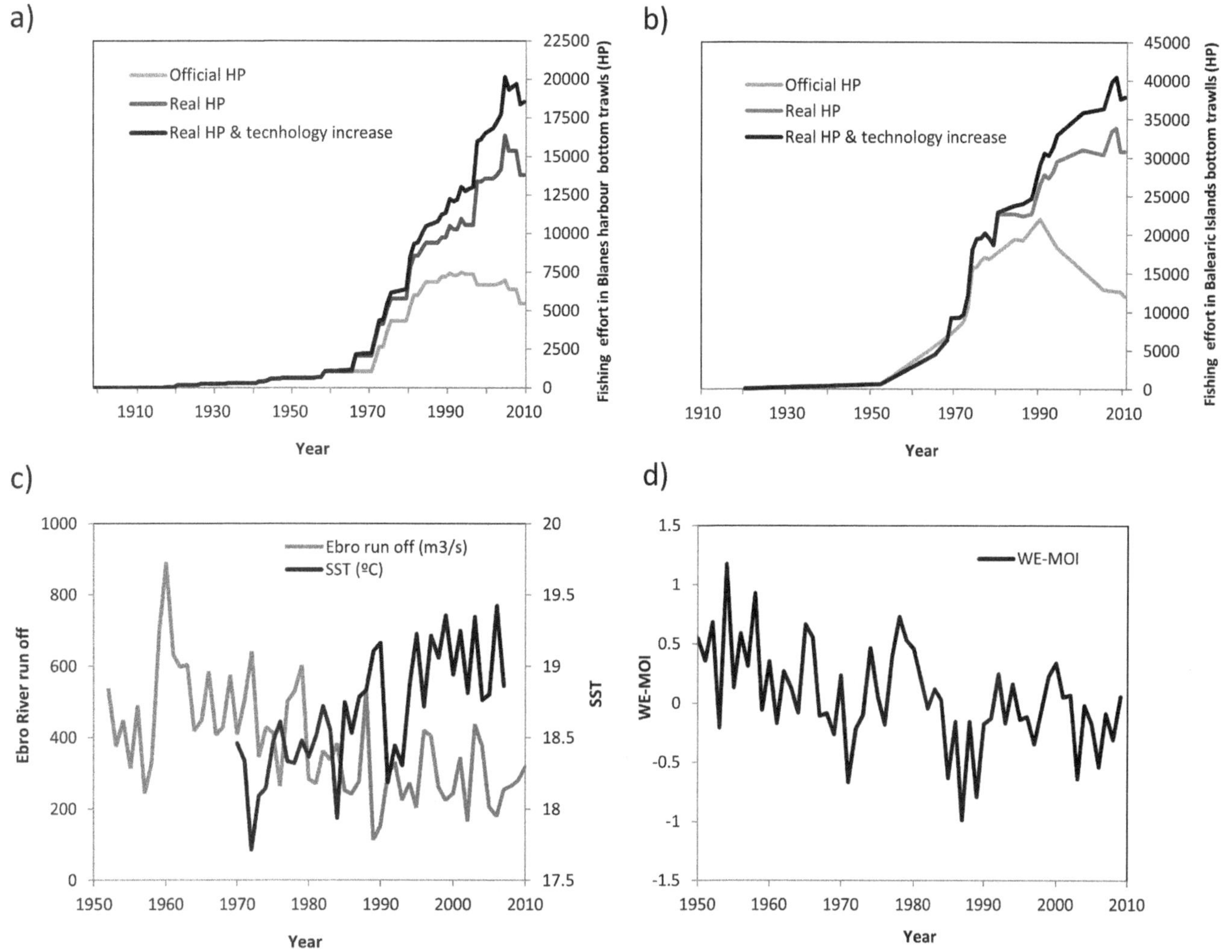

Figure 11. Temporal evolution of a) Bottom trawling fishing effort in the Blanes harbour (NW Mediterranean Sea) from 1900 to 2013 (HP; data sources: Annex S1), b) Bottom trawling fishing effort in the Balearic Islands (NW Mediterranean Sea) from 1920 to 2010 (HP; data sources: Annex S2), c) Ebro River water runoff from 1953 to 2010 (m3/s, annual mean; data sources: Ebro Hydrographical Confederation, Tortosa station, http://www.chebro.es/) and Sea Surface Temperature (SST) from 1970 to 2007 (°C, annual mean; data sources: data was collected near Tarragona, 40°N-2°E, Smith and Reynolds, 2004); and d) Western Mediterranean Oscillation Index from 1950 to 2009 (WEMOI; Error! Hyperlink reference not valid., [76]).

Balearic Islands may explain the fact that more species of cartilaginous fish are still listed in the islands since fishing effort is higher in the mainland [39,73,74] (Figure 11a–b and Annex S1 and S2). It is therefore important to highlight that regional dynamics are essential when analysing and interpreting biodiversity changes.

Overall, this study illustrates that traditional knowledge that can be retrieved from local fishers is useful to complement the information available regarding the evolution of fishing activities and changes in marine resources and biodiversity in the Mediterranean Sea. This information can complement the quantitative knowledge generated by fisheries assessments and can be useful to complement quantitative monitoring and evaluation surveys to keep track of fishing and climate change effects in the Mediterranean Sea. This knowledge is essential to set current data from the study area into a historical perspective, and prevent the shifting baseline syndrome [21–23]. Such recovery of information is also especially relevant in the context of the new Common Fisheries Policy (CFP) of the European Commission and the Marine Strategy Framework Directive (MSFD), which aims at achieving a Good Environmental Status (GES) in European marine waters by 2020, at the latest [75].

Supporting Information

Table S1 Complete list of targeted species listed by fishers.

Table S2 Complete list of taxa listed by fishers as being depleted (D), locally extinct (E), and as having proliferated (P). The classification of species in different international agreements or stock assessments is indicated.

Annex S1 Bottom trawling fishing effort reconstruction of the Blanes fishing harbour, Catalan Sea.

Annex S2 Bottom trawling fishing effort reconstruction of the Balearic Islands.

Acknowledgments

We would like to thank all fishers that participated in the interviews and all our colleagues that helped us collecting data from different regions, especially Sarah Camillieri, Macarena Delgado, Natalia Garcia, Juan Montes, Aitana Oltra, Fran Sobrado, Begoña Pique, and volunteers from *Ecologistas en Accion* NGO. We would also like to thank Enric Massutí for valuable help during the interview process in the Balearic Islands and Isabel Palomera, Francesc Sardà and Baptista Company for their help in the Tramontana region.

Author Contributions

Conceived and designed the experiments: M. Coll. Performed the experiments: M. Coll M. Carreras MJC CC GG EM RS. Analyzed the data: M. Coll M. Carreras GG. Contributed reagents/materials/analysis tools: M. Coll M. Carreras MJC CC GG EM RS. Wrote the paper: M. Coll M. Carreras EM RS.

References

1. Brown C, Corcoran E, Herkenrath P, Thonell J (2006) Marine and coastal ecosystems and human well-being: a synthesis report based on the findings of the Millennium Ecosystem Assessment; 92-807-2679-X UNEPNI, editor. 64 p.
2. Coll M, Libralato S, Tudela S, Palomera I, Pranovi F (2008) Ecosystem Overfishing in the Ocean. PLoS ONE 3: e3881.
3. Jackson JBC, Kirby MX, Berger WH, Bjorndal KA, Botsford LW, et al. (2001) Historical overfishing and the recent collapse of coastal ecosystems. Science 293: 629–638.
4. Lotze HK, Lenihan HS, Bourque BJ, Bradbury RH, Cooke RG, et al. (2006) Depletion, degradation, and recovery potential of estuaries and coastal seas. Science 312: 1806–1809.
5. Costello MJ, Coll M, Danovaro R, Halpin P, Ojaveer H, et al. (2010) A census of marine biodiversity knowledge, resources and future challenges PLoS One 5: e12110.
6. Halpern BS, Walbridge S, Selkoe KA, Kappel CV, Micheli F, et al. (2008) A global map of human impact on marine ecosystems. Science 319: 948–952.
7. Cheung WWL, Lam VWY, Sarmiento JL, Kearney K, Watson R, et al. (2009) Projecting global marine biodiversity impacts under climate change scenarios. Fish and Fisheries 10: 235–251.
8. Worm B, Barbier EB, Beaumont N, Duffy JE, Folke C, et al. (2006) Impacts of Biodiversity Loss on Ocean Ecosystem Services. Science 314: 787–790.
9. Lotze HK, Worm B (2009) Historical baselines for large marine animals. Trends in Ecology & Evolution 24: 254–262.
10. Roux J-P, van der Lingen CD, Gibbons MJ, Moroff NE, Shannon LJ, et al. (2013) Jellyfication of Marine Ecosystems as a Likely Consequence of Overfishing Small Pelagic Fishes: Lessons from the Benguela. Bulletin of Marine Science 89: 249–284.
11. Coll M, Libralato S (2012) Contributions of food-web modelling for an Ecosystem Approach of Marine Resource Management in the Mediterranean Sea. Fish and Fisheries 13: 60–88.
12. Coll M, Piroddi C, Kaschner K, Ben Rais Lasram F, Steenbeek J, et al. (2010) The biodiversity of the Mediterranean Sea: estimates, patterns and threats. PLoS ONE 5: doi:10.1371.
13. Coll M, Piroddi C, Albouy C, Ben Rais Lasram F, Cheung W, et al. (2012) The Mediterranean Sea under siege: spatial overlap between marine biodiversity, cumulative threats and marine reserves. Global Ecology and Biogeography 21: 465–480.
14. Lotze HK, Coll M, Dunne J (2011) Historical changes in marine resources, food-web structure and ecosystem functioning in the Adriatic Sea. Ecosystems 14: 198–222.
15. Colloca F, Cardinale M, Maynou F, Giannoulaki M, Scarcella G, et al. (2013) Rebuilding Mediterranean fisheries: a new paradigm for ecological sustainability. Fish and Fisheries 14: 89–109.
16. Abdul Malak D, Livingstone SR, Pollard D, Polidoro BA, Cuttelod A, et al. (2011) Overview of the Conservation Status of the Marine Fishes of the Mediterranean Sea. Gland, Switzerland and Malaga, Spain: IUCN. vii + 61pp.
17. Ferretti F, Myers RA, Serena F, Lotze HK (2008) Loss of large predatory sharks from the Mediterranean Sea. Conservation Biology 22: 952–964.
18. Lejeusne C, Chevaldonné P, Pergent-Martini C, Boudouresque C, Pérez T (2010) Climate change effects on a miniature ocean: the highly diverse, highly impacted Mediterranean Sea. Trends in Ecology & Evolution 25: 250–260.
19. Ben Rais Lasram F, Guilhaumon F, Somot S, Thuiller W, Mouillot D (2010) The Mediterranean Sea as a "cul-de-sac" for endemic fishes facing climate change. Global Change Biology 16: 3233–3245.
20. Narayanaswamy BE, Coll M, Danovaro R, Davidson K, Ojaveer H, et al. (2013) Synthesis of knowledge on marine biodiversity in European Seas: from census to sustainable management. PLoS ONE 8: e58909.
21. Pauly D (1995) Anecdotes and the Shifting Base-Line Syndrome of Fisheries. Trends in Ecology & Evolution 10: 430–430.
22. Pinnegar JK, Engelhard GH (2008) The 'shifting baseline'phenomenon: a global perspective. Reviews in Fish Biology and Fisheries 18: 1–16.
23. Guénette S, Gascuel D (2012) Shifting baselines in European fisheries: the case of the Celtic Sea and Bay of Biscay. Ocean & Coastal Management 70: 10–21.
24. Sala E (2004) The past and present topology and structure of Mediterranean subtidal rocky-shore food webs. Ecosystems 7: 333–340.
25. Fortibuoni T, Libralato S, Raicevich S, Giovanardi O, Solidoro C (2010) Coding early naturalists' accounts into long-term fish community changes in the Adriatic Sea (1800–2000). PLoS ONE 5: e15502.
26. Bertrand JA, De Sola LG, Papaconstantinou C, Relini G, Souplet A (2002) The general specifications of the MEDITS surveys. Scientia Marina 66: 9–17.
27. Sáenz-Arroyo A, Roberts CM, Torre J, Cariño-Olvera M, Enríquez-Andrade EA (2005) Rapidly shifting environmental baselines among fishers of the Gulf of California. Proceedings of the Royal Society B 22: 1957–1962.
28. Sáenz-Arroyo A, Roberts CM, Torre J, Cariño-Olvera M (2005) Using fishers' anecdotes, naturalists' observations and grey literature to reassess marine species at risk: the case of the Gulf grouper in the Gulf of California, Mexico. Fish and Fisheries 6: 121–133.
29. Sáenz-Arroyo A, Roberts CM, Torre J, Cariño-Olvera M, Hawkins JP (2006) The value of evidence about past abundance: marine fauna of the Gulf of California through the eyes of 16th to 19th century travellers. Fish and Fisheries 7: 128–146.
30. McClenachan L (2009) Documenting loss of large trophy fish from the Florida Keys with historical photographs. Conservation Biology 23: 636–643.
31. Boudreau SA, Worm B (2010) Top-down control of lobster in the Gulf of Maine: insights from local ecological knowledge and research surveys. Marine Ecology Progress Series 403: 181–191.
32. Maynou F, Sbrana M, Sartor P, Maravelias C, Kavadas S, et al. (2011) Estimating Trends of Population Decline in Long-Lived Marine Species in the Mediterranean Sea Based on Fishers' Perceptions. PLoS ONE 6: e21818.
33. Azzurro E, Moschella P, Maynou F (2011) Tracking Signals of Change in Mediterranean Fish Diversity Based on Local Ecological Knowledge. PLoS ONE 6: e24885.
34. Coll M, Carreras M, Cornax MJ, Massutí E, Morote E, et al. (2013) An estimate of the total catch in the Spanish Mediterranean Sea and Gulf of Cadiz regions (1950–2010). Fisheries Centre Research Report.
35. Álvarez Ereipa B (2008) Estrategias de comercialización de la pesca litoral. Caso de la Lonja de contratación del puerto pesquero de Palamós y sus lecciones para la pesca artesanal en Venezuela: MSc thesis en Economía y Gestión de la Actividad Pesquera. University of Barcelona.
36. MARM (2011) Ministerio de Medio Ambiente y Medio Rural y Marino: http://www.marm.es/es/pesca/estadisticas/ (accessed 01.2012).
37. FAO (2011) http://www.fao.org/fishery/statistics/software/fishstat/en: GFCM (Mediterranean and Black Sea) Capture Production 1970–2008 (Release date: December 2010).
38. Sea Around Us P (2011) Sea Around Us project: Large Marine Ecosystems (LME) NEW DATA: Mediterranean Sea Landings (visited 14.04.2011).
39. Carreras M, Coll M, Quetglas A, Massutí E, Oliver P, et al. (2013) Estimates of total fisheries removal for the Balearic Islands (1950–2010). Fisheries Centre Research Report.
40. Council of Europe (1979) Convention on the Conservation of European Wildlife and Natural Heritage. Bern, Switzerland. Available at:http://conventions.coe.int/Treaty/EN/Treaties/Html/104.htm.
41. Barcelona Convention (1976) Convention for the Protection Of The Mediterranean Sea Against Pollution. Signed 16 February 1976, in force 12 February 197, amended in 1995 by the Protocol Concerning specially Protected Areas and Biological Diversity in the Mediterranean (SAP-Bio). http://www.unep.ch/regionalseas/regions/med/t_barcel.htm.
42. Bonn Convention (1983) Convention on the Conservation of Migratory Species of Wild Animals. Signed 1979, in force 1983. http://www.cms.int.
43. CITES Convention (1975) Convention on International Trade in Endangered Species of Wild Fauna and Flora. Signed in 1973, in force in 1975. www.cites.org
44. SAC-GFCM (2011) General Fisheries Commission for the Mediterranean Sea - Scientific Advisory Committee GFCM:SAC13/2011/3.
45. Anderson MA, Gorley RN, Clarke KR (2008) PERMANOVA+ for PRIMER: Guide to software and statistical methods. Plymouth, UK: PRIMER-E.
46. Clarke KR, Gorley RN (2006) PRIMER v6: user manual/tutorial (Plymouth routines in multivariate ecological research). Plymouth: Primer-E Ltd. 192 p.
47. Froese R, Pauly D (2013) Fishbase, a Global Information System on Fishes: http://www.fishbase.org. World Wide Web electronic publication (accessed February 2013).

48. Morato T, Watson R, Pitcher TJ, Pauly D (2006) Fishing down the deep. Fish and Fisheries 7: 24–34.
49. Swartz W, Sala E, Tracey S, Watson R, Pauly D (2011) The Spatial Expansion and Ecological Footprint of Fisheries (1950 to Present). PLoS ONE 5: e15143.
50. Bas C, Maynou F, Sardà F, Lleonart J (2003) Variacions demogràfiques a les poblacions despècies demersals explotades als darrers quaranta anys a Blanes i Barcelona. Barcelona: Institut d'Estudis Catalans. Axius Secció Ciències. 202 p.
51. Coll M, Shannon LJ, Moloney CL, Palomera I, Tudela S (2006) Comparing trophic flows and fishing impacts of a NW Mediterranean ecosystem with coastal upwelling systems by means of standardized models and indicators. Ecological Modelling 198: 53–70.
52. Coll M, Palomera I, Tudela S, Dowd M (2008) Food-web dynamics in the South Catalan Sea ecosystem (NW Mediterranean) for 1978-2003. Ecological Modelling 217: 95–116.
53. Coll M, Palomera I, Tudela S, Sardà F (2006) Trophic flows, ecosystem structure and fishing impacts in the South Catalan Sea, Northwestern Mediterranean. Journal of Marine Systems 59: 63–96.
54. Piroddi C, Bearzi G, Christensen V (2010) Effects of local fisheries and ocean productivity on the northeastern Ionian Sea ecosystem. Ecological Modelling 221: 1526–1544.
55. Martin P, Carbonell A, Belcari P (2001) Estimation of trawl discards in the western Mediterranean. European hake (*Merluccius merluccius*) as case study. Final Report DG Fisheries Study 00/009, 135.
56. Hidalgo M, Massutí E, Guijarro B, Moranta J, Ciannelli L, et al. (2009) Population effects and changes in life history traits in relation to phase transitions induced by long-term fishery harvesting: European hake. Canadian Journal of Fisheries and Aquatic Sciences 66: 1355–1370.
57. FAO-GFCM (2013) Report of the fifteenth session of the Scientific Advisory Committee. Rome, 8-11 April 2013. FAO Fisheries and Aquaculture Report No. 1042. Rome, FAO. 2013. p. 96.
58. STECF (2013) 2012 Assessment of Mediterranean Sea stocks part II (STECF 13-05). Publications Office of the European Union, Luxembourg, EUR 25309 EN, JRC 81592, 618 pp.
59. Mayol J, Grau AM, Riera F, Oliver J (2000) Llista vermella dels peixos de les Balears. Documents tècnics de conservació, II època, núm, 7. Palma. Conselleria de Medi Ambient, Govern de les Illes Balears.
60. Durán J, Pastor E, Grau A, Valencia JM (2012) First results of embryonic development, spawning and larval rearing of the Mediterranean spider crab Maja squinado (Herbst 1788) under laboratory conditions, a candidate species for a restocking program. Aquaculture Research 43: 1777–1786.
61. Ferretti F, Worm B, Britten G, Heithaus M, Lotze H (2010) Patterns and ecosystem consequences of shark declines in the ocean. Ecology Letters 13: 1055–1071.
62. Baum JK, Myers RA, Kehler DG, Worm B, Harley SJ, et al. (2003) Collapse and conservation of shark populations in the Northwest Atlantic. Science 299: 389–392.
63. Aldebert Y (1997) Demersal resources of the Gulf of Lions (NW Mediterranean). Impact of exploitation on fish diversity. Vie Milieu 47: 275–284.
64. Coll M, Navarro J, Palomera I (2013) Ecological role of the endemic Starry ray *Raja asterias* in the NW Mediterranean Sea and management options for its conservation. Biological Conservation 157: 108–120.
65. Palomera I, Olivar MP, Salat J, Sabates A, Coll M, et al. (2007) Small pelagic fish in the NW Mediterranean Sea: An ecological review. Progress in Oceanography 74: 377–396.
66. Calvo E, Simó R, Coma R, Ribes M, Pascual J, et al. (2012) Effects of climate change on Mediterranean marine ecosystems: the case of the Catalan Sea. Climate Research 50: 1–29.
67. Hutchings JA, Festa-Bianchet M (2009) Canadian species at risk (2006-2008), with particular emphasis on fishes. Environmental Reviews 17: 53–65.
68. Coll M, Navarro J, Olson R, Christensen V (2013) Assessing the trophic position and ecological role of squids in marine ecosystems by means of food-web models. Deep Sea Research II: Topical Studies in Oceanography 95: 21–36.
69. Sabatés A, Martín P, Lloret J, Raya V (2006) Sea warming and fish distribution: the case of the small pelagic fish, *Sardinella aurita*, in the western Mediterranean. Global Change Biology 12: 2209–2219.
70. Purcell JE, Uye S, Lo W-T (2007) Anthropogenic causes of jellyfish blooms and their direct consequences for humans: a review. Marine Ecology Progress Series 350: 153–174.
71. Sala OE, Chapin FS, Armesto JJ, Berlow E, Bloomfield J, et al. (2000) Biodiversity - Global biodiversity scenarios for the year 2100. Science 287: 1770–1774.
72. Bosc E, Bricaud A, Antoine D (2004) Seasonal and interannual variability in algal biomass and primary production in the Mediterranean Sea, as derived from 4 years of SeaWiFS observations. Global Biogeochemical Cycles 18: doi:10.1029/2003GB002034.
73. Massuti E, Moranta J (2003) Demersal assemblages and depth distribution of elasmobrachs from the continental shelf and slope off the Balearic Islands (western Mediterranean) (vol 60, pg 753, 2003). ICES Journal of Marine Science 60: 1398–1398.
74. Quetglas A, Guijarro B, Ordines F, Massutí E (2012) Stock boundaries for fisheries assessment and management in the Mediterranean: the Balearic Islands as a case study. Scientia Marina 76: 17–28.
75. EC (2008) Directive of the European Parliament and the Council Establishing a Framework for Community Action in the Field of Marine Environmental Policy (Marine Strategy Framework Directive).
76. Martín-Vide J, Lopez-Bustins JA (2006) The Western Mediterranean Oscillation and rainfall in the Iberian Peninsula. International Journal of Climatology 26: 1455–1475.

21

Local Extinction and Unintentional Rewilding of Bighorn Sheep (*Ovis canadensis*) on a Desert Island

Benjamin T. Wilder[1]*, Julio L. Betancourt[2], Clinton W. Epps[3], Rachel S. Crowhurst[3], Jim I. Mead[4], Exequiel Ezcurra[1,5]

1 Department of Botany and Plant Sciences, University of California Riverside, Riverside, California, United States of America, **2** National Research Program, Water Mission Area, U.S. Geological Survey, Reston, Virginia, United States of America, **3** Department of Fisheries and Wildlife, Oregon State University, Corvallis, Oregon, United States of America, **4** Department of Geosciences, and Sundquist Center of Excellence in Paleontology, East Tennessee University, Johnson City, Tennessee, United States of America, **5** University of California Institute for Mexico and the United States (UC MEXUS), Riverside, California, United States of America

Abstract

Bighorn sheep (*Ovis canadensis*) were not known to live on Tiburón Island, the largest island in the Gulf of California and Mexico, prior to the surprisingly successful introduction of 20 individuals as a conservation measure in 1975. Today, a stable island population of ~500 sheep supports limited big game hunting and restocking of depleted areas on the Mexican mainland. We discovered fossil dung morphologically similar to that of bighorn sheep in a dung mat deposit from Mojet Cave, in the mountains of Tiburón Island. To determine the origin of this cave deposit we compared pellet shape to fecal pellets of other large mammals, and extracted DNA to sequence mitochondrial DNA fragments at the 12S ribosomal RNA and control regions. The fossil dung was ^{14}C-dated to 1476–1632 calendar years before present and was confirmed as bighorn sheep by morphological and ancient DNA (*a*DNA) analysis. 12S sequences closely or exactly matched known bighorn sheep sequences; control region sequences exactly matched a haplotype described in desert bighorn sheep populations in southwest Arizona and southern California and showed subtle differentiation from the extant Tiburón population. Native desert bighorn sheep previously colonized this land-bridge island, most likely during the Pleistocene, when lower sea levels connected Tiburón to the mainland. They were extirpated sometime in the last ~1500 years, probably due to inherent dynamics of isolated populations, prolonged drought, and (or) human overkill. The reintroduced population is vulnerable to similar extinction risks. The discovery presented here refutes conventional wisdom that bighorn sheep are not native to Tiburón Island, and establishes its recent introduction as an example of unintentional rewilding, defined here as the introduction of a species without knowledge that it was once native and has since gone locally extinct.

Editor: Michael Hofreiter, University of York, United Kingdom

Funding: Funding was provided by a UC MEXUS Dissertation Grant and a NSF Graduate Research fellowship to Wilder. The funders had no role in study design, data collection and analysis, decision to publish, or preparation of the manuscript.

Competing Interests: The authors have declared that no competing interests exist.

* E-mail: benjamin.wilder@email.ucr.edu

Introduction

As recorded in *Cmiique Iitom* —the language of the Seri people, an indigenous community of the coast of Sonora, Mexico and nearby Tiburón Island— Orion's belt, *Hapj*, consists of three stars. The middle star represents the mule deer, *hap*, and the two flanking stars are bighorn sheep, *mojet*, and pronghorn antelope, *haamoja*. When the great hunter of the sky, *Azoj Cmiique* (Scorpius), fired his arrow, it struck *hap* but missed the others. After dripping onto Tiburón Island, the mule deer's blood remained in the sky as the red star *Azoj haait* (Alpha or Betelgeuse). For the Seri, this myth explains why mule deer, but not bighorn sheep or pronghorn antelope, historically inhabited the island [1].

The events that have led to the formation of modern ecosystems, especially extinctions, are often cryptic in occurrence and causation. The anomalous absence of species in either the fossil record or on modern landscapes raises several questions. Did particular species once occur that are now lost? If so, what caused their extinctions, and are they reversible? How do we establish biological baselines to determine conservation priorities and strategies in the absence of historical data?

For example, controversial rewilding efforts to restore and even resurrect lost megafauna [2–5] at the very least demand accurate baselines. In this paper, we coin the term "unintentional rewilding" to mean the introduction of a species, deliberate or otherwise, without knowledge that it was once native and has since gone locally extinct. Here are a couple of examples. In North America, European horses were introduced during the Spanish conquest, subsequently went feral, and unwittingly replaced native horses that were genetically the same but became extinct at the end of the Pleistocene [6,7]. European domesticated horses gone wild are regarded by federal land management agencies in the USA as an exotic species that is harmful to native wildlife habitat and thus should be eradicated or reduced in numbers. Accepting these feral horses as native would challenge current management mandates within the federal government.

A second example involves bison. Extreme drought at the start of the twenty-first century drove bison from the adjacent Kaibab Plateau, where they were introduced and bred with cattle during the 1930s into Grand Canyon National Park, where bison are now trampling riparian areas and archeological sites. Holocene evidence for bison is scant in the Grand Canyon, and the modern herd may include bison-cattle hybrids, bolstering the National Park Service's case for removal. But what if the modern herd contains no hybrids and future paleontological evidence shows that bison occupied the Grand Canyon intermittently throughout the Holocene? Awareness of cases such as the North American horses and Grand Canyon bison will surely increase with expanded paleoecological studies and advances in genomics [8,9], and will continue to raise fundamental questions about conservation targets and measures.

Here we report on the unintentional rewilding of bighorn sheep (*Ovis canadensis*) on Tiburón Island in the Gulf of California, a few kilometers off the west coast of the State of Sonora and the largest island in Mexico (1,218 km^2; Fig. 1). The Canal del Infiernillo —a narrow (~2–10 km wide) and shallow (~5.5 m deep) channel — separates the island from the mainland and was submerged by rising sea level only 6,000–4,700 cal yrs B.P. [10–12].

Mountain ranges on the mainland (the Sierra Seri) and on the island (the Sierra Kunkaak) are very similar in area and suitable habitat. Mule deer (*Odocoileus hemionus*) and desert bighorn sheep (*Ovis canadensis nelsoni*, see end of methods for a discussion on bighorn sheep taxonomy) are found on the coastal mainland today, and pronghorn (*Antilocapra americana sonoriensis*) occurred along the coastal plains in the past [13]. Given that the island was part of the Sonoran mainland as recently as ca. 6,000 years ago, we expect the past occurrence of these species on the island. However, mule deer (*Odocoilus hemionus sheldoni*), an endemic subspecies, is the only non-introduced ungulate on Tiburón Island, and all previous accounts of bighorn sheep and pronghorn antelope correspond to modern attempts to introduce them [14].

The island has a diverse mammal fauna relative to other Gulf islands, but lacks top predators such as bobcat (*Lynx rufus*) and mountain lion (*Puma concolor*) that are common on the mainland. The reports of early explorers in the region confirm the historic absence of bighorn sheep on Tiburón. Charles Sheldon, an early 20th century naturalist and avid sportsman, hunted bighorn sheep on the mainland ranges of coastal Sonora and mule deer on Tiburón Island in 1921, in each case accompanied by Seri guides. Sheldon's detailed field notes record the absence of bighorn sheep on Tiburón, as indicated in this passage: "The chief, Buro Alesan, tells me there are no wild sheep on Tiburón Island, but a few are in the Sierra Seri" [15,16].

In 1975, sixteen female and four male desert bighorn sheep were introduced from the Sonoran mainland adjacent to Tiburón Island as a conservation measure [13,17,18]. The population grew rapidly to ~500 animals, where it seems to have reached its carrying capacity [13,19–21]. Low levels of hunting and other human disturbance, lack of mountain lion, absence of domesticated sheep and their contagious diseases, expansiveness of suitable habitat, and wetter conditions from 1976 to 1995 all probably contributed to introduction success. In agreement with early reports that bighorn sheep were not native to the island [15], all published studies about the transplanted population have treated the operation as an introduction of an alien species into a previously unoccupied ecosystem [13,22].

For wildlife biologists, Tiburón Island has become both a long-term field experiment and an object lesson in conservation. The introduction of bighorn sheep was not only successful in establishing a viable population but, through translocations back to Sonora, Tiburón animals also have contributed significantly to

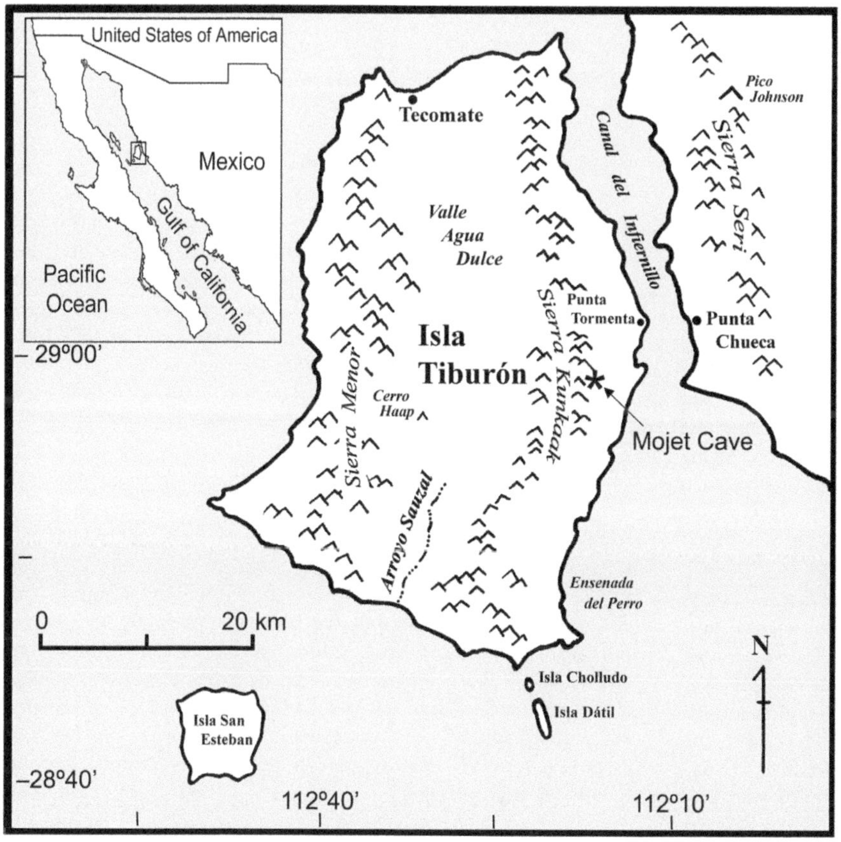

Figure 1. Tiburón Island in Gulf of California.

recovery efforts on the mainland. In 1995, a coalition of institutions initiated an innovative program to fund bighorn sheep research and conservation while providing needed income for the Seri through international auctioning of exclusive hunting tags on the island. Starting in 1999, hunting permits garnered six-figure auction bids [23]; most recent prices range from US $80,000–90,000 a tag [24,25]. Revenue from these auctions offers the Seri incentive to maintain Tiburón Island in an undisturbed state [13,14,26]. To date, this conservation story has been regarded as controversial due to the non-native status of bighorn sheep on the island. The impact of unchecked bighorn sheep herbivory on the island's Sonoran Desert flora, which includes several regional endemic species [26], was not considered prior to the introduction.

During a recent survey for fossil woodrat (*Neotoma*) middens on Tiburón Island, we discovered large pieces of an apparent sheep dung mat in Mojet Cave, a small rock shelter in the eastern foothills of the Sierra Kunkaak. Pellets from the recovered dung mat were ^{14}C-dated to 1476–1632 calendar years before present (cal yr B.P.). We used morphological and ancient DNA (*a*DNA) analyses to determine the identity of the species that deposited the pellets. The 'molecular caving' [27] approach taken here adds a new dimension to the study of paleoenvironments in aridlands, and an opportunity to link theory and observations that address long-standing questions of lost populations and future conservation strategies.

Figure 2. Dung from Mojet Cave and recent potential dung producers. A) dung pellet and mat (ETVP 4999) from Mojet Cave; identified as bighorn sheep, *Ovis canadensis* (see text). Note pellet in dung mat is featured below after removal. Opposite side of dung mat has a thin layer of bat guano adhering to the surface. Recent dung: B) bighorn sheep, *Ovis canadensis* (ETVP 6083); C) pronghorn antelope, *Antilocapra americana* (ETVP 6028); D) mule deer, *Odocoileus hemionus* (ETVP 6017). Scale bar equals 10 mm.

Results

Morphological Identification

The Mojet Cave dung deposit contained both isolated complete pellets and those incorporated into an amorphous mat of crushed pellets; all were consolidated with crystallized urine. The pellets analyzed were small (averaging 15.5×10.1 mm; n = 3), showing the characteristically blunt proximal ends and pointed distal ends that attribute them to bighorn sheep (Fig. 2). This morphology indicated that the pellets represented a dry-season diet, typical of desert vegetation throughout most of the year. Desert bighorn sheep diets focus on diverse array of desertscrub species, especially succulents [26,28]. Pellets of wapiti (*Cervus*; Cervidae), the extinct shrub-ox (*Euceratherium*; Ovibovinae), and the extinct Harrington's mountain goat (*Oreamnos harringtoni*; Rupicaprinae), all known to have inhabited mountains in the now arid Sonoran Desert during the late Pleistocene [29], are significantly larger in size, heavier, and have a conspicuously more robust pellet form [30,31]. Pellets of extant deer and pronghorn (including the extinct *Stockoceros*, which is known to have frequented shallow rock shelters) are similar in size with those of *Ovis*, yet they are characteristically and typically longer in form (Fig. 2). Dung pellets of the living mountain goat (*Oreamnos americanus*) are distinctly smaller than those of *Ovis*. Thus, the morphology of the pellets from Mojet Cave suggests that bighorn sheep produced them.

Ancient DNA

From exterior scrapings of several ancient pellets within dung mat sample B (see methods for sample descriptions), we sequenced one 90 base pair (bp) region of the 12S ribosomal RNA subunit of the mitochondrial genome (GenBank accession number KF769974) from DNA amplified in 12 replicate PCRs (7 forward, 5 reverse), and the three control region fragments from 2–4 replicate PCRs (78–117 bp each; KF769975 [3 forward, 4 reverse], KF769976 [3 forward, 3 reverse], KF769977 [3 forward, 2 reverse]). Neither the exterior scrapings of pellets from dung mat sample A nor the interior pellet material from either mat amplified successfully. Neither of the two extraction controls amplified at any locus, and none of the blank PCR controls amplified at the 12S or control region fragments. Despite the relatively short length of the target fragments (12S and control region), none of the PCR amplicons were successfully sequenced in their entirety in only one direction, but bidirectional reads could be aligned with each other. We saw no evidence of competing sequences (indicating contamination by multiple templates), and sequences were consistent over multiple PCR replicates, except for variation at one location in numbers of repeated base pairs (3 or 4) identified by different people reviewing the sequences.

From a BLAST search, the consensus sequence for the 12S region of the midden sample clearly matched known bighorn sheep sequences, with 1–2 differences from published *O. c. canadensis* (Rocky Mountain bighorn sheep) sequences, and 0–1 differences from published *O. c. nelsoni* (desert bighorn sheep) sequences (Fig. 3). The ancient pellet 12S sequence differed from reference sequences for *Antilocapra americana* by 8 bp, from *Odocoileus hemionus* by 6 bp, from *Odocoileus virginianus* (white-tailed deer) by 7 bp, from *Oreamnos americanus* by 4 bp, from *Bos taurus* (domestic cattle) by 7 bp, and from *Ovis aries* (domestic sheep) by 3 bp (Fig. 3).

Phylogenetic analysis of the 46 bp alignment of ancient and reference 12S sequences (after trimming primer sites) resulted in 9 equally parsimonious trees each with 24 steps, of which the majority rule consensus appears in Fig. 4. The ancient DNA sequence is identical to one of the modern bighorn samples, and clusters within a clade that includes all desert bighorn sheep samples as well as the Rocky Mountain bighorn, mountain goat, and domestic sheep individuals. The short length of the sequence alignment made it impossible to resolve relationships within this clade unambiguously, but the exact match of the ancient DNA to one of the modern desert bighorn sheep leaves no doubt about the identity of that sample.

In total, we sequenced 280 bp of three control region fragments from the ancient sample (Fig. S1). The ancient sequence exactly matched two published desert bighorn sheep haplotypes (*O. c. nelsoni* haplotypes B and V, GenBank accession numbers

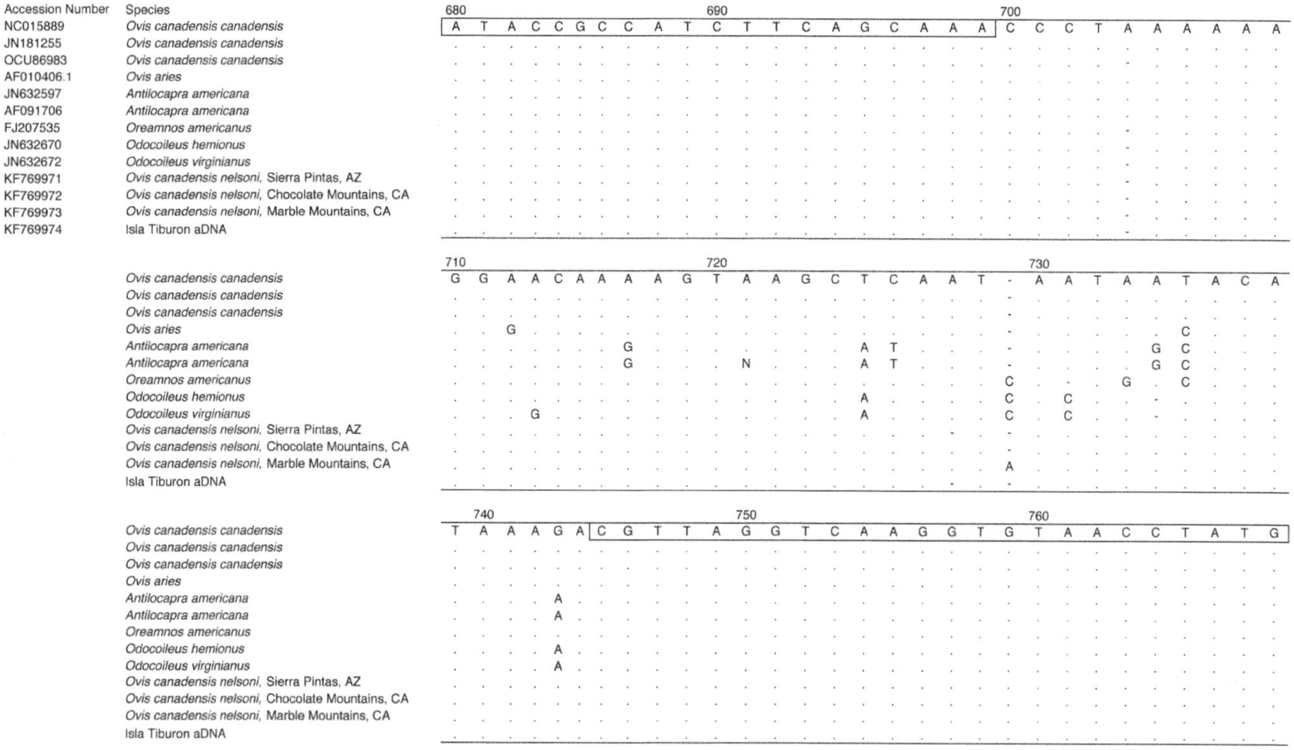

Figure 3. Isla Tiburón *a*DNA and Sonoran Desert ungulate sequences. Partial sequences of the 12S ribosomal RNA subunit of the mitochondrial genome for the Isla Tiburón *a*DNA sample and published sequences from all ungulate species thought to have existed in the area since the start of the Holocene, with GenBank accession numbers. Open box denotes primer region, dot indicates identical bases between sequences, N indicates unknown base. Sequence position numbers are derived from *O. c. canadensis* haplotype (GenBank Accession NC015889).

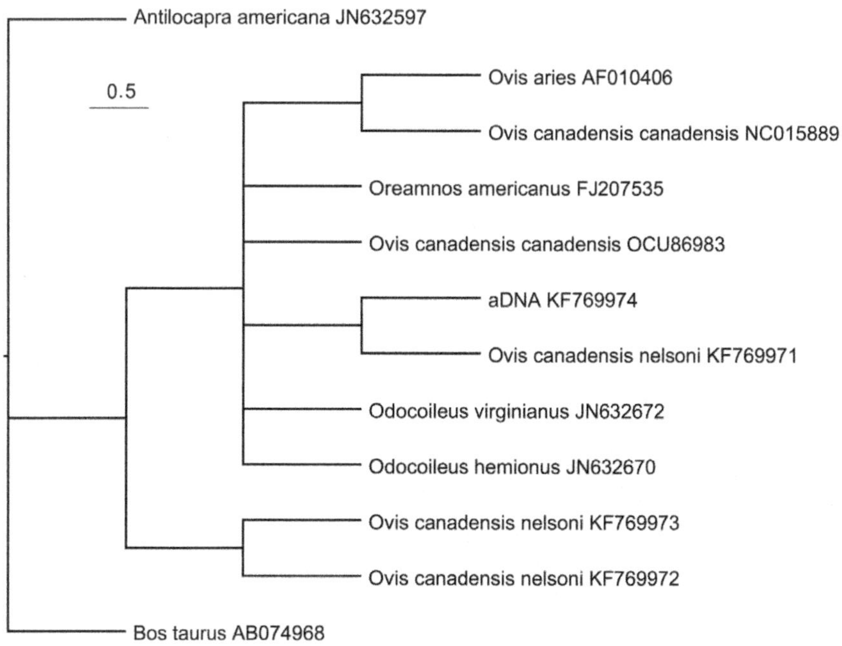

Figure 4. Phylogenetic tree for Isla Tiburón *a*DNA and ungulate sequences. Majority rule consensus tree from phylogenetic analysis of a 46 bp region of the 12S ribosomal RNA subunit of the mitochondrial genome, including the Isla Tiburón *a*DNA sample and published sequences from all ungulate species thought to have existed in the area since the start of the Holocene, with GenBank accession numbers. The phylogeny was inferred using parsimony and treating indels as a 5th character state.

AY903995, AY904015). Haplotypes B and V were differentiated by only 4 sites outside the three fragments that we sequenced for this project, and originated from southern Mojave (B) [32] and the Sierra Pinta mountains of southern Arizona (V). The ancient sequence varied from each of the two known modern Tiburón bighorn sheep control region haplotypes (by 5 and 2 bp), and by one bp from a published sequence from Baja California identified as *O. c. weemsi* (Weem's desert bighorn sheep from Baja California). The ancient sequence varied from a published *O. c. canadensis* sequence originating in the Canadian Rocky Mountains [33] at 11 sites. Lastly, there were 53–57 differences between the ancient DNA sequence and published haplotypes for *Ovis aries* (Fig. 4).

Discussion

The accumulation of plant and animal remains in aridland caves and rock shelters, sometimes imbedded in crystallized mammalian urine, offers unique opportunities for *a*DNA preservation and analysis. Although DNA molecules rapidly disintegrate after death and fossil DNA is highly fragmented [34], both the desiccation and protection within crystallized urine substantially slow DNA degradation [35]. Modern genetic techniques and diverse arrays of molecular primers for a multitude of taxa make it possible to analyze tiny amounts of DNA from dung, plant, and other material preserved in fossil middens [8]. In the case of Mojet Cave, the bighorn sheep pellets in the urine-hardened mat yielded low-quality DNA that could still be amplified via primers designed from extant material. However, extreme temperatures and temperature fluctuations, UV-radiation, and humidity in the area may have compromised the preservation of these samples, leading to fragmentation of DNA strands and the truncated sequences we observed.

We conducted extractions and PCR setup for ancient DNA samples in a laboratory with no history of mammalian genetic research, used new reagents, and observed standard *a*DNA practices [36]. We are confident that our *a*DNA sequences do not result from contamination. Sequence analysis of the *a*DNA recovered confirms that desert bighorn sheep deposited the pellets recovered from the Mojet Cave dung mat. Moreover, it is noteworthy that the *a*DNA control region haplotype does not match either of the two haplotypes recovered from 63 modern Tiburón sheep captured in 2007 [37] and previously reported in an unpublished study (GENBANK AY116622.1, AY116621.1).

We cannot specify whether bighorn sheep persisted or repeatedly colonized Tiburón Island during the Holocene, or pinpoint when or why they became locally extinct during the past 1500 years. Population model simulations under two future climate change scenarios show that extinction risk for the Tiburón bighorn sheep population increases more with mean drought severity than with drought variability [38]. According to the tree-ring record, mean drought severity and variability have varied substantially during the past two millennia. Gridded reconstructions of the July Palmer Drought Severity Index for the two grid points nearest Tiburón [39] (centered at 27.5N, 110W and 30N, 112.6W) show increased mean drought severity from AD 400 to 1400, encompassing the Medieval Climate Anomaly (MCA). The tree-ring record is not long enough to evaluate if this dry period was unprecedented in the Holocene, causing a unique extinction on Tiburón. Oceanic conditions can induce periods of extreme drought in the coastal deserts of the Gulf of California [40], which can greatly increase risk of population extinction of bighorn sheep in desert environments [41,42].

Bighorn sheep have a highly fragmented distribution through much of their current range, resembling a metapopulation in many regions, including the deserts of North America [43]. High rates of local population extinction in desert environments [41] and strong genetic drift [42] are counterbalanced to varying degrees by gene flow and recolonization, depending on population isolation and local climatic conditions [32,42,44]. In the absence of gene flow, the small population size characteristic of desert bighorn sheep populations results in very rapid genetic drift and likely inbreeding [44]. This situation is further exacerbated by the stochastic nature of precipitation and forage availability in Sonoran Desert systems, and potential hunting by indigenous peoples. The Seri lived off the seasonal bounty of the desert and sea [16,45], and it can be safely assumed that they occasionally hunted bighorn sheep. Bighorn sheep remains figure prominently in the archeological record of the Southwest, ranging from cremated remains in the Pinacate Mountains in northwestern Mexico [46] to widespread evidence of bighorn as a food resource in southern Arizona to the Colorado Plateau [47,48].

We hypothesize that isolation of the prehistoric Tiburón bighorn sheep population resulting from sea level rise, combined with subsequent drivers that act on small populations, including inbreeding, overharvesting by hunters, and megadroughts typical of Northern Mexico and the Southwestern U.S.A., figured in their local extinction. For similar reasons, the re-introduced population on Tiburón Island likewise is vulnerable to extinction. Genetic diversity of the island population is demonstrably low, attributed mainly to genetic drift and lack of gene flow with other populations [37,49]; low genetic diversity has been correlated with lower fitness in bighorn sheep [50]. The Tiburón population also was modeled to be susceptible to stochastic effects via increased aridity and warmer temperatures associated with climate variability and change [51], as well as and continued extraction of animals for repopulation efforts on the mainland [38].

Confirmation of the prior presence and local extinction of bighorn sheep on Tiburón Island refutes their status as a non-native species. This extended baseline anchors bighorn sheep within the changing ecology of Tiburón Island, furthers their importance as a focal point for conservation and management, and presents a cautionary tale. The introduction of bighorn sheep on Tiburón in 1975 is more justifiable than realized at the time. It raises other questions, however. Now that we know that bighorn sheep were native to the island, does this necessarily quell concern about the impacts of reintroduction? Did mountain lion, their main predator on the mainland, also occur on the island but suffer extinction at the same time as bighorn sheep? Does the absence of predators matter, or is it mimicked by current levels of trophy hunting that focuses on the healthiest animals, and not on the weak?

As evidenced by local extinction on Tiburón and throughout their range [44], bighorn sheep, as other large ungulate species, face increased extinction rates from future droughts exacerbated by warming [38,52]. The Tiburón bighorn sheep population is of critical significance as a source population for mainland introductions, great economic and cultural importance for the Seri community, and an indicator of environmental conditions. Careful management of trophy hunting and introductions of new animals from the mainland to broaden genetic diversity could help avoid another local extinction.

So how should we regard the reintroduction of bighorn sheep now that we know how recently it occurred naturally on the island? Is it a restoration or a biological invasion? This question should apply to most cases of rewilding and de-extinction efforts. Native plant communities on Tiburón Island clearly co-evolved with bighorn sheep and, given the short time since local extinction, should still be resilient to the perturbations caused by renewed herbivory. It is unlikely that local plants lost their defenses since the sheep's extinction, which happened sometime between 1500 and 100 years ago. Mountain lion may or may not have also occurred on the island, and predator pressure on the introduced population now may be greatly diminished.

In other cases of rewilding, exactly how long is required for the remaining native species to evolve and lose their defenses and resilience? Native horses have been gone from North America for more than 10,000 years, and so have most of their predators. This could have had evolutionary consequences in plant and animal communities that eroded their potential resilience to rewilding. Even if bighorn "belong" on Tiburón, as well as horses in North America, does this necessarily absolve wildlife managers and conservationists from considering any and all unintended impacts of rewilding?

Finally, we recognize that our discovery of the bighorn sheep deposit in a remote cave on Tiburon Island also was unintentional. We did not set out to test whether or not bighorn sheep were native to the island; we were actually looking for something else. We now envision strategic and purposeful 'molecular caving' to pinpoint the timing and circumstances for other Holocene extinctions, and thus inform conservation efforts, in the Gulf of California and other arid regions worldwide.

Methods

To better understand past plant communities and the origin of the modern Sonoran Desert, fossil packrat (*Neotoma* spp.) middens were collected on Tiburón Island in March 2012. Among the middens we found a different type of urine-hardened deposit containing ungulate dung in a rock shelter at 235 m elevation on Hast Coopol, a low-lying volcanic peak on the eastern foothills of the Sierra Kunkaak (Fig. 1).

Thick crusts of urine-hardened sheep dung commonly line the floors of caves and rock shelters occupied by both wild and domestic sheep worldwide. Domestic sheep (*Ovis aries*) are not known to have inhabited the island. Bighorn sheep (*O. canadensis*) are the only late Pleistocene wild sheep (*Ovis* sp.) known to inhabit the Intermountain West and south into Mexico [29,53,54]. Unlike other extant Sonoran Desert artiodactyls (e.g., pronghorn, and mule deer), bighorn sheep commonly use caves and rock overhangs to bed and escape the midday heat in summer [55] and their kidneys concentrate urine to conserve water [56]. As it evaporates, the viscous urine can crystallize and cement both sediment and dung on the cave floor, much like the process that forms packrat and other rodent middens [57].

The rock shelter on Tiburón Island named here Mojet Cave, (mojet ['moxɛt], the Seri name for bighorn sheep) [58] is about 8 m deep and 15 m wide (Video S1). Two blocks or mats of artiodactyl dung (sample A consisted of two small pieces both ~20×10×8 cm, and sample B one large mat ~40×30×10 cm) were collected amid roof fall on the cave floor, following the same protocols used for sampling packrat middens [59]. Five fecal pellets taken from both the top and bottom layers of dung mat sample B were sent to the UC, Irvine W. M. Keck Carbon Cycle Accelerator Mass Spectrometry Laboratory for radiocarbon dating. All pellets dated fell within the age range of 1595–1725±20 ^{14}C yr B.P., which was calibrated to a mean age of 1476–1632 calendar years before present (cal yr B.P.) using the Intcal13 calibration [60]. A pellet from the bottom of the midden dated to 1530±15 ^{14}C yr B.P. and one from the top as 1625±20 ^{14}C yr B.P. That the dates are stratigraphically reversed indicates the deposit likely formed at once. Two independent methods, morphological identification and ancient DNA analysis were used to identify the fossil pellets. The same pellets used for *a*DNA extraction were dated to 1720±20 and 1725±20 ^{14}C yr B.P. Those used for morphological identification were not dated to preserve the intact nature of the pellet and are deposited in the East Tennessee Vertebrate Paleontology Laboratory (Eastern Tennessee State University) under collection number ETVP 4999. All necessary permits were obtained for the described study, which complied with all relevant regulations. The material for this study was collected under collector's permit FAUT–0265 granted to Exequiel Ezcurra by the Secretaría de Medio Ambiente y Recursos Naturales (SEMARNAT; authorization document SGPA/DGVS/02213-13).

Morphological Identification

We compared identically sized and shaped pellets as the one dated with an extensive dung collection of both modern and extinct herbivores housed in the East Tennessee Vertebrate Paleontology Laboratory. Isolate dung pellets of adult *Cervus* and the extinct *Euceratherium*, *Symbos*, and *Oreamnos harringtoni* typically have a ratio of the width:length (measurements in mm) versus the weight (g) distinctly greater than that produced by *Ovis*, *Antilocapra*, and the living *Oreamnos americanus*; weights greater than 0.5 g readily distinguish these larger ungulates from *Ovis* and *Antilocapra* [31]. Isolated pellets of living adult *Ovis* spp., *Oreamnos*, *Antilocapra*, and the extinct *Stockoceros* can overlap in weight (all typically weigh less than 0.3 g in weight). When considering size/shape, pronghorns will more often have a longer pellet, with width:length ratios greater than 1.1, while most often *Ovis* spp. produce a pellet ratio between 0.9 to 0.5, creating their more cuboid appearance [30]. This can vary with a more boreal and green vegetation diet. The classification of the pellets from Mojet Cave as *Ovis* is based on gross morphology and should be viewed as a 'best fit' identification.

Ancient DNA

To substantiate our identification from pellet morphology we also sequenced DNA from the fossilized pellets. We observed basic tenets of ancient DNA handling [36,61,62] by 1) restricting all handling of material prior to PCR amplification to a laboratory at Oregon State University that had never been used for genetic research or mammalian research of any kind; 2) restricting laboratory equipment used in all pre-PCR operations to equipment that had never been used with mammalian genetic samples; 3) only allowing personnel to enter the ancient DNA laboratory if they had not previously entered the modern DNA laboratory that day. The modern DNA laboratory is on a different floor within the same building. We used only newly-purchased reagents, bleached surfaces between extractions, and autoclaved supplies. We attempted to extract and amplify DNA from pellets from two different dung deposits within the larger dung mat (sample B). We scraped the exterior of pellets (where epithelial cells are concentrated, [63]) with a bleached and flamed razor blade, but also collected dust from the interior of the pellets for a second extraction, in the event that the exterior surface had been contaminated or degraded. Due to the small size of the pellets, we used two pellets for each extraction of surface material; inner material for the second extraction was taken from a single pellet. We used the Aquagenomics/AquaPrecipi DNA extraction kit (Multitarget Pharmaceuticals, Utah), to extract DNA from 0.03 g of scraped dust from each replicate. We added 15 mAU of proteinase K (Qiagen, California) and approximately 0.2 mL of 1.0 mm zirconia beads (Biospec, Oklahoma) to the dust, vortexed briefly to lyse the cells, and incubated the samples for one hour at 60°C. Using the same protocol, we conducted two blank extractions to detect contamination of reagents or supplies. Samples were rehydrated with 100 uL of 1× TE buffer (pH 8.0) and stored at 4°C.

We attempted to amplify fragments of varying lengths to evaluate the *a*DNA principle that amplification success will increase with decreasing fragment size [36]. As expected, efforts to amplify a 350 bp fragment of the 16S ribosomal RNA gene (Table S1) that has previously been used for species identification of ungulate pellets were unsuccessful, with the exception of one contaminated PCR negative that was shown by BLAST-search to have amplified human DNA. Instead, we used Primer3 [64,65] to design primer pairs to amplify a ~90 bp fragment of the 12S ribosomal RNA gene on the mitochondrial genome (Table S1). Using sequences published on GenBank (see Table S2 for accession numbers), we designed primers for locations that we identified as largely conserved but with variation between priming sites across mule and white tailed deer (*Odocoileus hemionus* and *Odocoileus virginianus*, respectively), mountain goats (*Oreamnos americanus*), pronghorn antelope (*Antilocapra americana*), domestic sheep (*Ovis aries*), domestic cattle (*Bos taurus*), and bighorn sheep (*Ovis canadensis*). We used the polymerase chain reaction (PCR) to amplify this gene fragment in reactions consisting of 5× Qiagen Multiplex PCR master mix, 5 μg bovine serum albumin, 0.2 μM of each primer, and 0.5 μL of DNA extract and brought the reaction to a 10 μL reaction volume with nuclease-free molecular grade water. Thermalcycling conditions were as follows: 15 minutes at 95°C, followed by 35 cycles of [95°C for 30 s, 54°C for 90 s, 72°C for 60 s] and a final elongation of 30 minutes at 60°C. We included a negative control (molecular grade water) in each PCR run to monitor for contamination, and also attempted to amplify the two negative extraction controls. We verified amplification using an agarose gel stained with GelRed (Biotium, California) and used 2 μL shrimp alkaline phosphatase (ExoSAP-IT, Affymetrix, Santa Clara, CA) to prepare 4 μL of DNA from each amplicon for sequencing. PCR products were submitted to the Molecular Research Core Facility at Idaho State University for bidirectional sequencing on an ABI 3130XL DNA analyzer. We also sequenced the same section of the12S gene in one bighorn sheep sample from each of: Chocolate Mountains (CA), Marble Mountains (CA), Orocopia Mountains (CA), and the Sierra Pintas (AZ). These samples were chosen to assess potential variation in the 12S gene in desert bighorn sheep across a broad geographical area, and because this subspecies (*O. c. nelsoni*) was not represented in GenBank at this 12S gene.

To compare variation between the fossil haplotypes and those of modern bighorn sheep we also designed primers for three short sections of the mitochondrial DNA control region (~80–120 bp each; Table S1; Fig. S1) within a 515 bp region previously sequenced for numerous modern bighorn sheep populations [33]. Because of the highly degraded nature of ancient DNA, we did not attempt to sequence this entire region, but chose three sections that 1) showed variability in published sequences, and 2) supported reliable primers in conserved regions. We designed primers using sequences published on GenBank for bighorn sheep described as *O. c. nelsoni*, *O. c. weemsi*, and *O. c. mexicana* to capture variation across a wide range of desert bighorn sheep (Table S2), and included two published haplotypes from bighorn sheep descended from individuals translocated to Isla Tiburón during the 1970s (accession numbers AY116621 and AY116622, Table S2). Amplification of the control region fragments followed the same PCR recipe and cycling conditions as 12S.

Because of the degraded nature of most ancient DNA and the possibility that base modifications over time could alter the sequences, for the ancient pellets, we attempted to amplify and sequence each targeted region in a minimum of three separate PCR reactions. Raw chromatogram traces were trimmed to the length at which all bases could be unambiguously identified. We verified base calls, searched each sequence on NCBI BLAST, and aligned all replicate sequences in Geneious (v.6.1.2; Biomatters, available from http://www.geneious.com/) and had two other experienced researchers independently do the same. In case of a discrepancy between calls, we reviewed sequences again with a fourth experienced observer and corrected obvious errors, after which we selected the call made by the majority of observers. Discrepancies, when they arose, typically resulted from difficulty in interpreting strings of single repeated base pairs (e.g., AAAA). We generated consensus sequences, resolving discrepancies based on majority of calls and re-evaluation of the original sequences, and compared them with sequences published on Genbank for other candidate species (Table S2). In that comparison, we included 4 published sequences from domestic sheep (*O. aries*, GenBank AF010406, EF490455, NC001941, and HE577848) and two from domestic cattle (*Bos taurus*, GenBank AB074968 and AF492351) to rule out contamination from domestic animals in our lab reagents. Published sequences were selected from different studies to capture intraspecific diversity. However, as we observed no intraspecific diversity in these two species for this 12S fragment, we included only one example for each species in further analyses.

To confirm the species identity of the ancient DNA sample and rule out the possibility of contamination from other mammals, we conducted a phylogenetic analysis on the ancient 12S sequence and reference sequences for other wild and domestic artiodactyls that we downloaded from GenBank (a 46 bp region after trimming primers). While molecular data are most frequently analyzed with model-based likelihood or Bayesian methods, those methods treat alignment gaps due to insertions and deletions as missing data, and require moderately long sequence alignments in order to infer an appropriate model of nucleotide substitution. Our very short ribosomal alignment includes several indels that clearly demonstrate phylogenetic signal; treating these as missing data in a pilot likelihood analysis resulted in a loss of almost all phylogenetic resolution in the resulting consensus. To capture the signal in the indels, we inferred a phylogeny under parsimony in PAUP* 4.0 beta [66], while treating gaps as a fifth character state, using *Antilocapra americana* as the outgroup and conducting 1000 heuristic searches with random addition sequences.

The taxonomy of bighorn sheep subspecies remains somewhat confused. Cowan originally described seven subspecies of bighorn sheep [67]. However, recent genetic and morphometric evaluations [68–73] demonstrated that there are likely only three valid subspecies: *O. c. canadensis* (including bighorn sheep formerly described as *O. c. canadensis*, *O. c. auduboni* and *O. c. californiana* excepting populations in the Sierra Nevada mountains of California), *O. c. nelsoni* (desert bighorn sheep, including bighorn sheep formerly described as *O. c. nelsoni*, *O. c. mexicana*, *O. c. cremnobates*, and likely *O. c. weemsi* although insufficient specimens were available to confirm the validity of *O. c. weemsi*), and *O. c. sierrae* (populations in the Sierra Nevada Mountains of California, previously described as *O. c. californiana*).

Supporting Information

Figure S1 Control regions of Isla Tiburón *a*DNA and select bighorn sheep populations. Partial control region sequences of Isla Tiburón *a*DNA sample and published bighorn sheep samples, sequenced as three fragments (a–c). Open boxes denote primer region, dot indicates identical bases between sequences, dash indicates deletion, N indicates unknown base.

Sequence position numbers are derived from the reverse complement of *O. c. nelsoni* haplotype B [44] (GenBank Accession AY903995).

Table S1 Primer sequences and fragment size. The genes sequenced or attempted from ancient fecal pellets from Tiburón Island, Mexico.

Table S2 Accession numbers of sequences used. The sequences used for primer development and sequence alignment for reference sequences and ancient fecal DNA of unknown origin from Tiburón Island, Mexico.

Video S1 Movie of Mojet Cave, Tiburón Island. The video shows the fossil dung mat as encountered before any sampling or manipulation was made. Taken by Wilder 29 Mar 2012, featured in the video (in order of appearance) Andrew Semotiuk, Carlos Armando Mendes Romero, Jose Ramon Torres Molina.

Acknowledgments

Field assistance was provided by Carlos Armando Mendes Romero, Andrew Semotiuk, and Jose Ramon Torres Molina. Cathy Moser Marlett created the base maps for Fig. 1, Sandra Swift took the photographs for Fig. 2, and general assistance was provided by Kate A. Rylander. We thank Scott Heppell and Carl Schreck for access to alternative laboratory facilities and Brian Sidlauskas, Kendra Hoekzema, and Anne Davis for analytical assistance. Any use of trade, product, or firm names is for descriptive purposes only and does not imply endorsement by the U.S. Government.

Author Contributions

Conceived and designed the experiments: BTW JLB CWE RSC JIM. Performed the experiments: CWE RSC JIM. Analyzed the data: BTW JLB CWE RSC JIM EE. Contributed reagents/materials/analysis tools: JLB CWE JIM EE. Wrote the paper: BTW JLB CWE.

References

1. Montaño Herrera R (2010) Hapj Coi Iicp Hac. In: Herrera Casanova L, Astorga ML, Perales Torres O, Montaño Herrera R, Moreno Herrera X, editors. Comcaac Quih Ziix Quih Ocoaaj hac. Socaaix and Haxöl Iihom, Sonora. pp 31–32. Available: http://www.lengamer.org/admin/language_folders/seri/user_uploaded_files/links/File/Alfabetizacion/Enciclopedia_2010.pdf. Accessed 29 Oct 2013.
2. Donlan JC, Greene HW, Berger J, Bock CE, Bock JH, et al. (2005) Re-wilding North America. Nature 436(7053): 913–914.
3. Donlan JC, Berger J, Bock CE, Bock JH, Burney DA, et al. (2006) Pleistocene rewilding: An optimistic agenda for twenty-first century conservation. Am. Nat 168(5): 660–681.
4. Sherkow J, Greely HT (2013) What if extinction is not forever? Science 340(6128):32–33.
5. Sandler R (2013) The ethics of reviving long extinct species. Conservation Biology. doi: 10.1111/cobi.12198
6. Jansen T, Forster P, Levine MA, Oelke H, Hurles M, et al. (2002) Mitochondrial DNA and the origins of the domestic horse. Proc Natl Acad Sci USA 99(16): 10905–10910.
7. Hofreiter M, Serre D, Poinar HN, Kuch M, Pääbo S (2001) Ancient DNA. Nat Rev Genet 2(5): 353–359.
8. Murray DC, Pearson SG, Fullagar R, Chase BM, Houston J, et al. (2012) High-throughput sequencing of ancient plant and mammal DNA preserved in herbivore middens. Quat Sci Rev 58: 135–145.
9. Lorenzen ED, Nogués-Bravo D, Orlando L, Weinstock J, Binladen J, et al. (2011) Species-specific responses of Late Quaternary megafauna to climate and humans. Nature 479(7373): 359–364.
10. Richman AD, Case TJ, Schwaner TD (1988) Natural and unnatural extinction rates of reptiles on islands. Am Nat 131(5): 611–630.
11. Lambeck K, Chappell J (2001) Sea level change through the last glacial cycle. Science 292(5517): 679–686.
12. Graham MH, Dayton PK, Erlandson JM (2003) Ice ages and ecological transitions on temperate coasts. Trends Ecol Evol 18(1): 33–40.
13. Medellín RA, Manterola C, Valdéz M, Hewitt DG, Doan-Crider D, et al. (2005) History, ecology, and conservation of the pronghorn antelope, bighorn sheep, and black bear in Mexico. In: Cartron JLE, Ceballos G, Felger RS, editors. Biodiversity, Ecosystems, and Conservation in Northern Mexico. New York: Oxford University Press. pp. 387–404.
14. Ezcurra E, Bourillón L, Cantú A, Martínez ME, Robles A (2002) Ecological conservation. In: Case TJ, Cody ML, Ezcurra E, editors. A New Island Biogeography of the Sea of Cortés. New York: Oxford University Press. pp. 417–444.
15. Carmony NB, Brown DE (1983) The Wilderness of the Southwest: Charles Sheldon's Quest for Desert Bighorn Sheep and Adventures with the Havasupai and Seri Indians. Salt Lake City: University of Utah Press. 219 p.
16. Felger RS, Moser MB (1985) People of the Desert and Sea: Ethnobotany of the Seri Indians. Tucson: University of Arizona Press. 438 p.
17. Montoya B, Gates G (1975) Bighorn capture and transplant in Mexico. Desert Bighorn Council Transactions 19: 28–32.
18. Hernández-Alvídrez R., Campoy-Favela J (1989) Observaciones recientes de la población de Borrego cimarrón en Isla Tiburón, Sonora, México. Ecológica 1(1): 1–29.
19. DeForge JR, Valdez R (1984) Status of investigations of desert bighorn sheep in Sonora, Mexico. Desert Bighorn Council Transactions 28: 41–43.
20. Pallares E (1999) El Borrego cimarrón: beneficio para los indígenas. National Geographic (Spanish version, section Geografia America Latina) 3(4): 15–32.
21. Lee R, López-Saavedra EE (1994) A second helicopter survey of desert bighorn sheep in Sonora, Mexico. Desert Bighorn Council Transactions 38: 12–13.
22. Mellink E (2002) Invasive vertebrates on islands of the Sea of Cortés. In: Tellman B, editor. Invasive Species in the Sonoran Region. Tucson: The University of Arizona Press. pp. 112–125.
23. Navarro F (1999) Unidad de manejo y aprovechamiento sustentable de la vida silvestre (UMA) en Isla Tiburón: Un ejemplo de desarrollo sustentable. Insulario (Gaceta Informativa de la Reserva islas del Golfo de California) 8: 13–15.
24. Paulson K (2009) Tiburon Island desert sheep permit on auction! HuntingLife.com. Available: http://www.huntinglife.com/blog/detail/tiburon-island-desert-sheep-permit-on-auction. Accessed 29 Oct 2013.
25. Wild Sheep Foundation (2012) Tiburon Island Desert Bighorn Sheep Permit. Wild Sheep Foundation. Available: http://bid.wildsheepfoundation.org/Tiburon-Island-Desert-Bighorn-Sheep-Permit_i11627138. Accessed 29 Oct 2013.
26. Felger RS, Wilder BT, Romero-Morales H (2012) Plant Life of a Desert Archipelago: Flora of the Sonoran Islands in the Gulf of California. Tucson: University of Arizona Press. 624 p.
27. Hofreiter M, Mead JI, Martin PS, Poinar HN (2003) Molecular caving. Current Biology 13(18): R693–R695.
28. Tarango LA, Krausman PR, Valdez R, Kattnig RM (2002) Research observation: Desert bighorn sheep diets in Northwestern Sonora, Mexico. Journal of Range Management 55(6): 530–534.
29. Mead JI, Czaplewski NJ, Agenbroad LD (2005) Rancholabrean (Late Pleistocene) mammals and localities of Arizona. In: McCord RD, editor. Vertebrate Paleontology of Arizona. Mesa Southwest Museum Bulletin 11. pp. 139–180. Available: http://azmnh.org/paleo/pdf/Descrip11.pdf. Accessed 29 Oct 2013.
30. Mead JI, O'Rourke MK, Foppe TM (1986) Dung and diet of the extinct Harrington's mountain goat (*Oreamnos harringtoni*). J Mammal 67(2): 284–293.
31. Kropf J, Mead JI, Anderson RS (2007) Dung, diet, and the paleoenvironments of the extinct shrub-ox (*Euceratherium collinum*) on the Colorado Plateau, USA. Quaternary Res 67(1): 143–151.
32. Epps CW, Wehausen JD, Palsboll PJ, McCullough DR (2010) Using genetic tools to track desert bighorn sheep colonizations. J Wildl Manage 74(3): 522–531.
33. Hiendleder S, Kaupe B, Wassmuth R, Janke A (2002) Molecular analysis of wild and domestic sheep questions current nomenclature and provides evidence for domestication from two different subspecies. Proc R Soc Lond, B, Biol Sci 269(1494): 893–904.
34. Marota I, Rollo F (2002) Molecular paleontology. Cell Mol Life Sci 59(1): 97–111.
35. Chelomina GN (2006) Ancient DNA. Russ J Genet 42(3): 219–233.
36. Pääbo S, Poinar H, Serre D, Jaenicke-Després V, Hebler J, et al. (2004) Genetic analyses from ancient DNA. Annu Rev Genet 38: 645–679.
37. Gasca-Pineda J, Cassaigne I, Alonso RA, Eguiarte LE (2013) Effective population size, genetic variation, and their relevance for conservation: The bighorn sheep in tiburon Island and Comparisons with Managed Artiodactyls. PLoS ONE 8(10): e78120.
38. Colchero F, Medellín RA, Clark JS, Lee R, Katul GG (2009) Predicting population survival under future climate change: density dependence, drought and extraction in an insular bighorn sheep. J Anim Ecol 78(3): 666–673.
39. Cook ER, Woodhouse CA, Eakin CM, Meko DM, Stahle DW (2004) Long-term aridity changes in the western United States. Science 306(5698): 1015–1018.

40. Caso M, González-Abraham C, Ezcurra E (2007) Divergent ecological effects of oceanographic anomalies on terrestrial ecosystems of the Mexican Pacific coast. Proc Natl Acad Sci USA 104(25): 10530–10535.
41. Epps CW, McCullough DR, Wehausen JD, Bleich VC, Rechel JL (2004) Effects of climate change on population persistence of desert-dwelling mountain sheep in California. Conserv Biol 18(1): 102–113.
42. Epps CW, Palsboll PJ, Wehausen JD, Roderick GK, McCullough DR (2006) Elevation and connectivity define genetic refugia for mountain sheep as climate warms. Mol Ecol 15(14): 4295–4302.
43. Bleich VC, Wehausen JD, Ramey RRII, Rechel JL (1996) Metapopulation Theory and Mountain Sheep: Implications for conservation. In: McCullough DR, editor. Metapopulations and Wildlife Conservation. Washington D.C.: Island Press. pp. 353–373.
44. Epps CW, Palsboll PJ, Wehausen JD, Roderick GK, Ramey RRII, et al. (2005) Highways block gene flow and cause a rapid decline in genetic diversity of desert bighorn sheep. Ecol Lett 8(10): 1029–1038.
45. Nabhan GP (2000) Cultural dispersal of plants and reptiles to the Midriff Islands of the Sea of Cortés: integrating indigenous human dispersal agents into island biogeography. J Southwest 42(3): 545–558.
46. Hayden JD (1985) Food Animal Cremations of the Sierra Pinacate, Sonora, Mexico. Kiva 50(4): 237–248.
47. Grant C (1980) The desert bighorn and aboriginal man. In: Monson G, Summer L, Allen RW, editors. The Desert Bighorn: Its Life History, Ecology & Management. University of Arizona Press. pp. 7–39.
48. Nabhan GP (1993) When the spring runs dry. In: GP . Nabhan, editor. Counting Sheep: Twenty Ways of Seeing Desert Bighorn. pp. 7–26.
49. Hendrick PW, Gutierrez-Espeleta GA, Lee RN (2001) Founder effect in an island population of bighorn sheep. Mol Ecol 10(4): 851–857.
50. Johnson HE, Mills LS, Wehausen JD, Stephenson TR, Luikart G (2011) Translating effects of inbreeding depression on component vital rates to overall population growth in endangered bighorn sheep. Conserv Biol 25(6): 1240–1249.
51. Seager R., Ting M, Held I, Kushnir Y, Lu J, et al. (2007) Model projections of an imminent transition to a more arid climate in southwestern North America. Science 316(5828): 1181–1184.
52. Duncan C, Chauvenet ALM, McRae LM, Pettorelli N (2012) Predicting the future impact of droughts on ungulate populations in arid and semi-arid environments. PLoS ONE 7(12): e51490.
53. Valdez R, Krausman PR (1999) Mountain sheep of North America. Tucson: University of Arizona Press. 353 p.
54. Groves C, Grubb P (2011) Ungulate Taxonomy. Baltimore: Johns Hopkins University Press. 317 p.
55. Cain JW, Krausman PR, Morgart JR, Jansen BD, Pepper MP (2008) Responses of desert bighorn sheep to removal of water sources. Wildlife Monogr 171: 1–32.
56. Horst R (1971) Observations on the kidney of the desert bighorn sheep. Desert Bighorn Council Transactions 15: 24–37.
57. Mead JI, Coats LL, Schubert BW (2003) Late Pleistocene faunas from caves in the eastern Grand Canyon. In: Schubert W, Mead JI, Graham RW, editors. Ice Age Cave Faunas of North America. Bloomington: Indiana University Press. pp. 64–86.
58. Moser MB, Marlett SA (2010) Comcáac quih yaza quih hant ihíip hac; Diccionario Seri-Español-Inglés, 2nd ed. México, DF: Editorial UniSon, Hermosillo & Plaza y Váldés Editores. 937 p. Available: http://www-01.sil.org/mexico/seri/G004c-DiccionarioEd2-sei.pdf. Accessed 29 Oct 2013.
59. Spaulding WG, Betancourt JL, Croft LK, Cole KL (1990) Packrat Middens: Their composition and methods of analysis. In: Betancourt JL, Van Devender TR, Martin PS, editors. Packrat Middens: The last 40,000 years of biotic change. Tucson: University of Arizona Press. pp. 59–84.
60. Reimer PJ, Bard E, Bayliss A, Beck JW, Blackwell PG, et al. (2013) IntCal13 and Marine13 radiocarbon age calibration curves 0–50,000 years cal BP. Radiocarbon 55(4): 1869–1887.
61. Willserslev E, Cooper A (2005) Review Paper. Ancient DNA. Proc Royal Soc B 272(1558): 3–16.
62. Pääbo S (1990) Amplifying ancient DNA. In: Innis MA, Gelfand DH, Sninsky JJ, White TJ, editors. PCR protocols: A guide to methods and applications. San Diego: Academic Press. pp. 159–166.
63. Wehausen JD, Ramey RR, Epps CW (2004) Experiments in DNA extraction and PCR amplification from bighorn sheep feces: the importance of DNA extraction method. J Hered 95(6): 503–509.
64. Koressaar T, Remm M (2007) Enhancements and modifications of primer design program Primer3. Bioinformatics 23(10): 1289–1291.
65. Untergrasser A, Cutcutache I, Koressaar T, Ye J, Faircloth BC, et al. (2012) Primer3 – new capabilities and interfaces. Nucleic Acids Res 40(15): e115.
66. Swofford DL (2003) PAUP* Phylogenetic Analysis Using Parsimony (*and Other Methods). Version 4. Sinauer Associates, Sunderland, Massachusetts.
67. Cowan IM (1940) The distribution and variation in the native sheep of North America. Am Midl Nat 24(3): 505–580.
68. Ramey RRII (1993) Evolutionary genetics and systematics of North American mountain sheep: implications for conservation. Ph. D. Dissertation. Ithaca: Cornell University.
69. Wehausen JD, Ramey RRII (1993) A morphometric reevaluation of the Peninsular bighorn subspecies. Desert Bighorn Council Transactions 37: 1–10.
70. Ramey RRII (1995) Mitochondrial DNA variation, population structure, and evolution of mountain sheep in the southwestern United States and Mexico. Mol Ecol 4(4): 429–439.
71. Ramey RRII (1999) New perspectives on the evolutionary origins, historic phylogeography, and population structure of North American mountain sheep. Transactions of the North American Wild Sheep Conference 2: 9–20.
72. Wehausen JD, Ramey RRII (2000) Cranial morphometric and evolutionary relationships in the northern range of *Ovis canadensis*. J Mammal 81(1): 145–161.
73. Wehausen JD, Bleich VC, Ramey RRII (2005) Correct nomenclature for Sierra Nevada bighorn sheep. California Fish and Game 91(3): 216–218.

22

Conus: First Comprehensive Conservation Red List Assessment of a Marine Gastropod Mollusc Genus

Howard Peters[1]*, Bethan C. O'Leary[1], Julie P. Hawkins[1], Kent E. Carpenter[2], Callum M. Roberts[1]

1 Environment Department, University of York, York, United Kingdom, **2** International Union for Conservation of Nature, Global Marine Species Assessment, Biological Sciences, Old Dominion University, Norfolk, Virginia, United States Of America

Abstract

Marine molluscs represent an estimated 23% of all extant marine taxa, but research into their conservation status has so far failed to reflect this importance, with minimal inclusion on the authoritative Red List of the International Union for the Conservation of Nature (IUCN). We assessed the status of all 632 valid species of the tropical marine gastropod mollusc, *Conus* (cone snails), using Red List standards and procedures to lay the groundwork for future decadal monitoring, one of the first fully comprehensive global assessments of a marine taxon. Three-quarters (75.6%) of species were not currently considered at risk of extinction owing to their wide distribution and perceived abundance. However, 6.5% were considered threatened with extinction with a further 4.1% near threatened. Data deficiency prevented 13.8% of species from being categorised although they also possess characteristics that signal concern. Where hotspots of endemism occur, most notably in the Eastern Atlantic, 42.9% of the 98 species from that biogeographical region were classified as threatened or near threatened with extinction. All 14 species included in the highest categories of Critically Endangered and Endangered are endemic to either Cape Verde or Senegal, with each of the three Critically Endangered species restricted to single islands in Cape Verde. Threats to all these species are driven by habitat loss and anthropogenic disturbance, in particular from urban pollution, tourism and coastal development. Our findings show that levels of extinction risk to which cone snails are exposed are of a similar magnitude to those seen in many fully assessed terrestrial taxa. The widely held view that marine species are less at risk is not upheld.

Editor: David L. Roberts, University of Kent, United Kingdom

Funding: The Natural Environment Research Council (NERC) and Economic and Social Research Council (ESRC) generously supported the Red List research under a studentship grant to Howard Peters http://www.nerc.ac.uk/funding/. The authors gratefully acknowledge funding from the Biodiversity Synthesis Center of the Encyclopedia of Life (EOL) at the Field Museum of Natural History, Chicago for the Conus Synthesis Workshop http://synthesis.eol.org/center/funding; also Tom Haas and the New Hampshire Charitable Foundation and the Thomas W. Haas Foundation for their support of the Global Marine Species Assessment under the IUCN Global Species Programme. The funders had no role in study design, data collection and analysis, decision to publish, or preparation of the manuscript.

Competing Interests: The authors have declared that no competing interests exist.

* E-mail: hp510@york.ac.uk

Introduction

Extinction risk of marine organisms has attracted little attention compared to that of terrestrial taxa, with a widely held view that such risk is inconsequential due to high dispersal ability and large geographic ranges [1,2] especially when taking reference from the fossil record [1,3]. These beliefs are particularly prevalent when considering marine invertebrates, where a decline in abundance of the important phylum *Mollusca* has been overshadowed by the collapse in many exploited vertebrates, especially finfish [4]. This is primarily due to their relatively minor contribution to human protein requirements and the generally held belief that molluscs possess greater resilience to extinction through their perceived wide distribution and a likelihood of hidden pockets of survivors [5]. Marine invertebrates in general are seriously under-represented within the IUCN Red List [6]. Only cuttlefish, lobsters and scleractinian corals have been fully assessed and published [6,7]. Although limited research on the impact of habitat loss and fishing pressure on marine gastropod molluscs has been undertaken on a regional scale including for shell fisheries [8,9], there have been no comprehensive assessments of trends in species abundance, commercial and environmental impacts and extinction risk to any genera with a global biogeographical distribution.

Cone snails of the genus *Conus* offer an excellent opportunity to explore global threats to marine molluscs owing to their exceptional diversity [10], wide distribution, high degree of endemism, varied depth distribution [11], and an established global market in their trade from amateur shell collectors to commercial traders [12]. In addition, cone snails are used in some communities in the Pacific as an occasional foodstuff [13] but, more importantly, they are actively targeted by international drug companies and researchers as a potential pharmacological resource [14].

Cone snails constitute the family *Conidae*, which together with the *Turridae* (turrid snails) and *Terebridae* (auger snails) comprise the superfamily *Conoidea* otherwise known as *Toxoglossa* ('poison tongue') owing to the venom apparatus they deploy for immobilising prey [15]. The *Conoidea* form part of the order *Neogastropoda* in the sub-class *Prosobranchia* of the class *Gastropoda* of the phylum *Mollusca* [11].

Cone snails live throughout the world's tropical coastal waters with a steep latitudinal diversity gradient away from the tropics, extending into cooler regions that include southern California, northern Gulf of Mexico, Florida and the Carolinas, North Africa, the Mediterranean, South Africa, Australia, southern Japan and

China [16]. Distribution varies widely with some species occurring across the entire tropical Indo-Pacific but others restricted to a single bay or seamount [11,17].

The genus *Conus* is taxonomically challenging. Although morphological characteristics of the shell remain the initial means of species identification [11], more recently, other traits have also been employed to differentiate among species, in particular the radular teeth used in the capture of prey, whose shape and structure not only reflects the dietary preferences of the species [18] but may also be specific to a single species [19]. Separation of species through DNA sequence variations provides even greater reliability, but more recently, character-based DNA barcoding has been highly effective in distinguishing among closely-related species [20]. For this assessment we relied upon expertise from taxonomists in *Conus* to create a dataset of valid species.

The fossil record indicates that the first *Conus* appeared in a sea that covered what is now England and France during the Lower Eocene around 55 mya [21]. During subsequent radiations the genus expanded around the globe and by the Holocene had formed into four biogeographical regions: Indo-Pacific (IP), Eastern Pacific (EP), Western Atlantic (WA), and Eastern Atlantic (EA). Although widening of the Atlantic during the Cretaceous and Cenezoic has today created an impermeable barrier to *Conus* crossing the ocean, there have been some migrations in the past, as witnessed from the fossil record and more recently by *C. ermineus* extant in both the EA and WA and *C. chaldaeus, C. ebraeus* and *C. tessulatus* found in both the IP and EP [22].

The majority of the 632 species of *Conus* assessed (53.6%) occurs in the infralittoral zone of 5 m deep or less, with most of the remaining species (27.7%) at 50 m deep or less. However, there are some species such as *C. teramachii* that live in deeper parts of the continental shelf extending to 1,000 m where they may be brought to the surface as bycatch of demersal fisheries. The bathymetric ranges of individual species vary considerably with some shallow water species living within a one or two metre depth range and some deep-water species being found within a 500 m range or more [11].

Microhabitats vary by species and most often consist of sand or mud into which the cone snail may burrow, but may also include inter-tidal limestone benches (the smooth remains of reef structures from earlier geological periods when sea levels were higher [23]) with sand or algal turf, sub-tidal reef platforms with living and dead corals, or boulders with sandy layers [24]. They may also be found among coral rubble and occasionally among mangroves and seagrasses.

Cone snails are generally nocturnal in their feeding habit [25] and group-specific in their preference for worms, molluscs or fish (Fig. 1) although some species have a mixed diet [26]. The smallest groupings by diet are the obligate piscivores with around 50 species [11,27], and obligate molluscivores with approximately 80 [28]. The majority of *Conus* are vermivorous with polychaetes representing the largest dietary component, that can be the exclusive source of food for some species [11]. All cone snails use venom to immobilize their prey. The diversity of venoms employed by a particular species in the capture of prey is a reflection of the degree of specialisation in its diet [29].

From the earliest civilizations, people have prized cone shells for their exceptional beauty, with examples discovered among prehistoric artefacts used for personal adornment extending back 5,000 years [30]. Their striking patterns and wide range of colours and shades continue to attract collectors today with rare examples in perfect condition changing hands for thousands of dollars with common and abundant species traded for cents to a dollar or two each [12].

Over millions of years *Conus* has evolved a battery of peptide toxins (conopeptides/conotoxins) for immobilizing prey [31]. The venom of each species is a cocktail mixed from between 50 and 200 different peptides each of only 10 to 35 amino acids in length and is generally targeted at voltage-gated or ligand-gated ion channels [30]. These conopeptides have become a focus for biomedical research worldwide [14]. Indeed, with the probability that there are on average over 100 distinct toxins for each species [30], as a whole, the *Conidae* can probably synthesize in excess of 50,000 toxins with little, if any, replication [32].

Cone snails are therefore important to: a) biodiversity; they have evolved into one of the largest of all marine genera, b) biopharmaceutics; they offer unparalleled opportunities in the development of novel drugs, and c) economics; their shells provide income to poor fishing communities through sales to tourists, traders and a global business in the specimen shell trade.

Habitat loss is considered by many malacologists to be the primary risk factor facing tropical marine mollusc species (Bouchet pers. comm. 2011) and there is plenty of hard evidence to support this view. In Queensland, Australia, for example, abundance and species richness of mollusc assemblages have been shown to be adversely affected by removal of subtropical mangrove forests, with population declines of 83% recorded [33]. In San Diego, Southern California the endemic horn snail, *Cerithidia fuscata*, that lived along intertidal mudflats was last seen in 1935 after pollution and dredging had driven it to extinction [34]. Where coral cover has been extensively damaged or degraded through pollution, sedimentation, coastal development and destructive fishing, as witnessed throughout much of the tropics, coral-associated molluscs such as the *Conidae* are being usurped by bivalve crevice dwellers [35].

In this paper we report one of the first comprehensive extinction risk assessments of a taxonomically well-resolved marine taxon. Our research assesses the extinction risk to the global populations of *Conus*, examining each species' distribution, current and projected threats from disturbance to habitats, pollution, coastal development, and shell gathering. We have examined where possible the effects of fragmentation of populations and the likely impact of demersal fisheries on deeper water species. The assessment enables us to reappraise whether marine taxa are less extinction prone than terrestrial. In addition, our aim is to provide data in support of conservation measures for those species at the greatest risk of extinction over the short to medium term.

Methods

Red List Assessment

We used the assessment standards and procedures of the International Union for the Conservation of Nature (IUCN) Red List of Threatened Species to assess extinction risk to 632 species of *Conus*. This is the world's leading resource for describing the global conservation status of plants and animals and uses a standard methodology to classify species into one of nine categories, together with a codified set of criteria [36]. The assessment includes examination of the effects of both ecological change and commercial exploitation on the subject taxa. Data derived during the research and discovery process for each species is compiled to a standard format together with maps, images and other supporting documentation.

Following taxonomic review, we divided valid species into 12 biogeographical working sets for detailed assessment. A comprehensive assessment was not possible for those species where data was substantially absent. For example, species endemic to areas of protracted civil unrest such as the Horn of Africa may not have

Figure 1. Diet and toxicity. Left: *C. geographus* Linnaeus, 1758; piscivorous, 65–165 mm; intertidal to 20 m; significant fatality risk to humans. Centre: *C. textile* Linnaeus, 1758; molluscivorous, 40–150 mm; intertidal to 50 m; handle with extreme caution. Right: *C. betulinus* Linnaeus, 1758; vermivorous, 55–177 mm; intertidal to 20 m; minimal risk to humans; note operculum. All species Indo-Pacific.

been researched in the field for many years. Coincidentally, these regions are not generally subject to intensive coastal development, harbour works and refineries and so may offer a degree of protection to marine taxa. Similarly, species occurring in deep water, where recovery is most commonly through fisheries by-catch, often suffer a paucity of data including extent of distribution and habitat types. Furthermore, bathymetry data will often rely on the questionable estimation of fishers. Wherever possible for deep-water *Conus*, we have focussed our attention on the level of demersal fishing in the area, including destructive methods such as dredging that may seriously affect mollusc assemblages.

Most *Conus* species, however, occur in shallow water where impacts such as coastal development, pollution and habitat destruction can be more easily recorded. Such threats can give rise to population fragmentation leading to a serious decline in abundance which may be difficult to quantify until it has become extreme. However, indicators including market prices for specimen shells provide a useful guide to increasing scarcity. Knowledge voids are common for *Conus* but where they occur we have, where possible, used estimation or inference using suboptimal data permitted under Red List standards [36]. Despite this, 13.8% of *Conus* species were found to be so deficient in data we were unable to make an assessment with any degree of reliability.

Key indicators of risk

Distribution. A key indicator of potential risk to a species is the size of its geographical distribution. The Red List standard assessment uses two measures: Extent of Occurrence (EOO) and Area of Occupancy (AOO). EOO for marine species is the area within a polygon drawn around the boundary of the species' range, excluding land areas. This will include areas which may not be physically occupied by the taxon, e.g. deep water, but which could contribute to larval dispersal. AOO is the physical area within the EOO in which the taxon is known to occur. For shallow water species, this may be calculated from the perimeter of an island or length of coastline, extended by the width of habitat calculated from the known or inferred bathymetric range of the species over the area under review. However, for 'linear' habitats such as rivers and coastlines, IUCN suggests that their standard habitat width of 2 km should be used in computing AOO [36] and we have adopted this approach in the assessment for *Conus*. It should be noted that both the AOO and the EOO are only of major significance in assessing the level of threat if the species has a restricted range, and that for most wide-ranging *Conus* species that fall outside the parameters of the assessment criteria these range sizes are not calculated. However, spatial data, derived from range maps created for every species assessment, were projected using ArcGIS version 10.1 (Environmental Systems Research Institute) to generate species richness maps for a) all species and b) only those species with a range less than the median mapped range size. All data were standardised onto 1° grid cells and projected to world cylindrical equal area.

Number of Locations. It is possible that a catastrophic event could have a profound effect on the population size of some species. Although marine molluscs are resilient in being able to endure physical forces such as extreme weather events, small populations may be extirpated as a result of sudden habitat loss caused by catastrophic events such as major oil spills. The 'location' count indicates the number of areas in which a single catastrophic event could affect all individuals of the taxon present, events that may cumulatively drive a species into extinction. The value of this measure is another key factor in determining the level of risk a global population faces.

Literature search

We conducted a comprehensive search through published papers and other literature for data relating to *Conus* species' populations, depth, distribution, habitats, trade in animals and shells, use for foodstuff, pharmaceuticals, etc., together with any conservation measures in place, including indirect conservation as may be offered by marine protected areas. We sought information on current and possible future threats, including coastal development for tourism, industry or port construction, nutrient loading from agricultural run-off, pollution from domestic and industrial effluent, intensive trawling, siltation from land-based sources, dredging for shipping channels and mineral extraction. Data on activities such as these can often only be found in trade publications, contract award notifications etc.

We also examined the market in shells to determine 'collectability', pricing fluctuations, scarcity and demand. Some shells with exceptional colour and form will achieve iconic status, and if they are also rare like *C. gloriamaris* or *C. milneedwardsi*, it adds to their cachet. Species that live within a highly restricted range, within a single bay for example, are often at heightened risk from human activity. This particularly applies to shallow water species which may be gathered as curios in areas where new beach tourism projects are being developed or planned. We synthesised distribution data including observed fragmentation, location counts, marketability, population declines and threats for each species to apply one of the nine categories listed below.

Assessment categories

There are three categories of extinction risk: critically Endangered (CR), Endangered (EN) and Vulnerable (VU) that broadly define 'extremely high', 'very high, or 'high' risk of extinction respectively. In addition, there are two extinct categories, Extinct (EX) and Extinct in the Wild (EW), and three other categories: Near Threatened (NT) for species that will be elevated to a threatened category in the short term unless the potential risk is removed; Data Deficient (DD) where there is insufficient data to determine a category, and Least Concern (LC) where current and projected population levels indicate the species is not at risk. As this was a comprehensive assessment, we did not use the category Not Evaluated (NE), where the species has been recorded but no assessment has been carried out.

For *Conus*, the criteria in support of the selected category are primarily derived from a range of variables based on estimated population size and/or level of decline together with species range size and location count.

Synthesis and pre-publication checks

Following our research and assessment, the results were reviewed by a panel of fourteen international experts, each with specialist knowledge of the *Conus* species within their allotted biogeographical working sets. The review took the form of a five day synthesis workshop with teams comprising leading academics together with renowned specialists from the commercial sector with comprehensive field knowledge of species' distribution, scarcity and threats, and a facilitator experienced in Red List standards and procedures. This peer-review process confirmed or modified findings of the original assessment authors, and allowed inclusion of supplementary field-based knowledge from the participating experts. All reports were checked for consistency by the Mollusc Specialist Group of the IUCN Species Survival Commission before final approval and submission for publication through the IUCN Red List Unit.

Results

The greatest species richness for *Conus* is in the Philippines and countries to the south and east towards Papua New Guinea, the Solomon Islands, New Caledonia and Fiji (Fig. 2A). Species occurring in the Atlantic including West Africa and the Gulf of Mexico and Caribbean are fewer in number. However, when very wide-ranging species are omitted, i.e. those equal to or greater than the median mapped range size, centres of endemism are revealed in scattered locations including the Caribbean, the Marquesas, New Caledonia and the Eastern Atlantic, in particular Cape Verde, Senegal and Angola (Fig. 2B).

Global threats

Three of 632 *Conus* species assessed were considered to be Critically Endangered (CR), 11 Endangered (EN) and 27 Vulnerable (VU), which together represent 6.5% of all global species (Fig. 3), with a further 26 species (4.1%) categorised as Near Threatened (NT). Over one in ten of all *Conus* species is therefore considered at risk or may become so in the near future. 87 species (13.8%) were categorised as Data Deficient (DD) of which 75 (86.2%) occur in the Indo-Pacific.

All 14 CR and EN species occur in the waters off Cape Verde and Senegal, West Africa (Table 1). Of the 27 assessed as VU, eight are from Cape Verde and Senegal with three from Angola (Table 1), seven from the Western Atlantic (Table 2), and nine from the Indian Ocean, including two from Western Australia (Table 3). Only three threatened species occur east of longitude 60 (Oman to Mascarenes): *C. rawaiensis* from Western Thailand and *C. compressus* and *C. thevenardensis* from Western Australia – all VU. According to this assessment procedure, there are no threatened species in the Pacific (Fig. 4). Of the 26 Near-Threatened species (NT), Cape Verde and Senegal are again over-represented with 14 of the 17 species from the Eastern Atlantic. Of the remainder in this category, five are from the Western Atlantic, one from the Western Indian Ocean, and three from the Pacific (Fig. 4).

Analysis by Region

Marine molluscs that are wide-ranging are likely to be more resilient against threats than those that are range-restricted, with dispersed populations providing a reservoir for re-colonization in the event of local extirpations [2]. The Eastern Atlantic species occupy a limited length of coast with few islands when compared to the Western Atlantic and, more particularly, the Indo-Pacific. It is also intersected by large rivers draining the tropical land mass of Africa which render substantial areas of coastal water unsuitable for many marine molluscs. Conversely, islands of the tropical Indo-Pacific and Caribbean contribute substantial areas of shallow water habitat suitable for taxa such as *Conus* and do not generally suffer any significant flux of freshwater. Figure 5 shows the percentage distribution of species' range sizes within each of the four oceanic regions. This graphically illustrates that wide-ranging *Conus* species, i.e. AOO >2,000 km^2, are uncommon within the Eastern Atlantic compared to the other regions.

Eastern Atlantic. Ninety-eight species of *Conus* occur along the Eastern Atlantic seaboard from the Mediterranean and Morocco south to Namibia, with associated island archipelagos including the Canaries, Azores, and Cape Verde (plus one: *C. ermineus*, that also occurs in the Western Atlantic and was included in that region). There is one species from the island of St Helena, *C jourdani*, within this grouping although no live specimens have been observed and it is categorised as DD. With three CR, 11 EN and 11 VU species, representing 25.5% of the Eastern Atlantic species, and a further 17 species NT (Table 1), 42.9% of Eastern Atlantic

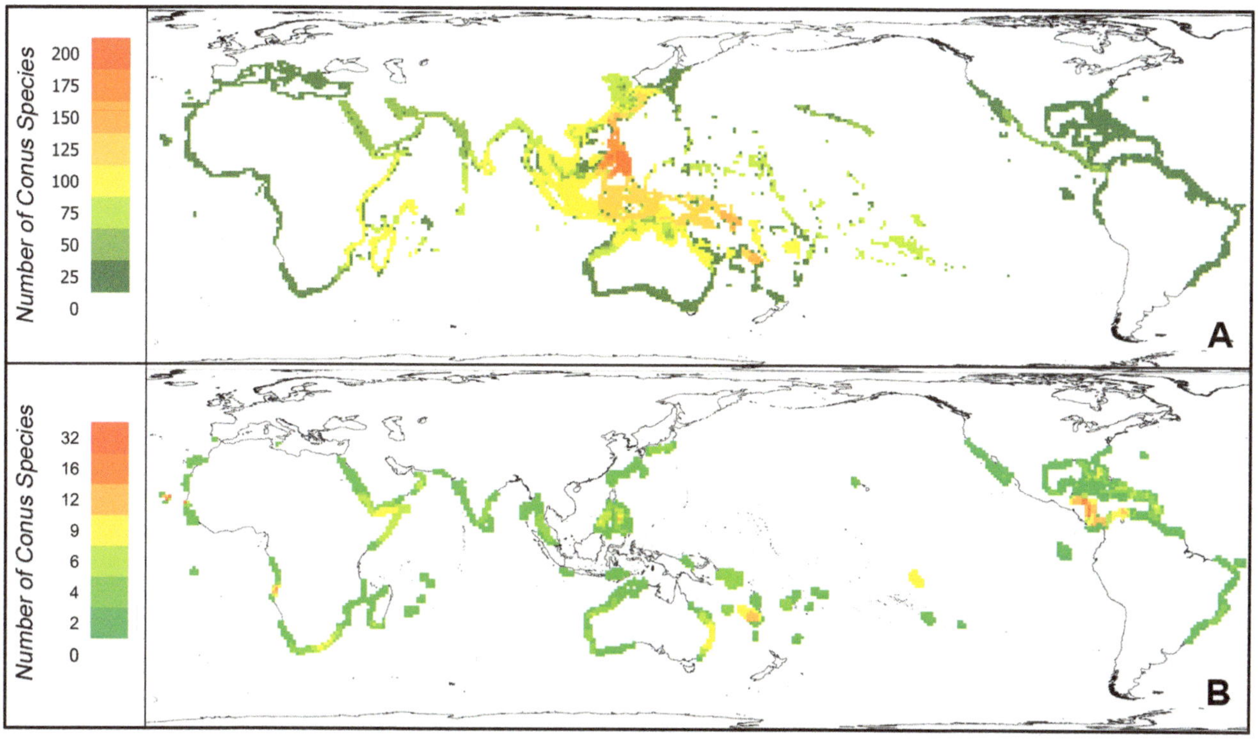

Figure 2. ***Conus*** **species distribution.** Species richness from a composite of individual species maps (Fig. 2A), and only species with mapped area less than the median indicating regions of potential endemism (Fig. 2B).

Conus are considered at risk of extinction or liable to become so. This exceptional concentration of threatened species is found nowhere else across the genus' wide distribution and the disproportionate contribution of species from Cape Verde and Senegal demands further explanation.

Cape Verde is home to 8.9% of all *Conus* species. With 53 species endemic from a total of 56 present in the archipelago, endemism is exceptionally high at 94.6%. Forty-three species are each restricted to a single island. All three CR species are found in Cape Verde, *C. lugubris*, *C. mordeirae*, and *C. salreiensis*, together with four EN and five VU (Table 1). There are also 12 NT species. With 24 species in either a threatened or near-threatened category, Cape Verde has 45.3% of its *Conus* diversity at risk compared to 7.4% for the remainder of the world. Angola and Senegal contribute the next largest numbers of endemic *Conus* species with 22 and 13 respectively, which together with 53 species endemic to Cape Verde account for 89.8% of all 98 species within the Eastern Atlantic. Senegal contributes seven EN and three VU species with Angola contributing three VU (Table 1).

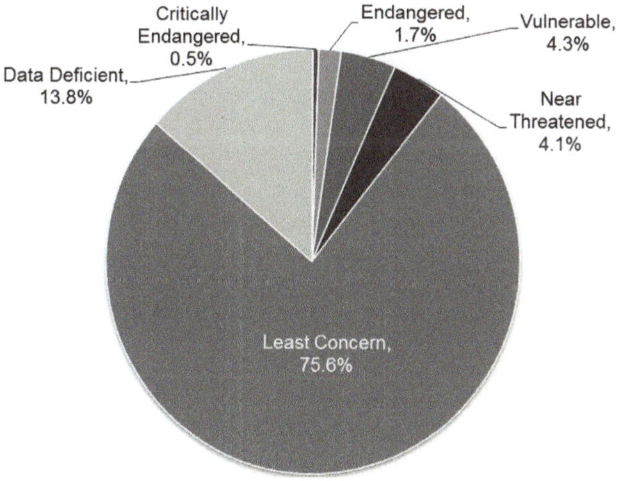

Figure 3. Global extinction risk to ***Conus.*** The percentage contribution for each assessed category to the global diversity of 632 *spp* of *Conus*. These are represented by 3 Critically Endangered species; 11 Endangered; 27 Vulnerable; 26 Near Threatened; 87 Data Deficient, and 478 of Least Concern.

Western Atlantic. We assessed 113 species of *Conus* from the Western Atlantic where they occur from the Carolinas and Bermuda south to Brazil and throughout the Gulf of Mexico and Caribbean. Species are widely variable in their distribution across the region. There are six threatened species, all categorised as VU (see Table 2), representing 5.3% of the total and a further four NT, together resulting in 8.8% of *Conus* species within this region considered at immediate or potential risk.

Indo-Pacific. We assessed 390 species of *Conus* from the Indo-Pacific where they occur across the tropics and subtropics, from East Africa south to South Africa and north to the Red Sea and the Persian Gulf and across the whole of the Indian Ocean and the Western and Central Pacific, south to Australia and New Zealand, north to Japan, east to French Polynesia and Easter Island and northeast to Hawaii.

Only nine species were found to be VU. All occur within the Indian Ocean with six species from the western flank: two from Oman, one from the southern Red Sea, two from the Mascarenes and one from South Africa. From the eastern flank there is one species from Thailand and two from Western Australia. There are also four NT species including one from Oman with the other

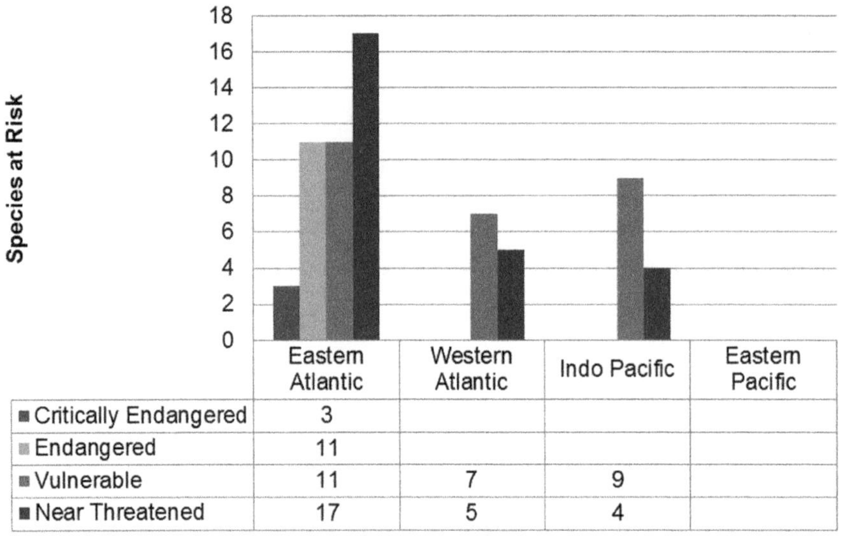

Figure 4. Number of *Conus* species at risk by ocean basin. The number of species at risk by ocean basin for each threatened category. There are no species at risk in the Eastern Pacific.

three being the only *Conus* species potentially at risk in the Pacific – one each from Queensland, the Philippines and the Marquesas.

Eastern Pacific. We assessed 31 species of *Conus* from the Eastern Pacific where they occur from Southern California south along the Pacific coast of Meso-America to Southern Ecuador including the Galapagos and other island groups of the region.

No species were assessed as threatened or near threatened in the Eastern Pacific.

Threats

The nature of threats to those species of *Conus* at risk of extinction are varied and depend primarily, but not exclusively, on the proximity and nature of human habitation and development adjacent to coastlines where the molluscs occur. This alone, however, will not normally create a scenario for species extinction. Wide-ranging species are capable of maintaining their viability through resilience from multiple sub-populations. Although most threatened *Conus* species are range-restricted, this is not always the case: two species from the USA, *C. anabathrum* and *C. stearnsii* occur along the west coast of Florida where their ranges are substantially fragmented by shoreline development. However, restricted range, coupled with shallow water habitat, magnifies the impact of stressors such as coastal development or pollution. Of the 41 *Conus* species globally assessed as threatened with extinction, 32 (78.0%) occur within an AOO of 250 km^2 and a minimum depth of 5 m or less. In the Eastern Atlantic, of the 25 threatened species, this rises to 100%.

Threats to those *Conus* species assessed within one of the three threatened categories can be classified into four causal groups: 1. pollution, either from proximity to actual or potential petrochemical spills, or urban and industrial effluent; 2. disturbance to habitat from coastal development either resulting from human population increases, e.g. sea defences, residential and commercial structures, including aquaculture facilities, and port construction, or tourism infrastructure. Also included in this group is damage to habitat caused by damaging and extensive demersal fishing; 3. shell gathering, and 4. environmental change e.g. elevated sea-surface temperatures (Fig. 6). There will frequently be a

Table 1. Threatened *Conus* of the Eastern Atlantic (EA).

Critically Endangered (CR)		**Endangered (EN)**		**Vulnerable (VU)**	
Cape Verde	*C. lugubris*	Cape Verde	*C. ateralbus*	Angola	*C. allaryi*
Cape Verde	*C. mordeirae*	Cape Verde	*C. crotchii*	Angola	*C. cepasi*
Cape Verde	*C. salreiensis*	Cape Verde	*C. cuneolus*	Angola	*C. xicoi*
		Cape Verde	*C. fernandesi*	Cape Verde	*C. decoratus*
		Senegal	*C. belairensis*	Cape Verde	*C. felitae*
		Senegal	*C. bruguieresi*	Cape Verde	*C. fontonae*
		Senegal	*C. cloveri*	Cape Verde	*C. regonae*
		Senegal	*C. echinophilus*	Cape Verde	*C. teodorae*
		Senegal	*C. hybridus*	Senegal	*C. cacao*
		Senegal	*C. mercator*	Senegal	*C. guinaicus*
		Senegal	*C. unifasciatus*	Senegal	*C. tacomae*

Table 2. Threatened *Conus* of the Western Atlantic (WA).

Vulnerable (VU)	
Aruba	*C. hieroglyphus*
Florida	*C. anabathrum*
Florida	*C. stearnsii*
Bahamas	*C. richardbinghami*
Brazil	*C. henckesi*
Martinique	*C. hennequini*
Venezuela	*C. duffyi*

Table 3. Threatened *Conus* of the Indo-Pacific (IP).

Vulnerable (VU)	
Oman	*C. ardisiaceus*
Oman	*C. melvilli*
S Red Sea	*C. cuvieri*
Mascarenes	*C. julii*
Réunion	*C. jeanmartini*
SE South Africa	*C. immelmani*
W Thailand	*C. rawaiensis*
Australia	*C. compressus*
Australia	*C. thevenardensis*

combination of causes, for example tourism infrastructure may also increase shell gathering. Similarly, the proximity of shanty towns devoid of planning regulations poses an elevated risk of effluent discharge into the marine environment. Finally habitat destruction from sand removal, beach nourishment works and recreational use of the sea may all result in disturbance to local mollusc populations.

Cape Verde is experiencing a major structural change from a largely services and fisheries based economy supported by development aid and remittances from its diaspora to one of beach tourism [37]. This is accompanied by a myriad of threats from road and resort construction, unlawful removal of beach sand for cement [38] and casual shell gathering by tourists. All three CR species occur in Cape Verde where their populations are already reduced. *C. lugubris* and *C. mordeirae* live in areas where habitat has already been lost to development and *C. salreiensis* which is restricted to a single bay has had observable declines in population since a harbour was constructed. Each is found in an area along a shallow coastal strip of less than 11 km in length. Harbour expansion and the accidental discharge of engine fuel increase the pressures on small, range-restricted *Conus* populations such as *C. fernandesi*, *C. fontonae* and *C. regonae*. With so many *Conus* species occupying highly restricted ranges within the archipelago, modest threats such as these could have a profound impact.

Around the Dakar peninsula, Senegal, it has been observed that species restricted to its highly polluted coastal waters are showing a marked decline in abundance coupled with an overall diminution of shell size including *C. echinophilus*, *C. hybridus*, *C. mercator* and *C. unifasciatus*. In common with many maritime cities in developing countries, Dakar suffers from a burgeoning population with largely inadequate waste-processing infrastructure. South in Angola, *Conus* species categorised as at risk face similar threats to those in Senegal.

In the Western Atlantic some disturbance to *Conus* can be traced to human migration to the Florida coast. Tourism and retirement have driven large-scale construction projects for condominiums and other coastal infrastructure leading to significant loss of habitat for *C. anabathrum* and *C. stearnsii*. Tourism also represents the underlying risk to the vulnerable species *C. hennequini* in Martinique and *C. hieroglyphus* in Aruba. Shell collecting in the Bahamas threaten *C. richardbinghami*. General coastal development in Bahia, Brazil threatens *C. henckesi* where it occurs only off two small islands. The Venezuelan government has voiced plans for substantial development on the islands of Los Roques which will place the shallow water species, *C. duffyi* at risk.

The *Conus* species of the Indo-Pacific are at less risk. In the north-western Indian Ocean, the Persian Gulf, the southern Red Sea including the Gulf of Aden and the Horn of Africa, civil wars, poverty, piracy and the security situation offer some degree of protection from coastal development. However, there are still concerns from oil spillage in the region and two scarce species from Oman, *C. ardisiaceus* and *C. melvilli* together with *C. cuvieri* from Djibouti are categorised as VU. In Southern Natal and the Mascarene islands of Mauritius and Réunion respectively, *C. immelmani* and *C. julii*, have both declined in numbers almost certainly from over-collecting, with *C. jeanmartini* also from Réunion being subject to intensive trawling in its deep-water habitat.

In the Eastern Indian Ocean, *C. rawaiensis* occurs only in an area estimated at less than 35 km^2 in a single location off the western shores of Thailand in a region zoned for tourism. In Western Australia, an extreme localized warm-water event in 2011 from La Niña, in the region around Geraldton to Shark Bay including the Abrolhos Islands, resulted in a catastrophic decline of marine molluscs including *Conidae*. *C. compressus*, a restricted range species, possibly suffered a 50% decline in abundance. Also in Western Australia, *C. thevenardensis*, already rare, is subject to a range of threats including a large oil installation, tourism and dredging.

Other Red List categories

The results for the three threatened categories paint an incomplete picture. There are also 87 species assessed to be Data Deficient and 26 as Near Threatened, together representing 17.9% of the global diversity. Many of the Data Deficient species are considered to be scarce in the wild even though the causes and extent of the threats they face cannot yet be determined with sufficient accuracy. Twice as many of the 87 DD species, (39.1%) occur only at depths of 50 m or more, i.e. below gleaning, scuba, and (many) artisanal fishing gears, compared to 19.4% for the remaining 545 species globally. Specimens may be brought to the surface from these depths as by-catch from fisheries. However, demersal gear such as dredges may also contribute substantially to the endangerment of the species recovered through destruction of their habitat, especially for those that are also of restricted range. Occurrence in deep water does not automatically result in a DD categorisation. Despite paucity of data, taxa with a known distribution greater than 2,000 km^2 but with no known threat would normally be assessed as Least Concern. At the other extreme, there are also a number of DD species where there is an almost total absence of recent sightings but extinction cannot be proven, i.e. there is reasonable doubt that the last individual has died [39]. This is exemplified by species such as *C. jourdani* from St Helena which is only known from 'dead' shells washed onto the beach in one small bay; *C. bellulus* and *C. luteus* which have not

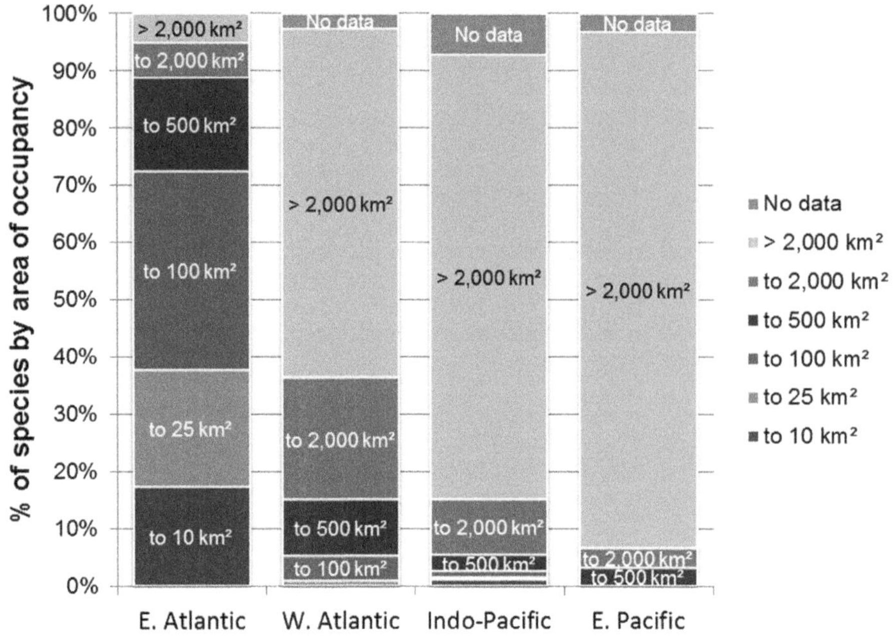

Figure 5. Contribution of range-restricted species to *Conus* biodiversity within each ocean basin. This illustrates by region the percentage of total species by area of occupancy, with wide-ranging species, i.e. >2,000 km^2 being minimal in the Eastern Atlantic but the major contributor to the Indo-Pacific and Eastern Pacific *Conus*. The abbreviated key describes the band sizes, e.g. to 10 km^2 = 0–10 km^2, to 25 km^2 = 11–25 km^2, to 100 km^2 = 26–100 km^2, etc.

been reported since the 1970s; *C. splendidulus* which has not been seen in 20 years and *C. sauros* which is possibly extinct.

Threatened species frequently have small ranges. Of 478 *Conus* species categorised as LC, 103 have an AOO of 2,000 km^2 of less (21.5%). However, of the 67 species with threatened and near-threatened categories, all except four occur within an AOO of 2,000 km^2 or less (94.0%), with 40 (59.7%) restricted to a range of 100 km^2 or less (Fig. 7). By comparison, 27 of the 59 DD species (45.8%) where it had been possible to approximate their AOO, were recorded at 2,000 km^2 or less (Fig. 7). Furthermore, many of

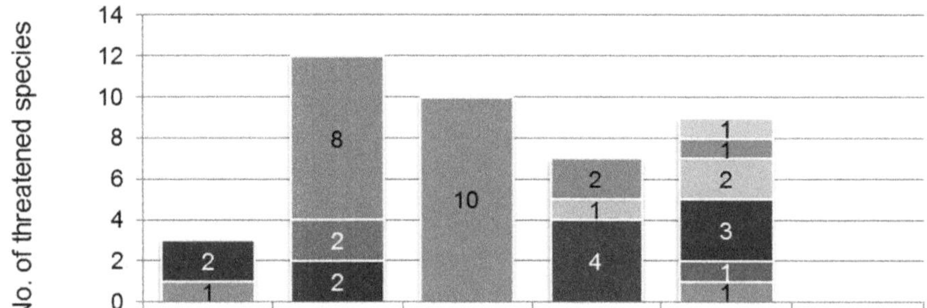

	Angola	Cape Verde	Senegal	Western Atlantic	Indo-Pacific	Eastern Pacific
	Eastern Atlantic					
Dem Fishing					1	
Tourism		8		2	1	
Shell Coll				1	2	
Ports etc		2				
Petro-chem	2	2			3	
Elev SST					1	
Effluent	1		10		1	
Coast Dev				4		

Figure 6. Main threats to *Conus* by ocean basin. The number of *Conus* species at risk (consolidation of CR, EN and VU) indicating primary causes of endangerment, being demersal fishing, tourism, shell collecting, ports and harbours, petro-chemical spills, elevated sea-surface temperatures, effluent discharge and runoff, and coastal development.

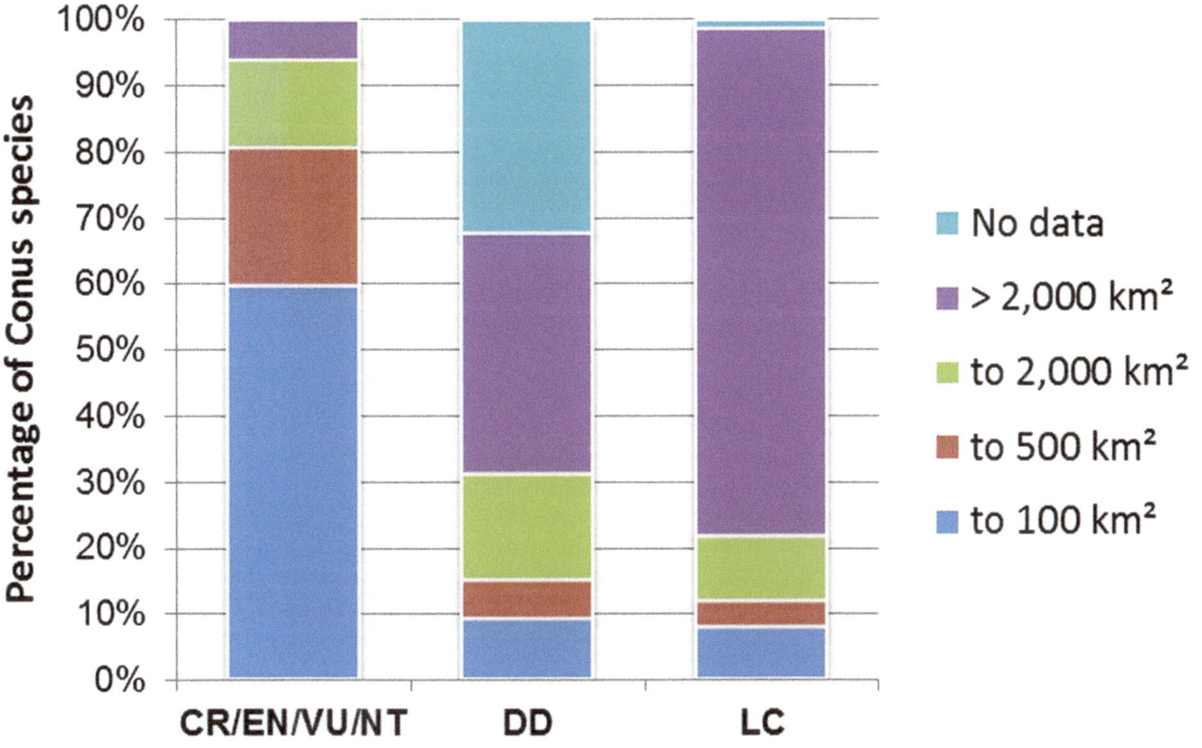

Figure 7. Contribution of range-restricted *Conus* species within assessment categories. This shows the percentage of species by assessment category, with DD species having a higher proportion of taxa with small ranges compared to LC species and also with a large number of species with no distribution data. All threatened and near-threatened categories have been grouped. Key: CR Critically Endangered, EN Endangered, VU Vulnerable, NT Near Threatened, DD Data Deficient and LC Least Concern

the 36.8% of DD species that are wider-ranging (AOO >2,000 km^2) are based on infrequent sightings. It would therefore be reasonable to suppose that a substantial proportion of the DD species are potential candidates for listing as threatened.

DD species are overwhelmingly found in the Indo-Pacific. There are three in the Eastern Atlantic, eight in the Western Atlantic, 75 in the Indo-Pacific and one in the Eastern Pacific. Outside the Indo-Pacific there is just a single species with a known AOO of more than 2,000 km^2.

Discussion

It is widely believed that extinction risk in the sea is less likely than in the terrestrial environment and that this is supported by the fossil record [2,3]. This view is based largely on perceived high fecundity, greater dispersal ability and geographic range size [40]. With 6.5% of *Conus* species at risk globally this would appear to follow this perception, however, in regions offering reduced dispersal opportunity, such as the whole of the Eastern Atlantic, 25.5% of species are threatened. Cone snails here have a similar level of extinction risk to species in well-assessed terrestrial taxa, such as freshwater invertebrates (34% of 7,784 species assessed at risk), lepidoptera (from 8.5% of butterflies in Europe to 17% in the U.S. at risk), European terrestrial molluscs (20% at risk) [41,42] and bryophyte flora from the Canaries (21% at risk) [43]. Contributing to the pattern seen, many cone snails have limited dispersal ability, small geographic ranges and/or are rare. The level of extinction risk is similar in other well assessed marine taxa, including corals (27% of species at risk) [7,44] and scombrid and billfish (11% of 61 species at risk) [45]. Given the rapid escalation of threats to the marine environment [46], if the pattern seen in these groups is typical of marine species generally, then there is a high risk that extinctions will soon become common in the sea, just as they now are on land.

Our global assessment of the conservation status of all 632 cone snails shows that three-quarters (75.6%) of species are classified as Least Concern under IUCN Red List standards. However, beneath this relatively optimistic result lies a picture of substantial regional variations with indicators signalling wider concerns. In the Eastern Atlantic along the shores of Senegal, Cape Verde and Angola, species restricted in their range and subject to the effects of industrialisation and urbanisation face an elevated risk of extinction. Endemism for marine species occurs most commonly in isolated island groups where the original dispersal was assisted by a pelagic larval stage or by transport on rafting matter [47]. Endemics may also be found where there may be non-reversing currents transporting water away from the tropics towards higher latitudes [40]. All Cape Verde endemic *Conus* have a non-planktonic larval stage having lost the ability during speciation to feed during larval dispersal [48]. This conforms to the hypothesis that non-planktonic, i.e. lecithotrophic, species of *Conus* commonly originate from planktotrophic species [49]. All three species assessed as Critically Endangered occur in the waters off Cape Verde where they are exposed to habitat degraded through coastal development primarily driven by tourism. Similarly, of the 11 Endangered species, four are found in Cape Verde with the remaining seven occurring off the coast of Senegal, in particular the Dakar peninsula, where high levels of pollution from industrial and residential effluent is thought to be the driver of declining abundance and observable reductions in body size. A further 11 species (40.7%) of the 27 assessed as Vulnerable occur in Cape Verde, Senegal and Angola, making West Africa home to 61% of the 41 *Conus* species threatened with extinction. Of the remaining 16 species categorised as Vulnerable, seven are found in the Western Atlantic where they are primarily exposed to coastal development, tourism and shell collecting. The remaining nine occur in the Indian Ocean where petrochemicals, shell-collecting and elevated sea-surface temperatures represent the principal causes of decline.

Threats

Overfishing. The effect of overfishing on the abundance of fish stocks has been extensively reported in both the scientific and general press over many years [50,51]. However, threats to invertebrates from fishing are seldom equated with extinction, especially marine molluscs. Although extremely unusual, near-extinctions in this group have occurred in the recent past; for example in the white abalone *Haliotus sorenseni* from southern California and Baja California by the mid-1990s had been fished to the edge of extinction [52]. Once counted in the millions there are now probably less than 1,600 individuals remaining.

Amongst marine molluscs, most species are sought by shell collectors [12]. Although this does not threaten the survival of the vast majority of molluscs, shell collecting has undoubtedly caused the decline and endangerment of some species, particularly 'trophy' shells. Throughout the Indo-Pacific, the spectacular giant triton (*Charonia tritonis*), has been extensively fished and in many areas has been extirpated [53]. Similarly, although primarily removed for its adductor muscles, the giant clam (*Tridacna gigas*) the largest of all bivalve molluscs, has met the same fate [54]. In Zanzibar, East Africa, the cowries *Cypraea tigris, C. histrio* and *C. lynx* were found to be up to 18 times less abundant in exploited tourist areas [8]. We identified three rare species of *Conus* threatened by shell collecting: *C. richardbinghami* from the Bahamas and *C. immelmani* and *C. julii* from the Mascarenes. Taxa already facing pressures from factors such as pollution may be pushed further towards extinction by gathering for shells, yet warning indicators such as sudden price inflation on the shell market may not alone warrant inclusion to a threatened category.

Bioprospecting. *Conus* is exceptionally important to biomedical science, although there is dispute about the number of animals taken for their bioactive compounds. To protect their intellectual property, pharmaceutical companies are silent on the issue, but researchers are adamant that volumes are negligible. In their dialogue in *Science* Chivian et al. (2003; 2004) raised important concerns about the quantity of cone snails taken from the wild, indicating that thousands were then collected to satisfy research demands [55,56]. This was forcefully rebutted by Duda et al. (2004) who reviewed recent conotoxin research from which they determined that a maximum of 20 research groups were working on *Conus* toxins at that time, and that any single characterisation required fewer than 21 animals to be sacrificed [57]. Regardless of where the true determinant lies, balancing the legitimate needs of medical research without further compromising natural resources is essential. Fortunately, alternative, more sustainable options are now available including milking venom without killing the animal [58], polymerase chain reaction (PCR) sequencing of DNA fragments that requires just one specimen [59] and more recently digital marine bioprospecting using massive parallel deep sequencing of transcriptomes that requires only minute samples of bioactive material [60].

Habitat loss. It has been shown that habit loss leads to declines in species richness, reduced biomass and loss of complexity [61,62], often accompanied by colonisation by species that inhibit recovery [63]. Virtually all of the world's 'trawlable' area of continental shelf has already been altered, and about half the area of all the continental shelves is hit by trawls every year [64], changing the structure and function of habitats, destroying assemblages and resulting in homogenisation of the seabed [65]. Of 133 marine species that have been recorded as having gone extinct either regionally or globally, 37% were attributed entirely or in part to habitat loss [66]. Extinctions of marine gastropod molluscs from loss of habitat are set to continue and include the horn snail *Cerithidea fuscata* from southern California last seen in 1935, the eelgrass limpet *Lottia alveus alveus* from the northwest USA last collected in 1929, and from the 19th century the rocky shore limpet '*Colisella*' *edmitchelli* also from southern California and the periwinkle *Littoraria flam*mea from China; all driven to extinction through loss of habitat from anthropogenic causes, with the possible exception of the eelgrass limpet that lost its habitat from a slime mould that may have been introduced from ships' ballast [2,34].

Our assessment found that with the exception of three species made vulnerable by shell collecting (see above), all 38 other *Conus* species threatened with extinction are impacted to some degree by habitat loss, either directly from coastal and port development or indirectly from pollution or from human exacerbated natural occurrences such as El Niño/La Niña–Southern Oscillation (ENSO) warm-water events (Fig. 6).

Red List comparatives

Our *Conus* assessment is the first global study for the IUCN Red List for any marine gastropod mollusc genus and one of the few for marine invertebrates. Other marine invertebrates that have been the subject of a global assessment include 845 reef-building corals, 247 lobsters and 195 cuttlefishes [44]. Data Deficiency is a common thread throughout each of these studies with 17%, 35% and 76% of species for each respective grouping [44] compared to 14% for *Conus*. Preliminary results available for oceanic squid show that 57% of this group are of Least Concern with the remaining 43% Data Deficient. As with the data deficient cone snails, many of these cephalopods are deep-water species that have only been captured on a few occasions [44].

Of the 845 corals that have been globally assessed 27.3% fall into a threatened category with a further 20.8% near threatened, although prior to the massive bleaching event of 1998 it has been estimated that 95.3% of non-DD species would have been categorised as Least Concern [7]. The exceptional ENSO event which resulted in this bleaching largely devalues any post-event comparison, although it has been shown that La Niña can impact some mollusc assemblages through stress, changes in productivity and availability of dietary preferences [67]. In Australia, the La Niña event of 2010-11 gave the highest monthly Southern Oscillation Index values on record accompanied by elevated sea-surface temperatures in Western Australia [68]. In the region around Geraldton to Shark Bay including the Abrolhos Islands, this coincided with an estimated 50% mortality in molluscs that included *Conus compressus* (H. Morrison pers. comm. 2011). Scleractinian corals and molluscs, including *Conus*, also share the threat of ocean acidification with the prospect of arrested development in their aragonite-forming structures [69,70]. Of all the threats faced by these fauna this is the most intractable and one that could even determine their continued existence.

For freshwater molluscs, Red List assessments have been completed for 1,500 of the 5,000 described species [6,42]. Results show that out of 7,784 freshwater invertebrates assessed to date, gastropods are the most threatened group with a threat range of 33% (if no DD species are threatened) to 68% (if all DD species are threatened) [42]. In common with *Conus*, the threatened species include range-restricted habitat specialists that are particularly at risk from loss of habitat and pollution.

Further Research and Conservation Priorities

As a global assessment for conservation has not been undertaken on any other marine gastropod it is not possible to explore relationships between different gastropod genera to identify commonality of risks. Further research is urgently needed to address this issue.

Data Deficient *Conus* species can normally be characterised by minimal sightings and a lack of data on distribution. This may result from their bathymetric profile (39% of DD species occur only at 50 m or more) and/or genuine rarity. However, DD species share a common trait with those categorised as threatened in having a higher percentage with restricted range than those from the general population, with the implication that there could be a significant proportion of DD species at risk. Bearing the title 'Data Deficient' or 'Near Threatened' places these 113 species jointly away from the spotlight afforded the 41 species in the three threatened categories. However, with the potential to double the number of species at risk it is essential that the taxa that make up these categories should not be ignored but instead benefit from further research into their true status.

One of the primary sources of information on species distribution, habitats, populations and threats for our Red List assessment has been the specimen shell trade. In many parts of the developing world, trade in shells provides valuable additional revenue to some of the poorest families living along tropical coastlines. Research is needed to assess the threat from rare shell collecting towards mollusc population decline to determine what measures should be taken to enable this activity to continue sustainably while at the same time allowing for protection of vulnerable species.

The need to identify conservation strategies for all species at risk is compelling, although for developing nations, snail conservation is unlikely to become a driver for environmental improvements. In the absence of in situ conservation measures, captive breeding programmes may ultimately be necessary, such as those undertaken for tree snails of the genus *Partula* from the Pacific Islands [71]. At present, except possibly for species such a *C. pennaceus* and *C. textile* that emerge as mature veligers, this is not a viable option, as the complexity of plankton essential for developing larvae cannot be easily replicated [28].

Over half (53.6%) of all *Conus* species occur at depths of 5 m or less where they are susceptible to gleaning, and nearly three-quarters (74.5%) occur at or above recreational SCUBA diving depths of 30 m or less. Marine Protected Areas (MPAs) offer one of the few sanctuaries, but regional authorities need encouragement to strengthen enforcement and to erect prominent signage against shell gathering within MPAs.

Cape Verde presents a special case in *Conus* conservation. With 45.3% of its species at risk there is a strong argument for legislating against export of both animals and shells, and with manageable borders the country is ideally suited to export controls. This small archipelago also signals a warning to other nations developing their coastal infrastructure: new roads bring visitors to areas previously protected by their isolation, and illegal sand removal for construction from beaches and shallow water of the littoral zone [72] pose a constant threat to habitat. Regional authorities should be required to undertake environmental impact assessments that take account of these issues when planning new developments.

The toxins that make *Conus* so successful are generally unique to each species [59] and any extinction in the genus could in turn deprive science of a potential pharmacological resource. The extraordinary number of species and the global distribution of these tropical snails make them an important contributor to marine biodiversity, and with the appeal of their shells they help support some of the world's poorest people.

Finally, there exists a well-defined community of cone snail aficionados who together are highly influential in the trade in cone shells. This includes leading academics as well as collectors and dealers. A positive first step from our Red Listing is that following a preliminary presentation of our findings at their international convention, a core of members has been motivated to explore a voluntary embargo in trade of critically endangered species and to consider this also for other *Conus* species at risk [73].

Acknowledgments

We would like to thank: Mark Westneat, Audrey Aronowsky, Sarah Kim and Beth Sanzenbacher at the Biodiversity Synthesis Center, Chicago for their organisation of the *Conus* synthesis workshop at the Field Museum in Chicago; Philippe Bouchet, José Coltro, Tom Duda, Alan Kohn, Eric Monnier, Hugh Morrison, Ed Petuch, Guido Poppe, Gabriella Raybaudi-Massilia, Sheila Tagaro, Manuel Jiménez Tenorio, Stephan Veldsman and Fred Wells for volunteering their time and expertise during the assessment and at the synthesis workshop; Monika Böhm, Heather Harwell, Andrew Hines, Suzanne Livingstone, Jonnell Sanciangco and Mary Seddon for facilitating at the synthesis workshop; Mike Filmer for helping to resolve the many taxonomic issues; Hannah Cubaynes, Zarozinia Sheriff and all the interns who assisted with species research; Klaus & Christina Groh of ConchBooks for use of images and maps for the Red List; Mia Theresa Comeros and Angela Goodpaster for their work on the Red List maps; and Caroline Pollock and Janet Scott of the IUCN Red List Unit for bringing the assessment to publication.

Author Contributions

Conceived and designed the experiments: HP BCO JPH CMR. Analyzed the data: HP BCO CMR. Contributed reagents/materials/analysis tools: KEC. Wrote the paper: HP JPH CMR. Revised the manuscript: JPH KEC CMR. Supported the synthesis workshop: HP KEC CMR.

References

1. Harnik PG, Simpson C, Payne JL (2012) Long-term differences in extinction risk among the seven forms of rarity. Proc Biol Sci 279: 4969–4976. doi:10.1098/rspb.2012.1902.
2. Roberts CM, Hawkins JP (1999) Extinction risk in the sea. Trends Ecol Evol 14: 241–246.
3. McKinney ML (1998) Is marine biodiversity at less risk? Evidence and implications. Divers Distrib 4: 3–8.
4. McManus JW (1997) Tropical marine fisheries and the future of coral reefs: a brief review with emphasis on Southeast Asia. Coral Reefs 16: S121–S127. doi:10.1007/s003380050248.
5. Jamieson GS (1993) Marine Invertebrate Conservation: Evaluation of Fisheries. Am Zool 33: 551–567.
6. IUCN (2013) IUCN Red List of Threatened Species. IUCN Red List Threat Species Version 20122. Available: http://www.iucnredlist.org. Accessed 8 February 2013.
7. Carpenter KE, Abrar M, Aeby G, Aronson RB, Banks S, et al. (2008) One-third of reef-building corals face elevated extinction risk from climate change and local impacts. Science 321: 560–563. doi:10.1126/science.1159196.
8. Newton LC, Parkes EVH, Thompson RC (1993) The Effects of Shell Collecting on the Abundance of Gastropods on Tanzanian Shores. Biol Conserv 63: 241–245.
9. Queensland Government Dept of Primary Industries and Fisheries (2007) Annual status report 2007 Queensland Marine Specimen Shell Collection Fishery. Available: http://www.daff.qld.gov.au/documents/Fisheries_SustainableFishing/AnnualStatusReport-QLDMarineSpecimen-ShellCollectionFishery-2007.pdf.
10. Bouchet P (1990) Turrid genera and mode of development: the use and abuse of protoconch morphology. Malacologia: 69–77.
11. Röckel D, Korn W, Kohn AJ (1995) Manual of the Living Conidae, Vol 1. Verlag Christa Hemmen.

12. Rice T (2007) A Catalog of Dealers' Prices for Shells: Marine, Land and Freshwater, 23rd edition. Of Sea and Shore Publications.
13. Chadwick A, Olivera BM (2009) Cone Shells and Human Culture. Cone Collect 12: 16–21. Available: http://www.theconecollector.com/.
14. Garber K (2005) Peptide leads new class of chronic pain drugs. Nat Biotechnol 23: 399. doi:10.1038/nbt0405-399.
15. Taylor JD, Kantor YI, Sysoev AV (1993) Foregut anatomy, feeding mechanisms, relationships and classification of the Conoidea (= Toxoglossa) (Gastropoda). Bull Nat Hist Museum London (Zoology) 59: 125–170.
16. Kohn AJ, Perron FE (1994) Life History and Biogeography – Patterns in Conus. Oxford Science Publications.
17. Monteiro A, Tenorio MJ, Poppe GT (2004) A Conchological Iconography. The Family Conidae. The West African and Mediterranean Species of Conus. Germany: ConchBooks, Hackenheim.
18. Tenorio MJ, Tucker JK, Chaney HW (2012) The Families Conilithidae and Conidae – The Cones of the Eastern Pacific. In: Poppe GT, Groh K, editors. A Conchological Iconography. ConchBooks, Hackenheim. p. 200.
19. Franklin JB, Fernando SA, Chalke BA, Krishnan KS (2007) Radular morphology of Conus (Gastropoda: Caenogastropoda: Conidae) from India. Molluscan Res 27: 111–122.
20. Zou S, Li Q, Kong L, Yu H, Zheng X (2011) Comparing the Usefulness of Distance, Monophyly and Character-Based DNA Barcoding Methods in Species Identification: A Case Study of Neogastropoda. PLoS One 6: e26619. doi:10.1371/journal.pone.0026619.
21. Kohn AJ (1990) Tempo and Mode of Evolution in Conidae. Malacologia 32: 55–67.
22. Duda TF, Kohn AJ (2005) Species-level phylogeography and evolutionary history of the hyperdiverse marine gastropod genus Conus. Mol Phylogenet Evol 34: 257–272. doi:10.1016/j.ympev.2004.09.012.
23. Kohn AJ (1983) Marine Biogeography and Evolution in the Tropical Pacific: Zoological Perspectives. Bull Mar Sci 33: 528–535.
24. Kohn AJ (1968) Microhabitats, Abundance and Food of Conus on Atoll Reefs in the Maldive and Chagos Islands. Ecology 49: 1046–1062.
25. Kohn AJ (1959) The Ecology of Conus in Hawaii. Ecol Monogr 29: 47–90.
26. Duda TF, Kohn AJ, Palumbi SR (2001) Origins of diverse feeding ecologies within Conus, a genus of venomous marine gastropods. Biol J Linn Soc 73: 391–409. doi:10.1006/bijl.2001.0544.
27. Olivera BM, Walker C, Cartier GE, Hooper D, Santos AD, et al. (1999) Speciation of cone snails and interspecific hyperdivergence of their venom peptides. Potential evolutionary significance of introns. Ann N Y Acad Sci 870: 223–237.
28. Livett BG, Gayler KR, Khalil Z (2004) Drugs from the sea: conopeptides as potential therapeutics. Curr Med Chem 11: 1715–1723.
29. Remigio EA, Duda TF (2008) Evolution of ecological specialization and venom of a predatory marine gastropod. Mol Ecol 17: 1156–1162. doi:10.1111/j.1365-294X.2007.03627.x.
30. Terlau H, Olivera BM (2004) Conus venoms: a rich source of novel ion channel-targeted peptides. Physiol Rev 84: 41–68. doi:10.1152/physrev.00020.2003.
31. Olivera BM (1997) E.E.Just Lecture, 1996. Conus venom peptides, receptor and ion channel targets, and drug design: 50 million years of neuropharmacology. Mol Biol Cell 8: 2101–2109.
32. Craig AG, Bandyopadhyay P, Olivera BM (1999) Post-translationally modified neuropeptides from Conus venoms. Eur J Biochem 264: 271–275.
33. Skilleter GA, Warren S (2000) Effects of habitat modification in mangroves on the structure of mollusc and crab assemblages. J Exp Mar Bio Ecol 244: 107–129. doi:10.1016/S0022-0981(99)00133-1.
34. Carlton JT (1993) Neoextinctions of Marine Invertebrates. Am Zool 33: 499–509.
35. Zuschin M, Hohenegger J, Steininger FF (2001) Molluscan assemblages on coral reefs and associated hard substrata in the northern Red Sea. Coral Reefs 20: 107–116. doi:10.1007/s003380100140.
36. IUCN Standards and Petitions Subcommittee (2010) Guidelines for Using the IUCN Red List Categories and Criteria. Version 8.0.: 85. Available: http://www.iucnssg.org/tl_files/Assets/pdf/RL Docs/RedListGuidelines.pdf.
37. AfDB OECD, UNDP UNECA (2012) African Economic Outlook 2012, Western African Countries. African Development Bank. Dakar, Senegal. Available: http://www.africaneconomicoutlook.org/en/countries/west-africa/cape-verde/.
38. Irwin A, Wilson C (2011) Cape Verde. Bradt Travel Guides Ltd.
39. IUCN (2012) IUCN Red List Categories and Criteria: Version 3.1. Second edition. Gland, Switzerland and Cambridge, UK: IUCN. iv.
40. Roberts CM, McClean CJ, Veron JEN, Hawkins JP, Allen GR, et al. (2002) Marine biodiversity hotspots and conservation priorities for tropical reefs. Science 295: 1280–1284. doi:10.1126/science.1067728.
41. Gerlach J, Hoffman Black S, Hochkirch A, Jepsen S, Seddon M, et al. (2012) Terrestrial invertebrate life. In: Collen B, Böhm M, Kemp R, Baillie JEM, editors. Spineless: Status and trends of the world's invertebrates. Zoological Society of London, United Kingdom. pp. 46–57.
42. Darwall W, Seddon M, Clausnitzer V, Cumberlidge N (2012) Freshwater invertebrate life. In: Collen B, Böhm M, Kemp R, Baillie JEM, editors. Spineless: Status and trends of the world's invertebrates. Zoological Society of London, United Kingdom. pp. 26–33.
43. González-Mancebo JM, Dirkse GM, Patiño J, Romaguera F, Werner O, et al. (2012) Applying the IUCN Red List criteria to small-sized plants on oceanic islands: conservation implications for threatened bryophytes in the Canary Islands. Biodivers Conserv 21: 3613–3636. doi:10.1007/s10531-012-0385-0.
44. Kemp R, Peters H, Allcock L, Carpenter KE, Obura D, et al. (2012) Marine invertebrate life. In: Collen B, Böhm M, Kemp R, Baillie JEM, editors. Spineless: Status and trends of the world's invertebrates. Zoological Society of London, United Kingdom. pp. 34–45.
45. Collette BB, Carpenter KE, Polidoro BA, Juan-Jorda MJ, Boustany A, et al. (2011) High Value and Long Life — Double Jeopardy for Tunas and Billfishes. Science (80-) 333: 291–292. doi:10.1126/science.1208730.
46. Roberts CM (2012) Ocean of Life. Allen Lane/Penguin Books.
47. Devantier LM (1992) Rafting of tropical marine organisms on buoyant coralla. Mar Ecol Prog Ser 86: 301–302.
48. Cunha RL, Castilho R, Rüber L, Zardoya R (2005) Patterns of cladogenesis in the venomous marine gastropod genus Conus from the Cape Verde islands. Syst Biol 54: 634–650. doi:10.1080/10635150591007471.
49. Duda TF, Palumbi SR (1999) Developmental shifts and species selection in gastropods. Proc Natl Acad Sci U S A 96: 10272–10277.
50. Myers RA, Hutchings JA, Barrowman NJ (1997) Why do fish stocks collapse? The example of cod in the Atlantic Canada. Ecol Appl 7: 91–106.
51. Thurstan RH, Brockington S, Roberts CM (2010) The effects of 118 years of industrial fishing on UK bottom trawl fisheries. Nat Commun 1: 15. doi:10.1038/ncomms1013.
52. National Marine Fisheries Service (2008) White Abalone Recovery Plan (Haliotis sorenseni). Available: http://www.nmfs.noaa.gov/pr/pdfs/recovery/whiteabalone.pdf.
53. Moore A, Ndobe S (2008) Reefs at risk in Central Sulawesi, Indonesia – status and Outlook. Proceedings of the 11th International Coral Reef Symposium, Ft. Lauderdale, Florida, 7-11 July 2008 Session number 18. pp. 840–844. Available: http://nova.edu/ncri/11icrs/proceedings/files/m18-35.pdf.
54. Wells S (1997) Giant clams: Status, trade and mariculture, and the role of CITES in management. IUCN—the World Conservation Union (Gland, Switzerland).
55. Chivian E, Roberts CM, Bernstein AS (2003) The threat to cone snails. Science 302: 391. doi:10.1126/science.302.5644.391b.
56. Chivian E, Roberts CM, Bernstein AS (2004) Response to: How much at risk are cone snails? Science 303: 955–957.
57. Duda TF, Bingham J-P, Livett BG, Kohn AJ, Raybaudi Massilia G, et al. (2004) How much at risk are cone snails? Science 303: 955–957.
58. Hopkins C, Grilley M, Miller C, Shon KJ, Cruz LJ, et al. (1995) A new family of Conus peptides targeted to the nicotinic acetylcholine receptor. J Biol Chem 270: 22361–22367.
59. Livett BG, Sandall DW, Keays D, Down J, Gayler KR, et al. (2006) Therapeutic applications of conotoxins that target the neuronal nicotinic acetylcholine receptor. Toxicon 48: 810–829. doi:10.1016/j.toxicon.2006.07.023.
60. Urbarova I, Karlsen BO, Okkenhaug S, Seternes OM, Johansen SD, et al. (2012) Digital marine bioprospecting: mining new neurotoxin drug candidates from the transcriptomes of cold-water sea anemones. Mar Drugs 10: 2265–2279. doi:10.3390/md10102265.
61. Airoldi L, Balata D, Beck MW (2008) The Gray Zone: Relationships between habitat loss and marine diversity and their applications in conservation. J Exp Mar Bio Ecol 366: 8–15. doi:10.1016/j.jembe.2008.07.034.
62. Munday PL (2004) Habitat loss, resource specialization, and extinction on coral reefs. Glob Chang Biol 10: 1642–1647. doi:10.1111/j.1365-2486.2004.00839.x.
63. Thrush SF, Dayton PK (2002) Disturbance to Marine Habitats by Trawling and Dredging: Implications for Marine Biodiversity. Annu Rev Ecol Syst 33: 449–473. doi:10.1146/annurev.ecolsys.33.010802.150515.
64. Watling L, Norse EA (1998) Effects of Mobile Fishing Gear on Marine Benthos. Conserv Biol 12: 1178–1179. doi:10.1046/j.1523-1739.1998.0120061178.x.
65. Gray JS, Dayton P, Thrush S, Kaiser MJ (2006) On effects of trawling, benthos and sampling design. Mar Pollut Bull 52: 840–843. doi:10.1016/j.marpolbul.2006.07.003.
66. Dulvy NK, Sadovy Y, Reynolds JD (2003) Extinction vulnerability in marine populations. Fish Fish 4: 25–64. doi:10.1046/j.1467-2979.2003.00105.x.
67. Riascos JM, Heilmayer O, Laudien J (2007) Population dynamics of the tropical bivalve Cardita affinis from Málaga Bay, Colombian Pacific related to La Niña 1999–2000. Helgol Mar Res 62: 63–71. doi:10.1007/s10152-007-0083-6.
68. Australian Government (2012) Record breaking La Niña events; An analysis of the La Niña life cycle and the impacts and significance of the 2010–11 and 2011–12 La Niña events in Australia. Melbourne, Australia: Bureau of Meteorology.
69. Doney SC, Fabry VJ, Feely RA, Kleypas JA (2009) Ocean Acidification: The Other CO2 Problem. Ann Rev Mar Sci 1: 169–192. doi:10.1146/annurev.-marine.010908.163834.
70. Rodolfo-Metalpa R, Houlbrèque F, Tambutté É, Boisson F, Baggini C, et al. (2011) Coral and mollusc resistance to ocean acidification adversely affected by warming. Nat Clim Chang 1: 308–312. doi:10.1038/nclimate1200.
71. Tonge S, Bloxam Q (1991) A review of the captive-breeding programme for Polynesian tree snails. Int Zoo Yearb 30: 51–59.
72. Lopes EP (2010) Recent data on marine bivalves (Mollusca, Bivalvia) of the Cape Verde Islands, with records of six species new to the archipelago. Zool Caboverdiana 1: 59–70.
73. Monteiro A, Malcolm G, Herndl G (2012) Endangered cone species. Cone Collect (Ed A Monteiro): 36. Available: http://www.theconecollector.com/. Accessed 1 January 2013.

PERMISSIONS

All chapters in this book were first published in PLOS ONE, by The Public Library of Science; hereby published with permission under the Creative Commons Attribution License or equivalent. Every chapter published in this book has been scrutinized by our experts. Their significance has been extensively debated. The topics covered herein carry significant findings which will fuel the growth of the discipline. They may even be implemented as practical applications or may be referred to as a beginning point for another development.

The contributors of this book come from diverse backgrounds, making this book a truly international effort. This book will bring forth new frontiers with its revolutionizing research information and detailed analysis of the nascent developments around the world.

We would like to thank all the contributing authors for lending their expertise to make the book truly unique. They have played a crucial role in the development of this book. Without their invaluable contributions this book wouldn't have been possible. They have made vital efforts to compile up to date information on the varied aspects of this subject to make this book a valuable addition to the collection of many professionals and students.

This book was conceptualized with the vision of imparting up-to-date information and advanced data in this field. To ensure the same, a matchless editorial board was set up. Every individual on the board went through rigorous rounds of assessment to prove their worth. After which they invested a large part of their time researching and compiling the most relevant data for our readers.

The editorial board has been involved in producing this book since its inception. They have spent rigorous hours researching and exploring the diverse topics which have resulted in the successful publishing of this book. They have passed on their knowledge of decades through this book. To expedite this challenging task, the publisher supported the team at every step. A small team of assistant editors was also appointed to further simplify the editing procedure and attain best results for the readers.

Apart from the editorial board, the designing team has also invested a significant amount of their time in understanding the subject and creating the most relevant covers. They scrutinized every image to scout for the most suitable representation of the subject and create an appropriate cover for the book.

The publishing team has been an ardent support to the editorial, designing and production team. Their endless efforts to recruit the best for this project, has resulted in the accomplishment of this book. They are a veteran in the field of academics and their pool of knowledge is as vast as their experience in printing. Their expertise and guidance has proved useful at every step. Their uncompromising quality standards have made this book an exceptional effort. Their encouragement from time to time has been an inspiration for everyone.

The publisher and the editorial board hope that this book will prove to be a valuable piece of knowledge for researchers, students, practitioners and scholars across the globe.

LIST OF CONTRIBUTORS

Yukai Chen, Xiaobo Yang, Qi Yang, Donghai Li, Wenxing Long and Wenqi Luo
Key Laboratory of Protection and Development Utilization of Tropical Crop Germplasm Resources, Ministry of Education, College of Horticulture and Landscapes, Hainan University, Haikou 570228, Hainan, China

Stephen Harris
School of Biological Sciences, University of Bristol, Bristol, United Kingdom

Steeves Buckland
School of Biological Sciences, University of Bristol, Bristol, United Kingdom
National Parks & Conservation Service, Reduit, Mauritius

Laura E. Gallagher, Sion M. Henshaw, Vishnu Bachraz and Kevin Ruhomaun
National Parks & Conservation Service, Reduit, Mauritius

Nik C. Cole
School of Biological Sciences, University of Bristol, Bristol, United Kingdom
Durrell Wildlife Conservation Trust, Jersey, Channel Islands
Mauritian Wildlife Foundation, Vacoas, Mauritius

Jesús Aguirre-Gutiérrez
Naturalis Biodiversity Center, Leiden, The Netherlands
Institute for Biodiversity and Ecosystems Dynamics (IBED), University of Amsterdam, Amsterdam, The Netherlands

Aurélien Besnard
Laboratoire de Biogeographie et d'Ecologie des Vertébrés, Centre d'Ecologie Evolutive et Fonctionnelle, Montpellier, France

Rachel M. Tucker
Mauritian Wildlife Foundation, Vacoas, Mauritius
Department of Animal and Plant Sciences, University of Sheffield, Sheffield, United Kingdom

Caroline E. Sogot
Department of Earth Sciences, University of Cambridge, Cambridge, United Kingdom
Department of Earth Sciences, Natural History Museum, London, United Kingdom

Elizabeth M. Harper
Department of Earth Sciences, University of Cambridge, Cambridge, United Kingdom

Paul D. Taylor
Department of Earth Sciences, Natural History Museum, London, United Kingdom

Ryan M. Ames and Simon C. Lovell
Computational and Evolutionary Biology, Faculty of Life Sciences, The University of Manchester, Manchester, United Kingdom

Daniel Money
Department of Ecology and Evolutionary Biology, University of Kansas, Lawrence, Kansas, United States of America

Gill T. Braulik
Sea Mammal Research Unit, Scottish Oceans Institute, University of St. Andrews, St. Andrews, Fife, United Kingdom
World Wildlife Fund-Pakistan, Lahore, Pakistan
Wildlife Conservation Society, Zanzibar, United Republic of Tanzania

Simon P. Northridge
Sea Mammal Research Unit, Scottish Oceans Institute, University of St. Andrews, St. Andrews, Fife, United Kingdom

Masood Arshad and Uzma Noureen
World Wildlife Fund-Pakistan, Lahore, Pakistan

Devin D. Bloom and Andrew E. Z. Short
Department of Ecology & Evolutionary Biology, University of Kansas, Lawrence, Kansas, United States of America
Division of Entomology, Biodiversity Institute, University of Kansas, Lawrence, Kansas, United States of America

Martin Fikàček
Department of Entomology, National Museum, Prague, Czech Republic
Department of Zoology, Faculty of Sciences, Charles University, Prague, Czech Republic

J. Gaspar Lorén
Departament de Microbiologia i Parasitologia Sanitàries, Facultat de Farmàcia, Universitat de Barcelona, Barcelona, Spain

Maribel Farfàn and M. Carmen Fusté
Departament de Microbiologia i Parasitologia Sanitàries, Facultat de Farmàcia, Universitat de Barcelona, Barcelona, Spain

Institut de Recerca de la Biodiversitat (IRBio), Universitat de Barcelona, Barcelona, Spain

Kerstin K. Zander
The Northern Institute, Charles Darwin University, Darwin NT, Australia

Gillian B. Ainsworth and Stephen T. Garnett
Research Institute for the Environment and Livelihoods, Charles Darwin University, Darwin NT, Australia

Jürgen Meyerhoff
Institute for Landscape Architecture and Environmental Planning, Technische Universität Berlin, Berlin, Germany

Logan Volkmann
Department of Biological Sciences, Simon Fraser University, Burnaby, British Columbia, Canada

Arne O. Mooers
Department of Biological Sciences, Simon Fraser University, Burnaby, British Columbia, Canada
Interdisciplinary Research in the Mathematical and Computational Sciences (IRMACS) Centre, Simon Fraser University, Burnaby, British Columbia, Canada

Iain Martyn
Interdisciplinary Research in the Mathematical and Computational Sciences (IRMACS) Centre, Simon Fraser University, Burnaby, British Columbia, Canada

Vincent Moulton
School of Computing Sciences, University of East Anglia, Norwich, United Kingdom

Andreas Spillner
Department of Mathematics and Computer Science, University of Greifswald, Greifswald, Germany

Adam K. Huttenlocker
Department of Biology and Burke Museum of Natural History and Culture, University of Washington, Seattle, Washington, United States of America

Frank A. La Sorte
Cornell Laboratory of Ornithology, Cornell University, Ithaca, New York, United States of America

Stuart H. M. Butchart
BirdLife International, Cambridge, United Kingdom

Walter Jetz
Department of Ecology and Evolutionary Biology, Yale University, New Haven, Connecticut, United States of America

Katrin Böhning-Gaese
Biodiversity and Climate Research Centre (BiK-F), Senckenberg Gesellschaft für Naturforschung, Frankfurt (Main), Germany
Department of Biological Sciences, Goethe Universität, Frankfurt (Main), Germany

Daniel L. Rabosky
Department of Ecology and Evolutionary Biology and Museum of Zoology, University of Michigan, Ann Arbor, Michigan, United States of America

Stephanie Dolrenry and Leela Hazzah
Nelson Institute for Environmental Studies, University of Wisconsin, Madison, Wisconsin, United States of America
Living With Lions, Mpala Research Centre, Nanyuki, Kenya

Jennifer Stenglein and R. Scott Lutz
Forest and Wildlife Ecology, University of Wisconsin, Madison, Wisconsin, United States of America

Laurence Frank
Living With Lions, Mpala Research Centre, Nanyuki, Kenya
Museum of Vertebrate Zoology, University of California, Berkeley, California, United States of America

Taku Kadoya, Akio Takenaka and Fumiko Ishihama
Center for Environmental Biology and Ecosystem Studies, National Institute for Environmental Studies, Onogawa, Tsukuba, Ibaraki, Japan

Taku Fujita
Nature Conservation Society of Japan, Shinkawa, Chuo-ku, Tokyo, Japan

Makoto Ogawa
Tokushima Prefectural Museum, Bunka-no-mori Park, Hachiman-cho, Tokushima-shi, Tokushima, Japan

Teruo Katsuyama
Kanagawa Prefectural Museum of Natural History, Odawara, Kanagawa, Japan

Yasuro Kadono5
Faculty of Science, Kobe University, Kobe, Japan

Nobumitsu Kawakubo
Faculty of Applied Biological Sciences, Gifu University, Yanagido, Gifu, Japan

Shunsuke Serizawa
Department of Biology, Aichi Kyoiku University, Igaya-cho, Kariya, Aichi, Japan

Hideki Takahashi
Hokkaido University Museum, Sapporo, Hokkaido, Japan

Masayuki Takamiya
Department of Environmental Science, Kumamoto University, Kumamoto, Japan

Shinji Fujii
Division of Human Environment, University of Human Environments, Okazaki, Aichi, Japan

Hiroyuki Matsuda and Kazuo Muneda
Faculty of Natural Environment and Information Sciences, Yokohama National University, Tokiwadai, Hodogaya-ku, Yokohama city, Kanagawa, Japan

Masatsugu Yokota
Faculty of Science, University of the Ryukyus, Senbaru, Nishihara, Nakagami, Okinawa, Japan

Koji Yonekura
Center for Academic Resources and Archives, Botanical Gardens, Tohoku University, Aoba, Aramaki, Aoba-ku, Sendai, Miyagi, Japan

Tetsukazu Yahara
Department of Biology, Kyushu University, Higashi-ku, Fukuoka, Japan

Claudio S. Quilodràn
Laboratory of anthropology, genetics and peopling history (AGP), Department of Genetics and Evolution, University of Geneva, Geneva, Switzerland
Laboratory ofmolecular phylogeny and evolution in vertebrates, Department of Genetics and Evolution, University of Geneva, Geneva, Switzerland

Mathias Currat
Laboratory of anthropology, genetics and peopling history (AGP), Department of Genetics and Evolution, University of Geneva, Geneva, Switzerland

Juan I. Montoya-Burgos
Laboratory ofmolecular phylogeny and evolution in vertebrates, Department of Genetics and Evolution, University of Geneva, Geneva, Switzerland

Diala Abu Awad, Sophie Gallina and Sylvain Billiard
UMR-CNRS 8198, Laboratoire Génétique et Évolution des Populations Végétales, Université Lille 1, Villeneuve d'Ascq, France

Cyrille Bonamy
Centre de Ressources Informatiques (CRI), Université Lille 1, Villeneuve d'Ascq, France

Martin Aberhan
Museum für Naturkunde, Leibniz Institute for Evolution and Biodiversity Science, Berlin, Germany

Wolfgang Kiessling
Museum für Naturkunde, Leibniz Institute for Evolution and Biodiversity Science, Berlin, Germany
GeoZentrum Nordbayern, Paläoumwelt, Universität Erlangen2 Nürnberg, Erlangen, Germany

Edwin T. Chester, Scott R. Strachan and Belinda J. Robson
Environmental and Conservation Sciences, Murdoch University, Perth, Western Australia, Australia

Ty G. Matthews, Travis J. Howson and Jonathon K. Mackie
School of Life and Environmental Sciences, Deakin University, Warrnambool, Victoria, Australia

Kerrylyn Johnston
Marine and Freshwater Research Laboratory, Murdoch University, Perth, Western Australia, Australia

Sebastian Höhna
Department of Mathematics, Stockholm University, Stockholm, Sweden

Marta Coll
Institut de Ciéncies del Mar (ICM-CSIC). Barcelona, Spain
UMR EME 212 Centre de Recherche Halieutique Méditerranéenne et Tropicale. IRD - IFREMER & Université Montpellier II. Séte Cedex, France

Giulia Gorelli and Raquel Sàez
Institut de Ciéncies del Mar (ICM-CSIC). Barcelona, Spain

Marta Carreras and Maria-José Cornax
OCEANA, Madrid, Spain

Cristina Ciércoles
Investigación, Planificación y Desarrollo S.A., Madrid, Spain

Elvira Morote
Universidad de Almería (UAL). Almeria, Spain

Benjamin T. Wilder
Department of Botany and Plant Sciences, University of California Riverside, Riverside, California, United States of America

Julio L. Betancourt
National Research Program, Water Mission Area, U.S. Geological Survey, Reston, Virginia, United States of America

Clinton W. Epps and Rachel S. Crowhurst
Department of Fisheries and Wildlife, Oregon State University, Corvallis, Oregon, United States of America

Jim I. Mead
Department of Geosciences, and Sundquist Center of Excellence in Paleontology, East Tennessee University, Johnson City, Tennessee, United States of America

Exequiel Ezcurra
Department of Botany and Plant Sciences, University of California Riverside, Riverside, California, United States of America
University of California Institute for Mexico and the United States (UC MEXUS), Riverside, California, United States of America

Howard Peters, Bethan C. O'Leary, Julie P. Hawkins and Callum M. Roberts
Environment Department, University of York, York, United Kingdom

Kent E. Carpenter
International Union for Conservation of Nature, Global Marine Species Assessment, Biological Sciences, Old Dominion University, Norfolk, Virginia, United States of America

Index